# Neurobiology of DOPA as a Neurotransmitter

# Neurobiology of DOPA as a Neurotransmitter

Yoshimi Misu
Yoshio Goshima

A CRC title, part of the Taylor & Francis imprint, a member of the Taylor & Francis Group, the academic division of T&F Informa plc.

Published in 2006 by
CRC Press
Taylor & Francis Group
6000 Broken Sound Parkway NW, Suite 300
Boca Raton, FL 33487-2742


Printed in the United States of America on acid-free paper
10 9 8 7 6 5 4 3 2 1

International Standard Book Number-10: 0-415-33291-5 (Hardcover)
International Standard Book Number-13: 978-0-415-33291-0 (Hardcover)
Library of Congress Card Number 2005050209



**Library of Congress Cataloging-in-Publication Data**

Neurobiology of DOPA as a neurotransmitter / editors, Yoshimi Misu, Yoshio Goshima.
p. cm.
Includes bibliographical references and index.
ISBN 0-415-33291-5 (alk. paper)
1. Dopa--Physiological effect. 2. Neurotransmitters. I. Misu, Yoshimi, 1934- II. Goshima, Yoshio.

QP364.7.N4414 2005
612.8'042--dc22 2005050209

Taylor & Francis Group
is the Academic Division of T&F Informa plc.

**Visit the Taylor & Francis Web site at**
**http://www.taylorandfrancis.com**

**and the CRC Press Web site at**
**http://www.crcpress.com**

# Preface

This book is concerned with L-3,4-dihydroxyphenylalanine (DOPA) as a biologically active substance acting on living bodies, including the brain, in addition to its being a precursor of dopamine. Historically, DOPA has been considered to be an inert amino acid that alleviates the symptoms of Parkinson's disease by its bioconversion to dopamine via the enzyme aromatic L-amino acid decarboxylase. In contrast to this generally accepted idea, it is highly probable that DOPA functions by itself as a neurotransmitter and/or neuromodulator in the hypothalamus, the striatum, the nucleus accumbens, the hippocampus, and the lower brainstem. DOPA itself appears to be involved in therapeutic and adverse effects on Parkinson's disease, the basic and clinical new aspects of which are surveyed in this book, and further appears to be involved in baroreflex neurotransmission and central regulation of blood pressure, and in pathogenesis of brain neurotoxicity. In general, several criteria of biosynthesis, metabolism, active transport, existence, physiological release, competitive antagonism, and physiological or pharmacological responses, including interactions with the other neurotransmitters such as dopamine, noradrenaline, acetylcholine, $\gamma$-aminobutyric acid (GABA), glutamate, and nitric oxide, must be satisfied before a substance is accepted as a neurotransmitter. Recent accumulating evidence suggests that DOPA fulfills these criteria. In a striatal system simultaneously monitoring DOPA release and the activity of tyrosine hydroxylase, the rate-limiting enzyme for biosynthesis of DOPA and the following catecholamines, $Ca^{2+}$-dependent DOPA release occurs during depolarization, whereas $Ca^{2+}$-dependent activation of tyrosine hydroxylase occurs after the end of depolarization. From inhibition or displacement studies with cold DOPA and its competitive antagonists against selective uptake or binding of labeled ligands, proposed DOPA recognition sites appear to differ from its transport sites, dopamine $D_1$ and $D_2$ receptors, $\alpha_2$- and $\beta$-adrenoceptors, GABA receptors, and ionotropic glutamate receptors. DOPA appears to be a neurotransmitter of the primary baroreceptor afferents terminating in the nucleus tractus solitarii and to be an upstream factor for glutamate release and resultant delayed neuron death in the striatum and the hippocampal CA1 region induced by transient brain ischemia.

It is hoped that this book concerning novel DOPA research will be a valuable addition to the libraries of students and scientists worldwide who are interested in DOPA and catecholamines, in particular, and in the other neurotransmitters, in general. We also believe that studies on neurobiology of DOPA will have implications for better understanding of the therapy of Parkinson's disease and that this book will be able to contribute new ideas for progress in pharmacology and toxicology.

**Yoshimi Misu**

**Yoshio Goshima**

# Acknowledgments

Many thanks go to Dr. Mannfred A. Hollinger, Professor Emeritus, Department of Pharmacology and Toxicology, University of California, Davis, for giving us the opportunity of publishing this DOPA book. We also greatly appreciate the cooperation of the staff of the Taylor & Francis Group.

# About the Editors

**Yoshimi Misu** was born in Japan in 1934, graduated in 1959 from the Faculty of Medicine, Kyoto University, Kyoto, and was awarded his M.D. He finished the postgraduate course, Department of Pharmacology, Faculty of Medicine, Kyoto University, and was awarded a Ph.D. in 1964. From 1964 to 1977, he served as a research associate and associate professor, Department of Pharmacology, Faculty of Pharmaceutical Sciences, Kyoto University. Also from 1965 to 1967, he was a postdoctoral fellow, Department of Pharmacology, State University of New York, Downstate Medical Center, under Professor R.F. Furchgott. From 1977 to 1999, he was Professor and Chairperson, Department of Pharmacology, Yokohama City University School of Medicine, Yokohama, and in 1999 Professor Emeritus, Yokohama City University, and Doctor, Shinobu Hospital, Fukushima.

**Yoshio Goshima** was born in Japan in 1956. After graduating in 1982 from Yokohama City University School of Medicine with his M.D., he served as a research associate, assistant professor, and associate professor, Department of Pharmacology, Yokohama City University School of Medicine from 1982 to 1999. In 1988, he received his Ph.D. From 1993 to 1994, he was a postdoctoral fellow, Department of Developmental Biology, Harvard Medical School, and Department of Neurology and Neurobiology, Yale University School of Medicine, under Professor Stephen M. Strittmatter. In 1999, he was appointed Professor and Chairperson, Department of Molecular Pharmacology and Neurobiology, Yokohama City University Graduate School of Medicine.

# About the Editors

# Contributors

**Akinori Akaike**
Department of Pharmacology
Graduate School of Pharmaceutical Sciences
Sakyo-ku
Kyoto University
Kyoto, Japan

**Muhammad Akbar**
Hasanuddin University
Indonesia

**Ibolya Bodnár**
Neuroendocrine Research Laboratory
Hungarian Academy of Sciences
Department of Human Morphology and Developmental Biology
Semmelweis University
Budapest, Hungary

**María José Casarejos**
Departamento de Neurobiología-Investigación
Hospital Ramón y Cajal
Madrid, Spain

**Stewart L. Chritton**
Harvard Medical School
Brigham and Women's Hospital
Boston, Massachusetts

**Marton I.K. Fekete**
Neuroendocrine Research Laboratory
Hungarian Academy of Sciences
Department of Human Morphology and Developmental Biology
Semmelweis University
Budapest, Hungary

**Justo García de Yébenes**
Departamento de Neurología
Hospital Ramón y Cajal
Madrid, Spain

**Yoshio Goshima**
Department of Molecular Pharmacology and Neurobiology
Yokohama City University School of Medicine
Yokohama, Japan

**Shinichi Hirose**
Department of Pediatrics
Fukuoka University
Fukuoka, Japan

**Kaneyoshi Honjo**
Director, Honjo Bioscience, Inc.
Takatsuki
Osaka, Japan

**Oleh Hornykiewicz**
Center for Brain Research
Medical University of Vienna
Spitalgasse
Vienna, Austria

**Susan P. Hume**
Biology and Modelling R&D
Hammersmith Imanet Ltd.
Hammersmith Hospital
London, U.K.

**Larry W. Hunter**
Department of Physiology and Biomedical Engineering
Mayo Clinic College of Medicine
Rochester, Minnesota

**Kumatoshi Ishihara**
Hiroshima International University
Kure, Japan

**Sunao Kaneko**
Department of Neuropsychiatry
Hirosaki University School of Medicine
Hirosaki, Japan

**Kunio Kitahama**
Physiologie Integrative Cellulaire et Moleculaire
Universite Claude Bernard
Villeurbanne, France

**Eric W. Kristensen**
Research Investigator and Analytical Chemist
Abbott Laboratories
North Chicago, Illinois

**Takao Kubo**
Department of Pharmacology
Showa Pharmaceutical University
Tokyo, Japan

**Li-Hsien Lin**
Laboratory of Neurobiology
Department of Neurology and Neuroscience Program
University of Iowa and Department of Veterans Affairs Medical Center
Iowa City, Iowa

**Stuart A. Lipton**
Center for Neuroscience and Aging
The Burnham Institute
La Jolla, California

**Satoru Masubuchi**
Division of Molecular Brain Science
Department of Brain Sciences
Kobe University Graduate School of Medicine
Kobe, Japan

**Takehiko Maeda**
Department of Pharmacology
Wakayama Medical University
Kimiidera
Wakayama, Japan

**María Angeles Mena**
Departamento de Neurobiología-Investigación
Hospital Ramón y Cajal
Madrid, Spain

**Jaime Menéndez**
Departamento de Neurobiología-Investigación
Hospital Ramón y Cajal
Madrid, Spain

**Yoshimi Misu**
Yokohama City University
Yokohama, Japan

**György M. Nagy**
Neuroendocrine Research Laboratory
Hungarian Academy of Sciences
Department of Human Morphology and Developmental Biology
Semmelweis University
Budapest, Hungary

**Taizo Nakazato**
Department of Neurophysiology
Juntendo University Graduate School of Medicine
Tokyo, Japan

**Motohiro Okada**
Department of Neuropsychiatry
Hirosaki University School of Medicine
Hirosaki, Japan

**Hitoshi Okamura**
Division of Molecular Brain Science
Department of Brain Sciences
Kobe University Graduate School of Medicine
Kobe, Japan

**Márk Oláh**
Neuroendocrine Research Laboratory
Hungarian Academy of Sciences
Department of Human Morphology and Developmental Biology
Semmelweis University
Budapest, Hungary

**Jolanta Opacka-Juffry**
School of Human and Life Sciences
Roehampton University
Whitelands College
London, U.K.

**Eulalia Rodríguez-Martín**
Departamento de Neurobiología-Investigación
Hospital Ramón y Cajal
Madrid, Spain

**Duane K. Rorie**
Department of Anesthesiology
Mayo Clinic College of Medicine
Rochester, Minnesota

**Masashi Sasa**
Hiroshima University
Hiroshima, Japan

**Yukio Sasaki**
Department of Molecular Pharmacology and Neurobiology
Yokohama City University School of Medicine
Yokohama, Japan

**Rosa María Solano**
Departamento de Neurobiología-Investigación
Hospital Ramón y Cajal
Madrid, Spain

**Miao-Kun Sun**
Blanchette Rockefeller Neurosciences Institute
Rockville, Maryland

**Daniel Szekács**
Neuroendocrine Research Laboratory
Hungarian Academy of Sciences
Department of Human Morphology and Developmental Biology
Semmelweis University
Budapest, Hungary

**William T. Talman**
Laboratory of Neurobiology
Department of Neurology and Neuroscience Program
University of Iowa and Department of Veterans Affairs Medical Center
Iowa City, Iowa

**Joakim Tedroff**
A. Carlsson Research AB
Biotech Center
Gothenburg, Sweden

**Gertrude M. Tyce**
Department of Physiology and Biomedical Engineering
Mayo Clinic College of Medicine
Rochester, Minnesota

**Tomoko Ueyama**
Division of Molecular Brain Science
Department of Brain Sciences
Kobe University Graduate School of Medicine
Kobe, Japan

**Miklós Vecsernyés**
Pharmacology and Biopharmacy
Medical School of Debrecen
Debrecen, Hungary

**Shukuko Yoshida**
Institute of Neuroscience and Molecular Biology
Shibata IRIKA
Hirosaki, Japan

**Gang Zhu**
Department of Neuropsychiatry
Hirosaki University School of Medicine
Hirosaki, Japan

# Table of Contents

# Part I

## History

# 1 L-DOPA: A Historical Perspective

*Oleh Hornykiewicz*

## CONTENTS



## 1.1 INTRODUCTION

In the history of drug research and development, it is not unusual to come across chemical compounds that, having been initially regarded as pharmacologically inert, unexpectedly turn out to be potent therapeutic agents. The aromatic amino acid L-3,4-dihydroxyphenylalanine (L-DOPA) offers an instructive example of this "transmutation." At the beginning of the last century — the period when L-DOPA was first found to occur in nature — the view prevailed that naturally occurring amino acids were by definition nontoxic compounds, lacking any pharmacological activity. Thus, for many years, the nonessential amino acid L-DOPA also did not attract much attention from laboratory experimentalists. This position changed when it was realized, in the late 1930s, that L-DOPA was the intermediate metabolite in the biosynthesis of the pharmacologically highly potent catecholamine neurotransmitters noradrenaline and adrenaline from the amino acid L-tyrosine. Today, L-DOPA is the most potent drug available for the treatment of Parkinson's disease. In this role, L-DOPA is a typical example of a prodrug, the pharmacologically effective agent being the dopamine formed from the amino acid in the brain. In contradistinction to its role as a prodrug, L-DOPA has been found, in the last decade, to have biological actions of its own, especially in the brain, displaying many characteristics of a neurotransmitter or neuromodulator substance. It is the aim of this chapter to trace some of the historically prominent stages of these exciting developments.

## 1.2 THE BEGINNINGS

In 1911, 3,4-dihydroxyphenylalanine (DOPA) was first synthesized in the laboratory by Casimir Funk in London as the D,L racemate [1]. Stimulated by the elucidation of the chemical structure of adrenaline as a catechol derivative, and its synthesis in the laboratory in 1904 by Stolz, Funk was searching, similarly to many others at that time, for a substance from which adrenaline could be formed in the body. The chemical similarity between adrenaline and his just-synthesized new catecholic amino acid suggested to Funk "the idea that it may be the parent substance of adrenaline" [1].

In a parallel development, adrenaline's chemical nature as a catechol derivative also attracted the interest of researchers who were investigating the occurrence of catecholic compounds in various plants. In the course of such studies, the Italian pharmacologist Torquato Torquati isolated and crystallized, in 1910 and 1911, a substance from the seedlings of *Vicia faba* (broad beans) that gave the same (pyro)catechol reaction as adrenaline [2]. Although Torquati was both aware of and intrigued by the similarity between his crystalline catecholic substance and adrenaline, he apparently did not pursue the matter further. Torquati's work was, however, immediately repeated by Markus Guggenheim in Switzerland. He proved, in 1913, that Torquati's substance, which he also isolated from *Vicia faba* and crystallized according to Torquati's procedures, was in fact the naturally occurring L-form of Funk's racemic DOPA [3]. Thus, Guggenheim was the first to show that the amino acid L-DOPA occurred in nature as a normal constituent of plant material.

Finally, in 1916, the Swiss dermatologist Bruno Bloch, to whom Guggenheim gave a sample of L-DOPA, published observations that suggested, albeit indirectly, the possibility that L-DOPA occurred normally in animal tissues. Studying histochemically the formation of melanin in the presence of L-DOPA, Bloch found that L-DOPA — but not L-tyrosine as he had believed — was converted (by a "dopaoxydase") to melanin in the melanocytes of mammalian skin [4]. (Today, it is established that L-DOPA is only an intermediate, the parent substance of skin melanin being in fact L-tyrosine, and the enzyme involved a "tyrosinase.") In the course of his studies, Bloch coined for the German name *Dioxyphenylalanin* the word *DOPA*, the acronym generally used today.

## 1.3 FIRST STUDIES OF L-DOPA BIOLOGICAL ACTIVITY

Funk, in his 1911 "Note on the probable formation of adrenaline in the animal body" [5], mentions that experiments with D, L-DOPA, which he had synthesized in the laboratory, "showed that the compound had practically no action on the blood pressure and was without any marked toxic power when injected." Funk does not mention the doses, the species used, or the exact route of administration.

Guggenheim's report in 1913 is more detailed [3]. He states that the L-DOPA that he isolated from *Vicia faba* was "like all hitherto known amino acids pharmacologically fairly indifferent, with 20 mg injected i.v. in a rabbit of average size resulting in no essential change in blood pressure or respiration." Similarly, L-DOPA had, in his hands, no effect on isolated smooth muscle preparations, such as the uterus or the intestine. Also, an oral dose of 1 g given to a rabbit weighing 2.2 kg "did not produce any unusual symptoms." However, when Guggenheim himself took 2.5 g of his L-DOPA orally, he felt very sick after 10 min and had to vomit twice "so that the substance was not completely resorbed." It is probable that, because of this Guggenheim missed some of the behavioral effects of this rather large dose of L-DOPA. Half a century later, in 1960, Ulf von Euler took a fifth of that dose orally and reported that "half a gram of dopa does disturb sleep ... it causes a disagreeable feeling of apprehension and uneasiness," and he thought that "it must have some effects on alertness, too" [6].

The view expressed by Funk and Guggenheim about L-DOPA's lack of biological activity seems to have prevailed for nearly two decades. However, in 1927, Hirai and Gondo in Nagasaki, Japan, published a report which clearly showed that in the rabbit, D, L-DOPA (200 to 300 mg, s.c.) caused a marked hyperglycemia [7]. Two years later, Chikano in Osaka, Japan, obtained (in rabbits) essentially the same result with similar doses of D, L-DOPA [8]. Finally, in 1930, Hasama in Nagasaki demonstrated that in the rabbit, D, L-DOPA produced "contrary to [his] expectations and in contrast to the [vasopressor] effect of adrenaline," a clear fall in the arterial blood pressure, the lowest effective dose given i.v. being 0.5 mg/kg [9]. Hasama thus disproved Guggenheim's negative result obtained in the same species (rabbit) using similar (or higher) i.v. doses of L-DOPA. The paradox that in the rabbit L-DOPA had an effect on the blood pressure opposite to that of adrenaline remained unexplained for many years (see the following text).

## 1.4 DISCOVERY OF DOPA DECARBOXYLASE

In 1911, Funk incubated his D, L-DOPA with suprarenal tissue but could not observe any formation of adrenaline [5]. He thought, very correctly, that this was because there were several intermediate enzymatic steps between L-DOPA and adrenaline. As an example, he points to the conversion, in the gut, of tyrosine to the corresponding amine, namely, *tyramine*, and seems to hint at the possibility that something similar might be happening, in the intestine with L-DOPA.

A quarter century later, in 1936, Bloch and Pinösch, following Guggenheim's advice, gave histidine (imidazolylalanine) — another amino acid with an alanine side chain — subcutaneously to guinea pigs and found a doubling of the amounts of the corresponding amine (histamine) in the animals' lung tissue [10]. This proved the existence of a one-step reaction, i.e., decarboxylation, in the *in vivo* transformation of a naturally occurring amino acid to the corresponding amine. Peter Holtz, who at that time was already studying the conversion of amino acids to amines in Germany, took up the lead provided by the Bloch–Pinösch experiment and demonstrated in 1938 that L-DOPA was indeed readily decarboxylated in kidney extracts directly to dopamine [11]. Holtz and others soon provided evidence for the occurrence of

DOPA decarboxylating activity in other organs (e.g., intestine, liver, adrenal medulla, and pancreas) as well as in nervous tissues (for references, cf. [12]).

The discovery of DOPA decarboxylase immediately suggested to both Holtz [13] and Blaschko [14] the still-valid biosynthetic pathway in the formation of catecholamines in the body (i.e., L-tyrosine → L-DOPA → dopamine → noradrenaline → adrenaline).

## 1.5 L-DOPA AS VASOACTIVE SUBSTANCE

The discovery that L-DOPA was converted to dopamine in animal tissues greatly increased the interest in L-DOPA as a substance of pharmacological and clinical importance. Holtz [15], as well as other research groups [16,17], soon demonstrated that L-DOPA produced an increase in the arterial blood pressure in laboratory animals (cats and rats). This was not too unexpected as dopamine, the reaction product of L-DOPA's decarboxylation, had been already shown in 1910, by Barger and Dale in London, to be (in the cat) a (low-potency) sympathomimetic (vasopressor) substance [18].

The demonstration by Holtz, that the conversion of L-DOPA to dopamine proceeded especially readily in kidney extracts, suggested to some investigators a possible association of the amino acid with the phenomenon of experimental renal (Goldblatt) hypertension, intensively studied in the 1930s and 1940s. Indeed, Bing demonstrated in 1941 that injection of L-DOPA into the circulation of the ischemic kidney of the cat (*in situ*) produced acute renal hypertension; this was not the case when the renal blood flow was maintained at normal levels [19]. Oster and Sorkin [20] obtained analogous results with i.v.-injected L-DOPA in the (Goldblatt) hypertensive cat. Bing, in addition, demonstrated for the first time that the perfused cat kidney indeed converted L-DOPA to dopamine [21].

Oster and Sorkin, in 1942, were the first to study the vasopressor effect of L-DOPA in human subjects [20]. They injected higher doses of L-DOPA (120 to 450 mg i.v.) in normal subjects and in patients with essential hypertension. They observed that in patients with essential hypertension, L-DOPA produced larger rises in arterial blood pressure than in healthy controls. As side effects of L-DOPA, they mention that the larger doses especially caused tachycardia, tinnitus, sweating, nausea, retching, and vomiting. They do not mention any effect on the patients' behavior. Holtz, also in 1942, injected a healthy volunteer with 50 mg of L-DOPA i.v. and noticed "a rather considerable tachycardia"; 150 mg taken orally remained "without any noteworthy effects" [15]. In this experiment, Holtz also provided, for the first time, evidence that L-DOPA was converted to dopamine in the human body; the 50 mg i.v. L-DOPA resulted in the excretion in the urine of nearly 40% of the equivalent amount of dopamine [15].

## 1.6 DOPAMINE: A BIOLOGICALLY ACTIVE SUBSTANCE IN ITS OWN RIGHT

In 1942, Holtz and Credner observed that dopamine had a paradoxical effect on the arterial blood pressure, specifically, in the guinea pig and the rabbit. In these species, dopamine, in contrast to its vasopressor effect in the cat, did not raise the blood pressure as adrenaline did, but lowered it [15]. Holtz explained this paradox by the presence of especially high amounts of the enzyme monoamine oxidase in the guinea pig and rabbit kidney. In the cat, the activity of kidney monoamine oxidase was relatively low, so that (the unmetabolized) dopamine could exert its full sympathomimetic (i.e., vasopressor) effect; in contrast, in the guinea pig and the rabbit, dopamine was assumed to be quickly metabolized by the high amounts of monoamine oxidase to the corresponding aldehyde (3,4-dihydroxyphenylaldehyde), which Holtz held responsible for the "unspecific" fall in blood pressure. However, when in 1957, this author (at that time in Blaschko's laboratory in Oxford) repeated Holtz's experiments on the guinea pig and used the *in vivo* monoamine oxidase inhibitor iproniazid, he found that the action of iproniazid (i.e., inhibition of dopamine's metabolism to the aldehyde), instead of abolishing the fall in blood pressure produced by dopamine, as Holtz had hypothesized, actually potentiated it [22]. Also, L-DOPA's blood-pressure-lowering effect in the guinea pig (first observed in the rabbit by Hasama in 1930 and confirmed later by Holtz and Credner [15], see the preceding text) was potentiated by inhibition of the enzymatic metabolism of dopamine [22]. This was the first clear experimental evidence that dopamine was a biologically active compound in its own right, independent of its function as an intermediate in the biosynthesis of noradrenaline and adrenaline or as a sympathomimetic agent. Later, this conclusion played an important role in this author's conception of the idea of L-DOPA as a dopamine-replenishing substance in the treatment of Parkinson's disease and other conditions accompanied by brain dopamine deficiency.

## 1.7 L-DOPA AS A CENTRALLY ACTING SUBSTANCE

It seems that Holtz was the first to clearly observe the central excitatory action of L-DOPA and relate it to the accumulation of dopamine in the brain. In a paper published in the fall of 1957, dealing with the effect of reserpine, iproniazid, and biogenic amines on the hexobarbital anesthesia, Holtz found that 20 mg/kg i.v. L-DOPA produced "central excitation" in mice pretreated with iproniazid [23]. He mentions the ineffectiveness of the (synthetic) direct noradrenaline precursor 3,4-dihydroxyphenylserine (DOPS) in these experiments; points out the rapid conversion of L-DOPA to dopamine; and concludes that of the two brain sites acted upon by 5-hydroxytryptamine and tryptamine, only one is the pharmacological site of action of "the dopamine formed in the brain from L-DOPA." In this study, Holtz also demonstrated that by its central excitatory action, L-DOPA completely abolished or greatly reduced the central depressant effect of hypnotic doses of hexobarbital [23].

Holtz's study was directly followed by A. Carlsson's report on the awakening effect of L-DOPA in animals "tranquilized" with reserpine at the end of November 1957. In a short communication, Carlsson reported that in reserpinized mice and rabbits, D, L-DOPA in doses of up to 1000 mg/kg i.p. (mice) and 200 mg/kg i.v. (rabbits), dramatically restored the animals' normal behavior, the effect being potentiated by pretreatment with iproniazid [24]. Similar to Holtz before, Carlsson noted that in normal animals (rabbits), "3,4-dihydroxyphenylalanine caused central stimulation which was, likewise, markedly potentiated by iproniazid pretreatment." In contradistinction to Holtz, who attributed L-DOPA's central stimulation to the accumulation of dopamine in the brain [23], Carlsson confined himself to the statement that the study "supports the assumption that the effect of 3,4-dihydroxyphenylalanine was due to an amine (dopamine, noradrenaline, adrenaline) formed from it" [24].

The central excitatory, awakening action of L-DOPA was reproduced in the reserpinized monkey in 1959 by Everett [25], and in 1960 in reserpinized mice by Blaschko and Chrusciel who, for the first time tested, in addition to L-DOPA, the antireserpine potential of several related amino acids [26]. In 1960, Degkwitz demonstrated L-DOPA's effectiveness in counteracting the reserpine "sedation" in human subjects [27].

L-DOPA's action on the brain's electrical activity (EEG) was for the first time examined in 1958 by Monnier and Tissot [28]. After confirming the central excitatory effect of L-DOPA in the normal animal (rabbit), they showed that the amino acid "activates … electrical brain activity through … increase of excitability of the reticular ascending system." In their study, they considered L-DOPA mainly as a precursor of brain noradrenaline.

## 1.8 OCCURRENCE OF L-DOPA AND DOPAMINE IN MAMMALIAN TISSUES

It was not until the beginning of the 1950s that L-DOPA and dopamine were actually found to occur in animal tissues. In 1951, Goodall became the first to identify L-DOPA as a normal tissue constituent of both the adrenal glands [29] and the heart [30]. In 1956, Weil-Malherbe found small amounts of L-DOPA in a pheochromocytoma [31]. In the brain of laboratory animals, L-DOPA was identified first by Montagu in 1957 [32] and later (in the brain of the rabbit) by Weil-Malherbe [33,34]; in the human brain, both Montagu in 1957 [32] and Sano in 1959 [35] found small amounts of L-DOPA.

The occurrence of dopamine in nature was first demonstrated by Schmallfuss in 1931; he identified the amine in the shoots of the yellow *Genista* [36]. In 1942, Holtz and Credner found evidence for the presence of dopamine in normal human urine, implying its occurrence in the (human) body [15]. In mammalian tissues, dopamine was first identified in 1951, together with L-DOPA, in the adrenal glands and the heart of sheep by Goodall [29,30]. Shepherd and West confirmed, in 1953, dopamine's presence in the adrenal medulla [37]. Schümann was the first to show, in 1956, that dopamine was a normal constituent of the adrenergic nerves [38]. Montagu, in 1957, reported for the first time on the occurrence of dopamine in the brain of several species (rabbit, guinea pig, rat, and chicken) as well as in the (whole) human brain [32]. This discovery was soon confirmed (in the rabbit brain) by Weil-Malherbe [33,34] and Carlsson [39]. However, the occurrence of a catecholamine different from noradrenaline and adrenaline, which judging by present knowledge must have been largely dopamine, had already been reported in 1951 by Raab and Gigee; they had found this substance in the brain of several mammalian species (rat, rabbit, dog, cow, bull, and hog), including primates (rhesus monkeys and humans) [40].

## 1.9 DOPAMINE AS PRINCIPAL METABOLITE OF EXOGENOUS L-DOPA IN THE BODY, INCLUDING THE BRAIN

In 1951, van Arman tested various theoretically possible catecholamine precursors for their ability to replenish the adrenal medulla depleted of catecholamines [41]. In this study, van Arman depleted the rat's adrenal glands of their catecholamines with subconvulsive doses of insulin and then treated the animals with (among other substances) phenylalanine, phenylserine, tyrosine, tyramine, and epinine, as well as L-DOPA. He found that of all these substances, only L-DOPA restored the adrenal catecholamine ("adrenaline") levels [41]. It can be assumed that in these experiments the major portion of the catecholamine formed from the administered L-DOPA was actually dopamine. This seems to follow from experiments performed in 1956 by Demis et al. [42] and Hagen and Welch [43]. These investigators incubated DOPA-α-C14 with adrenal homogenates, and recovered most of the radioactivity as dopamine and only a small fraction as radioactive noradrenaline. Also in a pharmacological study, Burn and Rand found, in 1960, that L-DOPA restored the functioning of the postganglionic adrenergic neurons abolished by reserpine treatment [44]; the fact that the restoration of function was only partial suggests that also in these *in vivo* experiments L-DOPA was mainly replenishing the neuronal transmitter stores with the less effective dopamine.

The first researchers to study, as early as in 1951, the effect of systemically given L-DOPA on brain catecholamines were Raab and Gigee [40]. In an attempt to change the concentration of the catecholamine-like substance they discovered in the brain in the late 1940s (see the preceding text), they injected i.p., among many other substances, 300 mg/kg (racemic?) DOPA. They found that only DOPA increased the brain level of their catecholamine-like substance. The time course of the change as described in their report [40] is, as we now know, identical with that of the accumulation of brain dopamine after parenteral administration of L-DOPA.

The studies by van Arman and by Raab and Gigee, both reported in 1951, demonstrated for the first time that L-DOPA had the potential to increase the (reduced or normal) catecholamine content in peripheral tissues (adrenals) and the brain. Therefore, it was not unexpected when both Carlsson et al. [39] and Weil-Malherbe and Bone [34] showed in 1958 that in rabbits whose brain had been depleted by reserpine of its monoamines, including dopamine, L-DOPA preferentially restored the depleted dopamine, with a much smaller effect on the likewise depleted noradrenaline. These two observations, namely, the depletion of brain dopamine by reserpine and the replenishment of the dopamine stores by L-DOPA, were soon to assume a special significance; however, in their published reports, neither Carlsson nor Weil-Malherbe were prepared to draw definite conclusions about dopamine's role in brain function. This changed dramatically upon the discovery of dopamine's specific localization in the brain.

## 1.10 DISCOVERY OF DOPAMINE'S REGIONAL DISTRIBUTION IN THE BRAIN

In their 1951 study, Raab and Gigee [40] also examined the regional distribution of the catecholamine-like compound in the brain of several larger domestic animals (dog, cow, bull, and hog) and in primates (rhesus monkeys and humans). They found the highest amounts concentrated in the caudate nucleus, the difference between this region and the other examined brain structures being most impressive in the bull and the rhesus monkey. In the human brain, the difference between the caudate nucleus and some other regions was much smaller, probably as the result of the postmortem decay of the analyzed catecholamine compound. The major drawback of this study was the very high tissue blanks, as shown by the values for corpus callosum, cerebellum, etc. Although Raab's studies were well known and referred to by notable investigators (e.g., Holtz [45], Vogt [46], Montagu [32], and Rothballer [47]), they did not have any impact on the actual progress in this research area. Thus, when Marthe Vogt in her pioneering 1954 study of the regional distribution of noradrenaline and adrenaline (called *sympathin*) in the brain of the dog also analyzed the caudate nucleus (along with the other brain regions studied) by paper chromatography, she did not try to look for Raab's catecholamine-like substance. She was satisfied with the statement that Raab's "catechol derivative … bears no resemblance to sympathin in its [brain] distribution" because it occurred [in Raab's study] in " … regions containing only minimal amounts of sympathin" [46].

In 1959, 8 yr after Raab's regional distribution study, Bertler and Rosengren [48] in Sweden and Sano [35] in Japan demonstrated, using specific chemical assay procedures, that the bulk of brain dopamine in the dog (Bertler) as well as in the human (Sano) was indeed highly concentrated in the caudate nucleus and putamen (the striatal component of the basal ganglia), thus, in principle, confirming Raab's observations.

This discovery, taken together with the observations about the effects of L-DOPA and reserpine on brain dopamine and the animals' motor behavior, immediately permitted scientists to assign a functional role to brain dopamine, specifically, the possibility that dopamine, as Bertler and Rosengren put it, "is concerned with the function of the corpus striatum and thus with the control of motor function." It was now possible to propose a biochemical explanation for the parkinsonism-like condition produced by reserpine and its reversal by L-DOPA in laboratory animals [48,49], as well as the reversible parkinsonism in human subjects treated with reserpine [35]. The observations also suggested, by analogy, a possible connection between dopamine and Parkinson's disease proper, an idea that immediately and independently occurred to several dopamine research laboratories [49,55,57,64].

## 1.11 L-DOPA IN THE TREATMENT OF PARKINSON'S DISEASE

L-DOPA's path from animal studies to patients with Parkinson's disease was straight and clear. The pertinent literature offers several historical accounts of this development [50–53]. As presented in detail in Section 1.7 to Section 1.10, the most important observations about brain dopamine obtained from animal experiments in the years 1957 to 1959 were, in chronological order, as follows: (1) in 1957, dopamine's occurrence in the mammalian brain was discovered [32] and L-DOPA's central awakening, antihypnotic [23], and antireserpine [24] effects were demonstrated; (2) in 1958, it was found that reserpine released brain dopamine, similar to noradrenaline and adrenaline, from its storage sites in the brain and L-DOPA replenished the depleted brain dopamine stores [34,39]; and (3) in 1959, dopamine's preferential localization in the brain's basal ganglia (striatum) was discovered [35,48].

This cluster of studies, reported by several research groups within a time span of not more than 18 months, set the stage for the subsequent dopamine/L-DOPA work in patients with Parkinson's disease. This research, which eventually led to the discovery of L-DOPA's high therapeutic efficacy in Parkinson patients, was started independently and simultaneously by two research groups in Vienna and Montreal; each of the two groups had its own research background and took its own approach to the question to be solved. As described in Section 1.6, this author became interested in dopamine as a substance of special biological significance when (in 1957) he observed, in Oxford, for the first time that dopamine qualified as a biologically active substance in its own right, independent from noradrenaline and adrenaline [22]. This author continued his interest in dopamine in 1958 in Vienna by studying the effect of cocaine, monoamine oxidase inhibitors, and the parkinsonism-inducing neuroleptic chlorpromazine and cataleptogenic bulbocapnine on the brain levels of dopamine [54] in the rat.

In 1959, immediately following Bertler and Rosengren's study on dopamine's localization in the basal ganglia, this author, together with his postdoctoral collaborator H. Ehringer, started a study on dopamine (and noradrenaline) in the postmortem brains of patients with Parkinson's disease and other basal ganglia disorders. They analyzed the brains of 17 nonneurological controls, 6 patients with clinical basal ganglia symptomatology of unknown etiology, 2 patients with Huntington's disease, and 6 patients with Parkinson's disease (4 with postencephalitic and 2 with idiopathic disease); 2 neonatal brains and 1 infant brain were also examined. Of the 14 cases with basal ganglia disease or symptomatology, only the 6 Parkinson's disease cases had a severe loss of dopamine in the caudate nucleus and putamen [55]. Ehringer and this author were quite aware of the significance and consequences of their observation, which in their view, if found reproducible, "could be regarded as comparable in significance to the histological changes in substantia nigra," so that "a particularly great importance would have to be attributed to dopamine's role in the pathophysiology and symptomatology of idiopathic Parkinson's disease ..." [55]. This discovery, published in 1960, can be regarded as the turning point in the history of brain dopamine; it has provided, until today, a rational basis for all the subsequent research into the mechanisms, causes, and new drug treatments of Parkinson's disease.

In Montreal, T. Sourkes, the discoverer, in 1954 at the Merck Institute in Rahway, N.J., of the DOPA decarboxylase inhibitory action of α-methyldopa continued his keen interest in catecholamine metabolism and found, while working with his Ph.D. student G. Murphy in 1959, that α-methyldopa reduced the levels of dopamine (and noradrenaline) in the brain [56]. Based on the reports on brain dopamine that were available by 1959 (see the preceding text), Sourkes, together with the neurologist A. Barbeau, decided to study, in 1960, the urinary excretion of catecholamines in a variety of extrapyramidal syndromes, including Parkinson's disease. They found that patients with Parkinson's disease, especially the postencephalitic variety, excreted significantly less dopamine in urine than control subjects [57]. In their report published in 1961, they concluded that "analyses of the catecholamine content of the brains of patients who have died with basal ganglia disorders … " should be carried out so as to determine "whether the concentration of cerebral dopamine itself undergoes major changes …" [57]. When making this suggestion, they could not have known that at the time such a study had already been carried out in Vienna (see the preceding paragraph).

The discovery, in Vienna, of the severe lack of dopamine in the striatum, specifically in patients with Parkinson's disease, immediately suggested to this author the next logical step, the step "from brain homogenate to treatment," in what became known as "the dopamine miracle" [58], namely, the therapeutic use of L-DOPA to replace the missing dopamine in the patient. In November 1960, this author suggested to the neurologist W. Birkmayer a clinical trial with i.v. L-DOPA; 8 months later, in July 1961, Birkmayer conducted a trial with 50 to 150 mg L-DOPA injected i.v. in a group of 20 patients with Parkinson's disease (both idiopathic and postencephalitic). The effect of the injected L-DOPA, especially in patients pretreated with a monoamine oxidase inhibitor, was spectacular. The first report published in November 1961 [59] reads as follows:

> The effect of a single i.v. administration of L-DOPA was, in short, a complete abolition or substantial relief of akinesia. Bedridden patients who were unable to sit up; patients who could not stand up when seated; and patients who when standing could not start walking, performed, after L-DOPA all these activities with ease. They walked around with normal associated movements, and they even could run and jump. The voiceless, aphonic speech, blurred by pallilalia and unclear articulation, became forceful and clear as in a normal person. For short periods of time the patients were able to perform motor activities which could not be prompted to any comparable degree by any other known drug [59].

At around the same time, in Montreal, Sourkes and Murphy suggested to Barbeau the use of orally administered L-DOPA in patients with Parkinson's disease [50]. In their first full report, published in 1962 (in French), the authors write:

> … In all cases L-DOPA ameliorated the rigidity, especially when combined with an inhibitor of monoamine oxidase. This amelioration was of the order of 50%. It started within half an hour after ingestion and lasted from 2 to 2.5 hr [60].

Today, it is generally recognized that the discovery of L-DOPA represents one of the triumphs of the pharmacology of our time, being "the defining finding for transmitter-based therapeutics" [61]. It has provided a stimulus for analogous investigations in many other brain disorders, both neurological and psychiatric. The discovery of L-DOPA as the most efficacious antiparkinson agent available "proved to be the culmination of a century and a half search for a treatment of Parkinson's disease" [62].

In the years between 1962 and 1967, several confirmatory and also a few negative L-DOPA trials were published; the negative results were mostly due to unsuitable trial conditions, uncritical patient selection, as well as problems with the clinical diagnoses (see Reference 50). An especially instructive case is that reported by Isamu Sano in Osaka, who in 1959 measured subnormal putamen dopamine in a single Parkinson brain, but " … had been reluctant to speculate, from that single experience, about the pathogenesis of Parkinson's disease" [63]. He then injected 200 mg of L-DOPA i.v. in two patients (pretreated with a monoamine oxidase inhibitor), but instead of testing the effect on the motor deficits (e.g., akinesia or rigidity), he made an electromyogram recording, with the patients lying on the examination table. He was "more interested in [the patients'] subjective complaints [L-DOPA's side effects]" [63]. In a printed (Japanese) version of a lecture in 1960, Sano concludes that " … treatment with DOPA has no practical therapeutic value" [64].

Six years after the original work in Vienna and Montreal, L-DOPA found its way into clinical practice. In 1967, Cotzias, well aware of the work in Vienna and Montreal [65], published the results of a study with gradually increasing large oral doses of D,L-DOPA. He showed that using this regimen, it was possible to achieve and maintain a strong and continuous L-DOPA effect in the patient [66]. Cotzias summarizes his observations as follows:

> Eight of the 16 patients treated orally with 3 to 16 g D,L-DOPA showed either complete, sustained disappearance or marked amelioration, of … tremor, cogwheel phenomenon, rigidity, loss of associated movements, muscular weakness, festination, salivation, and loss of facial expression [66].

Soon after Cotzias' clinical breakthrough study, Melvin Yahr published in 1969 the results of the first double-blind placebo-controlled study with L-DOPA, objectively establishing the amino acid's superior effectiveness as an antiparkinson drug [67].

Despite the unprecedented therapeutic success of L-DOPA, several distinguished neuroscientists expressed doubts and suspicions about its "miraculous" effect in the patient questioning the rationale on which its clinical use was based. Thus, for several years, Carlsson himself, so highly esteemed in the brain dopamine field, seems to have had doubts regarding dopamine's primary role in the central reserpine syndrome (as a model of parkinsonism). At a symposium in 1963, he felt that it was "not possible to draw any conclusions about the relative importance of dopamine and noradrenaline for the central actions of reserpine," substantiating this opinion by the claim that reserpine-treated animals, when injected with the direct noradrenaline precursor D,L-DOPS "are awakened and start to move around almost like normal animals" [68]. He reiterated the idea that "the actions of DOPA and DOPS are similar" [68] as late as 1965, adding that "the possibility should be kept in mind that dopamine can activate not only its own receptors [in the brain] but also those of noradrenaline, and *vice versa*" [69]. By these claims, L-DOPA's crucial dopamine-replacing potential as an antireserpine and antiparkinson drug was greatly undermined for a period of time. Similar doubts were expressed by Bertler and Rosengren, who stated in an important article published in 1966 in *Pharmacological Reviews* [70] that "the effect of L-DOPA was too complex to permit a conclusion about disturbances of the dopamine system in Parkinson's disease," concluding that "with the facts so far available, it is not possible to say what functions dopamine has in the central nervous system" [70]. (Ironically, in the same volume of *Pharmacological Reviews*, this author published his widely noted 1966 review article titled "Dopamine (3-hydroxytyramine) and brain function" [71]).

Echoing the early doubts, such highly distinguished neuroscientists as Herbert Jasper (Montreal Neurological Institute) and Arthur Ward (Neurosurgery Department, University of Washington, Seattle) cynically stated at a symposium on L-DOPA in 1969, that L-DOPA "was the right therapy for the wrong reason" [72,73]. Even as late as 1973, Marthe Vogt, one of the most competent British neuroscientists of the time, called in question the rational basis of L-DOPA's therapeutic use, by arguing that "since L-DOPA floods the brain with dopamine, to relate its [antiparkinson] effects to the natural function of dopamine neurons may be erroneous" [74]. However, none of the doubts and criticisms has withstood the test of time. Most of them were soon dispelled. In 1970, Everett [75] reinstated brain dopamine to its rightful place by bringing together all the experimental evidence disproving the claims of DOPS and noradrenaline. In 1975, Lloyd et al. [76] published a study (on postmortem brain) directly showing that patients treated with L-DOPA had more dopamine in their striatum than untreated patients, with the highest dopamine levels shortly after the last premortem dose of L-DOPA. The definite clinical proof of the dopamine-replenishing mechanism of action of L-DOPA was provided in 1974 by Donald Calne in the first study about the antiparkinson efficacy of bromocriptine, a direct agonist at the striatal postsynaptic dopamine receptors; bromocriptine had qualitatively the same therapeutic antiparkinson action as L-DOPA [77]. In today's clinical practice, L-DOPA is usually combined with substances that prevent its metabolism in the peripheral organs, thus allowing the maximal concentration of L-DOPA substance to penetrate into the brain; these L-DOPA adjuvants are: the peripheral DOPA decarboxylase inhibitors (e.g., carbidopa and benserazide) and the (peripheral) inhibitors of the enzyme catechol-*O*-methyl transferase (e.g., entacapone).

## 1.12 L-DOPA AS A BIOLOGICALLY ACTIVE SUBSTANCE IN ITS OWN RIGHT

In 1984, Reis and his colleagues in New York presented some evidence suggesting, for the first time, the existence of specific L-DOPA neurons in the brain. They found that in the rat, some neurons in the dorsal motor nucleus of the vagus contained tyrosine hydroxylase but lacked immunoreactivity to DOPA decarboxylase which suggested that "these cells … may be considered candidates for L-DOPA neurons" [78]. The actual existence of L-DOPA neurons was for the first time demonstrated in 1988 by Meister et al. in the arcuate nucleus of the rat hypothalamus [79] and by Tison et al. in 1989 in the dorsal vagal complex [80].

In the same time period, the first evidence for specific responses to L-DOPA independent of, and distinct from, dopamine was obtained in the rat brain by Misu, Goshima, and their colleagues in Japan. Their studies can be regarded as the beginning proper of the recognition of brain L-DOPA as a biologically active substance in its own right. In the course of these studies, Misu et al. adduced evidence that L-DOPA fulfills in the brain several criteria of a neurotransmitter, such as having specific localization, synthesis, active transport, release, recognition sites, and antagonists (for references, cf. Reference 81).

In their first studies, published in 1986, the Japanese investigators demonstrated that in the slice preparation of the rat striatum, in which no L-DOPA neurons have so far been found, and in the hypothalamus, L-DOPA had a dose-related, dual, facilitatory/inhibitory effect on the spontaneous as well as electrically evoked release of the endogenous catecholamine neurotransmitters [82,83]. Thus, in the striatum, dopamine release was facilitated by low (submicromolar) doses of L-DOPA, and inhibited by higher (micromolar) doses. The facilitatory effect of L-DOPA was mediated via presynaptic β-adrenoceptors (i.e., sensitive to propranolol), whereas the inhibitory effect was exerted via stimulation of the presynaptic $D_2$ dopamine receptors (i.e., sulpiride sensitive). A presynaptic site of action of exogenous L-DOPA has also been considered likely by Tedroff et al. in a recent positron emission tomography (PET) study in the primate (monkey) [84]. Misu and his group also demonstrated in 1988 that electrical field stimulation released in superfused striatal slices, *inter alia*, L-DOPA in a neurotransmitter-like manner [85]. They also found that in the striatum, L-DOPA potentiated the postsynaptic, $D_2$-receptor-related effects of dopamine. A rapid modulation (upregulation) by acute L-DOPA of the striatal postsynaptic $D_2$ dopamine receptors has also been observed by Hume et al. in a PET study in the rat [86]. Both the presynaptic and postsynaptic dopamine facilitatory effects of L-DOPA may play an auxiliary role in L-DOPA's high clinical efficacy in Parkinson's disease. In addition to its effects on dopamine neurotransmission, L-DOPA proved to be an effective releaser of glutamate, as studied in striatal (and hippocampal) preparations [87]. The various pharmacological effects of L-DOPA persisted after inhibition of the brain DOPA decarboxylase (with NSD-1015).

Summarizing this aspect of their work, Misu et al. propose that in the striatum, L-DOPA may be a neuromodulator substance, facilitating dopamine neurotransmission both by a presynaptic mechanism (i.e., via stimulation of the facilitatory β-adrenoceptors) and by a postsynaptic mechanism as a potentiator at the $D_2$ dopamine receptors (cf. [81,88]). On the other hand, L-DOPA, by releasing vesicular glutamate, is thought to directly contribute to some of the side effects of L-DOPA therapy in Parkinson's disease (especially dyskinesias) or possibly exert (excito)toxic effects (cf. Refernce 81 and Reference 87) indirectly.

Starting in the 1990s, Misu et al., in extensive experimental analysis of L-DOPA's physiological actions in the brain, presented evidence for the possibility that L-DOPA qualifies as a neurotransmitter in the primary baroreceptor afferents terminating in the nucleus tractus solitarii [89–91]. In the nucleus tractus solitarii, as in the neurons of the dorsal motor vagal nuclear complex [80], there exists an L-DOPA neuronal system [89–91]. During microdialysis of the nucleus tractus solitarii, L-DOPA's basal release was in part $Ca^{++}$ dependent and tetrodotoxin sensitive, and abolished by α-methyl-*p*-tyrosine. The authors also showed that after denervation of the primary baroreceptor afferents, L-DOPA (but not dopamine or dopamine-β-hydroxylase) immunoreactivity decreased in the nucleus tractus solitarii [90]. On the other hand, activation of the baroreceptors by i.v. phenylephrine released L-DOPA in this region [90]. Summarizing their experimental observations, Misu et al. [90] concluded:

> Taken together with previous evidence for the stereoselective involvement of exogenously applied L-DOPA itself in cardiovascular control in the NTS [nucleus tractus solitarii], it is probable that L-DOPA is a neurotransmitter of the primary baroreceptor afferents in rat. Endogenously released L-DOPA tonically functions to activate depressor neurons for regulation of BP [blood pressure] in the NTS.

Nothing needs to be added to this statement. There exist in the recent literature several excellent reviews, as well as critical evaluations, of this new, theoretically and clinically highly interesting, field of brain L-DOPA research [81,88,92–96].

## REFERENCES


1. Funk, C., Synthesis of dl-3:4-dihydroxyphenylalanine, *J. Chem. Soc.,* 99, 554, 1911.
2. Torquati, T., Sulla presenza di una sostanza azotata nei germogli dei semi di "Vicia faba," *Arch. Farmacol. Sper. Sci. Affini,* 15, 213, 1913.
3. Guggenheim, M., Dioxyphenylalanin, eine neue Aminosäure aus Vicia faba, *Hoppe-Seyler's Zeitschr. Physiol. Chem.,* 88, 276, 1913.
4. Bloch, B., Chemische Untersuchungen über das spezifische pigmentbildende Ferment der Haut, die Dopaoxydase, *Hoppe-Seyler's Zeitschr. Physiol. Chem.,* 98, 226, 1917.
5. Funk, C., Note on the probable formation of adrenaline in the animal body, *J. Physiol.,* 43, iv, 1911.
6. von Euler, U., Discussion, in *Adrenergic Mechanisms,* Vane, J.R., Wolstenholme, G.E.W., and O'Connor, M., Eds., Ciba Foundation Symposium, Churchill, London, 1960, 551–569.
7. Hirai, K. and Gondo, K., Über Dopa-Hyperglykämie, *Biochem. Zeitschr.,* 189, 92, 1927.
8. Chikano, M., Über den Einfluß von Aminosäuren und ihren Abkömmlingen auf die Adrenalinhyperglykämie, *Biochem. Zeitschr.,* 205, 154, 1929.
9. Hasama, B., Beiträge zur Erforschung der Bedeutung der chemischen Konfiguration für die pharmakologischen Wirkungen der adrenalinähnlichen Stoffe, *Arch. Exp. Path. Pharmak.,* 153, 161, 1930.
10. Bloch, W. and Pinösch, H., Die Umwandlung von Histidin in Histamin im tierischen Organismus, *Hoppe-Seyler's Zeitschr. Physiol. Chem.,* 239, 236, 1936.
11. Holtz, P., Heise, R., and Lüdtke, K., Fermentativer Abbau von L-Dioxyphenylalanin (Dopa) durch Niere, *Naunyn-Schmiedeberg's Arch. Exp. Pathol. Pharmak.,* 191, 87, 1938.
12. Holtz, P., Role of L-dopa decarboxylase in the biosynthesis of catecholamines in nervous tissue and the adrenal medulla, *Pharmacol. Rev.,* 11, 317, 1959.
13. Holtz, P., Dopa decarboxylase, *Naturwissenschaften,* 27, 724, 1939.
14. Blaschko, H., The specific action of L-dopa decarboxylase, *J. Physiol.,* 96, 50P, 1939.
15. Holtz, P. and Credner, K., Die enzymatische Entstehung von Oxytyramin im Organismus und die physiologische Bedeutung der Dopadecarboxylase, *Naunyn-Schmiedeberg's Arch. Exp. Pathol. Pharmak.,* 200, 256, 1942.
16. Schroeder, H.A., Arterial hypertension in rats, *J. Exp. Med.,* 75, 513, 1942.
17. Page, W.E. and Reed, R., Hypertensive effect of L-dopa and related compounds in the rat, *Am. J. Physiol.,* 143, 122, 1945.

18. Barger, G. and Dale, H.H., Chemical structure and sympathomimetic action of amines, *J. Physiol.,* 41, 19, 1910.
19. Bing, R.J. and Zucker, M.B., Renal hypertension produced by an amino acid, *J. Exp. Med.,* 74, 235, 1941.
20. Oster, K.A. and Sorkin, S.Z., Effects of intravenous injections of L-dopa upon blood pressure, *Proc. Soc. Exp. Biol. Med.,* 51, 67, 1942.
21. Bing, R.J., The formation of hydroxytyramine by extracts of renal cortex and by perfused kidneys, *Am. J. Physiol.,* 132, 497, 1941.
22. Hornykiewicz, O., The action of dopamine on the arterial blood pressure of the guinea pig, *Br. J. Pharmacol.,* 13, 91, 1958.
23. Holtz, P. et al., Beeinflussung der Evipannarkose durch Reserpin, Iproniazid und biogene Amine, *Arch. Exp. Pathol. Pharmak.,* 231, 333, 1957.
24. Carlsson, A., Lindqvist, M., and Magnusson, T., 3,4-Dihydroxyphenylalanine and 5-hydroxytryptophan as reserpine antagonists, *Nature,* 180, 1200, 1957.
25. Everett, G.M. and Toman, J.E.P., Mode of action of Rauwolfia alkaloids and motor activity, *Biol. Psychiatr.,* 1, 75, 1959.
26. Blaschko, H. and Chrusciel, T.L., The decarboxylation of amino acids related to tyrosine and their awakening action in reserpine-treated mice, *J. Physiol.,* 151, 272, 1960.
27. Degkwitz, R. et al., Über die Wirkungen des L-DOPA beim Menschen und deren Beeinflussung durch Reserpin, Chlorpromazin, Iproniazid und Vitamin B6, *Klin. Wochenschr.,* 38, 120, 1960.
28. Monnier, M. and Tissot, R., Action de la réserpine et de ses médiateurs (5-hydroxytryptophan – sérotonine et dopa – noradrénaline) sur le comportement et le cerveau du lapin, *Helv. Physiol. Acta,* 16, 255, 1958.
29. Goodall, Mc.G., Dihydroxyphenylalanine and hydroxytyramine in mammalian suprarenals, *Acta Chem. Scand.,* 4, 550, 1950.
30. Goodall, Mc.G., Studies of adrenaline and noradrenaline in mammalian hearts and suprarenals, *Acta Physiol. Scand.,* 24 (Suppl. 85), 7, 1951.
31. Weil-Malherbe, H., Phaeochromocytoma. Catechols in urine and tumour tissue, *Lancet,* ii, 282, 1956.
32. Montagu, K.A., Catechol compounds in rat tissues and in brains of different animals, *Nature,* 180, 244, 1957.
33. Weil-Malherbe, H. and Bone, A.D., Intracellular distribution of catecholamines in the brain, *Nature,* 180, 1050, 1957.
34. Weil-Malherbe, H. and Bone, A.D., Effect of reserpine on the intracellular distribution of catecholamines in the brain stem of the rabbit, *Nature,* 181, 1474, 1958.
35. Sano, I. et al., Distribution of catechol compounds in human brain, *Biochim. Biophys. Acta,* 32, 586, 1959.
36. Schmalfuss, H. and Heider, A., Tyramin und Oxytyramin, blutdrucksteigernde Schwarzvorstufen des Besenginsters Sarothamnus scoparius Wimm, *Biochem. Zeitschr.,* 236, 226, 1931.
37. Shepherd, D.M. and West, G.B., Hydroxytyramine and the adrenal medulla, *J. Physiol.,* 120, 15, 1953.
38. Schümann, H.J., Nachweis von Oxytyramin (Dopamin) in sympathischen Nerven und Ganglien, *Arch. Exp. Pathol. Pharmak.,* 227, 566, 1956.
39. Carlsson, A. et al., On the presence of 3-hydroxytyramine in brain, *Science,* 127, 471, 1958.
40. Raab, W. and Gigee, W., Concentration and distribution of "encephalin" in the brain of humans and animals, *Proc. Soc. Exp. Biol. Med.,* 76, 97, 1951.

41. van Arman, C.G., Amino acids and amines as precursors of epinephrine, *Am. J. Physiol.,* 164, 476, 1951.
42. Demis, D.J., Blaschko, H., and Welch, A.D., The conversion of dihydroxyphenylalanine-2-C14 (dopa) to norepinephrine by bovine adrenal medullary homogenates, *J. Pharmacol. Exp. Ther.,* 117, 208, 1956.
43. Hagen, P. and Welch, A.D., The adrenal medulla and the biosynthesis of pressor amines, *Recent Prog. Hormone Res.,* 12, 27, 1956.
44. Burn, J.H. and Rand, M.J., The effect of precursors of noradrenaline on the response to tyramine and sympathetic stimulation, *Br. J. Pharmacol.,* 15, 47, 1960.
45. Holtz, P., Über die sympathicomimetische Wirksamkeit von Gehirnextrakten, *Acta Physiol. Scand.,* 20, 354, 1950.
46. Vogt, M., The concentration of sympathin in different parts of the central nervous system under normal conditions and after the administration of drugs, *J. Physiol.,* 123, 451, 1954.
47. Rothballer, A.B., The effects of catecholamines on the central nervous system, *Pharmacol. Rev.,* 11, 494, 1959.
48. Bertler, A. and Rosengren, E., Occurrence and distribution of dopamine in brain and other tissues, *Experientia,* 15, 10, 1959.
49. Carlsson, A., The occurrence, distribution and physiological role of catecholamines in the nervous system, *Pharmacol. Rev.,* 11, 490, 1959.
50. Sourkes, T.L. and Gauthier, S., Levodopa and dopamine agonists in the treatment of Parkinson's disease, in *Discoveries in Pharmacology,* Vol. 1: *Psycho- and Neuropharmacology,* Parnham, M.J. and Bruinvels, J., Eds., Elsevier, Amsterdam, 1983, 249.
51. Hornykiewicz, O., From dopamine to Parkinson's disease: a personal research record, in *The Neurosciences: Paths of Discovery II,* Samson, F. and Adelman, G., Eds., Birkhäuser, Boston, 1992, 125.
52. Roe, D.L., From DOPA to Parkinson's disease: the early history of dopamine research, *J. Hist. Neurosci.,* 6, 291, 1997.
53. Hornykiewicz, O., How L-DOPA was discovered as a drug for Parkinson's disease 40 years ago, *Wien. Klin. Wochenschr.,* 113, 855, 2001.
54. Holzer, G. and Hornykiewicz, O., Über den Dopamin-(3-Hydroxytyramin-) Stoffwechsel im Gehirn der Ratte, *Arch. Exp. Pathol. Pharmak.,* 237, 27, 1959.
55. Ehringer, H. and Hornykiewicz, O., Verteilung von Noradrenalin und Dopamin (3-Hydroxytyramin) im Gehirn des Menschen und ihr Verhalten bei Erkrankungen des extrapyramidalen Systems, *Klin. Wochenschr.,* 38, 1236, 1960.
56. Murphy, G. and Sourkes, T.L., Effect of catecholamino acids on the catecholamine content of rat organs, *Rev. Can. Biol.,* 18, 379, 1959.
57. Barbeau, A., Murphy, G.F., and Sourkes, T.L., Excretion of dopamine in diseases of basal ganglia, *Science,* 133, 1706, 1961.
58. Hornykiewicz, O., Dopamine miracle: from brain homogenate to dopamine replacement, *Movement Disord.,* 17, 501, 2002.
59. Birkmayer, W. and Hornykiewicz, O., Der L-3,4-Dioxyphenylalanin (= DOPA)-Effekt bei der Parkinson-Akinese, *Wien. Klin. Wochenschr.,* 73, 787, 1961.
60. Barbeau, A., Sourkes, T.L, and Murphy, G.F., Les catécholamines dans la maladie de Parkinson, in *Monoamines et système nerveux central,* de Ajuriaguerra, J., Ed., Georg & Cie SA, Geneva, 1962, 247.
61. Hardy, J., Private communication to O. Hornykiewicz, Letter of September 2002.
62. Sourkes, T.L., How dopamine was recognised as a neurotransmitter: a personal view. *Parkinsonism Relat. Disord.,* 6, 63, 2000.

63. Sano I., Private communication to O. Hornykiewicz, Letter dated March 20, 1962.
64. Sano I., Biochemistry of the extrapyramidal system (in Japanese), *Shinkei Kenkyu no Shimpo,* 5, 42, 1960. (English translation: *Parkinsonism Relat. Disord.,* 6, 3, 2000.)
65. Cotzias, G.C. et al., Melanogenesis and extrapyramidal diseases, *Fed. Proc.,* 27, 713, 1964.
66. Cotzias, G.C., Van Woert, M.H., and Schiffer, L.M., Aromatic amino acids and modification of parkinsonism, *New Engl. J. Med.,* 276, 374, 1967.
67. Yahr, M.D. et al., Treatment of parkinsonism with levodopa, *Arch. Neurol.,* 21, 343, 1969.
68. Carlsson, A., Functional significance of drug-induced changes in brain monoamine levels, *Progr. Brain Res.,* 8, 9, 1964.
69. Carlsson, A., Drugs which block the storage of 5-hydroxytryptamine and related amines, in *Handbuch der experimentellen Pharmakologie, XIX: 5-Hydroxytryptamine and Related Indolealkylamines,* Eichler, O. and Farah, A., Eds., Springer, Berlin, Heidelberg, New York, 1965, 529.
70. Bertler, A. and Rosengren, E., Possible role of brain dopamine, *Pharmacol. Rev.,* 18, 769, 1966.
71. Hornykiewicz, O., Dopamine (3-hydroxytyramine) and brain function. *Pharmacol. Rev.,* 18, 925, 1966.
72. Jasper, H.H., Neurophysiological mechanisms in parkinsonism, in L-*DOPA and Parkinsonism,* Barbeau, A. and McDowell, F.H., Eds., FA Davies, Philadelphia, 1970, 408.
73. Ward, A.A., Physiological implications in the dyskinesias, in *L-DOPA and Parkinsonism,* Barbeau, A. and McDowell, F.H., Eds., FA Davies, Philadelphia, 1970, 151.
74. Vogt, M., Functional aspects of the role of catecholamines in the central nervous system, *Br. Med. Bull.,* 29, 168, 1973.
75. Everett, G.M., Evidence for dopamine as a central neuromodulator: neurological and behavioral implications, in *L-DOPA and Parkinsonism,* Barbeau, A. and McDowell, F.H., Eds., FA Davies, Philadelphia, 1970, 364.
76. Lloyd, K.G., Davidson, L., and Hornykiewicz, O., The neurochemistry of Parkinson's disease: effect of L-dopa therapy, *J. Pharmacol. Exp. Ther.,* 195, 453, 1975.
77. Calne, D.B. et al., Bromocriptine in parkinsonism, *Br. Med. J.,* 4, 442, 1974.
78. Jaeger, C.B. et al., Aromatic L-amino acid decarboxylase in the rat brain: immunocytochemical localization in neurons of the brain stem, *Neuroscience,* 11, 691, 1984.
79. Meister, B. et al., Do tyrosine hydroxylase-immunoreactive neurons in the ventrolateral arcuate nucleus produce dopamine or only L-dopa?, *J. Chem. Neuroanat.,* 1, 59, 1988.
80. Tison, F. et al., L-DOPA in the rat dorsal vagal complex: an immunochemical study by light and electron microscopy, *Brain Res.,* 497, 260, 1989.
81. Misu, Y., Kitahama, K., and Goshima, Y., L-3,4-Dihydroxyphenylalanine as a neurotransmitter candidate in the central nervous system, *Pharmacol. Ther.,* 97, 117, 2003.
82. Misu, Y., Goshima, Y., and Kubo, T., Biphasic actions of L-DOPA on the release of endogenous dopamine via presynaptic receptors in rat striatal slices, *Neurosci. Lett.,* 72, 194, 1986.
83. Goshima, Y., Kubo, T., and Misu, Y., Biphasic actions of L-DOPA on the release of endogenous noradrenaline and dopamine from rat hypothalamic slices, *Br. J. Pharmacol.,* 89, 229, 1986.
84. Tedroff, J. et al., L-DOPA modulates striatal dopaminergic function in vivo: evidence from PET investigations in nonhuman primates, *Synapse,* 25, 56, 1997.
85. Goshima, Y., Kubo, T., and Misu, Y., Transmitter-like release of endogenous 3,4-dihydroxyphenylalanine from rat striatal slices, *J. Neurochem.,* 50, 1725, 1988.

86. Hume, S.P. et al., Effect of L-dopa and 6-hydroxydopamine lesioning on [11C]raclopride binding in rat striatum, quantified using PET, *Synapse,* 21, 45, 1995.
87. Furukawa, N. et al., Endogenously released DOPA is a causal factor for glutamate release and resultant delayed neuronal cell death by transient ischemia in rat striata, *J. Neurochem.,* 76, 815, 2001.
88. Opacka-Juffry, J. and Brooks, D.J., L-Dihydroxyphenylalanine and its decarboxylase: new ideas on their neuroregulatory roles, *Movement Disord.,* 10, 241, 1995.
89. Kubo, T. et al., Evidence for L-DOPA systems responsible for cardiovascular control in the nucleus tractus solitarii of the rat, *Neurosci. Lett.,* 140, 153, 1992.
90. Yue, J.-L. et al., Baroreceptor-aortic nerve-mediated release of endogenous L-3,4-dihydroxyphenylalanine and its tonic depressor function in the nucleus tractus solitarii of rats, *Neuroscience,* 62, 145, 1994.
91. Yue, J.-L. et al., Altered tonic L-3,4-dihydroxyphenylalanine systems in the nucleus tractus solitarii and the rostral ventrolateral medulla of spontaneously hypertensive rats, *Neuroscience,* 67, 95, 1995.
92. Misu, Y., Ueda, H., and Goshima, Y., Neurotransmitter-like actions of L-DOPA, *Adv. Pharmacol.,* 32, 427, 1995.
93. Fisher, A. et al., Dual effects of L-3,4-dihydroxyphenylalanine on aromatic L-amino acid decarboxylase, dopamine release and motor stimulation in the reserpine-treated rat: evidence that behaviour is dopamine independent, *Neuroscience,* 95, 97, 2000.
94. Treseder, S.A., Rose, S., and Jenner, P., The central aromatic amino acid DOPA decarboxylase inhibitor, NSD-1015, does not inhibit L-DOPA-induced circling in unilateral 6-OHDA-lesioned rats, *Eur. J. Neurosci.,* 13, 162, 2001.
95. Misu, Y., Goshima, Y., and Miyamae, T., Is DOPA a neurotransmitter? *Trends Pharmacol. Sci.,* 23, 262, 2002.
96. Misu, Y. et al., DOPA causes glutamate release and delayed neuron death by brain ischemia in rats, *Neurotoxicol. Teratol.,* 24, 629, 2002.

# *Part II*

## *Biosynthesis, Metabolism, Active Transport, and Existence*

# 2 DOPA as a Neurotransmitter Candidate

*Yoshimi Misu and Yoshio Goshima*

## CONTENTS



## 2.1 INTRODUCTION

Since the 1950s [1,2], exogenously applied L-3,4-dihydroxyphenylalanine (levodopa) has been generally believed to be an inert amino acid that alleviates the symptoms of Parkinson's disease via its conversion to dopamine by the enzyme aromatic L-amino acid decarboxylase (AADC) [3,4]. This conversion initiates the resultant phenomena such as displacement of tritiated dopamine [5] and impulse-evoked release of radioactive dopamine [6] from brain slices. Both striatal accumulation of dopamine [7,8] and increase in 3-methoxytyramine, a metabolite formed from dopamine by the action of the enzyme catechol-*O*-methyltransferase (COMT) in the synaptic cleft [8], also occur. Dopamine converted from levodopa was thought to activate presynaptic inhibitory dopamine receptors leading to decreases in the firing rate of cells in the zona compacta and ventral tegmental area [9]. The converted dopamine also inhibits the cardiac acceleration caused by electrical sympathetic nerve stimulation, via presynaptic dopamine receptors [10]. These effects following the administration of levodopa are abolished by an AADC inhibitor and are mimicked by dopamine [10] and apomorphine, a dopamine agonist [9].

In 1986 [11,12], we proposed that endogenous L-3,4-dihydroxyphenylalanine (DOPA) is a neurotransmitter or neuromodulator in its own right, in addition to being a precursor of dopamine [13–15]. Recently accumulated evidence suggests that DOPA fulfills the classical criteria related to biosynthesis, metabolism, active transport, presence, physiological release, competitive antagonism, and physiological or pharmacological responses, including interactions with the other neurotransmitter systems [16,17]; these criteria must be satisfied before a compound is accepted as a neurotransmitter [18].

In the course of our studies on the presynaptic regulatory mechanisms of endogenous catecholamine release from rat hypothalamic slices [19–26], we found an unexpected peak, increased by electrical field stimulation, on chromatographic charts. We identified it with DOPA in striatal slices [27]. We further found that the evoked release of DOPA is in part tetrodotoxin sensitive and $Ca^{2+}$ dependent (see Chapter 5).

The first evidence for responses to levodopa *per se* (and not to the dopamine converted from levodopa) is the presynaptic biphasic regulatory actions of levodopa on the impulse-evoked release of endogenous noradrenaline from rat hypothalamic slices (see Chapter 8). Nanomolar levodopa facilitates this via activation of presynaptic β-adrenoceptors in a propranolol-sensitive manner in the presence of intact activity of AADC. This facilitation is neither inhibited by the AADC inhibitor [11] nor mimicked by nanomolar dopamine and apomorphine [26]. In contrast, micromolar levodopa inhibits noradrenaline release via presynaptic dopamine $D_2$ receptors in a sulpiride-sensitive manner under inhibition of AADC [11]. Such a biphasic concentration–response relationship for nanomolar and micromolar levodopa is completely opposite to that for dopamine agonists regulating noradrenaline release [26]. Nanomolar dopamine elicits the sulpiride-sensitive inhibition via dopamine $D_2$ receptors, whereas micromolar dopamine elicits the propranolol-sensitive facilitation via β-adrenoceptors. Nanomolar apomorphine also inhibits noradrenaline release in a manner similar to dopamine. In rat striatal slices, nanomolar levodopa also facilitates dopamine release via presynaptic β-adrenoceptors in the absence and presence of AADC inhibition [12]. Furthermore, noneffective picomolar levodopa potentiates the β-agonist-induced facilitation of noradrenaline release via presynaptic β-adrenoceptors in hypothalamic slices [28].

DOPA methyl ester (DOPA ME) was the first competitive antagonist of levodopa to be identified [29] (See Chapter 8). The facilitation of noradrenaline release by nanomolar levodopa in hypothalamic slices is antagonized by DOPA ME in a competitive manner and is also antagonized by propranolol in a noncompetitive manner, suggesting that DOPA recognition sites differ from β-adrenoceptors. DOPA ME antagonizes the potentiation induced by picomolar levodopa of β-agonist-iduced facilitation completely (to the level of facilitation obtained by β-agonist alone [28]). Among the DOPA ester compounds, DOPA cyclohexyl ester is the most potent and a relatively stable competitive antagonist [30,31]. Competitive antagonism [29,30,32] suggests the existence of DOPA recognition sites on which DOPA or levodopa acts to elicit physiological or pharmacological responses. DOPA ME microinjected into depressor sites of the nucleus tractus solitarii (NTS) antagonizes the hypotension and bradycardia elicited by either electrical stimulation of the aortic depressor nerve (ADN) or microinjected exogenous levodopa [33,34] (see Chapter 11). The ADN is one of the primary baroreceptor afferents terminating in the NTS. The NTS is the gate of baroreflex neurotransmission in the lower brainstem.

Responses to DOPA or levodopa include the various types of interactions with the other neurotransmitter systems. From binding studies, however, DOPA recognition sites appear to differ from dopamine $D_1$ and $D_2$ receptors, $\alpha_2$- and β-adrenoceptors [29,31], γ-aminobutyric acid receptors [35], and ionotropic glutamate receptors [31,36]. From uptake studies, DOPA recognition sites appear to differ from $Na^+$-dependent DOPA transport sites [37,38].

Misu and Goshima [13], Misu et al. [14–17], Opacka-Juffry and Brooks [39], Sun [40], and Tedroff [41] have published review articles related to DOPA.

## 2.2 BIOSYNTHESIS AND METABOLISM OF DOPA

DOPA is synthesized from tyrosine by the action of tyrosine hydroxylase (TH), the rate-limiting enzyme of catecholamine biosynthesis, and is converted to dopamine by the action of AADC (Figure 2.1) [13–18]. Dopamine is converted to noradrenaline by the action of the enzyme dopamine-β-hydroxylase. Dopamine and noradrenaline are widely distributed neurotransmitters in the central and peripheral nervous systems [18]. Noradrenaline is further converted to adrenaline by the action of the enzyme phenylethanolamine-*N*-methyltransferase. Adrenaline appears to play a limited but significant role as a neurotransmitter in the central nervous system [23–25,42,43] and the peripheral adrenal–sympathetic nervous system [44]. These catecholamines are metabolized by the actions of COMT and the enzyme monoamine oxidase (MAO).

On the other hand, the involvement of several metabolic or degradation pathways of DOPA, and dopamine converted from DOPA, in excitotoxic neuronal damages [45,46] have been proposed. These are autoxidation or enzymatic oxidation of DOPA and dopamine into the reactive free radicals, DOPA quinones and dopamine quinones [47–53] (see Chapter 18 and Chapter 19), and a 3,4,6-trihydroxyphenylalanine (TOPA)-quinone, a non-*N*-methyl-D-aspartate agonist and excitotoxin [54–56] (Figure 2.1). Apart from these degradation pathways, excitotoxicity due to increases in extracellular glutamate, a representative excitatory amino acid neurotransmitter, has been proposed to be one of the major causal factors for degenerative neuronal disorders [57–60]. Furthermore, DOPA appears to be, by itself, an upstream factor for glutamate release and the resulting delayed neuron death in rat striata *in vitro* [46,61] and *in vivo* [62] (see Chapter 18 and Chapter 20). The interactions of the released glutamate and degradation products resulting from DOPA oxidation might start amplifying cycles of neurotoxic cascades.

## 2.3 IDENTIFICATION OF DOPA PEAK

As shown in Figure 2.2, putative DOPA and dopamine (DA) peaks in superfusates and dialysates are consistently detectable by high-performance liquid chromatography (HPLC) with electrochemical detection (ECD) in striata *in vitro* [27] and *in vivo* [63]. On chromatographic charts, the peak of a putative DOPA in dialysate comigrates with that of authentic levodopa when the pH of the buffer solution is slightly changed.

As shown in Figure 2.3, the retention time of putative DOPA differs from that of dopamine, noradrenaline, and the major metabolites produced following bioconversion

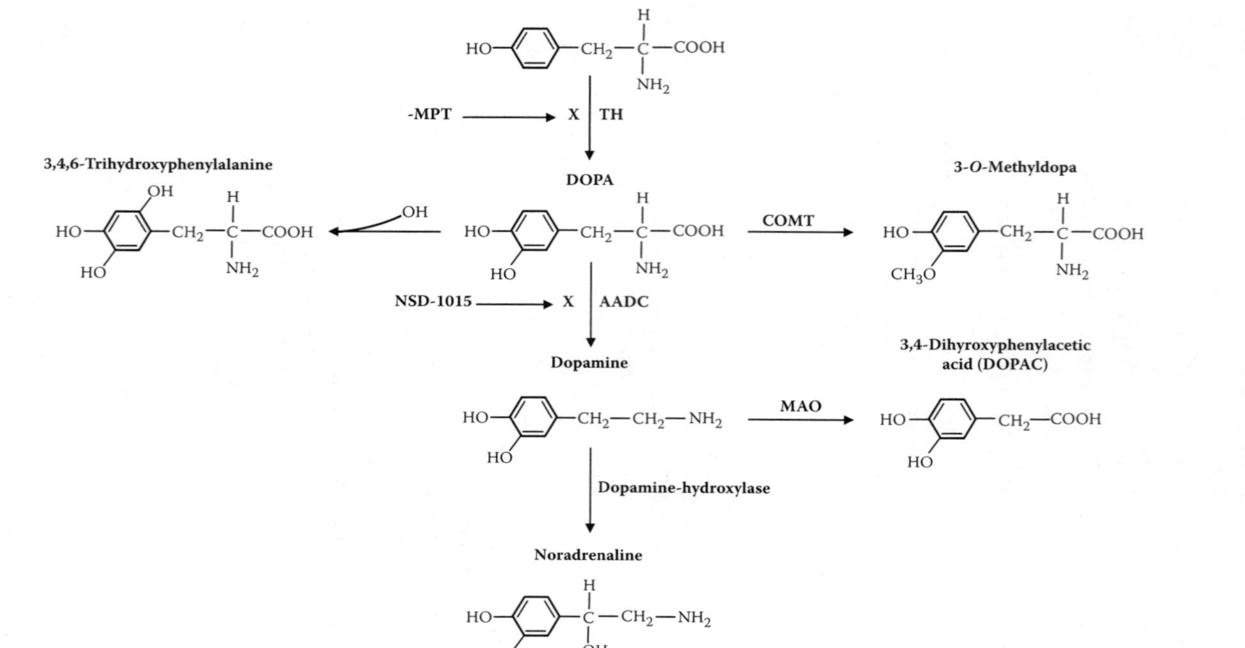


**FIGURE 2.1** Major pathways for the biosynthesis and metabolism of DOPA. DOPA is produced from tyrosine by the rate-limiting enzyme of catecholamine biosynthesis, TH. α-MPT inhibits the action of TH. DOPA is metabolized to 3-*O*-methyldopa by the action of COMT or is converted to dopamine by the action of AADC. NSD-1015 inhibits central AADC. DOPA autoxidation yields 3,4,6-trihydroxyphenylalanine and a neurotoxic quinone derivative. (From Misu, Y., Kitahama, K., and Goshima, Y., L-3,4-Dihydroxyphenylalanine as a neurotransmitter candidate in the central nervous system, *Pharmacol. Ther.*, 97, 117, 2003. With permission.)

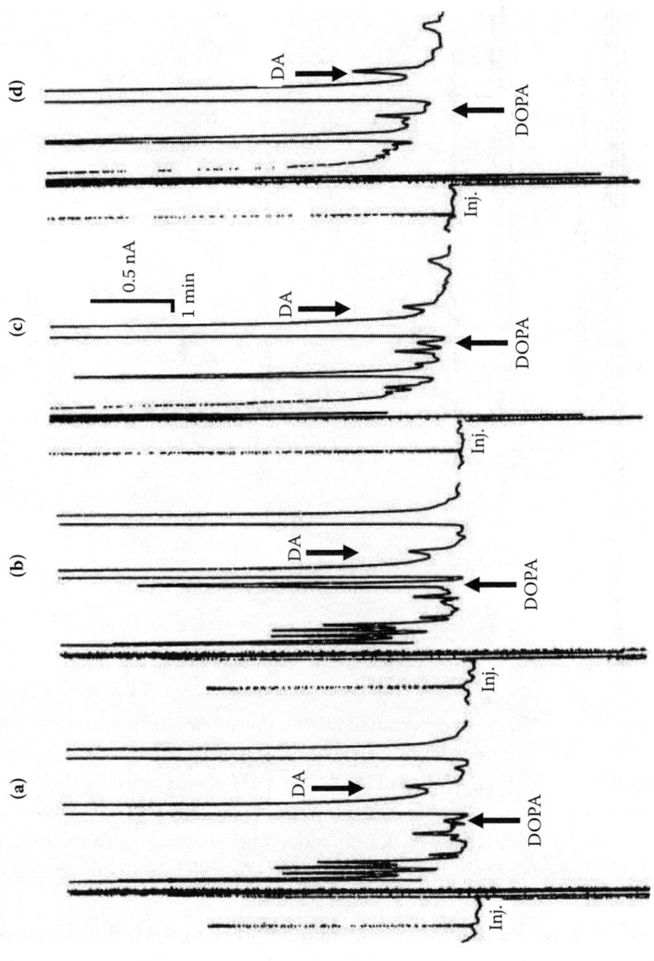


**FIGURE 2.2** Typical chromatograms of DOPA and dopamine obtained by HPLC with ECD during striatal microdialysis with Ringer solution in conscious rats. A dialysate was collected for 20 min 3 h after the start of perfusion immediately before (a) and 20 min after the i.p. injection of 100 mg/kg NSD-1015 (b) and was directly injected at Inj. Another dialysate was collected for 30 min as in (a), incubated without (c), or with (d), an excess of purified hog AADC, adsorbed onto alumina, and then injected at Inj. Limits of sensitivity were 5 fmol for DOPA at upward arrows and dopamine (DA) at downward arrows. The mean absolute values (fmol) in (a) were 38.4 ± 5.2 for DOPA and 73.3 ± 10.0 for dopamine (n = 28). (From Nakamura, S. et al., Transmitter-like basal and $K^+$-evoked release of 3,4-dihydroxyphenylalanine from the striatum in conscious rats studied by microdialysis, *J. Neurochem.*, 58, 270, 1992. With permission.)

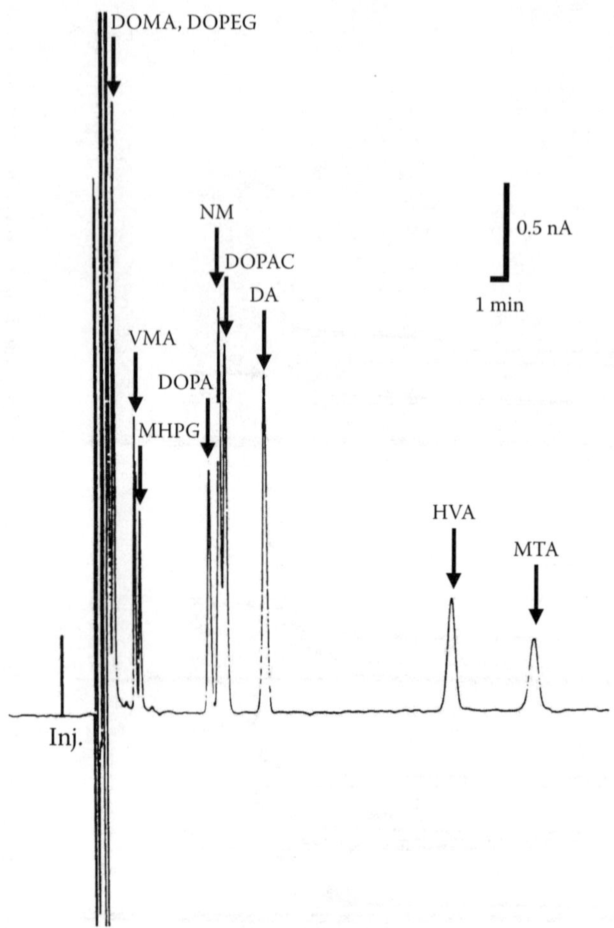


**FIGURE 2.3** A typical HPLC–ECD chromatogram of DOPA, dopamine, and major metabolites of dopamine and noradrenaline. A 20 ml solution containing 2 p*M* of DOMA, DOPEC, DOPA, and dopamine, 200 p*M* of VMA, MHPG, HVA, and MTA, and 1 n*M* of NM was directly injected at Inj. Downward arrows show the peaks of standard substances. (From Nakamura, S. et al., Transmitter-like basal and $K^+$-evoked release of 3,4-dihydroxyphenylalanine from the striatum in conscious rats studied by microdialysis, *J. Neurochem.,* 58, 270, 1992. With permission.)

by COMT and MAO such as 3,4-dihydroxymandelic acid (DOMA), 3,4-dihydroxyphenylglycol (DOPEG), 3-methoxy-4-hydroxymandelic acid (VMA), 3-methoxy-4-hydroxyphenylglycol (MHPG), normetanephrine (NM), 3,4-dihydroxyphenylacetic acid (DOPAC), homovanillic acid (HVA), and 3-methoxytyramine (MTA).

As shown in Figure 2.2a and Figure 2.2c, the peaks of putative basal DOPA before treatment are approximately half of that of dopamine during striatal microdialysis in conscious rats [63]. The ratio of the absolute basal release of DOPA to dopamine is 1:2 on an average in a series of striatal microdialysis experiments [63–66]. This is a surprising phenomenon because the striatal tissue content of DOPA is lower than that of dopamine by three orders of magnitude [67]. These findings suggest that the turnover rate of DOPA is much higher than that of dopamine. Indeed, the ratio of the basal release of DOPA to dopamine is also 1:2 to 3 in striatal slices [27,68]. The amount of DOPA released by depolarizing stimuli as a proportion of the tissue content after the end of experiments is 2 to 3 orders of magnitude higher than that of dopaminc. DOPA appears to be more efficiently synthesized and released to elicit function compared with dopamine [14–17].

As shown in Figure 2.2b, the central AADC inhibitor such as 3-hydroxybenzylhydrazine (NSD-1015) markedly increases the basal release of putative DOPA [62,63,65,66]. Figure 2.2d shows the important finding that purified hog AADC converts putative DOPA to dopamine [63]. On the other hand, the usual dose (200 mg/kg, i.p.) of α-methyl-*p*-tyrosine (α-MPT) required to inhibit the activity of TH decreases the peaks of both putative DOPA and dopamine. Importantly, α-MPT decreases both basal release [34,63,66,69–71] and impulse-evoked release [27,71] of putative DOPA in the striatum and in the sites of baroreflex neurotransmission in the lower brainstem of rats. These findings indicate that the putative DOPA identified on chromatographic charts is indeed DOPA.

## 2.4 FURTHER EVIDENCE FOR RESPONSES TO LEVODOPA *PER SE* BUT NOT FOLLOWING ITS BIOCONVERSION

In an *in vitro* experimental system, NSD-1015 (20 μ*M*) is estimated to inhibit AADC activity by 99.6% in a noncompetitive manner [72]. In the presence of NSD-1015, a COMT inhibitor does not inhibit the levodopa-induced facilitation of noradrenaline release via presynaptic β-adrenoceptors. This facilitation is neither mimicked by 3-*O*-methyldopa, a metabolite of DOPA following bioconversion by the action of COMT, nor by DOPAC, a metabolite of dopamine produced by the action of MAO [29]. The responses we obtained to levodopa appear to be those of its own and not those resulting from its bioconversion to these metabolites.

Furthermore, nanomolar levodopa (10 to 100 ng) microinjected into depressor sites of the NTS [33,73] and the caudal ventrolateral medulla (CVLM) [69,71] elicits hypotension and bradycardia. The NTS sends predominantly excitatory projections to the CVLM. Inhibitory neurons of the CVLM project directly to the rostral ventrolateral medulla (RVLM) to reduce vasomotor tone. The RVLM is the exit of baroreflex pathways from the lower brainstem to the thoracic spinal cord. Nanomolar levodopa (30 to 600 ng) microinjected into pressor sites of the RVLM elicits hypertension and tachycardia [70,73]. These responses to nanomolar levodopa in the NTS, CVLM, and RVLM are seen in the absence, and even in the presence, of NSD-1015 and are not mimicked by dopamine, noradrenaline, and adrenaline at the same low dose ranges as levodopa [33,69,70]; the effective doses of catecholamines microinjected to elicit responses in the NTS, CVLM, and RVLM are in the micromolar ranges. These findings suggest that the responses to levodopa that we obtained are mediated by levodopa on its own but not following its bioconversion to catecholamines and support the idea that DOPA is a neurotransmitter in its own right.

## 2.5 CONCLUSION

DOPA fulfills the classical criteria to be identified as a neurotransmitter such as biosynthesis, metabolism, active transport, existence, physiological release, competitive antagonism, and physiological or pharmacological responses, including interactions with the other neurotransmitter systems. On chromatographic charts, purified AADC converts DOPA to dopamine. The peak of DOPA is decreased by TH inhibition and is increased by AADC inhibition. Turnover rate of DOPA is higher by two to three orders of magnitude compared with that of dopamine. Responses to dopamine converted from levodopa are abolished by AADC inhibition and are mimicked by dopamine agonists. In contrast, responses to levodopa itself as a neurotransmitter or neuromodulator are seen in the absence and presence of AADC inhibition. In addition, these responses are not decreased by COMT inhibition and are not mimicked by a metabolite of DOPA following COMT activity, by a metabolite of dopamine following MAO activity and, further, by dopamine, noradrenaline, and adrenaline at the same low dose ranges to levodopa. These findings support the idea that DOPA is a neurotransmitter or neuromodulator in its own right, in addition to being a precursor of dopamine.

The effectiveness of picomolar to nanomolar levodopa on its own in the striatum and the sites of baroreflex neurotransmission in the lower brainstem of rats suggests that DOPA recognition sites have extremely high affinity.

## ACKNOWLEDGMENTS

This work was mainly supported by Grants-in-Aid for Developmental Scientific Research (No. 06557143); General Scientific Research (No. 61480119, 03454146, 07407003, 10470026) from the Ministry of Education, Science, Sports, and Culture, Japan; and by grants from the Mitsubishi Foundation, Japan; the Uehara Memorial Foundation, Japan; Japanese Heart Foundation and Mitsui Life Social Welfare Foundation, Japan; and SRF, Japan. Many thanks go to Sanae Sato, Department of Pharmacology, Fukushima Medical University School of Medicine, Fukushima, Japan, for remaking the figures.

## REFERENCES

1. Carlsson, A., Lindqvist, M., and Magnusson, T., 3,4-Dihydroxyphenylalanine and 5-hydroxytryptophan as reserpine antagonists, *Nature,* 180, 1200, 1957.
2. Ehringer, H. and Hornykiewicz, O., Verteilung von Noradrenaline und Dopamine (3-Hydroxytyramin) im Gehirn des Menschen und ihr Verhalten bei Erkrankungen des extrapyramidalen Systems, *Klin. Wochenschr.,* 38, 1236, 1960.
3. Bartholini, G. et al., Increase of cerebral catecholamines by 3,4-dihydroxyphenylalanine after inhibition of peripheral decarboxylase, *Nature,* 215, 852, 1967.
4. Hefti, F. and Melamed, E., L-DOPA's mechanism of action in Parkinson's disease, *Trends Neurosci.,* 3, 229, 1980.
5. Ng, K.Y. et al., L-Dopa–induced release of cerebral monoamines, *Science,* 170, 76, 1970.
6. Ng, K.Y. et al., Dopamine: stimulation-induced release from central neurons, *Science,* 172, 487, 1971.

7. Lloyd, K.G., Davidson, L., and Hornykiewicz, O., The neurochemistry of Parkinson' disease: effect of L-dopa therapy, *J. Pharmacol. Exp. Ther.,* 195, 453, 1975.
8. Ponzio, F. et al., Does acute L-DOPA increase active release of dopamine from dopaminergic neurons?, *Brain Res.,* 273, 45, 1983.
9. Bunney, B.S., Aghajanian, G.K., and Roth, R.H., Comparison of effects of L-DOPA, amphetamine and apomorphine on firing rate of rat dopaminergic neurons, *Nature (New Biol.),* 245, 123, 1973.
10. Lokhandwala, M.F. and Buckley, J.P., The effect of L-DOPA on peripheral sympathetic nerve function: role of presynaptic dopamine receptors, *J. Pharmacol. Exp. Ther.,* 204, 362, 1978.
11. Goshima, Y., Kubo, T., and Misu, Y., Biphasic actions of L-DOPA on the release of endogenous noradrenaline and dopamine from rat hypothalamic slices, *Br. J. Pharmacol.,* 89, 229, 1986.
12. Misu, Y., Goshima, Y., and Kubo, T., Biphasic actions of L-DOPA on the release of endogenous dopamine via presynaptic receptors in rat striatal slices, *Neurosci. Lett.,* 72, 194, 1986.
13. Misu, Y. and Goshima, Y., Is L-dopa an endogenous neurotransmitter?, *Trends Pharmacol. Sci.,* 14, 119, 1993.
14. Misu, Y., Ueda, H., and Goshima, Y., Neurotransmitter-like actions of L-DOPA, *Adv. Pharmacol.,* 32, 427, 1995.
15. Misu, Y. et al., Neurobiology of L-DOPAergic systems, *Prog. Neurobiol.,* 49, 415, 1996.
16. Misu, Y., Goshima, Y., and Miyamae, T., Is DOPA a neurotransmitter?, *Trends Pharmacol. Sci.,* 23, 262, 2002.
17. Misu, Y., Kitahama, K., and Goshima, Y., L-3,4-Dihydroxyphenylalanine as a neurotransmitter candidate in the central nervous system, *Pharmacol. Ther.,* 97, 117, 2003.
18. Hoffman, B.B. and Taylor, P., Neurotransmission: the autonomic and somatic motor nervous systems, in *Goodman & Gilman's The Pharmacological Basis of Therapeutics,* Hardman, J.G., Limbird, L.E., and Gilman, A.G., Eds., McGraw-Hill, New York, 2001, 115.
19. Misu, Y. and Kubo, T., Presynaptic β-adrenoceptors, *Trends Pharmacol. Sci.,* 4, 506, 1983.
20. Misu, Y. and Kubo, T., Presynaptic β-adrenoceptors, *Med. Res. Rev.,* 6, 197, 1986.
21. Ueda, H., Goshima, Y., and Misu, Y., Presynaptic mediation by $\alpha_2$-, $\beta_1$- and $\beta_2$-adrenoceptors of endogenous noradrenaline and dopamine release from slices of rat hypothalamus, *Life Sci.,* 33, 371, 1983.
22. Ueda, H., Goshima, Y., and Misu, Y., Presynaptic $\alpha_2$- and dopamine-receptor-mediated inhibitory mechanisms and dopamine nerve terminals in the rat hypothalamus, *Neurosci. Lett.,* 40, 157, 1983.
23. Ueda, H. et al., Adrenaline involvement in the presynaptic β-adrenoceptor-mediated mechanism of dopamine release from slices of the rat hypothalamus, *Life Sci.,* 34, 1087, 1984.
24. Ueda, H. et al., Involvement of epinephrine in the presynaptic beta adrenoceptor mechanism of norepinephrine release from rat hypothalamic slices, *J. Pharmacol. Exp. Ther.,* 232, 507, 1985.
25. Goshima, Y., Kubo, T., and Misu, Y., Autoregulation of endogenous epinephrine release via presynaptic adrenoceptors in the rat hypothalamic slices, *J. Pharmacol. Exp. Ther.,* 235, 248, 1985.
26. Misu, Y. et al., Presynaptic inhibitory dopamine receptors on noradrenergic nerve terminals: analysis of biphasic actions of dopamine and apomorphine on the release of endogenous norepinephrine in rat hypothalamic slices, *J. Pharmacol. Exp. Ther.,* 235, 771, 1985.

27. Goshima, Y., Kubo, T., and Misu, Y., Transmitter-like release of endogenous 3,4-dihydroxyphenylalanine from rat striatal slices, *J. Neurochem.,* 50, 1725, 1988.
28. Goshima, Y. et al., Picomolar concentrations of L-DOPA stereoselectively potentiate activities of presynaptic β-adrenoceptors to facilitate the release of endogenous noradrenaline from rat hypothalamic slices, *Neurosci. Lett.,* 129, 214, 1991.
29. Goshima, Y., Nakamura, S., and Misu, Y., L-Dihydroxyphenylalanine methyl ester is a potent competitive antagonist of the L-dihydroxyphenylalanine-induced facilitation of the evoked release of endogenous norepinephrine from rat hypothalamic slices, *J. Pharmacol. Exp. Ther.,* 258, 466, 1991.
30. Misu, Y. et al., L-DOPA cyclohexyl ester is a novel stable and potent competitive antagonist against L-DOPA, as compared to L-DOPA methyl ester, *Jpn. J. Pharmacol.,* 75, 307, 1997.
31. Furukawa, N. et al., L-DOPA cyclohexyl ester is a novel potent and relatively stable competitive antagonist against L-DOPA among L-DOPA ester compounds in rats, *Jpn. J. Pharmacol.,* 82, 40, 2000.
32. Goshima, Y. et al., L-DOPA induces $Ca^{2+}$-dependent and tetrodotoxin-sensitive release of endogenous glutamate from rat striatal slices, *Brain Res.,* 617, 167, 1993.
33. Kubo, T. et al., Evidence for L-DOPA systems responsible for cardiovascular control in the nucleus tractus solitarii of the rat, *Neurosci. Lett.,* 140, 153, 1992.
34. Yue, J.-L. et al., Baroreceptor-aortic nerve-mediated release of endogenous L-3,4-dihydroxyphenylalanine and its tonic function in the nucleus tractus solitarii of rats, *Neuroscience,* 62, 145, 1994.
35. Honjo, K. et al., GABA may function via $GABA_A$ receptors to inhibit hypotension and bradycardia by L-DOPA microinjected into depressor sites of the nucleus tractus solitarii in anesthetized rats, *Neurosci. Lett.,* 261, 93, 1999.
36. Miyamae, T. et al., Some interactions of L-DOPA and its related compounds with glutamate receptors, *Life Sci.,* 64, 1045, 1999.
37. Ishii, H. et al., Involvement of rBAT in $Na^+$-dependent and -independent transport of the neurotransmitter candidate L-DOPA in *Xenopus laevis* oocytes injected with rabbit small intestinal epithelium poly $A^+$ RNA, *Biochim. Biophys. Acta,* 1466, 61, 2000.
38. Sugaya, Y. et al., Autoradiographic studies using L-[$^{14}$C]DOPA and L-[$^{3}$H]DOPA reveal regional $Na^+$-dependent uptake of the neurotransmitter candidate L-DOPA in the CNS, *Neuroscience,* 104, 1, 2001.
39. Opacka-Juffry, J. and Brooks, D.J., L-Dihydroxyphenylalanine and its decarboxylase: new ideas on their neuroregulatory roles, *Movement Disord.,* 10, 241, 1995.
40. Sun, M.-K., Pharmacology of reticulospinal vasomotor neurons in cardiovascular regulation, *Pharmacol. Rev.,* 48, 465, 1996.
41. Tedroff, J.M., The neuroregulatory properties of L-DOPA. A review of the evidence and potential role in the treatment of Parkinson's disease, *Rev. Neurosci.,* 8, 195, 1998.
42. Hökfelt, T. et al., Evidence for adrenaline neurons in the rat brain, *Acta Physiol. Scand.,* 89, 286, 1973.
43. Ward-Routledge, C. and Marsden, C.A., Adrenaline in the CNS and the action of antihypertensive drugs, *Trends Pharmacol. Sci.,* 9, 209, 1988.
44. Misu, Y. et al., Evidence for tonic activation of prejunctional β-adrenoceptors in guinea-pig pulmonary arteries by adrenaline derived from the adrenal medulla, *Br. J. Pharmacol.,* 98, 45, 1989.
45. Fahn, S., Is levodopa toxic?, *Neurology,* 47 (Suppl. 3), S184, 1996.
46. Cheng, N.-N. et al., Differential neurotoxicity induced by L-DOPA and dopamine in cultured striatal neurons, *Brain Res.,* 743, 278, 1996.

47. Graham, D.G., Oxidative pathways for catecholamines in the genesis of neuromelanin and cytotoxic quinones, *Mol. Pharmacol.,* 14, 633, 1978.
48. Halliwell, B., Reactive oxygen species and the central nervous system, *J. Neurochem.,* 59, 1609, 1992.
49. Fahn, S. and Cohen, G., The oxidant stress hypothesis in Parkinson's disease. Evidence supporting it, *Ann. Neurol.,* 32, 804, 1992.
50. Smith, T.S., Parker, W.D., and Bennett, J.P., L-DOPA increases nigral production of hydroxyl radicals in vivo: potential role of L-DOPA toxicity?, *Neuroreport,* 5, 1009. 1994.
51. Basma, A.N. et al., DOPA cytotoxicity in PC12 cells in culture is via its autoxidation, *J. Neurochem.,* 64, 718, 1995.
52. Metodiewa, D. and Koska, C., Reactive oxygen species and reactive nitrogen species: relevance to cyto(neuro)toxic events and neurologic disorders. An overview, *Neurotoxicity Res.,* 1, 197, 2000.
53. Asanuma, M., Miyazaki, I., and Ogawa, N., Dopamine or L-DOPA-induced neurotoxicity: the role of dopamine quinone formation and tyrosinase in a model of Parkinson's disease, *Neurotoxicity Res.,* 5, 165, 2003.
54. Olney, J.W. et al., Excitotoxicity of L-DOPA and 6-OH-DOPA: implications for Parkinson's and Huntington's diseases, *Exp. Neurol.,* 108, 269, 1990.
55. Rosenberg, P.A. et al., 2,4,5-Trihydroxyphenylalanine in solution forms a non-*N*-methyl-D-aspartate glutamatergic agonist and neurotoxin, *Proc. Natl. Acad. Sci. U.S.A.,* 88, 4865, 1991.
56. Newcomer, T.A., Rosenberg, P.A., and Aizenman, E., TOPA quinone, a kainate-like agonist and excitotoxin, is generated by a catecholaminergic cell line, *J. Neurosci.,* 15, 3172, 1995.
57. Olney, J.W. and Sharpe, L.G., Brain lesions in an infant rhesus monkey treated with monosodium glutamate, *Science,* 166, 386, 1969.
58. Choi, D.W., Glutamate neurotoxicity and diseases of the nervous system, *Neuron,* 1, 623, 1988.
59. Lipton, S.A. and Rosenberg, P.A., Excitatory amino acids as a final common pathway for neurologic disorders, *New Engl. J. Med.,* 330, 613, 1994.
60. Rossi, D.J., Oshima, T., and Attwell, D., Glutamate release in severe brain ischaemia is mainly by reversed uptake, *Nature,* 403, 316, 2000.
61. Maeda, T. et al., L-DOPA neurotoxicity is mediated by glutamate release in cultured rat striatal neurons, *Brain Res.,* 771, 159, 1997.
62. Furukawa, N. et al., Endogenously released DOPA is a causal factor for glutamate release and resultant delayed neuronal cell death by transient ischemia in rat striata, *J. Neurochem.,* 76, 815, 2001.
63. Nakamura, S. et al., Transmitter-like basal and $K^+$-evoked release of 3,4-dihydroxyphenylalanine from the striatum in conscious rats studied by microdialysis, *J. Neurochem.,* 58, 270, 1992.
64. Nakamura, S. et al., Transmitter-like 3,4-dihydroxyphenylalanine is tonically released by nicotine in striata of conscious rats, *Eur. J. Pharmacol.,* 222, 75, 1992.
65. Nakamura, S. et al., Non-effective dose of exogenously applied L-DOPA itself stereoselectively potentiates postsynaptic $D_2$-receptor-mediated locomotor activities of conscious rats, *Neurosci. Lett.,* 170, 22, 1994.
66. Yue, J.-L. et al., Endogenously released L-DOPA itself tonically functions to potentiate $D_2$ receptor-mediated locomotor activities of conscious rats, *Neurosci. Lett.,* 170, 107, 1994.

67. Westerink, B.H.C., Van Es, T.P., and Spaan, S.J., Effects of drugs interfering with dopamine and noradrenaline biosynthesis on the endogenous 3,4-dihydroxyphenylalanine levels in rat brain, *J. Neurochem.,* 39, 44, 1982.
68. Misu, Y. et al., Nicotine releases stereoselectively and $Ca^{2+}$-dependently endogenous 3,4-dihydroxyphenylalanine from rat striatal slices, *Brain Res.,* 520, 334, 1990.
69. Yue, J.-L., Goshima, Y., and Misu, Y., Transmitter-like L-3,4-dihydroxyphenylalanine tonically functions to mediate vasodepressor control in the caudal ventrolateral medulla of rats, *Neurosci. Lett.,* 159, 103, 1993.
70. Yue, J.-L. et al., Evidence for L-DOPA relevant to modulation of sympathetic activity in the rostral ventrolateral medulla of rats, *Brain Res.,* 629, 310, 1993.
71. Miyamae, T. et al., L-DOPAergic components in the caudal ventrolateral medulla in baroreflex neurotransmission, *Neuroscience,* 92, 137, 1999.
72. Goshima, Y., Nakamura, S., and Misu, Y., L-DOPA facilitates the release of endogenous norepinephrine and dopamine via presynaptic $\beta_1$ and $\beta_2$-adrenoceptors under essentially complete inhibition of L-aromatic amino acid decarboxylase in rat hypothalamic slices, *Jpn. J. Pharmacol.,* 53, 47, 1990.
73. Yue, J.-L. et al, Altered tonic L-3,4-dihydroxyphenylalanine systems in the nucleus tractus solitarii and the rostral ventrolateral medulla of spontaneously hypertensive rats, *Neuroscience,* 67, 95, 1995.

# 3 Active Uptake System of L-DOPA in CNS

*Yukio Sasaki, Yoshio Goshima, and Yoshimi Misu*

## CONTENTS



## 3.1 INTRODUCTION

Once neurotransmitters are released from presynapses, they must be readily removed from synaptic clefts for the termination of synaptic neurotransmission. There are three mechanisms of removal of neurotransmitters: diffusion, enzymatic degradation, and reuptake. Among them, reuptake of the neurotransmitter is the most common mechanism for inactivation. High-affinity uptake of the released neurotransmitters is mediated by transporter molecules in the plasma membrane of nerve terminals and glial cells (for review, see [1]). One of the most important aspects of these transporters is that reuptake is coupled to transmembrane ion gradients which provide the energy for the transport [1,2]. The plasma membrane transporters of neurotransmitters can be divided into two families depending on their ionic dependence: (1) the $Na^+/Cl^-$-dependent transporters, including monoamine and $\gamma$-aminobutyric acid (GABA) transporters, and (2) the $Na^+/K^+$-dependent transporters, including glutamate transporters [1–4]. Therefore, existence of $Na^+$-dependent transporter(s) would be expected at sites where L-3,4-dihydroxyphenylalanine (L-DOPA)ergic systems are localized.

Until now, systems for $Na^+$-dependent L-DOPA transport have not been well characterized. In the central nervous system (CNS), L-DOPA transport in rat cerebral cortex slices [5] and cultured rat and mouse astrocytes [6] has been shown to be $Na^+$ independent. Our previous biochemical analysis, however, demonstrated that the uptake of L-[$^3$H]DOPA into rat hypothalamic slices is partially $Na^+$ dependent [7]. The L-DOPA uptake is not modified by 3-hydroxybenzylhydrazine (NSD-1015), a central aromatic L-amino acid DOPA decarboxylase (AADC) inhibitor, suggesting that the observed tritium incorporation is not due to uptake of [$^3$H]dopamine converted from L-[$^3$H]DOPA. A large number of studies on L-DOPA uptake have been done using autoradiography and radiolabeled compounds administered to rats [8,9] and humans [10]. However, uptake studies of L-[$^3$H]DOPA in *in vivo* rat brains appear to represent the uptake sites of dopamine because the labeled cell bodies show a pattern similar to that of the dopamine transporter detected by the *in situ* hybridization method [11–14] and by immunocytochemistry [15]. Very few attempts have been made so far at determining the cellular sites of uptake of L-DOPA itself in the brain.

From another point of view, L-DOPA is one of aromatic amino acids possessing two hydroxyl groups on its aromatic ring. Therefore, it is possible that L-DOPA uptake is mediated by amino acid transport systems. Although amino acid transport systems have been investigated as uptake systems for nutrients in nonneural tissues like small intestine and kidney, some of the amino acid transport systems (e.g., system $X_{AG}^-$ [glutamate transporters] and system Gly [glycine transporters]) have been found to function for amino acid neurotransmitters in the CNS [for review, see [16]]. Until now, 21 kinds of amino acid transport systems have been classified by substrate specificity and sodium dependency, and about 40 kinds of amino acid transporter cDNA have been cloned [16]. Among these systems, aromatic amino acids like L-tyrosine and L-phenylalanine can be taken up by system $B^0$, system $B^{0,+}$, system L, system T, and system $b^{0,+}$. System $B^0$ and system $B^{0,+}$ are $Na^+$ dependent transport systems, whereas system L, system T, and system $b^{0,+}$ are $Na^+$ independent. Therefore, these amino acid transport systems could be candidates for L-DOPA uptake.

In this chapter, we will show the properties of L-DOPA uptake in the *Xenopus* oocyte expression system and rat brain slices, and further elucidate the uptake sites of the L-DOPA in the CNS using microautoradiography techniques.

## 3.2 CHARACTERIZATION OF L-DOPA UPTAKE USING *XENOPUS* OOCYTE EXPRESSION SYSTEM

Since the GABA transporter became the first neurotransmitter transporter to be characterized and cloned using the *Xenopus laevis* oocyte expression system [17,18], many transporters have been examined with this useful tool. In an attempt to identify the transporter for L-DOPA, we investigated whether or not an active L-DOPA transport system with high affinity is functionally expressed in *Xenopus* oocytes, by using poly $A^+$ RNA from various animal tissues, and further characterized the transport activity of L-DOPA in detail [19].

At first, we examined the L-DOPA uptake activities expressed by injection of various poly $A^+$ RNA from the rat CNS. Microinjection of the hypothalamus, the cerebral cortex, the cerebellum, and the brainstem poly $A^+$ RNA into oocytes significantly increased the uptake of L-DOPA by approximately 40% of that due to the endogenous activity. However, these levels of increase in L-DOPA uptake were quite low and, hence, we were unable to characterize the uptake properties in detail. We then tested rabbit small intestinal epithelium because amino acid uptake activity is known to be relatively high in this tissue. As expected, oocytes injected with rabbit intestinal epithelium poly $A^+$ RNA showed the highest L-DOPA uptake activity (an approximately fivefold increase over the endogenous activity of oocytes) among the tissue poly $A^+$ RNAs tested. The L-DOPA transport in the oocytes showed saturable uptake with a high affinity. The apparent $K_m$ value (38 $\mu M$ in the presence of $Na^+$) was comparable to that of the transporters for the neurotransmitter glutamate [20]. These data suggest that uptake carrier molecule(s), but not diffusion across the membrane, were involved in L-DOPA transport. The uptake experiments in the presence or absence of $Na^+$ revealed that the L-DOPA transport was composed of $Na^+$-dependent (56%) and $Na^+$-independent (44%) fractions. The inhibition study using nonradioactive competitors showed that the L-DOPA uptake system is stereoselective and is not displaced by excess amounts of dopamine. These results suggest that L-DOPA uptake in small intestine, at least in part, is responsible for $Na^+$-dependent transporter(s) that are different from dopamine transporters.

The uptake of L-DOPA was inhibited by neutral amino acids such as L-tyrosine, L-phenylalanine, and L-leucine. The uptake was also inhibited by the basic amino acid L-lysine but was not attenuated by the acidic amino acid L-glutamate. These data suggest that the uptake of L-DOPA is mediated via basic as well as neutral amino acid transporter(s). Among the aromatic amino acid transport systems listed in Section 3.1, L-lysine can be taken up by system $B^{0,+}$ and system $b^{0,+}$. To clarify which transport systems are involved in the L-DOPA uptake system, we used an antisense oligonucleotide derived from a part of related $b^{0,+}$ amino acid transporter (rBAT) cDNA sequence, which is one of the subunits of heterodimers forming transporter units of system $b^{0,+}$. An antisense, but not sense, oligonucleotide for rBAT almost completely suppressed the L-DOPA uptake, suggesting that system $b^{0,+}$ was responsible for the L-DOPA transport. However, $b^{0,+}$ amino acid transport has been characterized as the system for $Na^+$-independent neutral and basic amino acid transport and, hence, this property appears to be different from that of L-DOPA uptake in the oocytes injected with rabbit poly $A^+$ RNA. In this regard, it is noteworthy that in rBAT-expressing *Xenopus* oocytes, a partially $Na^+$-dependent transport activity for L-histidine is observed [21]. It is thus possible that rBAT and rBAT-associated transporter(s) may also be involved in the $Na^+$-dependent transport of amino acids such as L-DOPA. The first rBAT-associated transporter molecules, termed $b^{0,+}$AT or BAT1, have been identified [22,23]. Heterodimerization of rBAT and BAT1 results in expression of $Na^+$-independent $b^{0,+}$ activity. Further studies are required to investigate whether or not $b^{0,+}$AT/BAT1 is involved in L-DOPA transport in oocytes injected with rabbit intestinal poly $A^+$ RNA.

## 3.3 $NA^+$-DEPENDENT UPTAKE OF L-DOPA IN THE CNS

To examine the properties and localizations of the L-DOPA transport system in the CNS, we performed autoradiographic uptake studies by L-[$^{14}$C]DOPA and L-[$^{3}$H]DOPA using rat brain slice preparations [24]. At first, we estimated the level of L-[$^{14}$C]DOPA uptake in the whole brain. The density of L-[$^{14}$C]DOPA uptake was relatively high in the cerebral cortex, the hippocampus, the hypothalamus, and the cerebellum, but was moderate or even low in the striatum. Therefore, we characterized the properties of L-DOPA transport in the CNS mainly using the cortex and the hippocampus slices. As a result, we clarified the following properties of L-DOPA transport in the CNS: (1) partially $Na^+$- and $Cl^-$-dependent, such as monoamine transporters, (2) no displacement by excess amounts of dopamine, (3) inhibition by L-phenylalanine, and (4) partial inhibition by either L-lysine or 2-amino-2-norbornate-carboxylic acid (BCH), a substrate for system L and system $B^{0,+}$ [24]. The candidates of amino acid transport systems, which are responsible for uptake of aromatic amino acids like L-DOPA, would be the $Na^+$-dependent system $B^0$ and system $B^{0,+}$ and the $Na^+$-independent system L, system T, and system $b^{0,+}$ mentioned earlier. Taken together, the uptake of L-DOPA is suggested to be, at least in part, mediated via the $Na^+/Cl^-$-dependent neutral and basic amino acid transporter system $B^{0,+}$ [25], although expression of ATB, which is a transporter protein responsible for system $B^{0,+}$, in the CNS remains unclear [26]. It is also noteworthy that the $Na^+/Cl^-$ dependency and the portion inhibited by L-lysine are only 40 to 50% of total L-DOPA uptake, suggesting that other transport systems could also be involved in L-DOPA uptake.

In Section 3.2, we showed the role of rBAT and rBAT-associate protein(s) in the $Na^+$-dependent uptake of L-[$^{14}$C]DOPA in *Xenopus* oocytes injected with poly $A^+$ RNA from rabbit intestinal epithelium. The properties of system $b^{0,+}$ described in the *Xenopus* system seem to differ from the properties of L-DOPA uptake observed in the cortex and hippocampus slices. However, it is still possible that system $b^{0,+}$ is involved in L-DOPA transport in the CNS because several systems could be responsible for this, depending on the brain region. It is noteworthy that rBAT expression [27–29] is localized at the target regions of L-DOPA in the CNS such as the hypothalamus and the lower brainstem [7,30,31].

## 3.4 L-DOPA UPTAKE SITES IN THE CNS

We analyzed the uptake sites of L-DOPA in detail by the use of L-[$^{3}$H]DOPA in microautoradiography. Overall, diffuse grain accumulation was seen in the entire brain area, which could be displaced by an excess amount of nonlabeled L-DOPA. In addition, dense grain accumulation was observed in some brain areas. As mentioned in the text that follows, high-density accumulation was also seen in the hypothalamus, the cerebral cortex, the nucleus tractus solitarii (NTS), the area postrema (AP), and the dorsal motor nucleus of the vagus (DMV) in the medulla oblongata, as well as in the Purkinje cell layer and the molecular layer of the cerebellum. Accumulation was moderate or even low in the striatum in microautoradiography of L-[$^{3}$H]DOPA uptake. Our study showing relatively high uptake activity in the cerebellum appeared to be consistent with a substantial level of uptake for L-[$^{3}$H]-DOPA in the cerebellum after i.v. administration of the compound [9]. In addition, studies of fluoroDOPA accumulation in the healthy human brain demonstrate accumulation in extrastriatal cerebral structures such as the amygdala and the hippocampal formation [10], a finding consistent with that of the present study. Our principal result obtained with both L-[$^{14}$C]DOPA and L-[$^{3}$H]DOPA, however, was in marked contrast to similar autoradiographic data showing extensive accumulation of radioactivity in the striatum and the other dopaminergic neurons after *in vivo* i.v. administration of L-[$^{3}$H]DOPA [8,9]. Our study was also distinct from studies of dopamine transporters, also being highly expressed in dopaminergic neurons [11,13–15,32]. Although the reasons for the differences between these studies are unclear, it is probably due to the L-DOPA administered peripherally being more readily converted to dopamine by extraneuronal AADC in glial and/or endothelial cells during its transport from blood to the brain and/or species or regional differences in the levels of tissue AADC activity. Although decarboxylation of exogenous L-DOPA to dopamine has also been demonstrated in serotonergic neurons [33], the pattern of L-DOPA uptake sites obtained appeared to differ from that of serotonergic neurons.

In the following subsections, we will discuss uptake sites of L-[$^{3}$H]DOPA in rat brain slices individually, and discuss their significance in detail.

### 3.4.1 The Hypothalamus

Heavy to moderate silver grain accumulation was observed in the lateral portion of the median eminence (ME) and the hypothalamic supraoptic nucleus (SON) in microautoradiography of L-[$^{3}$H]DOPA uptake (Figure 3.1A, [24]). This distribution pattern in the ME coincided well with the L-DOPA-immunoreactive neuronal terminals observed in this area (Figure 3.1B, [24]). L-DOPA-immunoreactive neurons were located in the dorsomedial and the ventrolateral parts of the arcuate nucleus.

Dense accumulation of L-[$^{3}$H]DOPA grains in the lateral part of the ME is consistent with previous findings of immunoreactive neuronal terminals in the ME at light microscopic and electron microscopic levels [30]. Furthermore, the lateral part of the ME is innervated by hypothalamic neurons located in the dorsomedial and the ventrolateral parts of the arcuate nucleus [34,35]. These neurons are tyrosine hydroxylase (TH)-positive and growth hormone releasing hormone (GHRH) positive [36]. Consistently, TH-positive but AADC-negative cells are located in the ventrolateral part of the arcuate nucleus, whereas dopamine-immunoreactive cells are localized exclusively in the dorsomedial arcuate nucleus [37]. The GHRH-immunoreactive terminals are located in the central regions of the ME, with the highest density on the two sides of the ME [38], the distribution pattern being reminiscent of that of L-[$^{3}$H]DOPA grain accumulation. Because these neurons are suspected to secrete corticotropin releasing hormone or GHRH, endogenous L-DOPA in the ME region may have some role in the production and/or release of these pituitary hormones.

### 3.4.2 The Cerebral Cortex

Diffuse silver grain accumulation of L-[$^{3}$H]DOPA was observed in the molecular layer (layer I), the external granular layer (layer II), the external pyramidal layer (layer III), and the internal granular layer (layer IV) in the cerebrum [24]. Grain density was

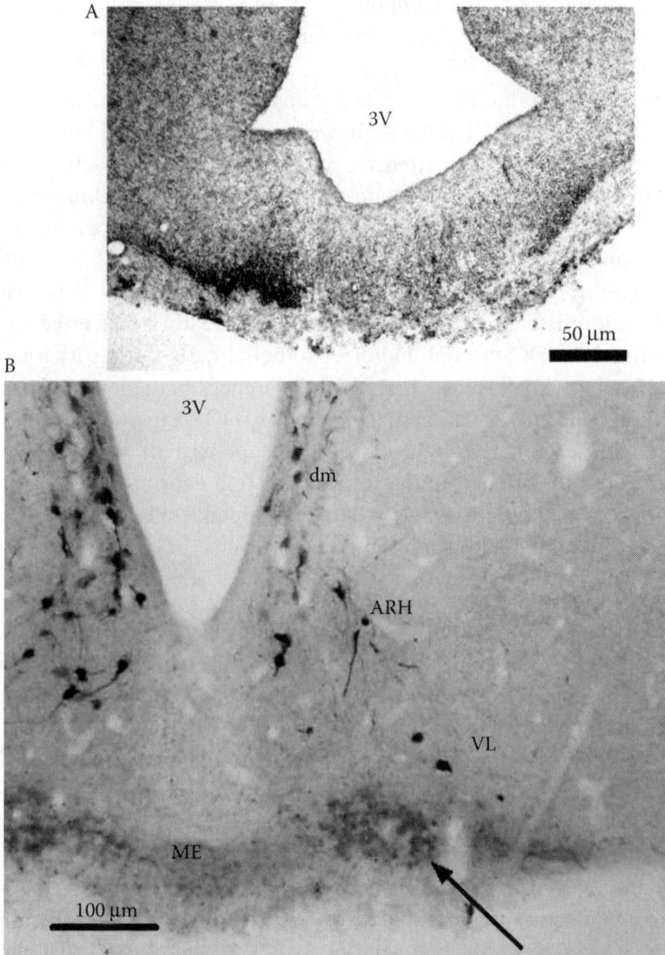


**FIGURE 3.1** Microautoradiogram of the uptake of L-[$^3$H]DOPA and L-DOPA-immunoreactive neurons in a coronal section through the hypothalamus. (A) Silver grains heavily accumulated on the median eminence (ME). The most dense accumulation was on both sides of the ME. (B) L-DOPA-immunoreactive cell bodies in the dorsomedial (dm) arcuate nucleus of the hypothalamus (ARH). Some were scattered in the ventrolateral part (VL). They contained only L-DOPA. Labeled terminal fibers were abundant in the ME, in particular, its lateral portion indicated by an arrow. 3V, the third ventricle. (From Sugaya, Y. et al., Autoradiographic studies using L-[$^{14}$C]DOPA and L-[$^3$H]DOPA reveal regional $Na^+$-dependent uptake of the neurotransmitter candidate L-DOPA in the CNS, *Neuroscience*, 104, 1, 2001. With permission.)

relatively high in layer I. Grains accumulated on the cell bodies of some subsets of neurons and/or nonneuronal cells. These cells with grain accumulation were relatively small and, hence, are thought to belong to nonneuronal cell bodies, probably of glial origin. This pattern of grain accumulation with higher density was also observed in the amygdala and the piriform cortex.

It is noteworthy that the L-[$^3$H]DOPA uptake was mainly observed in layer I of the cerebral cortex, the amygdala, and the piriform cortex. TH-immunoreactive neurons have been described in the rodent cortex during development [39] and in the human and monkey neocortex [40,41]. These TH-immunoreactive neurons probably contain L-DOPA as an end product because they lack the catecholaminergic synthesizing enzymes, AADC, dopamine-β-hydroxylase, and phenylethanolamine-*N*-methyl-transferase [39–41]. On the other hand, the amygdala and the piriform cortex have been reported to receive projections from the ventral tegmental area (VTA) dopaminergic neurons. Although these neurons contain dopamine, L-DOPA immunostaining in the VTA is more intense than in the lateral part of the substantia nigra [42]. Moreover, our recent study demonstrated that some neuronal fibers show intense L-DOPA-immunoreactivity in the amygdala and the piriform cortex (unpublished observations). It is thus possible that these TH-positive and/or L-DOPA-immunoreactive neurons may be responsible for the L-[$^3$H]DOPA uptake in layer I of the cerebral cortex, the amygdala, and the piriform cortex.

### 3.4.3 The Medulla Oblongata

Silver grains of L-[$^3$H]DOPA were accumulated in the AP, the ependymal cell layer of the central canal, the DMV, and the NTS of the medulla oblongata. The grains in the NTS and the DMV appeared to accumulate in cell bodies of some neurons and/or nonneuronal cells [24]. This pattern resembled that observed in the cerebral cortex. We have previously reported that the NTS is one of the main regions with L-DOPA as a neurotransmitter [30,31]. Therefore, it was expected that $Na^+$-dependent active transport systems for L-DOPA would be localized in the NTS. However, the L-[$^3$H]DOPA uptake of the NTS and DMV region did not show any $Na^+$ dependency. This negative result may be related to our observation that the uptake of L-DOPA was composed of $Na^+$-dependent and -independent transport systems as discussed earlier. If the $Na^+$-dependent L-DOPA uptake constitutes only a minor component in the NTS, the whole uptake activity in the NTS would apparently exhibit $Na^+$-independent properties. The property of L-DOPA uptake in the NTS may be attributed mainly to the $Na^+$-independent uptake system described in rat brain slices [5] and cultured rat or mouse astrocytes [6].

### 3.4.4 The Hippocampus

In the hippocampus, the distribution pattern of the silver grains of L-[$^3$H]DOPA was rather diffuse, with the cell bodies and their surrounding area being equally labeled with L-[$^3$H]DOPA [24]. The grains of L-[$^3$H]DOPA were not associated with a specific subset of neuronal and/or nonneuronal cells. In the hippocampus, L-DOPA-immunoreactive axons have not been described [30]. However, the hippocampal areas are innervated by noradrenergic nerve fibers, which should contain L-DOPA as a precursor of dopamine and noradrenaline. In addition, in the hippocampus, our observations suggest a modulatory action of L-DOPA: nanomolar concentrations of L-DOPA inhibited the population spikes elicited by electrical stimuli applied to the Schaffer collateral commissural fibers in the hippocampal CA1 region [43]. This action of L-DOPA was antagonized by L-DOPA cyclohexyl ester (DOPA CHE), a competitive antagonist for L-DOPA [44], thereby suggesting a role for L-DOPA as a neuroactive substance in the hippocampus as well as in the striatum and the hypothalamus [45]. Furthermore, DOPA CHE antagonized glutamate release from the CA1 region during transient ischemia and suppressed the resultant delayed cell death of CA1 pyramidal neurons [46]. These results suggest that L-DOPA could act as a neurotransmitter in the hippocampus under physiological and pathophysiological conditions, although further studies are required to elucidate the existence and the role of endogenous L-DOPA in the hippocampus.

### 3.4.5 The Cerebellum

The silver grains of L-[$^3$H]DOPA formed clusters and surrounded the Purkinje cells [24]. The grains in the molecular layer of the cerebellum formed lines extending upward to the pial surface. This pattern of grain accumulation was quite similar to that of the immunoreactivity of glutamate aspartate transporter (GLAST), a glial type transporter for glutamate [24]. Because GLAST is known to be highly expressed in Bergmann cells [47], these cells may be those responsible for the uptake of L-[$^3$H]DOPA in the cerebellum. This glial cell type has been implicated as a $Na^+$-dependent reuptake system for glutamatergic neurotransmission in the cerebellum. Our preliminary observations suggest that the L-[$^3$H]DOPA uptake in the cerebellum is $Na^+$ dependent (unpublished observation). Further studies are required to clarify the relation between L-DOPA and glutamate uptakes and the physiological significance of L-DOPA uptake in the cerebellum.

### 3.4.6 The Choroid Plexus

Heavy grains also accumulated in the choroid plexus of the lateral ventricle, the third and fourth dorsal ventricles, the ependymal cell layer in the third ventricle, and the central canal of the medulla oblongata [24]. The grain density in the choroid plexus was not modified in the absence of $Na^+$. This finding, together with results from the inhibition studies with BCH and L-lysine, suggests that system L was responsible for the L-[$^3$H]DOPA transport in the choroid plexus [25]. A similar $Na^+$-independent transport system exhibiting system L properties has been described for small neutral amino acids in the sheep choroid plexus [48]. Recently, 4F2hc and LAT1, which form a transporter complex for system L, have been reported to transport L-DOPA across the blood–brain barrier in the CNS [49]. As uptake of L-DOPA mediated by system L exhibits an essentially complete $Na^+$-independent property [49], it is unlikely that the neuronal or glial L-DOPA transporter(s) described here, other than choroid plexus, might have some relevance to the L-DOPA transporters in the blood–brain barrier responsible for uptake of L-DOPA from blood into the brain.

## 3.5 CONCLUSION

We characterized the properties of L-DOPA uptake using the *Xenopus* expression system and rat brain slices. We also demonstrated the uptake sites of L-DOPA in the CNS. Some of the uptake sites of L-DOPA coincide well with target areas for neurons that contain L-DOPA as a neurotransmitter and/or neuromodulator. This L-DOPA uptake system may be involved in L-DOPAergic neurotransmission. We also make a new suggestion: that the hippocampus and the cerebellum are potential target regions for L-DOPA.

## ACKNOWLEDGMENTS

This study was in part supported by Grants-in-Aid for Developmental Scientific Research (No. 06557143) and Scientific Research (No. 07407003, 09877022, 09280280, 09480224, 10176229, 10470026) from the Ministry of Education, Science, Sports, and Culture, Japan; and by grants from the Mitsubishi Foundation, the Uehara Memorial Foundation, and SRF, all in Japan.

## REFERENCES

1. Masson, J. et al., Neurotransmitter transporters in the central nervous system, *Pharmacol. Rev.,* 51, 439, 1999.
2. Nelson, N., The family of $Na^+/Cl^-$ neurotransmitter transporters, *J. Neurochem.,* 71, 1785, 1998.
3. Torres, G.E., Gainetdinov, R.R., and Caron, M.G., Plasma membrane monoamine transporters: structure, regulation and function, *Nat. Rev. Neurosci.,* 4, 13, 2003.
4. Gegelashvili, G. and Schousboe, A., High affinity glutamate transporters: regulation of expression and activity, *Mol. Pharmacol.,* 52, 6, 1997.
5. Garcia-Sancho, F.J. and Herreros, B., Characterization of transport systems for the transfer of 3,4-L-dihydroxyphenylalanine into slices of rat cerebral cortex, *Biochim. Biophys. Acta,* 406, 538, 1975.
6. Tsai, M.J. and Lee, E.H., Characterization of L-DOPA transport in cultured rat and mouse astrocytes, *J. Neurosci. Res.,* 43, 490, 1996.
7. Goshima, Y., Nakamura, S., and Misu, Y., L-Dihydroxyphenylalanine methyl ester is a potent competitive antagonist of the L-dihydroxyphenylalanine-induced facilitation of the evoked release of endogenous norepinephrine from rat hypothalamic slices, *J. Pharmacol. Exp. Ther.,* 258, 466, 1991.
8. Cumming, P. et al., The effect of unilateral neurotoxic lesions to serotonin fibres in the medial forebrain bundle on the metabolism of [$^3$H]DOPA in the telencephalon of the living rat, *Brain Res.,* 747, 60, 1997.
9. Melega, W.P. et al., Comparative in vivo metabolism of 6-[$^{18}$F]fluoro-L-dopa and [$^3$H]L-dopa in rats, *Biochem. Pharmacol.,* 39, 1853, 1990.
10. Brown, W.D. et al., FluoroDOPA PET shows the nondopaminergic as well as dopaminergic destinations of levodopa, *Neurology,* 53, 1212, 1999.
11. Augood, S.J. et al., Co-expression of dopamine transporter mRNA and tyrosine hydroxylase mRNA in ventral mesencephalic neurons, *Brain Res. Mol. Brain Res.,* 20, 328, 1993.
12. Cerruti, C. et al., Dopamine transporter mRNA expression is intense in rat midbrain neurons and modest outside midbrain, *Brain Res. Mol. Brain Res.,* 18, 181, 1993.
13. Lorang, D., Amara, S.G., and Simerly, R.B., Cell-type-specific expression of catecholamine transporters in the rat brain, *J. Neurosci.,* 14, 4903, 1994.
14. Shimada, S. et al., Dopamine transporter mRNA: dense expression in ventral midbrain neurons, *Brain Res. Mol. Brain Res.,* 13, 359, 1992.

15. Ciliax, B.J. et al., The dopamine transporter: immunochemical characterization and localization in brain, *J. Neurosci.,* 15, 1714, 1995.
16. Hyde, R., Taylor, P.M., and Hundal, H.S., Amino acid transporters: roles in amino acid sensing and signalling in animal cells, *Biochem. J.,* 373, 1, 2003.
17. Guastella, J. et al., Cloning and expression of a rat brain GABA transporter, *Science,* 249, 1303, 1990.
18. Nelson, H., Mandiyan, S., and Nelson, N., Cloning of the human brain GABA transporter, *FEBS Lett.,* 269, 181, 1990.
19. Ishii, H. et al., Involvement of rBAT in $Na^+$-dependent and -independent transport of the neurotransmitter candidate L-DOPA in *Xenopus laevis* oocytes injected with rabbit small intestinal epithelium poly $A^+$ RNA, *Biochim. Biophys. Acta,* 1466, 61, 2000.
20. Gegelashvili, G. and Schousboe, A., Cellular distribution and kinetic properties of high-affinity glutamate transporters, *Brain Res. Bull.,* 45, 233, 1998.
21. Ahmed, A. et al., Electrogenic L-histidine transport in neutral and basic amino acid transporter (NBAT)-expressing *Xenopus laevis* oocytes. Evidence for two functionally distinct transport mechanisms induced by NBAT expression, *J. Biol. Chem.,* 272, 125, 1997.
22. Chairoungdua, A. et al., Identification of an amino acid transporter associated with the cystinuria-related type II membrane glycoprotein, *J. Biol. Chem.,* 274, 28845, 1999.
23. Feliubadalo, L. et al., Non-type I cystinuria caused by mutations in SLC7A9, encoding a subunit ($b^{o,+}$AT) of rBAT. International Cystinuria Consortium, *Nat. Genet.,* 23, 52, 1999.
24. Sugaya, Y. et al., Autoradiographic studies using L-[$^{14}$C]DOPA and L-[$^{3}$H]DOPA reveal regional $Na^+$-dependent uptake of the neurotransmitter candidate L-DOPA in the CNS, *Neuroscience,* 104, 1, 2001.
25. McGivan, J.D. and Pastor-Anglada, M., Regulatory and molecular aspects of mammalian amino acid transport, *Biochem. J.,* 299 (Pt. 2), 321, 1994.
26. Kekuda, R. et al., Cloning of the sodium-dependent, broad-scope, neutral amino acid transporter $B^o$ from a human placental choriocarcinoma cell line, *J. Biol. Chem.,* 271, 18657, 1996.
27. Hisano, S. et al., The basic amino acid transporter (rBAT)-like immunoreactivity in paraventricular and supraoptic magnocellular neurons of the rat hypothalamus, *Brain Res.,* 710, 299, 1996.
28. Nirenberg, M.J. et al., Immunocytochemical localization of the renal neutral and basic amino acid transporter in rat adrenal gland, brainstem, and spinal cord, *J. Comp. Neurol.,* 356, 505, 1995.
29. Pickel, V.M. et al., Regional and subcellular distribution of a neutral and basic amino acid transporter in forebrain neurons containing nitric oxide synthase, *J. Comp. Neurol.,* 404, 459, 1999.
30. Misu, Y. et al., Neurobiology of L-DOPAergic systems, *Prog. Neurobiol.,* 49, 415, 1996.
31. Yue, J.L. et al., Baroreceptor-aortic nerve-mediated release of endogenous L-3,4-dihydroxyphenylalanine and its tonic depressor function in the nucleus tractus solitarii of rats, *Neuroscience,* 62, 145, 1994.
32. Freed, C. et al., Dopamine transporter immunoreactivity in rat brain, *J. Comp. Neurol.,* 359, 340, 1995.
33. Arai, R. et al., Immunohistochemical evidence that central serotonin neurons produce dopamine from exogenous L-DOPA in the rat, with reference to the involvement of aromatic L-amino acid decarboxylase, *Brain Res.,* 667, 295, 1994.

34. Lechan, R.M., Nestler, J.L., and Jacobson, S., The tuberoinfundibular system of the rat as demonstrated by immunohistochemical localization of retrogradely transported wheat germ agglutinin (WGA) from the median eminence, *Brain Res.,* 245, 1, 1982.
35. Wiegand, S.J. and Price, J.L., Cells of origin of the afferent fibers to the median eminence in the rat, *J. Comp. Neurol.,* 192, 1, 1980.
36. Okamura, H. et al., Coexistence of growth hormone releasing factor-like and tyrosine hydroxylase-like immunoreactivities in neurons of the rat arcuate nucleus, *Neuroendocrinology,* 41, 177, 1985.
37. Okamura, H. et al., L-dopa-immunoreactive neurons in the rat hypothalamic tuberal region, *Neurosci. Lett.,* 95, 42, 1988.
38. Merchenthaler, I. et al., Immunocytochemical localization of growth hormone-releasing factor in the rat hypothalamus, *Endocrinology,* 114, 1082, 1984.
39. Berger, B. et al., Transient expression of tyrosine hydroxylase immunoreactivity in some neurons of the rat neocortex during postnatal development, *Brain Res.,* 355, 141, 1985.
40. Gaspar, P. et al., Tyrosine hydroxylase-immunoreactive neurons in the human cerebral cortex: a novel catecholaminergic group?, *Neurosci. Lett.,* 80, 257, 1987.
41. Lewis, D.A. et al., The distribution of tyrosine hydroxylase-immunoreactive fibers in primate neocortex is widespread but regionally specific, *J. Neurosci.,* 7, 279, 1987.
42. Kitahama, K. et al., Endogenous L-dopa, its immunoreactivity in neurons of midbrain and its projection fields in the cat, *Neurosci. Lett.,* 95, 47, 1988.
43. Akbar, M. et al., Inhibition by L-3,4-dihydroxyphenylalanine of hippocampal CA1 neurons with facilitation of noradrenaline and gamma-aminobutyric acid release, *Eur. J. Pharmacol.,* 414, 197, 2001.
44. Furukawa, N. et al., L-DOPA cyclohexyl ester is a novel potent and relatively stable competitive antagonist against L-DOPA among several L-DOPA ester compounds, *Jpn. J. Pharmacol.,* 82, 40, 2000.
45. Misu, Y. and Goshima, Y., Is L-dopa an endogenous neurotransmitter?, *Trends Pharmacol. Sci.,* 14, 119, 1993.
46. Arai, N. et al., DOPA cyclohexyl ester, a competitive DOPA antagonist, protects glutamate release and resultant delayed neuron death by transient ischemia in hippocampus CA1 of conscious rats, *Neurosci. Lett.,* 299, 213, 2001.
47. Storck, T. et al., Structure, expression, and functional analysis of a $Na^+$-dependent glutamate/aspartate transporter from rat brain, *Proc. Natl. Acad. Sci. U.S.A.,* 89, 10955, 1992.
48. Segal, M.B. et al., Kinetics and Na independence of amino acid uptake by blood side of perfused sheep choroid plexus, *Am. J. Physiol.,* 258, F1288, 1990.
49. Kageyama, T. et al., The 4F2hc/LAT1 complex transports L-DOPA across the blood-brain barrier, *Brain Res.,* 879, 115, 2000.

# 4 Morphology of DOPAergic Neurons in Mammals

*Hitoshi Okamura, Tomoko Ueyama, Satoru Masubuchi, and Kunio Kitahama*

## CONTENTS



## 4.1 INTRODUCTION

Attempts to visualize amine first succeeded in the late 1950s. Falck et al. [1] created an ingenius method, applying the vapor of paraformaldehyde to tissue sections of the adrenal gland and then observing the fluorescence produced upon exposure to ultraviolet light. The development of this procedure not only announced the birth of the morphological survey of amines but also the birth of the histochemical method in general, by which most neurotransmitters and bioactive substances are localized at cellular levels. Using this amine-fluorescence method, the morphology of the brain's catecholaminergic system was clarified and classified (from A1 to A15) according to the location of its cell bodies in brain nuclei [2].

This amine-fluorescence method had been the standard for many years, but the successes in the purification of the enzymes involved in the synthesis of biogenic amines in the 1960s enabled the production of antisera against these enzymes. Immunocytochemistry of amine synthetic enzymes, including tyrosine hydroxylase

(TH); the rate-limiting enzyme of catecholamine biosynthesis, aromatic L-amino acid decarboxylase (AADC), which converts L-3,4-dihydroxyphenylalanine (DOPA) to dopamine; and dopamine β-hydroxylase (DBH), which converts dopamine to noradrenaline, enabled the discrimination of dopaminergic and noradrenergic neurons. Sensitive and specific antisera to TH, especially, were widely used for detecting amine-synthesizing neurons [3,4] because TH immunocytochemistry could easily and sharply visualize the shape of cell bodies and nerve terminals, which was difficult to do using the amine-fluorescence method. Thus, TH immunocytochemistry superceded the amine-fluorescence method after the mid-1970s.

In the pioneering age of TH immunocytochemistry, concurrence with the amine-fluorescence method was emphasized. In due course of the time, however, it was noted that TH-immunoreactive (-ir) neurons are more widely distributed in the brain than catecholamine-fluorescent neurons [3,5,6]. For a long period, the difference had been explained only from a technical standpoint: TH immunocytochemistry is more sensitive than the amine-fluorescence method. However, there was also the possibility that the two substances, amine (stained by the amine-histofluorescence method) and TH (by TH immunocytochemistry), really do have different distributions. When considering the latter possibility, introduction of TH-positive but AADC-negative cells outside the classical aminergic neurons (A1 to A15) revealed a new type of neuronal group called D cells [7]. These lines of evidence support the existence of the aminergic cells not classically defined as dopaminergic or noradrenergic neurons. However, the elucidation of these new types of cells was obliged to wait for more sensitive techniques.

In 1980s, Geffard and his coworkers [8] succeeded in producing antibody to amine itself, which was suitable for immunocytochemistry, by utilizing the glutaraldehyde condensation reaction. In this method, tissue amine was detected with high specificity and high sensitivity, thus overcoming a weakness of the amine-fluorescence method. Using this technique, it was demonstrated that the immunoreactions of DOPA and dopamine are heterogeneous even within each catecholamine cell group, reflecting the differences in metabolism in each neuron [9]. This evidence suggested the presence of nonclassical amine producing neurons lacking essential enzymes for producing amines in the brain [10,11]. As an example of one of these nonclassical amine neurons, we describe here a neuronal system, containing only TH as a synthesized enzyme and lacking AADC, which produces DOPA but not dopamine. Here, we refer to these morphologically "TH-positive/DOPA-positive/AADC-negative/and dopamine-negative neurons" as *DOPA neurons*, and detail the distribution of these neurons in the following sections.

## 4.2 DOPA IMMUNOREACTIVITY IN RAT BRAIN

Before describing DOPA neurons, we first survey how DOPA immunoreactivity occurs in the brain. The content of DOPA is below 0.1 pmol/mg tissue, which is much lower than dopamine and/or noradrenaline contents. This suggests that most of the brain DOPA may exist as the intermediate product in catecholamine neurons. In morphology studies using DOPA antiserum, two types of neurons were detected. One had strongly immunoreactive cell bodies with long immunostained processes and the other had weakly immunoreactive cell bodies. Interestingly, the distribution

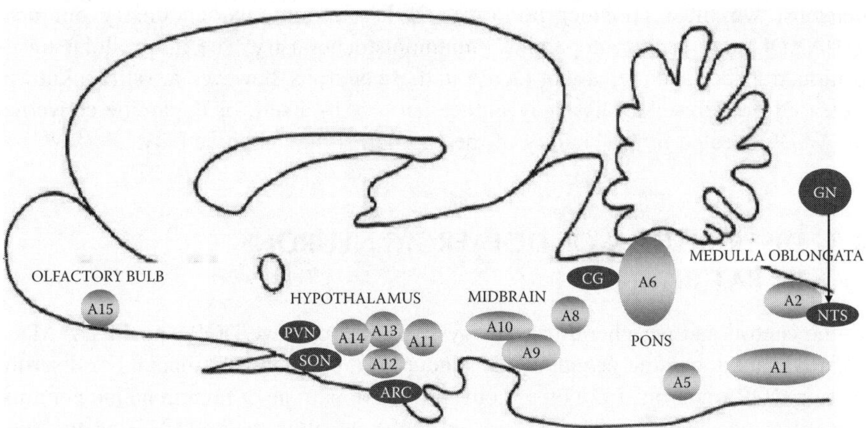


**FIGURE 4.1** Central DOPAergic neurons in relation to the classical noradrenergic and dopaminergic neuronal systems. Closed areas show the location of DOPAergic neurons and open areas show the noradrenergic (A1, A2, A5, and A6) and dopaminergic (A8, A9, A10, A11, A12, A13, A14, and A15) neurons. DOPAergic neurons in the ganglion nodosum (GN) innervate the nucleus tractus solitarius (NTS). A1 to A15 are from Dahlström and Fuxe (1964). A6 attributes to the locus coeruleus, A8 to the ventral tegmental area, and A9 to the substantia nigra. ARC, the hypothalamic arcuate nucleus; CG, the midbrain central gray; PVN, the hypothalamic paraventricular nucleus; SON, the hypothalamic supraoptic nucleus.

of strong DOPA-ir neurons was completely in concordance with that of TH-positive/AADC-negative dopamine-negative neurons, and the distribution of weakly stained neurons was in concordance with classical TH-positive/AADC-positive and/or catecholamine-fluorescent neurons. The distribution of catecholamine-fluorescent cell bodies described by the Falk-Hillarp method was confirmed by immunocytochemistry using antisera specific to noradrenaline and dopamine. A1 to A7 were stained with antinoradrenaline serum, and A8 to A15 were stained with antidopamine serum [12] (Figure 4.1). Strong DOPA immunoreaction may indicate that DOPA is not decarboxylated and remains as DOPA itself in DOPA neurons. In classical amine neurons, however, high amounts of DOPA produced by TH may be rapidly decarboxylated to dopamine by AADC, and only a weak DOPA immunoreactivity remains at cell body levels. Indeed, DOPA immunostaining was very low or absent in the rat striatum, to which the abundant midbrain dopaminergic terminals project, under normal conditions. The increase of DOPA staining in the striatal terminals following the central administration of an AADC inhibitor also proves the effectiveness of DOPA immunocytochemistry for detecting DOPA.

The glutaraldehyde condensation method provides evidence that metabolism of each monoamine is regulated heterogeneously in each monoaminergic neuron even within the same neuronal group. Recently available guanosine triphosphate (GTP) cyclohydrolase 1 (GCH), which synthesizes tetrahydrobiopterine (BH4) as a cofactor of TH, may also provide functional insights into the differences in the production of DOPA in each TH expressing neuron [13]. At present, for the definition of DOPA neurons, we must consider not only DOPA immunohistochemistry but also TH/AADC and, further, dopamine immunohistochemistry. The intracellular localization, transport, and release of DOPA in these neurons, however, are still unknown. Released or diffused DOPA may induce activity by itself, or dopamine converted by AADC located in the vicinity of the capillaries may elicit activity.

## 4.3 DISTRIBUTION OF DOPAERGIC NEURONS IN RAT BRAIN

In the central and peripheral nervous systems, TH-positive/DOPA-positive/AADC-negative/and dopamine-negative DOPA neurons are widely distributed. For discriminating DOPA neurons from other conventional dopamine or noradrenaline neurons, we utilize two methods. The first is the double-labeling method utilizing the anti-TH and anti-AADC method, with which we defined TH-positive/AADC-positive neurons as classical catecholaminergic neurons and TH-positive/AADC-negative neurons as DOPA neurons. The second is the immunocytochemistry using the glutaraldehyde condensation method described by Mons and Geffard [12], with which we defined DOPA-positive/dopamine-negative neurons as DOPA neurons. In Figure 4.1, we compare the DOPA neurons with the classical catecholaminergic neurons in the rat brain.

### 4.3.1 Hypothalamic Arcuate Nucleus

The hypothalamic arcuate nucleus contains the A12 dopaminergic cell group [14], which is suspected to act on prolactin cells in the anterior pituitary. The amine-fluorescence method demonstrates the dopaminergic neurons in the dorsomedial part of the arcuate nucleus and the adjacent periventricular nucleus (DMAR/PV) in the vicinity of the third ventricle. However, with TH immunocytochemistry, TH-immunostained neurons were detected not only in this DMAR/PV but also in the ventrolateral part of the arcuate nucleus and the periarcuate region just dorsal to the ventral surface of the brain (VLAR/PA) [15] (Figure 4.2). Other catecholamine markers show characteristic distribution patterns in these two (DMAR/PV and VLAR/PA) parts. AADC-ir and dopamine-ir neurons were detected in the DMAR/PV but not in the VLAR/PA [16–18]. On the other hand, strong DOPA-ir neurons were detected in the VLAR/PA [10] (Figure 4.3). These findings suggest that TH in the VLAR/PA has a capacity to produce DOPA which will not be converted to dopamine because there exists no AADC. Recently we demonstrated strongly DOPA-ir terminals in the external layer of the median eminence at light microscopic and electron microscopic levels (Okamura, unpublished observation). DOPA immunoreactivity was detected throughout the nerve terminals with vesicle-like aggregates. These findings suggest the possibility that DOPA produced in the VLAR/PA is transported to the external layer of the median eminence and secreted into the portal vessels. Recently, it has been demonstrated that the application of monosodium glutamate in the neonatal period primarily injures DOPA neurons in the VLAR/PA, which seem to contain growth hormone releasing factor [15], but

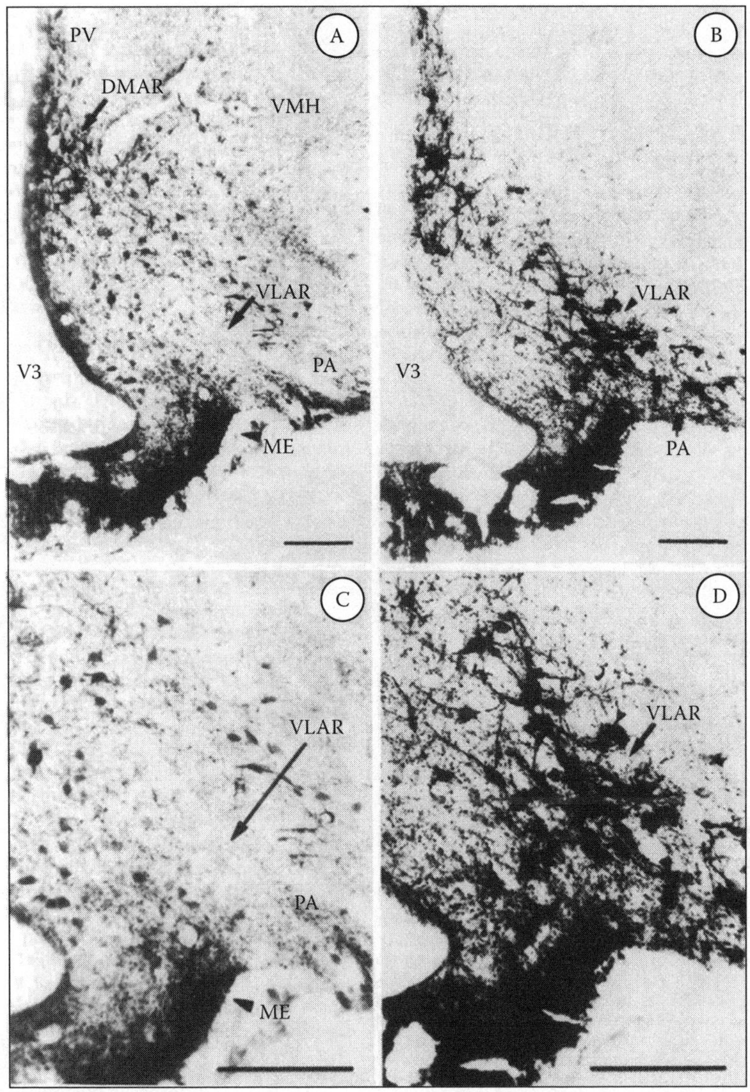


**FIGURE 4.2** Concordance and discordance of AADC and TH immunoreactivity. AADC-ir cells (A) and TH-ir cells (B) in the rat hypothalamic arcuate nucleus. C and D are high-power photomicrographs of a part of A and B, respectively. In the dorsomedial part of the arcuate nucleus (DMAR) which attributes to A12 dopaminergic cell group, both AADC-ir and TH-ir cells are visualized. However, there are no AADC-ir cells equivalent to medium-sized TH-ir cells in the ventrolateral part of the arcuate nucleus (VLAR) and its neighboring periarcuate region (PA). ME, the median eminence; VMH, the ventromedial hypothalamic nucleus; V3, the third ventricle. Bars = 100 μm. (From Okamura, H. et al., Aromatic L-amino acid decarboxylase (AADC)-immunoreactive cells in the tuberal region of the rat hypothalamus, *Biomed. Res.*, 9, 261, 1988. With permission.)

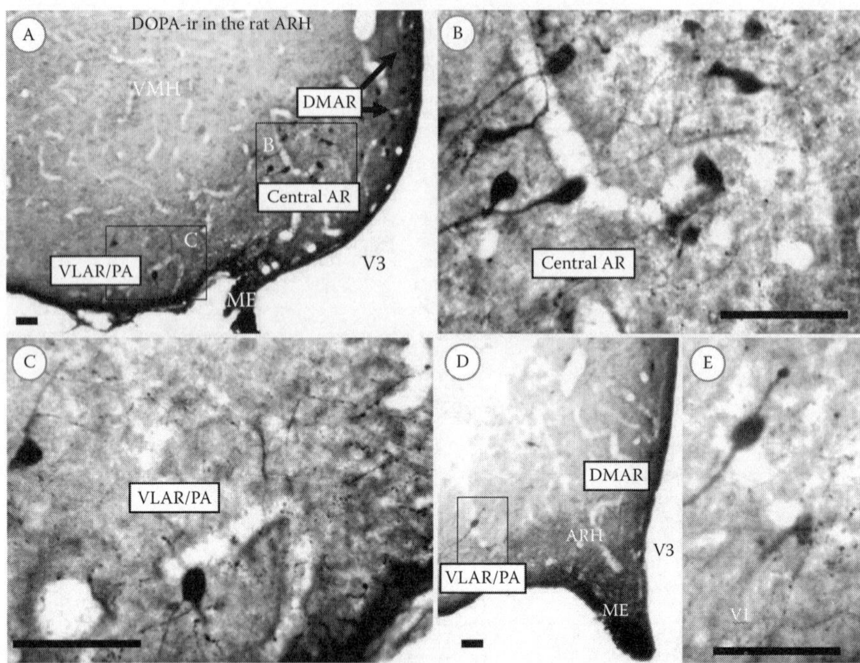


**FIGURE 4.3** DOPA immunoreactive neurons in the rat hypothalamic arcuate nucleus (ARH). In A, DOPA immunoreactivity (DOPA-ir) was observed not only in neurons of the dorsomedial part of the arcuate nucleus (DMAR) in dopaminergic A12 group but also in the central part of the arcute nucleus (Central AR) and the ventrolateral part of the arcuate nucleus and its neighboring periarcuate region (VLAR/PA). B and C are high-power photomicrographs of the square regions of A. D and its high-power photomicrograph E are taken from the section at the more rostral level of the arcuate nucleus. Note DOPA immunoreactive neurons with full well-developed processes (C and E). ME, the median eminence; V3, the third ventricle. Bars = 50 μm.

does not affect dopamine neurons in the DMAR/PV, which elicit the inhibitory effect on pituitary prolactin secretion [19].

### 4.3.2 Nucleus Tractus Solitarius

Jaeger et al. [7] demonstrated that some neurons, A2 catecholaminergic cells, in the rat nucleus tractus solitarius (NTS) and the dorsal motor nucleus of the vagus (DMV), in which the primary baroreceptor afferents terminate, express TH but not AADC. Tison et al. [20] showed that there are strong DOPA-ir neurons in this NTS/DMV complex. These findings are in agreement with those obtained by other investigators [21,22]. Yue et al. [22] demonstrated that the denervation of the aortic nerve, one of the primary baroreceptor afferents, peripheral to the ganglion nodosum decreases TH-ir neurons and DOPA-ir neurons without causing decreases in dopamine-ir neurons and DBH-ir neurons in the NTS/DMV. These DOPAergic neurons may be involved in cardiovascular [22] and/or respiratory [23] functions. In fact, delayed increase in TH activity was reported after long-term hypoxia in the rat caudal A2 medullary neurons [23].

### 4.3.3 Other Brain Regions

TH-ir neurons were detected outside of classical catecholaminergic cell groups. In the rat spinal cord, TH-ir neurons were detected, particularly, at lumbosacral levels [24], which was also observed in the monkey spinal cord. In the developmental stage of the rat brain, TH immunoreactivity was transiently expressed in neurons of the cerebral cortex, the amygdala, and the inferior colliculus [25]. In the tottering and leaner mutant [26] and osmotic stressed [13] mouse cerebellum, Purkinje cells express TH immunoreactivity and/or TH mRNA. However, in the case of the osmotic-stimulated Purkinje cells, AADC was also expressed [13]; thus they may not meet our criteria for DOPA neurons. Strongly DOPA-ir neurons were sometimes detected in the periaqueductal gray of the midbrain [9], the interfascicular nucleus, and the anterior hypothalamic and preoptic areas of the hypothalamus [27] in rats.

## 4.4 DISTRIBUTION OF DOPA IN BRAIN OF OTHER SPECIES

Except in the case of rodents, existence of TH-ir neurons outside of classical catecholaminergic cell groups was not rare. In the human [28] and monkey [29] cerebral cortex, not a few TH-ir neurons were distributed. In the thalamus, TH-ir neurons were detected abundantly in the cat periventricular nucleus [11], and the house shrew lateral habenular nucleus [30]. In the case of the cat, they are all TH-ir/AADC-negative neurons, and in the case of house shrew, they seem to be TH-positive/AADC-negative/and DOPA-positive DOPA neurons [30].

In the human hypothalamic arcuate-periventricular zone (A12 and A14), TH-ir and AADC-ir cells were abundant [31]. The strong immunostaining in these cells suggests that they might contain a high concentration of dopamine, allowing for their visualization by the aldehyde-induced histofluorescence technique [32–35]. In various animal species, most of the cells in the DMAR/PV show strong catecholamine fluorescence and dopamine immunoreactivity. Small TH-ir and AADC-negative neurons could be identified in the ventrolateral part of the human arcuate nucleus [36,31]. As in the rat [10,16], these neurons may be DOPA neurons.

In the hypothalamus, in addition to the A12 dopaminergic arcuate cell group, we confirmed the existence of a great number of TH-ir cell bodies along the entire extent of the human paraventricular, supraoptic, and accessory nuclei, as described by many authors [37–39]. TH-ir cells account for nearly 40% of all paraventricular cells, and most of them are magnocellular [39–40]. On the other hand, we failed to recognize detectable AADC immunoreactivity in magnocellular cells of the human paraventricular nucleus and supraoptic nucleus. Thus, the magnocellular

TH may not be capable of synthesizing dopamine. Indeed, in the monkey *Macaca fuscata*, paraventricular TH-ir large cell bodies showed no immunoreactivity for dopamine [31].

In the cat hypothalamus, dopamine-ir neurons showed a distribution pattern similar to TH-ir ones, although the intensity of the immunostaining varied. DOPA-ir cells showed a similar distribution pattern, although they were less in number. In the cat brain, the ventrolateral part of the arcuate nucleus showed only a few TH-ir neurons, as a result of which no, or very weak, dopamine immunoreactivity could be detected. However, there are some neurons that display very strong DOPA immunoreactivity in this nucleus.

In the cat dorsal vagal complex, in particular, in the caudal A2 noradrenergic medullary group located in the NTS, dopamine immunoreactivity was absent or very weak in TH-ir neurons, whereas DOPA immunoreactivity was quite strong in the cells as well as in their dendritic arbors (Figure 4.4).

Although not through DOPA immunocytochemistry, DOPA neurons have also been reported in many species in recent years using immunocytochemistry of TH

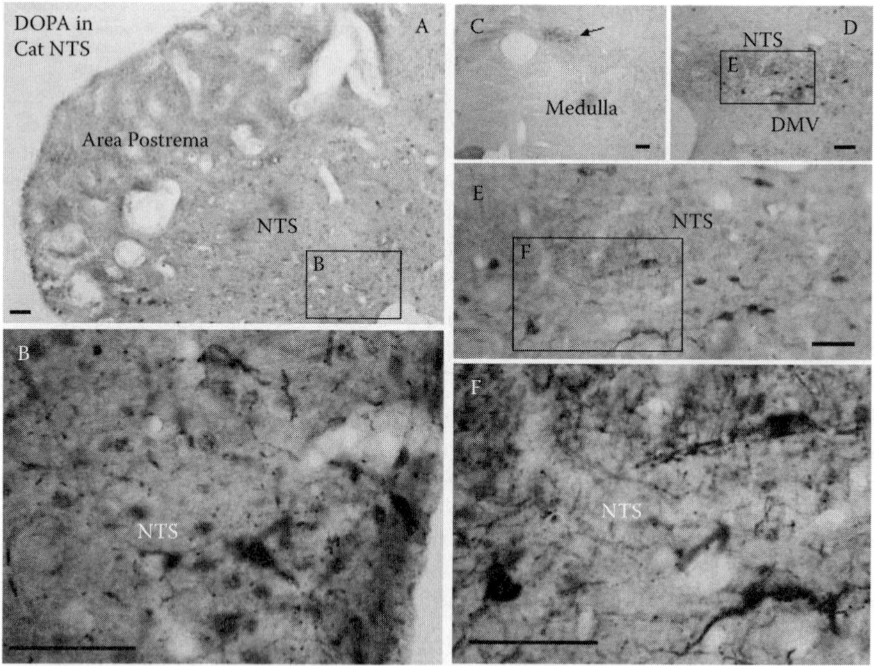


**FIGURE 4.4** DOPA immunoreactivity in the cat NTS. At the level of the area postrema (A), a variety of DOPA-ir neurons and processes were observed in the NTS. B is a high-power photomicrograph of the squared region of A. C shows the DOPA immunoreactivity in the NTS at the caudal level of the medulla oblongata adjacent to the central canal. D, E, and F are high-power photomicrographs of C, D, and E, respectively. Bars = 50 μm in A, B, D, E, and F, and 100 μm in C.

and AADC. TH-positive and AADC-negative neurons were recognized in several discrete regions of the hamster brain [41]. In the lamprey *Lampetra fluviatilis*, large numbers of this type of neurons were found in the olfactory bulb and the preoptic lobe [42].

## 4.5 INCREASE OF DOPAERGIC NEURONS AFTER SALT LOADING

Although TH immunoreactivity is found in the magnocellular hypothalamic neurons in the postmortem human brain [39], this immunoreactivity is virtually absent in other species such as rat, cat, sheep, and reptiles [43,44]. In control rats, TH-ir neurons were hardly ever detected, and only sporadic weakly immunoreactive neurons (1 cell per 2 to 4 sections) were observed in the supraoptic and paraventricular nuclei. On the seventh day of salt loading, however, many weakly to strongly immunoreactive neurons were detected (Figure 4.5) [45–47] without the induction of AADC [48]. Because these TH-positive magnocellular neurons are suspected to trigger secretion of vasopressin [49] or oxytocin [45], the produced DOPA may have some role in the production or release of these posterior pituitary hormones. After rehydration, the number of TH-ir neurons in the supraoptic nucleus gradually decreased, and by the fourth day after rehydration, their number decreased to about 30% of that at the seventh day of salt loading [47]. On the seventh day of rehydration, almost all TH-ir neurons became invisible, as they were before dehydration (Figure 4.5). These salt-sensitive neurons fulfill the criteria for TH-positive/DOPA-positive/AADC-negative/and dopamine-negative DOPA neurons. Recently, Marsais and Calas [50] suggested that TH-positive magnocellular neurons under hyperosmotic conditions lack GCH, which is involved in the synthesis of BH4, an important cofactor of TH, AADC, tryptophan hydroxylase, and nitric oxide synthase (NOS). Because most dopaminergic and noradrenergic neurons express GCH, they speculated that the TH-positive GCH-negative magnocellular neurons, without accompanying BH4, may lack the enzyme activity of TH. However, Hwang et al. [51] suggest that the olfactory dopaminergic neurons (A15) in the periglomerular cells lack GCH, which is necessary for detection by immunocytochemistry and *in situ* hybridization. Moreover, most cells expressing neuronal NOS express no GCH [51]. To explain this discrepancy, it is speculated that these GCH-negative neurons will receive BH4 released from the surrounding GCH-containing nerve terminals [51, 52]. Thus, at present, it is still possible that TH in the magnocellular neurons is enzymatically active and produces DOPA, but this requires further research.

## 4.6 DOPA IN PRIMARY SENSORY NEURONS

Classically, in the peripheral nervous system, catecholaminergic phenotypic characteristics have been restricted to the sympathetic neurons. However, many investigators have shown the presence of catecholamine traits in primary sensory neurons such as the dorsal root ganglion [53] and the ganglion nodosum [54–56]. The nodose ganglion, namely, the vagal inferior ganglion, is a primary sensory ganglion that conveys sensory

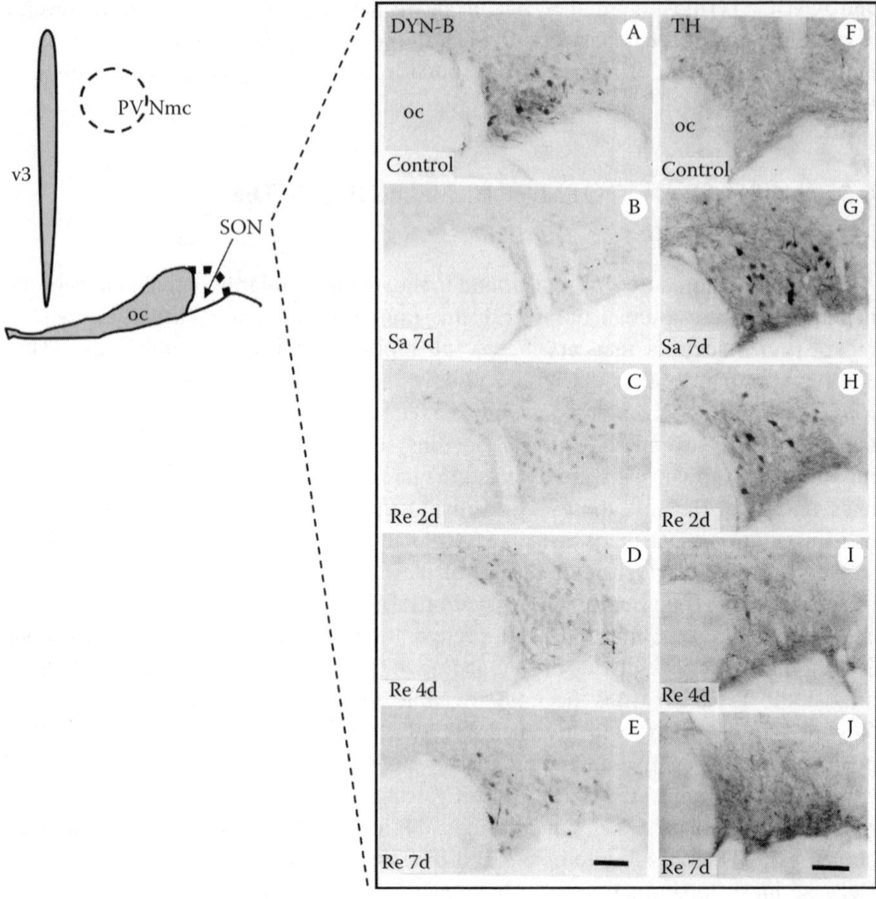


**FIGURE 4.5** Salt-loading transiently induces TH traits in the supraoptic nucleus (SON). Location of the SON is shown in the left schema. Immunocytochemistry of TH (F–J) in the SON was shown in comparison with that of dynorphin B (DYN-B) (A–E), in control (A, F), seventh day of salt loading (Sa 7d) (B, G), second day of rehydration (Re 2d) (C, H), fourth day of rehydration (Re 4d) (D, I), and seventh day of rehydration (Re 7d) (E, J). Note TH-ir neurons appeared only during salt loading and in the second day of rehydration stages, whereas the immunoreactive intensity of DYN-B was decreased during salt loading. oc, the optic chiasma; PVNmc, the magnocellular part of the paraventricular nucleus; V3, the third ventricle. Bars = 100 μm. (From Yagita, K., Okamura, H., and Ibata, Y., Rehydration process from salt-loading of vasopressin and coexisting galanin, dynorphin and tyrosine hydroxylase immunoreactives in the supraoptic and paraventicular nuclei, *Brain Res.,* 667, 13, 1994. With permission.)

information from neck and chest organs to the NTS of the medulla oblongata [57]. Primary sensory neurons in this nucleus abundantly express substance P and calcitonin gene-related peptide [58,59] in a manner similar to the other primary sensory ganglia. Although this nucleus is not a noradrenergic sympathetic ganglion, some neurons in this nucleus are reported to contain TH [55,60]. TH immunoreactivity was also found in the petrosal and geniculate ganglia but rarely found in the other sensory ganglia such as the dorsal root ganglia. However, in the developmental stage, TH-ir cells were common in the trigeminal, geniculate, jugular superior, petrosal, nodose, and cervical dorsal root ganglia in rats between 10.5 and 15.5 d of gestation [61]. This suggests that the TH in the adult nodose ganglion is not an artifact but the preservation of developmental traits in the adult stage.

Furthermore, after the section of the aortic nerve peripheral to the ganglion nodosum, we detected the decrease of TH and DOPA immunoreactivity in the host cell bodies in the ganglion nodosum and its projecting terminals in the NTS/DMV [22]. Because direct injury of peripheral processes and the denervation of the target organ may decrease the neurotransmitter-synthesizing enzyme activity in cell bodies, these findings strongly suggest that peripheral aortic nerve denervation decreases DOPA production in the ganglion, and the influence extends to the centrally projecting terminals in the NTS/DMV.

Recently, nitric oxide synthase (NOS) was identified in neurons of the nodose ganglion [62,63]. We demonstrated that NOS and/or TH expressing neurons in the rat nodose ganglion are centrally projecting primary sensory neurons by a combination of three methods: neuronal tracer, rhodamine-labeled latex microspheres injected into the rostral part of the NTS, NADPH-diaphorase histochemistry, and TH immunocytochemistry [64] (Figure 4.6). This evidence is in concordance with the observed reduction of NOS immunoreactivity in the medial NTS induced by nodose ganglionectomy [65] or with the reduction of TH and DOPA immunoreactivity in the nodose ganglion and the NTS/DMV elicited by aortic nerve denervation [22]. Thus, it is most likely that the primary sensory neurons of the nodose ganglion express NOS and/or TH/DOPA.

## 4.7 CONCLUSION

Morphological techniques have now become available to localize DOPA in brain neurons. Independent of previously described biochemical and pharmacological works, the idea of neurons containing DOPA as an end product is also advocated as "DOPA neurons." However, there are still many hurdles to present morphological evidence for DOPA as an end product. The most convincing data were obtained by the detection of DOPA itself by immunocytochemistry using DOPA antiserum. However, we are not sure what percentage of DOPA in the brain is detected by the present glutaraldehyde fixation method. By the glutaraldehyde condensation method utilized here, DOPA obtained antigenicity after the conjugation of tissue protein with glutaraldehyde fixation and is detected by antiserum against DOPA and bovine serum albumin (BSA) conjugated with glutaraldehyde [12]. With this immunocytochemical reaction we could not detect unconjugated DOPA. Thus, the success of the method was dependant on whether DOPA was fixed at the first fixation process. Because the DOPA antiserum was highly specific to DOPA-G-BSA (DOPA and BSA conjugated with glutaraldehyde) without ability to recognize dopamine-G-BSA (dopamine and BSA conjugated with glutaraldehyde), there may not be any cross-reaction, and, therefore, little overestimation of DOPA

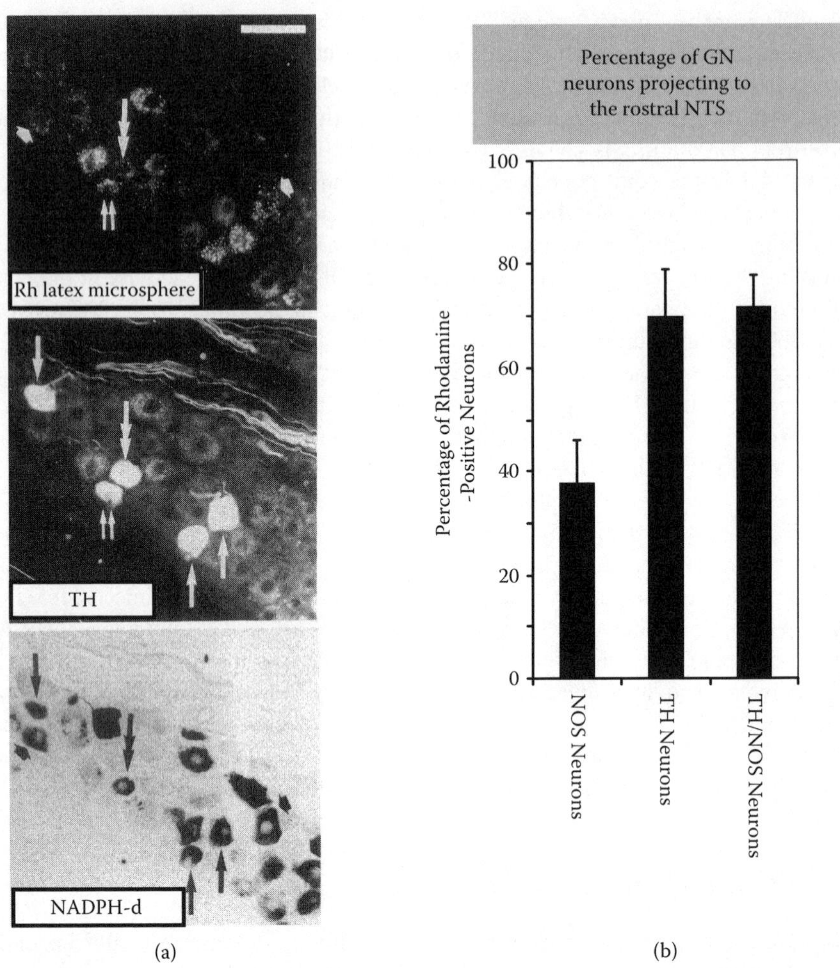


**FIGURE 4.6** Projection of TH/DOPA neurons in the ganglion nodosum (GN) to the rat medulla oblongata. In A, retrograde neuronal tracer rhodamine (Rh) latex microspheres were injected into the rostral part of the NTS, and 7 d later, animals were processed for TH immunocytochemistry and NADPH-diaphorase (NADPH-d) histochemistry. NADPH-d is a marker of nitric oxide synthase (NOS) expressing neurons. Rh latex microspheres were first photographed, then TH immunocytochemistry was visualized by fluorescein isothiocyanate, and finally NADPH-d reaction was performed. Arrows indicate neurons bearing two markers, and double arrows indicate triply-labeled neurons. Short thick arrows indicate neurons having microspheres and NADPH-d but no TH immunoreactivity. Bar = 70 μm. B shows percentage of the GN neurons projecting to the rostral NTS in TH neurons, NOS neurons, and TH/NOS neurons. Note that 40–60% of NOS neurons and TH/DOPA neurons project to the NTS. We calculated 102 neurons expressing both NADPH-d and TH traits from 513 NADPH-d-positive and 156 TH-positive cells in 3 rats (Mean ± SE). (From Nishiyama, K. et al., Tyrosine hydroxylase and NADPH-diaphorase in the rat nodose ganglion: Colocalization and central projection, *Acta Histochem. Cytochem.*, 34, 135, 2001. With permission.)

immunoreactivity. Thus, underestimation of DOPA immunoreactivity due to the first fixation process may be a main problem. Indeed, by DOPA immunocytochemistry, very faint or no immunoreactivity was detected in dopaminergic terminals innervating the striatum, whereas biochemical techniques utilizing HPLC or other methods, can detect the fair amounts of DOPA there [66]. Ultrastructural localization of DOPA, which may provide key knowledge for understanding the function and metabolism of DOPA, may face great difficulties because of this weak sensitivity in immunoreaction.

## ACKNOWLEDGMENTS

We are grateful to M. Geffard for donating anti-DOPA serum. This study was in part supported by 21st Century COE Program "Center of Excellence for Signal Transduction Disease: Diabetes Mellitus as Model" from the Ministry of Education, Culture, Sports, Science, and Technology of Japan, SRF Japan, and UMR CNRS 5123, France.

## REFERENCES

1. Falck, B. et al., Fluorescence of catecholamines and related compounds condensed with formaldehyde, *J. Histochem. Cytochem.,* 10, 348, 1962.
2. Dahlström, A. and Fuxe, K., Evidence of the existence of monoamine containing neurons in the central nervous system. 1. Demonstration of monoamines in the cell bodies of brain stem neurons, *Acta Physiol. Scand.,* 62, Suppl. 232, 1, 1964.
3. Hökfelt, T. et al., Immunohistochemical studies on the localization and distribution of monoamine neuron system in the rat brainstem. 1. Tyrosine hydroxylase in the mes- and diencephalon, *Med. Biol.,* 54, 427, 1976.
4. Nagatsu, I. et al., Immunofluorescent studies on tyrosine hydroxylase: application for its axonal transport, *Acta Histochem. Cytochem.,* 10, 494, 1977.
5. Chan-Palay, V. et al., Distribution of tyrosine-hydroxylase immunoreactive neurons in the hypothalamus of rats, *J. Comp. Neurol.,* 227, 467, 1984.
6. Hökfelt, T. et al., Immunohistochemical studies on the localization and distribution of monoamine neuron system in the rat brainstem. 2. Tyrosine hydroxylase in the telencephalon, *Med. Biol.,* 55, 21, 1977.
7. Jaeger, C.B. et al., Aromatic L-amino acid decarboxylase in the rat brain: immunocytochemical localization in neurons of the brain stem, *Neuroscience,* 11, 691, 1984.
8. Geffard, M. et al., First demonstration of highly specific and sensitive antibodies against dopamine, *Brain Res.,* 294, 161, 1984.
9. Okamura, H. et al., Heterogeneous distribution of L-DOPA immunoreactivity in dopaminergic neurons of the rat midbrain, in *Basic, Clinical, and Therapeutic Aspects of Alzheimer's and Parkinson's Diseases, Vol.* 1, Nagatsu, T., Fischer, S., and Yoshida, M. Eds., Plenum Press, New York, 1990, 423.
10. Okamura, H. et al., L-DOPA immunoreactive neurons in the rat hypothalamic tuberal region, *Neurosci. Lett.,* 95, 42, 1988.
11. Kitahama, K. et al., A new group of tyrosine hydroxylase-immunoreactive neurons in the cat thalamus, *Brain Res.,* 478, 156, 1989.

12. Mons, N. and Geffard, M., Specific antisera against the catecholamines: L-3,4-dihydroxyphenylalanine, dopamine, noradrenaline and octopamine tested by an enzyme-linked immunosorbent assay, *J. Neurochem.*, 48, 1826, 1987.
13. Sakai, M. et al., Enhanced expression of tyrosine hydroxylase and aromatic L-amino acid decarboxylase in cerebellar Purkinje cells of mouse after hyperosmotic stimuli, *Neurosci. Lett.*, 194, 142, 1995.
14. Björklund, A. and Lindvall, O., Dopamine-containing system in the CNS, in: *Handbook of Chemical Neuroanatomy. Vol. 2. Classical Transmitter in the CNS, Part I,* Björklund, A. and Hökfelt, T., Eds., Elsevier, Amsterdam, 1984, 55.
15. Okamura, H. et al., Coexistence of growth hormone releasing factor-like and tyrosine hydroxylase-like immunoreactivities in neurons of the rat arcuate nucleus, *Neuroendocrinology,* 41, 177, 1985.
16. Meister, B. et al., Do tyrosine hydroxylase-immunoreactive neurons in the ventral arcuate nucleus produce dopamine or only L-DOPA?, *J. Chem. Neuroanat.*, 1, 59, 1988.
17. Okamura, H. et al., Aromatic L-amino acid decarboxylase (AADC)-immunoreactive cells in the tuberal region of the rat hypothalamus, *Biomed. Res.*, 9, 261, 1988.
18. Okamura, H. et al., Comparative topography of dopamine and tyrosine hydroxylase-immunoreactive neurons in the rat arcuate nucleus, *Neurosci. Lett.*, 95, 347, 1988.
19. Bodnar, I. et al., Effect of neonatal treatment with monosodium glutamate on dopaminergic and L-DOPAergic neurons of the medial basal hypothalamus and on prolactin and MSH secretion of rats, *Brain Res. Bull.*, 55, 767, 2001.
20. Tison, F. et al., Endogenous L-DOPA in the rat dorsal vagal complex: an immunocytochemical study by light and electron microscopy, *Brain Res.*, 497, 260, 1989.
21. Manier, M. et al., Evidence for the existence of L-dopa- and dopamine-immunoreactive nerve cell bodies in the caudal part of the dorsal motor nucleus of the vagus nerve, *J. Chem. Neuroanat.*, 3, 193, 1990.
22. Yue, J.-L. et al., Baroreceptor-aortic nerve-mediated release of endogenous L-3,4-dihydroxyphenylalanine and its tonic depressor function in the nucleus tractus solitarii of rats, *Neuroscience,* 62, 145, 1994.
23. Soulier, V. et al., Delayed increase of tyrosine hydroxylation in the rat A2 medullary neurons upon long-term hypoxia, *Brain Res.*, 674, 188, 1995.
24. Uda, K. et al., Regional distribution of tyrosine hydroxylase immunoreactive neuronal elements in the rat and monkey spinal cord, *Biogen. Amines,* 4, 153, 1987.
25. Berger, B. et al., Transient expression of tyrosine hydroxylase immunoreactivity in some neurons of the rat neocortex during postnatal development, *Dev. Brain Res.*, 23, 141, 1985.
26. Hess, E.J. and Wilson, M.C., Tottering and leaner mutations perturb transient developmental expression of tyrosine hydroxylase in embryologically distinct Purkinje cell, *Neuron,* 6, 123, 1991.
27. Mons, N., Tison, F., and Geffard, M., Existence of L-DOPA immunoreactive neurons in the rat preoptic area and anterior hypothalamus, *Neuroendocrinology,* 51, 425, 1990.
28. Gaspar, P. et al., Tyrosine hydroxylase-immunoreactive neurons in the human cerebral cortex: a novel catecholamine group?, *Neurosci. Lett.*, 80, 257, 1987.
29. Köhler, C. et al., Immunohistochemical evidence for a new group of catecholamine-containing neurons in the basal forebrain of the monkey, *Neurosci. Lett.*, 37, 161, 1983.
30. Karasawa, N., Isomura, G., and Nagatsu, I., Production of specific antibody against L-DOPA and its ultrastructural localization of immunoreactivity in the house shrew (*Suncus murinus*) lateral habenular nucleus, *Neurosci. Lett.*, 143, 267, 1992.

31. Kitahama, K. et al., Aromatic L-amino acid decarboxylase and tyrosine hydroxylase immunohistochemistry in the adult human hypothalamus, *J. Chem. Neuroanat.,* 16, 43, 1998.
32. de la Torre, J.C., Catecholamines in the human diencephalon: a histochemical fluorescence study, *Acta Neuropathol.,* 21, 165, 1972.
33. Hyyppa, M., Hypothalamic monoamines in human fetuses, *Neuroendocrinology,* 9, 257, 1972.
34. Olson, L., Boreus, L.O., and Seiger, A., Histochemical demonstration and mapping of 5-hydroxytryptamine- and catecholamine-containing neuron systems in the human fetal brain, *Z. Anat. Entwickl. Grsch.,* 139, 259, 1973.
35. Nobin, A. and Bjorklund, A., Topography of the monoamine neuron systems in the human brain as revealed in fetuses, *Acta Physiol. Scand.,* Suppl. 388, 1, 1973.
36. Komori, K., Fujii, T., and Nagatsu, I., Do some tyrosine hydroxylase-immunoreactive neurons in the human ventrolateral arcuate nucleus and globus pallidus produce only L-DOPA?, *Neurosci. Lett.,* 133, 203, 1991.
37. Spencer, S. et al., Distribution of catecholamine-containing neurons in the normal human hypothalamus, *Brain Res.,* 328, 73, 1985.
38. Gaspar, P. et al., Catecholaminergic innervation of the septal area in man: immunocytochemical study using TH and DBH antibodies, *J. Comp. Neurol.,* 241, 12, 1985.
39. Li, Y.W. et al., Tyrosine hydroxylase-containing neurons in the supraoptic and paraventricular nuclei of the adult human, *Brain Res.,* 461, 75, 1988.
40. Panayotacopoulou, M.T. and Swaab, D.F., Development of tyrosine hydroxylase-immunoreactive neurons in the human paraventricular and supraoptic nucleus, *Dev. Brain Res.,* 72, 145, 1993.
41. Vincent, S.R. and Hope, B.T., Tyrosine hydroxylase containing neurons lacking aromatic amino acid decarboxylase in the hamster brain, *J. Comp. Neurol.,* 295, 290, 1990.
42. Pierre, J. et al., Immunohistochemical localization of dopamine and its synthetic enzymes in the central nervous system of the Lamprey *Lampeta fluviatilis, J. Comp. Neurol.,* 380, 119, 1997.
43. Hökfelt, T. et al., Distributional maps of tyrosine hydroxylase-immunoreactive neurons in the rat brain, in *Handbook of Chemical Neuroanatomy. Vol. 2. Classical Transmitter in the CNS, Part I,* Björklund, A. and Hökfelt, T., Eds., Elsevier, Amsterdam, 1984, 277.
44. Tillet, Y., Catecholaminergic neuronal systems in diencephalon of mammals, in *Phylogeny and Development of Catecholamine Systems in the CNS of Vertebrates,* Smeets, J.A.J. and Reiner, A., Eds., Cambridge University Press, Cambridge, 1994, 207.
45. Kiss, J.Z. and Mezey, E., Tyrosine hydroxylase in magnocellular neurosecretory neurons; response to physiological manipulations, *Neuroendocrinology,* 43, 519, 1986.
46. Young, W.S., Warden, M., and Mezey, E., Tyrosine hydroxylase mRNA is increased by hyperosmotic stimuli in the paraventricular and supraoptic nuclei, *Neuroendocrinology,* 46, 439, 1987.
47. Yagita, K., Okamura, H., and Ibata, Y., Rehydration process from salt-loading of vasopressin and coexisting galanin, dynorphin and tyrosine hydroxylase immunoreactives in the supraoptic and paraventicular nuclei, *Brain Res.,* 667, 13, 1994.
48. Okamura, H. et al., Immunocytochemistry of aromatic L-aminoacid decarboxylase in the rat preoptic area and anterior hypothalamus, with special reference to tyrosine hydroxylase immunocytochemistry, *Biogen. Amines,* 7, 351, 1990.

49. Tanaka, M. et al., L-DOPA-immunoreactivity in hypothalamic magnocellular neurons of hypothalamo-neurohypophysial system, *Neurosci. Res.,* 16, Suppl. S29, 1991.
50. Marsais F. and Calas A., Ectopic expression of non-catecholaminergic tyrosine hydroxylase in rat hypothalamic magnocellular neurons, *Neuroscience,* 94, 151, 1999.
51. Hwang, O. et al., Localization of GTP cyclohydrolase in monoaminergic but not nitric oxide-producing cells, *Synapse,* 28, 140, 1998.
52. Nagatsu, I. et al., Transient tyrosine hydroxylase expression in the non-catecholaminergic neurons of pre- and post-natal mice, in *Advances in Behavioral Biology, Vol. 53, Catecholamine Research From Molecular Insights to Clinical Medicine,* Nagatsu, T., McCarty, B., and Goldstein, D.S., Eds., Kluwer Academic/Plenum Publishers, New York, 2002, 111.
53. Price, J. and Mudge, A.W., A subpopulation of rat dorsal root ganglion neurones is catecholaminergic, *Nature,* 301, 241, 1983.
54. Yoshida, M. et al., Immunohistochemical localization of catecholamine-synthesizing enzymes in suprarenal, superior cervical and nodose ganglia of dogs, *Acta Histochem. Cytochem.,* 14, 588, 1981.
55. Katz, D.M. et al., Expression of catecholaminergic characteristics by primary sensory neurons in the normal adult rat in vivo, *Proc. Natl. Acad. Sci. U.S.A.,* 80, 3526, 1983.
56. Helke, C.J. and Niederer, A.J., Studies on the coexistence of substance P with other putative transmitters in the nodose and petrosal ganglia, *Synapse,* 5, 144, 1990.
57. Hatai, S., Number and size of the spinal ganglion cells and dorsal root fibers in the white rat at different ages, *J. Comp. Neurol.,* 12, 107, 1902.
58. Ayer-LeLievre, C.S. and Seiger, A., Development of substance P-immunoreactive neurons in cranial sensory ganglia of the rat, *Int. J. Dev. Neurosci.,* 2, 451, 1984.
59. Hisa, Y. et al., Neuropeptide participation in canine laryngeal sensory innervation. Immunohistochemistry and retrograde labeling, *Ann. Otol. Rhinol. Laryngol.,* 103, 767, 1994.
60. Uno, T. et al., Tyrosine hydroxylase-immunoreactive cells in the nodose ganglion for the canine larynx, *Neuroreport,* 7, 1373, 1996.
61. Katz, D.M. and Erb, M., Development regulation of tyrosine hydroxylase expression in primary sensory neurons of the rat, *Dev. Biol.,* 137, 233, 1990.
62. Alm, P. et al., Nitric oxide synthase-containing neurons in rat parasympathetic and sensory ganglia: a comparative study, *Histochem. J.,* 27, 819, 1995.
63. Zhang, X., et al., Expression of peptides, nitric oxide synthase and NPY receptor in trigeminal and nodose ganglia after nerve lesions, *Exp. Brain Res.,* 111, 393, 1996.
64. Nishiyama, K. et al., Tyrosine hydroxylase and NADPH-diaphorase in the rat nodose ganglion: colocalization and central projection, *Acta Histochem. Cytochem.,* 34, 135, 2001.
65. Lawrence, A.J., Krstew, E., and Jarrott, B., Actions of nitric oxide and expression of the mRNA encoding nitric oxide synthase in rat vagal afferent neurons, *Eur. J. Pharmacol.,* 127, 315, 1996.
66. Westerink, B.H.C., Van Es, T.P., and Spaan, S.J., Effects of drugs interfering with dopamine and noradrenaline biosynthesis on the endogenous 3,4-dihydroxyphenylalanine levels in the rat brain, *J. Neurochem.,* 39, 44, 1982.

# *Part III*

## *Release*

# 5 Physiological Release of DOPA

*Yoshimi Misu, Kaneyoshi Honjo, and Yoshio Goshima*

**CONTENTS**

## 5.1 INTRODUCTION

Exogenously applied L-3,4-dihydroxyphenylalanine (levodopa) has been traditionally thought to be an inert amino acid that alleviates the symptoms of Parkinson's disease via its conversion to dopamine by the enzyme aromatic L-amino acid decarboxylase (AADC) [1–4]. On the other hand, since 1986, we have been proposing that endogenous L-3,4-dihydroxyphenylalanine (DOPA) is a neurotransmitter and neuromodulator in its own right, in addition to being a precursor of dopamine [5,6].

The evidence for physiological release of an endogenous substance in response to nerve stimulation is one of the important classical criteria to be accepted as a neurotransmitter. However, extracellular presence of DOPA had been thought to be only an indicator of the biosynthesis of dopamine in striata *in vivo* [7,8]. On the other hand, in the course of studies on presynaptic regulatory mechanisms of endogenous catecholamine release from rat hypothalamic slices [9–16], we found an unexpected peak increased by electrical field stimulation on chromatographic charts. We identified it with DOPA [17]. We further found that this evoked release of DOPA is abolished by tetrodotoxin (TTX), a blocker of $Na^+$ channels, and by deprivation of extracellular $Ca^{2+}$ in striatal slices in a manner similar to that seen for the established neurotransmitter dopamine [18]. High $K^+$ [18] and nicotine [19] also release DOPA as well as dopamine in a $Ca^{2+}$-dependent manner.

At that time [17], we thought that we would start DOPA studies in our laboratory, if we could get two pieces of evidence. In general, $Ca^{2+}$ plays essential roles in the depolarization-induced processes of both the release of neurotransmitters [20,21] and the activation of tyrosine hydroxylase (TH) [22], the rate-limiting enzyme of catecholamine biosynthesis. Thus, one piece of evidence was proof that $Ca^{2+}$ is involved primarily in the process of DOPA release, which was shown by Goshima et al. (1988) [18]. The other piece of evidence is that DOPA elicits responses by itself, independent of its bioconversion to dopamine.

The first evidence for responses to levodopa in its own right, and not through its conversion to dopamine, is the presynaptic biphasic regulatory actions of levodopa on the impulse-evoked release of endogenous noradrenaline from rat hypothalamic slices [5]. Nanomolar levodopa facilitates this via activation of presynaptic β-adrenoceptors in a propranolol-sensitive manner under intact activity of AADC. This facilitation is neither inhibited by an AADC inhibitor [5] nor mimicked by nanomolar dopamine and apomorphine [16]. In contrast, micromolar levodopa inhibits noradrenaline release via activation of presynaptic dopamine $D_2$ receptors in a sulpiride-sensitive manner under inhibition of AADC [5]. Such a biphasic concentration–response relationship for nanomolar and micromolar levodopa is completely opposite to that for dopamine agonists regulating noradrenaline release [16]. Nanomolar dopamine elicits the sulpiride-sensitive inhibition via presynaptic dopamine $D_2$ receptors, whereas micromolar dopamine elicits the propranolol-sensitive facilitation via presynaptic β-adrenoceptors. Nanomolar apomorphine also inhibits noradrenaline release in a manner similar to that of dopamine. In rat striatal slices, nanomolar levodopa also facilitates dopamine release via presynaptic β-adrenoceptors in the absence and presence of AADC inhibition, whereas micromolar levodopa inhibits it via presynaptic $D_2$ receptors in the presence of AADC inhibition [6].

We have accumulated evidence suggesting that DOPA fulfills the classical criteria, such as biosynthesis, metabolism, active transport, presence, physiological release, competitive antagonism, and physiological or pharmacological responses including interactions with the other neurotransmitter systems [23–28], which must be satisfied before a compound is accepted as a neurotransmitter [29].

We have further explored DOPA release during microdialysis of the striatum [30–34] and the shell compartment of the nucleus accumbens [35,36] in conscious rats. Microdialysis has also been performed in the blood pressure regulatory centers of the lower brainstem in anesthetized rats, such as the nucleus tractus solitarii (NTS) [37,38], the caudal ventrolateral medulla (CVLM) [39–41], and the rostral ventrolateral medulla (RVLM) [38,42,43].

Neurons showing immunocytochemically TH-positive, DOPA-positive, AADC-negative, and dopamine-negative reactivity exist in some regions of the central nervous system, including the NTS [37,44,45], that may contain DOPA as an end product [25,28]. We have accumulated evidence to support the idea that DOPA is a neurotransmitter of the aortic depressor nerve (ADN), one of the primary baroreceptor afferents terminating in the NTS [37,38,46,47]. During microdialysis of the NTS, basal DOPA release is partially TTX sensitive and $Ca^{2+}$ dependent, suggesting that DOPA is released, at least in part, via some spontaneous neuronal activity [37,38]. Indeed, electrical stimulation of the ADN releases DOPA in a TTX-sensitive manner [37]. Furthermore, stimulation of baroreceptors by the i.v. infusion of phenylephrine also releases DOPA, an effect that is abolished by bilateral sinoaortic denervation.

In contrast, no immunocytochemical evidence for neurons having DOPA as an end product has been found in the striatum, the nucleus accumbens, the CVLM, and the RVLM [25,28]. Notwithstanding this, during microdialysis of the striatum [30,31] and the nucleus accumbens [35], basal DOPA release is TTX sensitive and $Ca^{2+}$ dependent in a manner similar to that of dopamine. High $K^+$ [30] and nicotine [31] causes release of DOPA and dopamine in a $Ca^{2+}$-dependent manner in the striatum. Nicotine systemically applied and locally injected into the ventral tegmental area (VTA) releases not only dopamine but also DOPA in the nucleus accumbens [35]. Extracellular levels of DOPA as well as dopamine are elevated in the nucleus accumbens in response to the discontinuation of electrical foot-shock stress [36]. In addition, partially TTX-sensitive basal DOPA release and high-$K^+$-evoked $Ca^{2+}$-dependent DOPA release are seen during microdialysis of the CVLM [39] and the RVLM [42]. Furthermore, baroreceptor activation by phenylephrine selectively releases DOPA in the CVLM, which is suppressed by an acute lesion of the ipsilateral NTS [41]. Electrical stimulation of the ADN also selectively releases DOPA in the CVLM, which is suppressed by inhibition of DOPA biosynthesis with $\alpha$-methyl-*p*-tyrosine ($\alpha$-MPT), a TH inhibitor, locally infused into the CVLM. Electrical stimulation of the posterior hypothalamic nucleus (PHN) selectively releases TTX-sensitive DOPA in the RVLM [43]. Thus, we can safely say that the sensitivity of immunocytochemical analysis to identify DOPAergic neurons appears to be lower as compared to biochemical approaches in combination with denervation experiments [26–28].

In this chapter, we describe the characteristics of the basal and evoked release of DOPA in these regions.

## 5.2 STRIATUM

An important target region for DOPA is the striatum. Parkinson's disease is an example of a neurological disorder in which development of an effective symptomatic therapy has evolved rationally from basic neurochemical studies in animals and in postmortem human brains. The akinetic–rigid syndrome elicited by reserpine could be antagonized by dopamine [1]. A dopamine deficiency was discovered in the brains of patients dying with Parkinson's disease [2]. Because dopamine does not cross the blood–brain barrier, levodopa was administered to parkinsonian patients in an attempt at replacement therapy [48]. Levodopa therapy, especially in combination with a peripheral AADC inhibitor, revolutionized the treatment of Parkinson's disease, improving the quality of life. Since then, dopaminergic therapy, including the use of levodopa and dopamine $D_1$ and $D_2$ agonists, has made major advances. Disease progression, however, is not altered and adverse effects such as decreased control of symptoms, increased dyskinesia, alterations in mentation, increased diurnal fluctuations, episodes of akinetic freezing and crisis, increased fatigue, and neurasthenia during long-term levodopa therapy are common [49]. These adverse effects, in turn, markedly disturb the quality of life with the progress of Parkinson's disease [24,25]. In addition, micromolar levodopa appears to release, by itself, vesicular glutamate from the rat striatum *in vitro* independent of its bioconversion to dopamine [50]. Glutamate release evoked by levodopa but not by dextrodopa is also evident in the striatum *in vivo* [51], which may indicate that it is related to the neuroexcitatory side effects, at least, for example, the dyskinesia encountered during chronic therapy of Parkinson's disease. Furthermore, levodopa causes dopaminergic neuron death at least in neuron cultures and in experiments using normal animals [27,51–57]. Endogenously released DOPA appears to be an upstream causal factor for glutamate release and the delayed neuron death elicited by transient ischemia in rat striata [51].

Today, levodopa still remains the most effective drug for the reversal of the symptoms of Parkinson's disease [58]. But most of the physician's efforts in providing optimum care to patients with Parkinson's disease is in trying to overcome the all too common adverse effects of levodopa.

In general, nicotine has the ability to release various kinds of neurotransmitters and/or neuromodulators via activation of nicotinic receptors located on neuronal axon terminals or cell bodies. Experimental evidence for the neuroprotective effect of nicotine has been shown against *N*-methyl-D-aspartate receptor-mediated glutamate neurotoxicity using primary cultures of rat cortical neurons [59] and mouse striatal neurons [60]. Epidemiological studies have repeatedly shown that Parkinson's disease occurs less frequently in persons who have smoked cigarettes for many years, compared with nonsmokers [61–63]. Cigarette smoking is effective in relief of symptoms for 10 to 30 min in the case of patients with early-onset Parkinson's disease [64]. These neuroprotective and therapeutic effects of cigarette smoking in Parkinson's disease have been attributed mainly to nicotine-induced dopamine release in the striatum. However, nicotine releases, repetitively and simultaneously, DOPA as well as dopamine from the rat striatum *in vitro* [19] and *in vivo* [31].

### 5.2.1 *In Vitro* DOPA Release from Rat Striatal Slices

Using high-performance liquid chromatography with electrochemical detection, the basal efflux of both DOPA and dopamine is consistently detectable in superfusate samples from striatal slices [18,19]. Identification of putative DOPA peaks on chromatographic charts is described in detail in Chapter 2, Section 2.3 titled Identification of DOPA Peak.

The ratio of the absolute value of the basal efflux of DOPA to dopamine is 1:10 in the presence of cocaine [18] and 1:2 to 1:3 in the absence of cocaine [18,19]. Cocaine was used in the earlier stages of our studies in an effort to reduce as much as possible the chromatographic interference with DOPA peaks by the metabolites of dopamine [18]. Dopamine but not DOPA is a substrate for the cocaine-sensitive reuptake of catecholamines into catecholaminergic neurons [65]. Neither TTX application nor $Ca^{2+}$ deprivation inhibits the basal efflux of both DOPA and dopamine. α-MPT (0.2 m*M*) produces no effects on the basal efflux of both DOPA and dopamine [18].

Electrical field stimulation with bipolar square wave pulses (2 Hz) releases, repetitively, both DOPA and dopamine over a similar time course in the presence of cocaine [18]. This release is completely TTX sensitive and $Ca^{2+}$ dependent, suggesting that DOPA is released by some neuronal activity, which is the first evidence to be identified. α-MPT (0.2 m*M*) inhibits the evoked release of DOPA but not that of dopamine, suggesting that this concentration of TH inhibitor is not enough to inhibit the second step of catecholamine biosynthesis in our experimental conditions. An important finding is that the amount of DOPA released by 2 Hz pulses, expressed as a fraction of the tissue content after the end of experiments, is approximately 20 times higher than that of the established neurotransmitter dopamine in the presence of cocaine.

Depolarization induced by high $K^+$ (15 m*M*) releases simultaneously both DOPA and dopamine in a $Ca^{2+}$-dependent manner. Levodopa appears to be taken up by neurons and to be released from rat striatal slices by high $K^+$ in a neurotransmitter-like manner [66]. The $Ca^{2+}$-dependent evoked release of DOPA is also seen in peripheral tissues such as dog sympathetic ganglia [67], portal vein [68], and adrenal medulla [69].

In our striatal slice preparations [18], frequency–release relationship at 0.5 to 5 Hz pulses and concentration–release relationship at 10 to 30 m*M* $K^+$ show a characteristic pattern for DOPA and dopamine. Dopamine release is completely frequency dependent and concentration dependent. The peak release of dopamine occurs at 5 Hz pulses and 30 m*M* $K^+$. On the other hand, the peak release of DOPA is seen at 2 Hz pulses and 15 m*M* $K^+$. DOPA release decreases from the peak at 5 Hz pulses and 30 m*M* $K^+$. This decrease appears to result from the inhibition of TH activity, following a negative feedback control via activation of presynaptic dopamine $D_2$ autoreceptors [70,71] induced by the higher amounts of simultaneously released dopamine (see Figure 5.1). Even if decreases in biosynthesis and release of DOPA occur following the negative feedback control, the ratio of DOPA to dopamine released at 5 Hz pulses and 30 m*M* $K^+$ is as low as 1:100 to 1:1000, compared with 1:10 to 1:40 at lower frequencies and

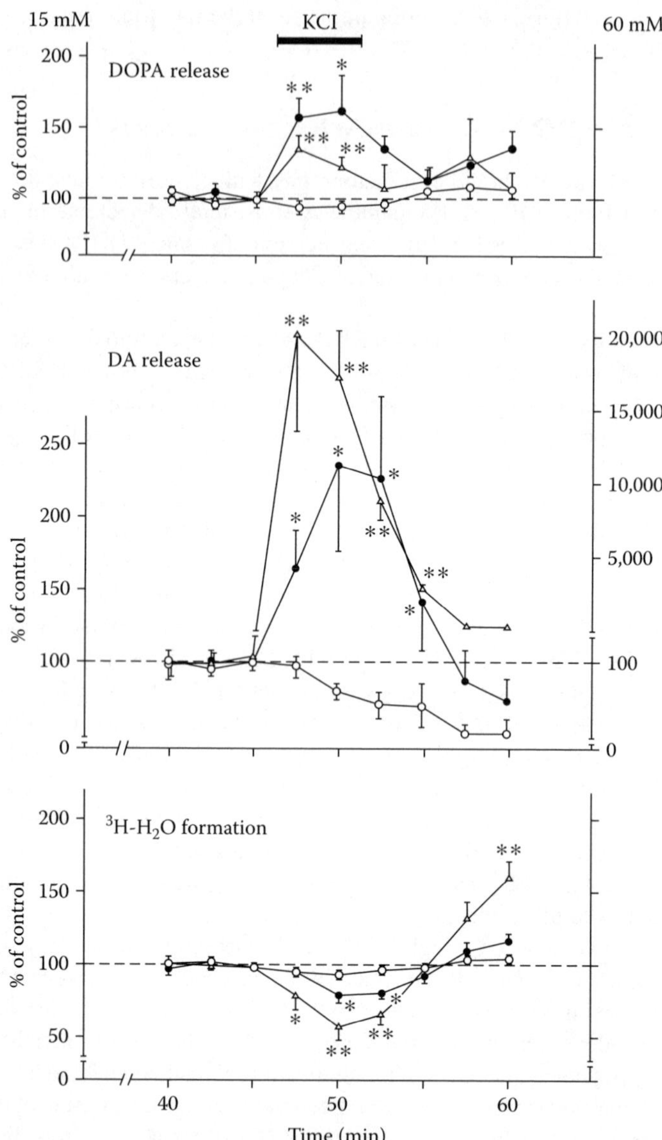


**FIGURE 5.1** Time courses of the release of both endogenous DOPA and dopamine (DA) and changes of TH activity, [$^3$H]H$_2$O formation, evoked by high K$^+$ from rat striatal slices superfused with [$^3$H]tyrosine-containing medium. Slices (25 mg) were superfused with normal Krebs medium for 20 min and then with the medium containing L-[3,5-$^3$H]tyrosine, 50 μCi/ml, in the absence of cocaine. Fractions of superfusates were collected every 2.5 min, 40 min after the start of superfusion. Slices were depolarized by high K$^+$ 15 m*M* (●) and 60 m*M* (△) for 5 min at a horizontal bar. Only dopamine concentrations evoked by 60 m*M* K$^+$ are indicated by the ordinate on the right. Respective ordinate shows percentage of control calculated from the means of the absolute values for the three fractions before depolarization.

lower $K^+$ concentrations. This could not occur if DOPA is released from dopamine-containing vesicles. Thus, DOPA appears to be released from some cytoplasmic compartment other than dopamine-containing vesicles.

Nicotine releases, repetitively and simultaneously, DOPA as well as dopamine from striatal slices [19]. This release is concentration dependent (0.1 to 10 μ*M*), stereoselective, sensitive to mecamylamine (a central nicotinic receptor antagonist), and $Ca^{2+}$ dependent. The ratio of the absolute value of DOPA to dopamine released by nicotine is 1:2 to 1:3 in the absence of cocaine. An important finding is that the amount of DOPA released by nicotine (10 μ*M*) expressed as a fraction of the tissue content after the end of the experiments is approximately 300 times higher than that of established neurotransmitter dopamine. We can safely say that although the tissue content of DOPA in the striatum is lower than that of dopamine by three orders of magnitude [7], the turnover rate of DOPA is higher than that of dopamine by two to three orders of magnitude [18,19]. It appears likely that DOPA is more efficiently synthesized and released to elicit function as compared with dopamine [24,25,28].

### 5.2.2 Time Courses of DOPA Release and TH Activation by High $K^+$

DOPA release evoked by depolarizing stimuli such as electrical field stimulation and high $K^+$ and by nicotine is always $Ca^{2+}$ dependent in a manner similar to that in dopamine release [18,19]. We attempted to clarify if this $Ca^{2+}$ dependency is primarily involved in the process of neurotransmitter release [20,21] rather than the process of TH activation [22]. Thus, we compared the time course of the release of both DOPA and dopamine with that of TH activation, tritiated $H_2O$ formation from superfused tritiated tyrosine simultaneously induced by high $K^+$ in striatal slices in the absence of cocaine [18]. Indeed, as shown in Figure 5.1, DOPA release occurs prior to the enhancement of TH activity. High $K^+$ (15 and 60 m*M*) releases DOPA and dopamine at the same time that the concentration-dependent decreases in TH activity are observed during depolarization. The decreases in TH activity are probably due to the negative feedback control via activation of presynaptic dopamine $D_2$ autoreceptors [70,71] following the simultaneously evoked concentration-dependent release of dopamine. In contrast, concentration-dependent increases in TH activity were observed after the release of both DOPA and dopamine ended. It is evident that the primary mechanism for the $Ca^{2+}$-dependency of DOPA release is in an excitation–secretion coupling process for neurotransmitter release [20,21]. In addition, depolarization induced by 60 m*M* $K^+$ releases less amounts of DOPA, compared with that by 15 m*M* $K^+$, which appears to occur following the negative feedback control elicited by the concentration-dependent

---

**FIGURE 5.1 (Continued)** Control values were: DOPA, 0.226 ± 0.07 pmol; dopamine, 0.696 ± 0.236 pmol; and $[^3H]H_2O$ formation, 26.8 ± 5.8 nCi (n = 3 to 6). Data are the means ± SE. $^*P < .05$, $^{**}P < .01$, vs. the corresponding value in nondepolarized slices (○) (a two-tailed Student's *t*-test). (From Goshima, Y., Kubo, T., and Misu, Y., Transmitter-like release of endogenous 3,4-dihydroxyphenylalanine from rat striatal slices, *J. Neurochem.*, 50, 1725, 1988. With permission.)

release of dopamine. Indeed, released dopamine can inhibit TH activity, DOPA biosynthesis, and further DOPA release via activation of presynaptic dopamine $D_2$ autoreceptors [70,71].

### 5.2.3 *In Vivo* DOPA Release during Striatal Microdialysis of Conscious Rats

The basal release of both DOPA and dopamine is also consistently detectable in dialysate samples collected during striatal microdialysis of conscious rats [30]. The ratio of the basal release of DOPA to dopamine is 1:2 on an average in the series of our striatal microdialysis experiments [30–34]. The various ratios of DOPA to dopamine reported by other investigators are 1:15 [8] and 1:4 [72] during striatal microdialysis in conscious rats, and 1:0.4 in anesthetized rats [73].

In contrast to the basal efflux of DOPA and dopamine from striatal slices [18,19], basal DOPA release is inhibited by TTX application (1 μ*M*) and by $Ca^{2+}$ removal plus 12.5 m*M* $Mg^{2+}$ addition (which antagonizes $Ca^{2+}$-dependent dopamine release) [74] via microdialysis probes [30]. It was found that purified hog AADC converts DOPA to dopamine and the DOPA peak is markedly increased by 3-hydroxybenzylhydrazine (100 mg/kg, i.p.), a central AADC inhibitor (see Figure 2.2 in Chapter 2). The DOPA peak is markedly decreased by α-MPT (200 mg/kg, i.p.). The maximal degrees of TTX sensitivity and $Ca^{2+}$ dependency for basal DOPA release are approximately 50 to 60% of control. This means that at least some part of DOPA is physiologically released in a manner similar to that of many neurotransmitters via the spontaneous firing of some striatal neurons. On the other hand, the maximal degrees of TTX sensitivity and $Ca^{2+}$ dependency for basal dopamine release are substantially complete, approximately 80 to 90% of control. These findings suggest that there are two components of extracellular DOPA level in the striatum [24–26,28]. One is a TTX-sensitive and $Ca^{2+}$-dependent physiological release of DOPA as a neurotransmitter or neuromodulator. The other is a TTX-insensitive and $Ca^{2+}$-independent efflux of DOPA as a precursor of dopamine.

Infusion of high $K^+$ (50 m*M*) markedly releases both DOPA and dopamine over a similar time course [30]. The ratio of DOPA to dopamine is 1:3.5 during depolarization. The evoked release of both DOPA and dopamine is $Ca^{2+}$ dependent, which is supported by other finding [72].

Nicotine at 10 to 300 μ*M* perfused via microdialysis probes releases both DOPA and dopamine in a dose-dependent and repetitive manner [31]. The ratio of the evoked release of DOPA to dopamine is 1:2 to 1:5. The evoked release of DOPA and dopamine is stereoselective, $Ca^{2+}$ dependent and mecamylamine (500 μ*M*) sensitive. These findings indicate that nicotine releases DOPA via a process similar to that of the neurotransmitter dopamine, by activation of central nicotinic receptors; this is consistent with the observations in striatal slices [19]. A new finding is that the release of both DOPA and dopamine is TTX sensitive [31]; this is contrary to the observations *in vitro* [19]. Depolarization of some kinds of neurons via activation of the $Na^+$ channels may occur in intact neuronal networks in striata and may be involved in the nicotine-induced release of DOPA and dopamine in experimental systems *in vivo*.

A further important finding is that mecamylamine alone (500 μ*M*) inhibits the basal release of DOPA but not that of dopamine in striata of conscious rats [31], suggesting that DOPA release, but not dopamine release, is tonically regulated via activation of nicotinic acetylcholine receptors. Endogenously released acetylcholine appears to play an important role in DOPA release, compared with dopamine release.

## 5.3 NUCLEUS ACCUMBENS

The VTA located in the midbrain projects the mesolimbic dopamine neurons to the nucleus accumbens. Several lines of evidence have suggested that the modifications of this system play an essential role in the behavioral effects, including locomotor activity, elicited by nicotine [75–77]. Indeed, nicotine locally infused into the ventral tegmental area (VTA) induces locomotor activity, which is antagonized by systemic mecamylamine [78]. In addition, systemic nicotine-induced dopamine release in the nucleus accumbens is antagonized by mecamylamine locally infused into the VTA but not by the infusion into the nucleus accumbens [79].

### 5.3.1 Basal and Nicotine-Evoked Release of DOPA during Microdialysis of the Nucleus Accumbens of Conscious Rats

The basal release of not only dopamine but also DOPA is consistently detectable during microdialysis of the shell compartment in the unilateral nucleus accumbens of conscious rats [35]. The ratio of the absolute value of the basal release of DOPA to dopamine is 1:1 to 1:3 [35,36]. The basal release of both DOPA and dopamine is inhibited by TTX perfusion (1 μ*M*) and $Ca^{2+}$ deprivation with 12.5 m*M* $Mg^{2+}$ addition. The maximal degrees of TTX sensitivity and $Ca^{2+}$ dependency for DOPA are approximately 70% of control, which is nearly similar to that for dopamine (approximately 80% of control) [35].

Systemic nicotine (1 mg/kg, s.c.) releases DOPA as well as dopamine over a similar time course accompanied with locomotor activity. The evoked release of DOPA is slightly less than that of dopamine. We confirmed previous findings that nicotine-induced dopamine release in the nucleus accumbens is greater than that in the striatum [75,76]. This is also the case for DOPA.

Nicotine (30 μg) injected into the left VTA releases DOPA as well as dopamine in the nucleus accumbens accompanied with locomotor activity over a time course roughly similar to that elicited by systemic nicotine [35]. All of these responses are antagonized by mecamylamine (100 μg) injected into the VTA. The mecamylamine-sensitive dopamine release and the locomotor activity elicited by nicotine are both consistent with previous findings [78,79]. Systemic nicotine also expresses Fos-like immunoreactivity in the nucleus accumbens, which is antagonized by mecamylamine infused locally into the VTA [80]. These findings suggest that nicotinic receptors in the somatodendritic region may have greater importance than those located in the terminal area for the stimulatory effects of nicotine on the mesolimbic dopamine system [78–80]. In addition, our findings suggest a probable role for the mesolimbic DOPA system [35]. An interesting finding related to nicotine-induced activation of a probable mesolimbic DOPA system [35] is demonstrated in the human VTA [81,82]. Some neuronal cell bodies show immunocytochemically TH-positive, AADC-negative, DOPA-positive, and dopamine-negative reactivity, and further the reactivity of guanosine triphosphate cyclohydrolase I, the first and rate-limiting enzyme for the biosynthesis of DOPA.

## 5.4 LOWER BRAINSTEM

The most important target region for DOPA is the blood pressure regulatory centers such as the NTS, the CVLM, and the RVLM because it is highly probable that DOPA is a neurotransmitter of the baroreflex neurotransmission pathways in the lower brainstem [24–26,28]. Baroreflex is the principal neuronal mechanism by which the cardiovascular system is regulated under a negative feedback control. Figure 5.2 represents the survey of these events [24–26,28,83–89]. Arterial baroreceptors are located both in the aortic arch and the carotid sinus, whereas chemoreceptors are located only in the carotid bodies in rats [90]. Glutamate, a representative

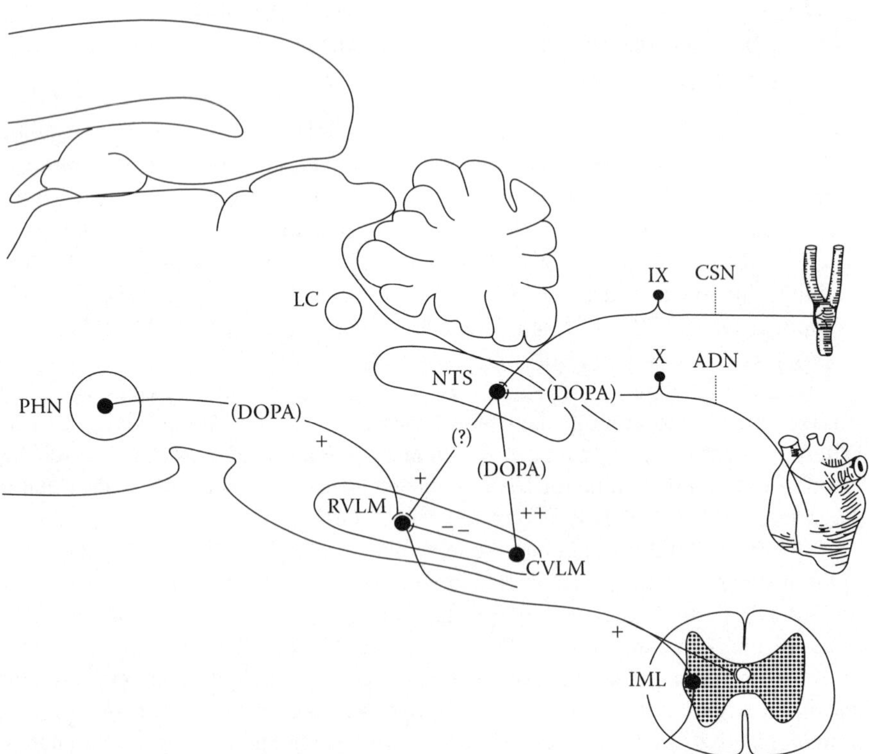


**FIGURE 5.2** Negative feedback control of baroreflex neurotransmission, central regulatory mechanisms of arterial blood pressure, and hypothesized DOPA-mediated pathways in the

of excitatory amino acids, has been regarded as the most probable neurotransmitter of the primary sensory baroreceptor afferents [83], the excitatory pathway from the NTS to the CVLM [91–93], and the excitatory pathway from the RVLM to the thoracic intermediolateral cell column [94] among many neurotransmitter candidates. The neurotransmitter of the inhibitory neurons projecting from the CVLM to the RVLM is generally accepted to be γ-aminobutyric acid (GABA) [87,95,96].

On the other hand, in the rat NTS, there exist neurons that may contain DOPA as an end product [37,44,45]. Our data further suggest that tonically functioning DOPA is a highly probable neurotransmitter of the ADN [37,38,46]. A tonically functioning DOPA-mediated baroreceptor–ADN–NTS–CVLM depressor pathway [39–41] and a tonically functioning DOPA-mediated pressor pathway from the PHN to the RVLM [38,42,43] appear to exist.

---

**FIGURE 5.2 (Continued)** lower brainstem of anesthetized rats. The primary baroreceptor afferents, the ADN from the aortic arch and the CSN from the carotid sinus, terminate in depressor sites of the NTS. Glutamate has been regarded as the most probable neurotransmitter of the primary baroreceptor afferents, the second-order interneurons and/or microcircuit-neurons within the depressor sites, and of the direct excitatory pathway from the NTS to the depressor sites of the CVLM. The RVLM receives excitatory inputs from various regions of the CNS, including the PHN and the NTS. The total tonicity of these excitatory pathways, shown as (+), appears to be less than the tonicity of the predominant excitatory NTS–CVLM pathway, shown as (++). Vasomotor tone is reduced by the neuronal activity of the CVLM, the integration of which constitutes the predominant inhibitory pathway for baroreflex neurotransmission. GABA-containing neurons in the CVLM project directly to the RVLM to inhibit excitatory activity. The tonicity of GABA-containing neurons is shown as (− −). The excitatory neurons of the RVLM project directly to the intermediolateral cell column (IML) of the thoracic spinal cord, the main origin of the sympathetic discharge. The RVLM plays an important role in controlling tonic resting and reflex integration of arterial blood pressure. Electrical stimulation of the ADN releases TTX-sensitive DOPA in the NTS and also α-MPT-sensitive DOPA in the CVLM, accompanied with hypotension and bradycardia. Baroreceptor stimulation by phenylephrine elicits DOPA release in the NTS, which is abolished by bilateral sinoaortic denervation, and also elicits DOPA release in the CVLM, which is abolished by lesion of the ipsilateral NTS. Denervation of the ADN distal to the ganglion nodosum decreases, ipsilaterally and selectively, the imumunocytochemical reactivity of DOPA in the ganglion nodosum and the NTS. Lesion of the NTS decreases selectively the tissue content of DOPA in the ipsilateral CVLM. Electrical stimulation of the unilateral PHN releases selectively TTX-sensitive DOPA in the ipsilateral RVLM accompanied with hypertension and tachycardia. Lesion of the bilateral PHN decreases selectively the tissue content of DOPA in the unilateral RVLM. DOPA appears to be a highly probable neurotransmitter of the primary ADN. A tonically functioning DOPA-mediated baroreceptor–ADN–NTS–CVLM depressor pathway and a tonically functioning DOPA-mediated PHN–RVLM pressor pathway appear to exist. Additional details are shown in the caption of Figure 11.1. LC, locus coeruleus. (Modified from Misu, Y., Kitahama, K., and Goshima, Y., L-3,4-Dihydroxyphenylalanine as a neurotransmitter candidate in the central nervous system, *Pharmacol. Ther.*, 97, 117, 2003. With permission.)

### 5.4.1 Basal DOPA Release during Microdialysis of NTS, CVLM, and RVLM in Anesthetized Rats

Male Wistar rats anesthetized with urethane (1.2 g/kg, i.p.) were usually used. Basal DOPA release is consistently detectable during microdialysis of the NTS [37,38], the CVLM [39–41], and the RVLM [38,42,43]. Even in the anesthetized condition, basal DOPA release is in part inhibited by 1 μ*M* TTX perfusion and by $Ca^{2+}$ deprivation plus 12.5 m*M* $Mg^{2+}$ addition. Basal DOPA is at least partially released by the spontaneous firing of some neurons within neuronal networks in these areas. The maximal degrees of both TTX sensitivity and $Ca^{2+}$ dependency of basal DOPA release are commonly approximately 30% of control in these three areas, which is quantitatively lower than the 50 to 60% in the striatum [30] and 70% in the nucleus accumbens [35] of conscious rats. This may be due to a difference between the anesthetized and conscious conditions or some regional differences between the NTS, the CVLM, or the RVLM and the striatum or the nucleus accumbens. Some part of TTX-insensitive and $Ca^{2+}$-independent basal DOPA release appears to reflect efflux of DOPA through a mechanism other than the proposed excitation–secretion coupling process [20,21] for DOPA [24,25]. α-MPT (200 mg/kg, i.p.) markedly and similarly decreases basal DOPA release in the NTS, the CVLM, and the RVLM, which is similar to the findings in the striatum.

### 5.4.2 High-$K^+$-Evoked DOPA Release during Microdialysis of NTS, CVLM, and RVLM in Anesthetized Rats

Local perfusion of high-$K^+$ 40-, 60-, and 100-m*M* solution via microdialysis probes increases DOPA release by approximately 50, 100, and 160% of control, respectively, in the NTS [37]. Fifty m*M* $K^+$ repetitively and constantly evokes DOPA release, and this release is $Ca^{2+}$ dependent. Similar findings are obtained in the CVLM [39] and the RVLM [42]. The $Ca^{2+}$ may be primarily involved in the release process of DOPA rather than the activation process of TH (Figure 5.1) [18]. Thus, DOPA appears to be evoked via a depolarization-dependent neurotransmitter release process from some neurons of the NTS, the CVLM, and the RVLM in anesthetized rats.

### 5.4.3 DOPA Release from Baroreceptor Activation by Phenylephrine and Electrical Stimulation of the ADN during Microdialysis of NTS in Anesthetized Rats

As shown in Figure 5.3, baroreceptor stimulation induced by strong and sustained infusion of phenylephrine (50 μg/kg/min, for 20 min, i.v.) triggers DOPA release in the NTS and a reflex bradycardia temporally associated with the rise and recovery of arterial blood pressure [37]. Both DOPA release and reflex bradycardia are abolished by acute bilateral denervation of the carotid sinus nerve (CSN) and ADN, which contain the primary baroreceptor afferents; hypertension is similarly seen after denervation. This finding is a piece of direct evidence that reflex information is carried from baroreceptors to the NTS via the sinoaortic nerves to result in DOPA release in the NTS and, further, supports the idea that DOPA is a neurotransmitter

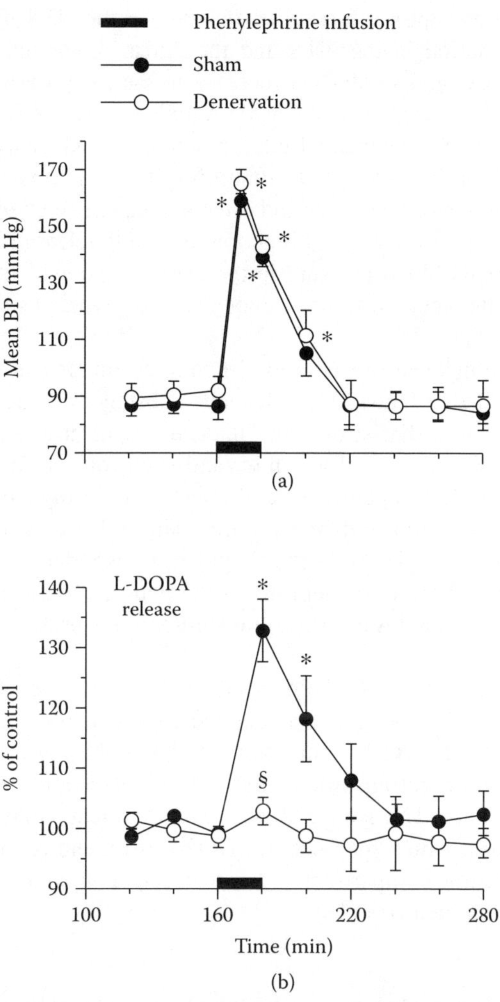


**FIGURE 5.3** Time course of phenylephrine-induced hypertension and DOPA (L-DOPA) release during microdialysis of the NTS in anesthetized rats and abolition of DOPA release by bilateral sinoaortic denervation. Phenylephrine infusion (50 μg/kg/min, i.v.) was done for 20 min at the horizontal bars 160 min after the start of perfusion. In denervation, the bilateral carotid sinus and aortic depressor nerves were acutely cut, before microdialysis, at their junctions with the glossopharyngeal and superior laryngeal nerves, respectively. Sham operation was done in sham. In (a), changes of mean blood pressure (BP) are shown. In (b), DOPA release is shown as percentage of control, the mean of the initial stable three absolute values before phenylephrine infusion. Each value represents mean ± SE. *$P < .01$ vs. the value immediately before phenylephrine infusion (Dunnett's multiple comparison test). §$P < .01$ vs. sham (unpaired *t*-test). Control (fmol) was 5.8 ± 0.6 for sham (n = 10) and 5.4 ± 0.6 for denervation (n = 5). (From Yue, J.-L. et al., Baroreceptor-aortic nerve-mediated release of endogenous L-3,4-dihydroxyphenylalanine and its tonic function in the nucleus tractus solitarii of rats, *Neuroscience*, 62, 145, 1994. With permission.)

of the primary baroreceptor afferents terminating in the NTS. DOPA released by baroreceptor stimulation in the NTS and the dorsal motor nucleus of the vagus (DMV) elicits reflex bradycardia via increase in the peripheral vagal tone and/or decrease in the sympathetic tone in the heart. A tonic function of DOPA for baroreflex integration in the NTS is evident. Reflex bradycardia induced by mild and brief infusion of phenylephrine (15 μg/10 μl/min, for 10 sec, i.v.) is antagonized by the bilateral microinjection of DOPA methyl ester, a competitive DOPA antagonist [97], into the depressor sites [37]. A tonic function of basally released DOPA for integration of resting arterial blood pressure in the NTS is also evident because bilateral microinjection of the antagonist alone elicits marked hypertension and tachycardia [37] (See Chapter 11).

The ADN contains sensory A- and C-fibers terminating in the NTS [98] and may utilize at least two kinds of endogenous substances as neurotransmitters. It is an important finding that strong and intermittent electrical stimulation of the ADN (100 Hz, 8 V, 0.1 msec, for 4 min; and such four episodes over a 20 min interval) releases DOPA repetitively and constantly during microdialysis of the NTS, causing hypotension and bradycardia. This release is abolished by TTX perfusion (1 μ*M*) [37]. This shows that impulses transported from the ADN to the NTS release DOPA from some neurons in the neuronal networks of the NTS. It is highly probable that DOPA is a neurotransmitter of the primary ADN terminating in the NTS.

The release of endogenous DOPA from the NTS *in vivo*, evoked by three kinds of stimulation (i.e., high-$K^+$-induced depolarization, baroreceptor activation by phenylephrine, and electrical stimulation of the ADN) is the first evidence in support of being a neurotransmitter among a lot of candidates for the primary baroreceptor afferents. The source of this DOPA release may be derived from DOPAergic neurons, which may contain DOPA as an end product [37,44,45], or from some cytoplasmic compartment other than catecholamine-containing vesicles in catecholaminergic neurons [18].

### 5.4.4 Immunocytochemistry of DOPA-Related Neuronal Markers in NTS and DMV Area of Rats

Figure 5.4 shows the effect of unilateral ADN denervation peripheral to the ganglion nodosum, which contains the cell bodies. Consistent with the present understanding of the physiological release of DOPA, 7 d after denervation the immunocytochemical reactivity of TH and DOPA, but not dopamine and dopamine-β-hydroxylase (DBH), decreased ipsilaterally in the medial and intermediate subdivisions of the NTS and the DMV [37]. In the ganglion nodosum, this denervation also decreased the staining of TH-immunoreactive cells and the number and staining of DOPA-immunoreactive cells. These findings suggest that the ADN may utilize DOPA, but not dopamine and noradrenaline, as a neurotransmitter and that it possesses a DOPAergic monosynaptic property. DOPAergic long axons appear to transmit baroreflex information directly to depressor sites or indirectly to some secondary neurons within the neuronal microcircuits of depressor sites in the NTS.

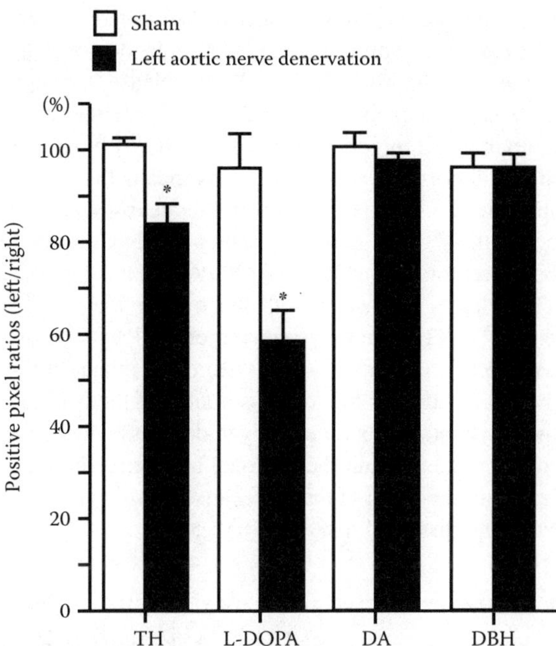


**FIGURE 5.4** Morphometic analysis of TH-, DOPA (L-DOPA)-, dopamine (DA)-, and DBH-immunoreactivity in the NTS and DMV area following the left ADN denervation peripheral to the ganglion nodosum in rats. Specific antisera against these catecholamine-related neuronal markers were treated 7 d after denervation. The ordinate shows positive pixel ratios. The total value of respective immunoreactivity in the left side is expressed as a percentage of that in the right side in sham and denervated rats (respective n = 5). Each value represents mean ± SE. $^{*}P < .01$ (unpaired *t*-test) vs. sham. (From Yue, J.-L. et al., Baroreceptor-aortic nerve-mediated release of endogenous L-3,4-dihydroxyphenylalanine and its tonic function in the nucleus tractus solitarii of rats, *Neuroscience*, 62, 145, 1994. With permission.)

### 5.4.5 DOPA Release due to Baroreceptor Activation by Phenylephrine and Electrical Stimulation of the ADN during Microdialysis of CVLM in Anesthetized Rats

During microdialysis of the unilateral CVLM, the basal release of DOPA, 3,4-dihydroxyphenylacetic acid (DOPAC), a metabolite of dopamine, noradrenaline, and adrenaline are consistently detectable [41]. Extracellular levels of dopamine, however, are lower than the assay limit of sensitivity defined as a signal-to-noise ratio (2 fmol). Baroreceptor activation by intense and sustained infusion of phenylephrine (i.v.) selectively releases DOPA without increasing dopamine, DOPAC, noradrenaline, and adrenaline. In addition, this DOPA release is suppressed by an acute electrolytic lesion of the ipsilateral NTS, suggesting that baroreflex information is transmitted from baroreceptors via the the primary baroreceptor afferents, the NTS, and then to the CVLM to release DOPA in the depressor sites of the CVLM.

Furthermore, strong and intermittent electrical stimulation of the unilateral ADN (20 Hz, 3 V, 0.3 msec, for 3 min; and six such episodes over a 30-min interval) elicits repetitively and constantly selective DOPA release, hypotension, and bradycardia without increasing dopamine, DOPAC, noradrenaline, and adrenaline. This DOPA release is interrupted by local inhibition of DOPA biosynthesis with α-MPT (30 μ*M*) infused into the ipsilateral CVLM. This means that DOPA, newly synthesized by TH within the CVLM, is involved in this neurotransmission.

Thus, it appears likely that a neurotransmitter involved in the neurotransmission of signals from baroreceptors to the CVLM is DOPA, and not dopamine, noradrenaline, or adrenaline. This idea is consistent with the finding that 10 d after a unilateral electrolytic lesion of the NTS, the tissue content of DOPA in the dissected ipsilateral CVLM decreases selectively by 45% without decreases in that of the three catecholamines. This finding also suggests that depressor sites of the NTS project DOPAergic neurons with a monosynaptic property directly to depressor sites, or indirectly to some neurons near depressor sites, within the neuronal microcircuits of the CVLM. On the other hand, this electrolytic lesion of the NTS causes hardly any decrease in the tissue content of DOPA in the dissected ipsilateral RVLM.

### 5.4.6 DOPA Release by Electrical Stimulation of the PHN during Microdialysis of RVLM in Anesthetized Rats

The RVLM receives, neuroanatomically, inputs from various regions of the central nervous system, such as the CVLM, the NTS, the hypothalamic paraventricular nucleus, the periaqueductal gray, and the lateral hypothalamic area [24,25,84,85,99]. However, there is no evidence for DOPAergic inputs to this area [24,25,28]. In the posterior and dorsal hypothalamus, immunocytochemically DOPA-reactive cells are generally fewer in number, compared with dopamine-reactive cells [25,28]. In the rat PHN (A11 cell group), however, some cells display TH-positive, weakly AADC-positive [100], and markedly DOPA-positive immunoreactivity with a very weak dopamine-positive immunoreactivity [101]. The rat PHN neurons send descending projections to the RVLM [86]. On the basis of these findings, we attempted to clarify whether or not electrical PHN stimulation releases DOPA during unilateral microdialysis of the RVLM.

Extracellular basal levels of DOPA, dopamine, noradrenaline, and adrenaline are consistently detectable. TTX (1 μ*M*) perfused into the ipsilateral RVLM gradually decreases the basal levels of DOPA (Figure 5.5) and the three types of catecholamines by 25 to 40%, suggesting some roles for these endogenous substances as neurotransmitters or neuromodulators in this area. As shown in Figure 5.5, intensive electrical stimulation of the ipsilateral PHN (50 Hz, 0.3 mA, 0.1 msec; twice for 5 min at an interval of 5 min) selectively releases DOPA in a repetitive and constant manner accompanied by hypertension and tachycardia. However, no catecholamines are released. TTX suppresses evoked DOPA release but partially inhibits hypertension with a slight inhibition of tachycardia. These findings suggest that no catecholamines are involved in the neurotransmission from the PHN to the RVLM, and that the unilateral PHN projects a DOPAergic pressor relay at least to the ipsilateral RVLM (see Chapter 11). These ideas are consistent with the observation

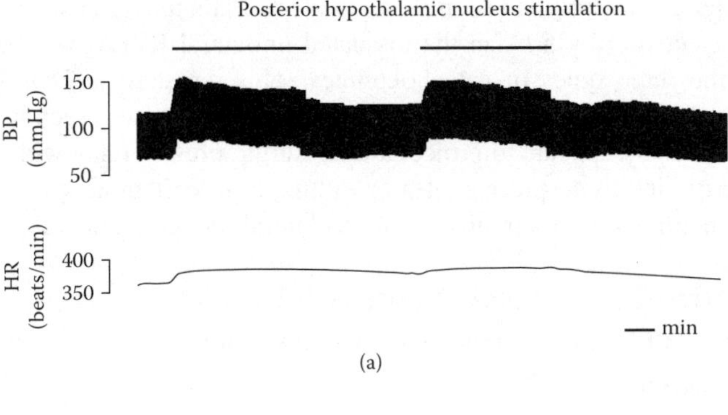


(a)

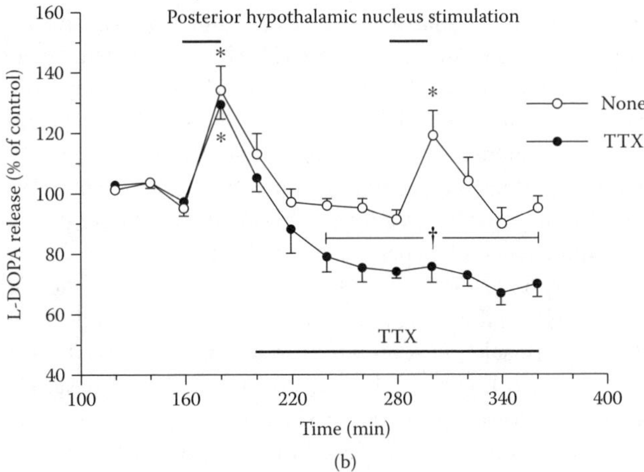


(b)

**FIGURE 5.5** Effects of TTX on DOPA (L-DOPA) released basally and evoked by intense electrical stimulation of the right PHN during microdialysis of the ipsilateral RVLM in anesthetized rats. The RVLM was perfused at a rate of 2 ml/min. (a) Shows a typical trace of mean blood pressure (BP) and heart rate (HR) during the first period of PHN stimulation ($S_1$). The PHN was stimulated twice for 5 min (50 Hz, 0.3 mA, 0.1 msec duration) at an interval of 5 min. (b) Shows time courses of DOPA released basally and repetitively evoked by stimulation of the PHN during microdialysis with Ringer solution without TTX (none), and inhibition of the basal release and suppression of the evoked release by TTX perfusion (TTX). Perfusates were collected every 20 min. The electrical stimulation shown in (a) was repeated twice 160 min ($S_1$) and 280 min ($S_2$) after the start of perfusion, as indicated by the upper horizontal bars. TTX (1 $\mu M$) was perfused 80 min before $S_2$ and continued throughout the experiments, as shown by a lower horizontal bar. The ordinate shows DOPA release expressed as percentage of control, the mean absolute value of three stable successive samples 120 to 160 min after the start of perfusion. Control (fmol) was 8.9 ± 0.7 (n = 7) for none and 10.8 ± 1.8 (n = 6) for TTX, respectively. Values are mean ± S.E. $^*P < .05$ vs. the value immediately before $S_1$ and $S_2$ and $^\dagger P < .05$ vs. the value immediately before $S_1$ in TTX (paired *t*-test). (From Nishihama, M. et al., An L-DOPAergic relay from the posterior hypothalamic nucleus to the rostral ventrolateral medulla and its cardiovascular function in anesthetized rats, *Neuroscience*, 92, 123, 1999. With permission.)

that 10 d after an electrolytic lesion of the bilateral PHN the tissue content of DOPA decreases selectively by 50% in the dissected unilateral RVLM without decreases in that of the three types of catecholamines. This decrease in DOPA content is selective for the RVLM because no evidence for such decrease is seen in the CVLM. The unilateral PHN appears to project a DOPAergic pressor relay with a monosynaptic property directly to pressor sites or indirectly to certain neurons near pressor sites in the neuronal microcircuits of the ipsilateral and contralateral RVLM.

### 5.4.7 Altered Tonic DOPA Release in NTS, CVLM, and RVLM of Anesthetized, Spontaneously Hypertensive Rats

The maximal degree of TTX sensitivity to inhibit basal DOPA release is usually low, approximately 30% of control, in the NTS, the CVLM, and the RVLM in normotensive anesthetized Wistar rats [37,39,42]. However, when the absolute value of basal DOPA release and of TTX-sensitive tonic neuronal activity to release basal DOPA are observed during microdialysis of these three areas, clear differences are seen between adult spontaneously hypertensive rats (SHR), a widely used genetic model of essential hypertension, and age-matched Wistar Kyoto rats (WKY). Therefore, we will briefly survey the alterations in basal DOPA release that appears to be involved in the maintenance of hypertension in SHR.

During microdialysis of the NTS [38], the CVLM [40], and the RVLM [38] in anesthetized adult SHR and age-matched WKY, basal DOPA release is constantly detectable [37,39,42]. Basal DOPA release is in part inhibited by TTX except in the CVLM of SHR [40] (see Figure 11.4). The absolute value of basal DOPA release during microdialysis is lower in the depressor NTS [38] and the depressor CVLM [40] and is higher in the pressor RVLM [38] of SHR compared with WKY. Locally perfused TTX (1 $\mu M$) reduces basal DOPA release to the same levels in these three regions of the two strains. TTX-sensitive tonic neuronal activity to release basal DOPA is lower in the depressor NTS [38], is lost in the depressor CVLM [40], and is higher in the pressor RVLM [38] of SHR compared with WKY. All of these alterations of basal DOPA release, and TTX-sensitive neuronal activity to release basal DOPA, appear to be involved in the maintenance of hypertension in SHR [25,26,28,102,103].

## 5.5 CONCLUSION

Evidence for TTX-sensitive and $Ca^{2+}$-dependent basal release and release evoked by nerve impulses is one of the most important classical criteria that must be fulfilled by an endogenous substance for its identification as a neurotransmitter. DOPA appears to fulfill this criterion in the striatum and the nucleus accumbens in conscious rats; and this is also the case for baroreflex neurotransmission and central blood pressure regulation in the NTS, the CVLM, and the RVLM in the lower brainstem of anesthetized rats.

In the striatum, $Ca^{2+}$ appears to be involved primarily in the process of DOPA release rather than the process of TH activation elicited by high $K^+$ depolarization.

Although the tissue content of DOPA in the striatum is lower than that of the established neurotransmitter dopamine by three orders of magnitude, the turnover rate of DOPA released by electrical nerve impulses and nicotine is higher than that of dopamine by two to three orders of magnitude. DOPA appears to be more efficiently synthesized and released to elicit function as compared with dopamine. A DOPA system from the VTA to the nucleus accumbens appears to function in addition to the mesolimbic dopamine system.

It is highly probable that DOPA is a tonically functioning neurotransmitter of the ADN, one of the primary sensory baroreceptor afferents, terminating in the NTS. In addition, DOPA appears to be a tonically functioning neurotransmitter of a baroreceptor–ADN–NTS–CVLM depressor relay and a PHN–RVLM pressor relay. Furthermore, alterations of basal DOPA release and TTX-sensitive tonic neuronal activity to release basal DOPA in the NTS, the CVLM, and the RVLM appear to be involved in the maintenance of hypertension in adult SHR.

## ACKNOWLEDGMENTS

This work was mainly supported by Grants-in-Aid for Developmental Scientific Research (No. 06557143) and General Scientific Research (No. 61480119, 03454146, 07407003, 10470026) from the Ministry of Education, Science, Sports, and Culture, Japan, and by grants from the Mitsubishi Foundation, the Uehara Memorial Foundation, Japanese Heart Foundation, Mitsui Life Social Welfare Foundation, and SRF, all in Japan. Many thanks go to Sanae Sato, Department of Pharmacology, Fukushima Medical University School of Medicine, Fukushima, Japan, for remaking the figures.

## REFERENCES

1. Carlsson, A., Lindqvist, M., and Magnusson, T., 3,4-Dihydroxyphenylalanine and 5-hydroxytryptophan as reserpine antagonists, *Nature,* 180, 1200, 1957.
2. Ehringer, H. and Hornykiewicz, O., Verteilung von Noradrenaline und Dopamine (3-Hydroxytyramin) im Gehirn des Menschen und ihr Verhalten bei Erkrankungen des extrapyramidalen Systems, *Klin. Wochenschr.,* 38, 1236, 1960.
3. Bartholini, G. et al., Increase of cerebral catecholamines by 3,4-dihydroxyphenylalanine after inhibition of peripheral decarboxylase, *Nature,* 215, 852, 1967.
4. Hefti, F. and Melamed, E., L-DOPA's mechanism of action in Parkinson's disease, *Trends Neurosci.,* 3, 229, 1980.
5. Goshima, Y., Kubo, T., and Misu, Y., Biphasic actions of L-DOPA on the release of endogenous noradrenaline and dopamine from rat hypothalamic slices, *Br. J. Pharmacol.,* 89, 229, 1986.
6. Misu, Y., Goshima, Y., and Kubo, T., Biphasic actions of L-DOPA on the release of endogenous dopamine via presynaptic receptors in rat striatal slices, *Neurosci. Lett.,* 72, 194, 1986.
7. Westerink, B.H.C., Van Es, T.P., and Spaan, S.J., Effects of drugs interfering with dopamine and noradrenaline biosynthesis on the endogenous 3,4-dihydroxyphenylalanine levels in the rat brain, *J. Neurochem.,* 39, 44, 1982.

8. Carboni, E., Tanda, G., and Di Chiara, G., Extracellular striatal concentrations of endogenous 3,4-dihydroxyphenylalanine in the absence of a decarboxylase inhibitor: a dynamic index of dopamine synthesis in vivo, *J. Neurochem.,* 59, 2230, 1992.
9. Misu, Y. and Kubo, T., Presynaptic β-adrenoceptors, *Trends Pharmacol. Sci.,* 4, 506, 1983.
10. Misu, Y. and Kubo, T., Presynaptic β-adrenoceptors, *Med. Res. Rev.,* 6, 197, 1986.
11. Ueda, H., Goshima, Y., and Misu, Y., Presynaptic mediation by $\alpha_2$-, $\beta_1$- and $\beta_2$-adrenoceptors of endogenous noradrenaline and dopamine release from slices of rat hypothalamus, *Life Sci.,* 33, 371, 1983.
12. Ueda, H., Goshima, Y., and Misu, Y., Presynaptic $\alpha_2$- and dopamine-receptor-mediated inhibitory mechanisms and dopamine nerve terminals in the rat hypothalamus, *Neurosci. Lett.,* 40, 157, 1983.
13. Ueda, H. et al., Adrenaline involvement in the presynaptic β-adrenoceptor-mediated mechanism of dopamine release from slices of the rat hypothalamus, *Life Sci.,* 34, 1087, 1984.
14. Ueda, H. et al., Involvement of epinephrine in the presynaptic beta adrenoceptor mechanism of norepinephrine release from rat hypothalamic slices, *J. Pharmacol. Exp. Ther.,* 232, 507, 1985.
15. Goshima, Y., Kubo, T., and Misu, Y., Autoregulation of endogenous epinephrine release via presynaptic adrenoceptors in the rat hypothalamic slices, *J. Pharmacol. Exp. Ther.,* 235, 248, 1985.
16. Misu, Y. et al., Presynaptic inhibitory dopamine receptors on noradrenergic nerve terminals: analysis of biphasic actions of dopamine and apomorphine on the release of endogenous norepinephrine in rat hypothalamic slices, *J. Pharmacol. Exp. Ther.,* 235, 771, 1985.
17. Goshima, Y., Kubo, T., and Misu, Y., Output of endogenous dopa evoked by depolarization from slices of rat striatum and hypothalamus, *Jpn. J. Pharmacol.,* 40 (Suppl. I), 149P, 1986.
18. Goshima, Y., Kubo, T., and Misu, Y., Transmitter-like release of endogenous 3,4-dihydroxyphenylalanine from rat striatal slices, *J. Neurochem.,* 50, 1725, 1988.
19. Misu, Y. et al., Nicotine releases stereoselectively and $Ca^{2+}$-dependently endogenous 3,4-dihydroxyphenylalanine from rat striatal slices, *Brain Res.,* 520, 334, 1990.
20. Douglas, W.W. and Rubin, R.P., The mechanism of catecholamine release from the adrenal medulla and the role of calcium in stimulus-secretion coupling, *J. Physiol.,* 167, 288, 1963.
21. Kirpekar, S.M. and Misu, Y., Release of noradrenaline by splenic nerve stimulation and its dependence on calcium, *J. Physiol.,* 188, 219, 1967.
22. El Mestikawy, S., Glowinski, J., and Hamon, M., Tyrosine hydroxylase activation in depolarized dopaminergic terminals: involvement of $Ca^{2+}$-dependent phosphorylation, *Nature,* 302, 830, 1983.
23. Misu, Y. and Goshima, Y., Is L-dopa an endogenous neurotransmitter?, *Trends Pharmacol. Sci.,* 14, 119, 1993.
24. Misu, Y., Ueda, H., and Goshima, Y., Neurotransmitter-like actions of L-DOPA, *Adv. Pharmacol.,* 32, 427, 1995.
25. Misu, Y. et al., Neurobiology of L-DOPAergic systems, *Prog. Neurobiol.,* 49, 415, 1996.
26. Misu, Y., Goshima, Y., and Miyamae, T., Is DOPA a neurotransmitter?, *Trends Pharmacol. Sci.,* 23, 262, 2002.
27. Misu, Y. et al., DOPA causes glutamate release and delayed neuron death by brain ischemia in rats, *Neurotoxicol. Teratol.,* 24, 629, 2002.

28. Misu, Y., Kitahama, K., and Goshima, Y., L-3,4-Dihydroxyphenylalanine as a neurotransmitter candidate in the central nervous system, *Pharmacol. Ther.,* 97, 117, 2003.
29. Hoffman, B.B. and Taylor, P., Neurotransmission: the autonomic and somatic motor nervous systems, in *Goodman & Gilman's The Pharmacological Basis of Therapeutics,* Hardman, J.G., Limbird, L.E., and Gilman, A.G., Eds., McGraw-Hill, New York, 2001, 115.
30. Nakamura, S. et al, Transmitter-like basal and $K^+$-evoked release of 3,4-dihydroxyphenylalanine from the striatum in conscious rats studied by microdialysis, *J. Neurochem.,* 58, 270, 1992.
31. Nakamura, S. et al., Transmitter-like 3,4-dihydroxyphenylalanine is tonically released by nicotine in striata of conscious rats, *Eur. J. Pharmacol.,* 222, 75, 1992.
32. Nakamura, S. et al., Endogenously released DOPA is probably relevant to nicotine-induced increases in locomotor activities of rats, *Jpn. J. Pharmacol.,* 62, 107, 1993.
33. Nakamura, S. et al., Non-effective dose of exogenously applied L-DOPA itself stereoselectively potentiates postsynaptic $D_2$-receptor-mediated locomotor activities of conscious rats, *Neurosci. Lett.,* 170, 22, 1994.
34. Yue, J.-L. et al., Endogenously released L-DOPA itself tonically functions to potentiate $D_2$ receptor-mediated locomotor activities of conscious rats, *Neurosci. Lett.,* 170, 107, 1994.
35. Goshima, Y. et al., Ventral tegmental injection of nicotine induces locomotor activity and L-DOPA release from nucleus accumbens, *Eur. J. Pharmacol.,* 309, 229, 1996.
36. Yamanashi, K. et al., Tonic function of nicotinic receptors in stress-induced release of L-DOPA from the nucleus accumbens in freely moving rats, *Eur. J. Pharmacol.,* 424, 199, 2001.
37. Yue, J.-L. et al., Baroreceptor-aortic nerve-mediated release of endogenous L-3,4-dihydroxyphenylalanine and its tonic function in the nucleus tractus solitarii of rats, *Neuroscience,* 62, 145, 1994.
38. Yue, J.-L. et al., Altered tonic L-3,4-dihydroxyphenylalanine systems in the nucleus tractus solitarii and the rostral ventrolateral medulla of spontaneously hypertensive rats, *Neuroscience,* 67, 95, 1995.
39. Yue, J.-L., Goshima, Y., and Misu, Y., Transmitter-like L-3,4-dihydroxyphenylalanine tonically functions to mediate vasodepressor control in the caudal ventrolateral medulla of rats, *Neurosci. Lett.,* 159, 103, 1993.
40. Miyamae, T. et al., Loss of tonic neuronal activity to release L-DOPA in the caudal ventrolateral medulla of spontaneously hypertensive rats, *Neurosci. Lett.,* 198, 37, 1995.
41. Miyamae, T. et al., L-DOPAergic components in the caudal ventrolateral medulla in baroreflex neurotransmission, *Neuroscience,* 92, 137, 1999.
42. Yue, J.-L. et al., Evidence for L-DOPA relevant to modulation of sympathetic activity in the rostral ventrolateral medulla of rats, *Brain Res.,* 629, 310, 1993.
43. Nishihama, M. et al., An L-DOPAergic relay from the posterior hypothalamic nucleus to the rostral ventrolateral medulla and its cardiovascular function in anesthetized rats, *Neuroscience,* 92, 123, 1999.
44. Jaeger, C.B. et al., Aromatic L-amino acid decarboxylase in the rat brain: immunocytochemical localization in neurons of the brain stem, *Neuroscience,* 11, 691, 1984.
45. Tison, F. et al., Endogenous L-DOPA in the rat dorsal vagal complex: an immunocytochemical study by light and electron microscopy, *Brain Res.,* 497, 260, 1989.
46. Kubo, T. et al., Evidence for L-DOPA systems responsible for cardiovascular control in the nucleus tractus solitarii of the rat, *Neurosci. Lett.,* 140, 153, 1992.

47. Yamanashi, K. et al., Involvement of nitric oxide production via kynurenic acid-sensitive glutamate receptors in DOPA-induced depressor responses in the nucleus tractus solitarii of anesthetized rats, *Neurosci. Res.,* 43, 231, 2002.
48. Birkmayer, W. and Hornykiewicz, O., Der L-2,3-dioxyphenylalanin (= DOPA)-Effekt bei der Parkinson-Akinese, *Wien. Klin. Wochenschr.,* 73, 787, 1961.
49. Yahr, M.D., Parkinson's disease. The L-dopa era, *Adv. Neurol.,* 60, 11, 1993.
50. Goshima, Y. et al., L-DOPA induces $Ca^{2+}$-dependent and tetrodotoxin-sensitive release of endogenous glutamate from rat striatal slices, *Brain Res.,* 617, 167, 1993.
51. Furukawa, N. et al., Endogenously released DOPA is a causal factor for glutamate release and resultant delayed neuronal cell death by transient ischemia in rat striata, *J. Neurochem.,* 76, 815, 2001.
52. Olney, J.W. et al., Excitotoxicity of L-DOPA and 6-OH-DA: implications for Parkinson's and Huntington's disease, *Exp. Neurol.,* 108, 269, 1990.
53. Newcomer, T.A. et al., TOPA quinone, a kainate-like agonist and excitotoxin, is generated by a catecholaminergic cell line, *J. Neurosci.,* 15, 3172, 1995.
54. Fahn, S., Is levodopa toxic?, *Neurology,* 47 (Suppl. 3), S184, 1996.
55. Cheng, N.-N. et al., Differential neurotoxicity induced by L-DOPA and dopamine in cultured striatal neurons, *Brain Res.,* 743, 278, 1996.
56. Maeda, T. et al., L-DOPA neurotoxicity is mediated by glutamate release in cultured rat striatal neurons, *Brain Res.,* 771, 159, 1997.
57. Jenner, P.G. and Brin, M.F., Levodopa neurotoxicity: experimental studies versus clinical relevance, *Neurology,* 50 (Suppl. 6), S39, 1998.
58. Fahn, S., Parkinson disease, the effect of levodopa, and the ELLDOPA trial, *Arch. Neurol.,* 56, 529, 1999.
59. Akaike, A. et al., Nicotine-induced protection of cultured cortical neurons against N-methyl-D-aspartate receptor-mediated glutamate cytotoxicity, *Brain Res.,* 644, 181, 1994.
60. Marin, P. et al., Nicotine protects cultured striatal neurons against N-methyl-D-aspartate receptor-mediated neurotoxicity, *Neuroreport,* 5, 1977, 1994.
61. Morens, D.M. et al., Cigarette smoking and protection from Parkinson's disease: false association or etiologic clue?, *Neurology,* 45, 1041, 1995.
62. Fratiglioni, L. and Wang, H.X., Smoking and Parkinson's and Alzheimer's disease: review of the epidemiological studies, *Behav. Brain Res.,* 113, 117, 2000.
63. Ross, G.W. and Petrovitch, H., Current evidence for neuroprotective effects of nicotine and caffeine against Parkinson's disease, *Drug. Aging,* 18, 797, 2001.
64. Ishikawa, A. and Miyatake, T., Effects of smoking in patients with early-onset Parkinson's disease, *J. Neurol. Sci.,* 117, 28, 1993.
65. Iversen, L.L., *The uptake and storage of noradrenaline in sympathetic nerves,* Cambridge University Press, London, 1967, 157.
66. Chang, W.Y. and Webster, R.A., Effects of 3-*O*-methyl dopa on L-dopa-facilitated synthesis and efflux of dopamine from rat striatal slices, *Br. J. Pharmacol.,* 116, 2637, 1995.
67. Kristensen, E.W. et al., Precursors and metabolites of norepinephrine in sympathetic ganglia of the dog, *J. Neurochem.,* 54, 1782, 1990.
68. Hunter, L.W., Rorie, D.K., and Tyce, G.M., Dihydroxyphenylalanine and dopamine are released from portal vein together with noradrenaline and dihydroxyphenylglycol during nerve stimulation, *J. Neurochem.,* 59, 972, 1992.
69. Chritton, S.L. et al., Adrenomedullary secretion of DOPA, catecholamines, catechol metabolites, and neuropeptides, *J. Neurochem.,* 69, 2413, 1997.

70. Roth, R.H., CNS dopamine autoreceptors: distribution, pharmacology, and function, *Ann. N.Y. Acad. Sci.,* 430, 27, 1984.
71. Koeltzow, T.E. et al., Alterations in dopamine release but not dopamine autoreceptor function in dopamine D3 receptor mutant mice, *J. Neurosci.,* 18, 2231, 1998.
72. Okada, M. et al., Effects of $Ca^{2+}$ channel antagonists on striatal dopamine and DOPA release, studied by in vivo microdialysis, *Br. J. Pharmacol.,* 123, 805, 1998.
73. Brannan, T., Martinez-Tica, J., and Yahr, M.D., Catechol-*O*-methyltransferase inhibition increases striatal L-dopa and dopamine: an in vivo study in rats, *Neurology,* 42, 683, 1992.
74. Westerink, B.H.C. et al., Use of calcium antagonism for the characterization of drug-evoked dopamine release from the brain of conscious rats determined by microdialysis, *J. Neurochem.,* 52, 722, 1989.
75. Imperato, A., Mulas, A., and Di Chiara, G., Nicotine preferentially stimulates dopamine release in the limbic system of freely moving rats, *Eur. J. Pharmacol.,* 132, 337, 1986.
76. Clarke, P.B. et al., Evidence that mesolimbic dopaminergic activation underlies the locomotor stimulant action of nicotine in rats, *J. Pharmacol. Exp. Ther.,* 246, 701, 1988.
77. O'Neill, M.F., Dourish, C.T., and Iversen, S.D., Evidence for an involvement of D1 and D2 dopamine receptors in mediating nicotine-induced hyperactivity in rats, *Psychopharmacology,* 104, 343, 1991.
78. Reavill, C. and Stolerman, I.P., Locomotor activity in rats after administration of nicotinic agonists intracerebrally, *Br. J. Pharmacol.,* 99, 273, 1990.
79. Nisell, M., Nomikos, G.G., and Svensson, T.H., Systemic nicotine-induced dopamine release in the rat nucleus accumbens is regulated by nicotinic receptors in the ventral tegmental area, *Synapse,* 16, 36, 1994.
80. Schilstrom, B. et al., Nicotine-induced Fos expression in the nucleus accumbens and the medial prefrontal cortex of the rat: role of nicotinic and NMDA receptors in the ventral tegmental area, *Synapse,* 36, 314, 2000.
81. Komori, K. et al., Identification of L-DOPA immunoreactivity in some neurons in the human mesencephalic region: a novel DOPA neuron group?, *Neurosci. Lett.,* 157, 13, 1993.
82. Nagatsu, I. et al., Specific localization of the guanosine triphosphate (GTP) cyclohydrolase I-immunoreactivity in the human brain, *J. Neural Transm.,* 106, 607, 1999.
83. Talman, W.T., Perrone, M.H., and Reis, D.J., Evidence for L-glutamate as the neurotransmitter of baroreceptor afferent nerve fibers, *Science,* 209, 813, 1980.
84. Urbanski, R.W. and Sapru, H.N., Evidence for a sympathoexcitatory pathway from the nucleus tractus solitarii to the ventrolateral medullary pressor area, *J. Auton. Nerv. Syst.,* 23, 161, 1988.
85. Sun, M.-K., Pharmacology of reticulospinal vasomotor neurons in cardiovascular regulation, *Pharmacol. Rev.,* 48, 465, 1996.
86. Vertes, R.P. and Crane, A.M., Descending projections of the posterior nucleus of the hypothalamus: *Phaseolus vulgaris* leucoagglutinin analysis in the rat, *J. Comp. Neurol.,* 354, 607, 1996.
87. Chan, R.K.W. and Sawchenko, P.E., Organization and transmitter specificity of medullary neurons activated by sustained hypertension: implications for understanding baroreceptor reflex circuitry, *J. Neurosci.,* 18, 371, 1998.
88. Aicher, S.A. et al., Anatomical substrates for baroreflex sympathoinhibition in the rat, *Brain Res. Bull.,* 51, 107, 2000.

89. Andresen, M.C. et al., Cellular mechanisms of baroreceptor integration at the nucleus tractus solitarius, *Ann. N.Y. Acad. Sci.,* 940, 132, 2001.
90. Ciriello, J., Brainstem projections of aortic baroreceptor afferents fibers in the rat, *Neurosci. Lett.,* 36, 37, 1983.
91. Gordon, F.J., Aortic baroreceptor reflexes are mediated by NMDA receptors in caudal ventrolateral medulla, *Am. J. Physiol.,* 252, R628, 1987.
92. Kubo, T, Kihara, M., and Misu, Y., Ipsilateral but not contralateral blockade of excitatory amino acid receptor in the caudal ventrolateral medulla inhibits aortic baroreceptor reflex in rats, *Naunyn-Schmiedeberg's Arch. Pharmacol.,* 343, 46, 1991.
93. Miyawaki, T. et al., Role of AMPA/kainate receptors in transmission of the sympathetic baroreflex in rat CVLM, *Am. J. Physiol.,* 272, R800, 1997.
94. Morrison, S.F. et al., Rostral ventrolateral medulla: a source of the glutamatergic innervation of the sympathetic intermidiolateral nucleus, *Brain Res.,* 562, 126, 1991.
95. Kubo, T. et al., Cardiovascular effects of L-glutamate and γ-aminobutyric acid injected into the rostral ventrolateral medulla in normotensive and spontaneously hypertensive rats, *Arch. Int. Pharmacod. T.,* 279, 150, 1986.
96. Minson, J.B. et al., c-fos identifies GABA-synthesizing barosensitive neurons in caudal ventrolateral medulla, *Neuroreport,* 8, 3015, 1997.
97. Goshima, Y., Nakamura, S., and Misu, Y., L-Dihydroxyphenylalanine methyl ester is a potent competitive antagonist of the L-dihydroxyphenylalanine-induced facilitation of the evoked release of endogenous norepinephrine from rat hypothalamic slices, *J. Pharmacol. Exp. Ther.,* 258, 466, 1991.
98. Numao, Y. et al., Physiological properties of the three subtypes constituting the aortic nerve-renal sympathetic reflex in rabbits, *J. Auton. Nerv. Syst.,* 9, 361, 1983.
99. Sved, A.F. et al., Excitatory inputs to the RVLM in the context of the baroreceptor reflex, *Ann. N.Y. Acad. Sci.,* 940, 247, 2001.
100. Skagerberg, G. et al., Studies on dopamine-, tyrosine hydroxylase- and aromatic L-amino acid decarboxylase-containing cells in the rat diencephalon: comparison between formaldehyde-induced histofluorescence and immunofluorescence, *Neuroscience,* 24, 605, 1988.
101. Tison, F. et al., Immunohistochemistry of endogenous L-DOPA in the rat posterior hypothalamus, *Histochemistry,* 93, 655, 1990.
102. Misu, Y., Yue, J.-L., and Goshima, Y., L-DOPA systems for blood pressure regulation in the lower brainstem, *Neurosci. Res.,* 23, 147, 1995.
103. Misu, Y. et al., Altered L-DOPA systems for blood pressure regulation in the lower brainstem of spontaneously hypertensive rats, *Hypertens. Res.,* 18, 267, 1995.

# 6 Mechanisms of Calcium-Associated Exocytosis of Striatal Dopamine and DOPA Release Studied by *In Vivo* Microdialysis

*Motohiro Okada, Gang Zhu, Shukuko Yoshida, Shinichi Hirose, and Sunao Kaneko*

## CONTENTS



## 6.1 INTRODUCTION

Over the past decade, a considerable number of investigations have explored the mechanisms of neurotransmitter-like release of L-3,4-dihydroxyphenylalanine (DOPA) in various secretory tissues. In superfused striatal slices from the brain of rats, electrical field stimulation produces DOPA release, which is tetrodotoxin (TTX) sensitive, $Ca^{2+}$ dependent, and α-methyl-*p*-tyrosine sensitive [1]. Similarly, both $K^+$-evoked stimulation [1] and superfusion with nicotine [2] also produce $Ca^{2+}$-dependent DOPA release. DOPA appears to be taken up by neurons [3]. The $Ca^{2+}$-dependent evoked release of DOPA has also been demonstrated in peripheral tissues such as dog sympathetic ganglia [3], portal vein [4], and adrenal medulla [4]. These lines of evidence strongly suggest that the output of striatal DOPA from presynaptic terminals might be regulated by exocytosis mechanisms, similar to that seen with striatal dopamine; however, in spite of these efforts, the detailed exocytosis mechanisms of striatal DOPA release remain to be demonstrated.

## 6.2 CALCIUM-ASSOCIATED EXOCYTOSIS MECHANISM

The importance of $Ca^{2+}$ in central nervous system neurotransmission as a second messenger that participates in the triggering and regulation of many neuronal processes, including neurotransmitter exocytosis [5–8], has been well established. Arrival of the action potential at synaptic terminals activates an influx of $Ca^{2+}$ through voltage-sensitive $Ca^{2+}$ channels (VSCCs), containing L-, N-, O-, P-, Q-, R-, and T-types VSCCs [9,10]. The influx of $Ca^{2+}$ through VSCC, which increases the intracellular $Ca^{2+}$ level, activates the $Ca^{2+}$-activated protein kinases. This activation involves $Ca^{2+}$ phospholipid-dependent protein kinase C (PKC), $Ca^{2+}$ calmodulin-dependent protein kinase II (CaMK-II), and endoplasmic reticulum-associated $Ca^{2+}$-induced $Ca^{2+}$-releasing systems (CICR), which act through the ryanodine receptor- and inositol 1,4,5-trisphosphate receptor-sensitive channels [5,6] (Figure 6.1A to Figure 6.1C). The supply of $Ca^{2+}$ via both VSCCs and CICRs increases the $Ca^{2+}$ level from a basal level of 100 n*M* to more than 100 μ*M* in the presynaptic active zone and then activates $Ca^{2+}$-dependent interaction between synaptic proteins, containing soluble *N*-ethylmaleimide-sensitive factor attachment protein receptors (SNAREs) and VSCC [6,7] (Figure 6.1D). This $Ca^{2+}$-dependent interaction between synaptic proteins has been considered to play an important role in the docking, fusion, and exocytosis processes (Figure 6.1E).

The binding of the synprint site of N-type VSCC with syntaxin, SNAP-25, the syntaxin/SNAP-25 dimer, or the synaptic core complex of syntaxin/SNAP-25/synaptobrevin has a biphasic dependence on the $Ca^{2+}$ concentration in the active zone but direct interaction with synaptobrevin does not occur [11,12] (Figure 6.1A to Figure 6.1C). This binding is observed in the absence of $Ca^{2+}$; however, maximal binding occurs at approximately 20 μ*M* of free $Ca^{2+}$, which is near the threshold for maximal neurotransmitter release, and reduced binding occurs at 100 μ*M* $Ca^{2+}$ [8,11,12]. This interaction is regulated by PKC but not PKA [13,14]. The synprint site of N-type VSCC binds to synaptotagmin in a $Ca^{2+}$-independent manner [8,11,12]; however, maximum binding of syntaxin to synaptotagmin requires a higher concentration of $Ca^{2+}$, in the range from 100 μ*M* to 1 m*M* [6,8,11,12] (Figure 6.1D and Figure 6.1E). Therefore, when the $Ca^{2+}$ concentration increases beyond 30 μ*M*, the binding of syntaxin with N-type VSCC is weakened but the binding of syntaxin with synaptotagmin is strengthened [6,8,11,12] (Figure 6.1D and Figure 6.1E).

The $\alpha_{1A}$ subunit of P-type VSCC has multiple isoforms detected by cDNA sequencing and analysis with sequence-specific antibodies [15,16]. Binding of SNARE to rbA and BI isoforms of the P-type VSCC has a different dependence on $Ca^{2+}$ concentration compared to that of N-type VSCC [11]. The BI isoform binds to syntaxin, SNAP-25, and synaptotagmin $Ca^{2+}$ independently. The rbA isoform has

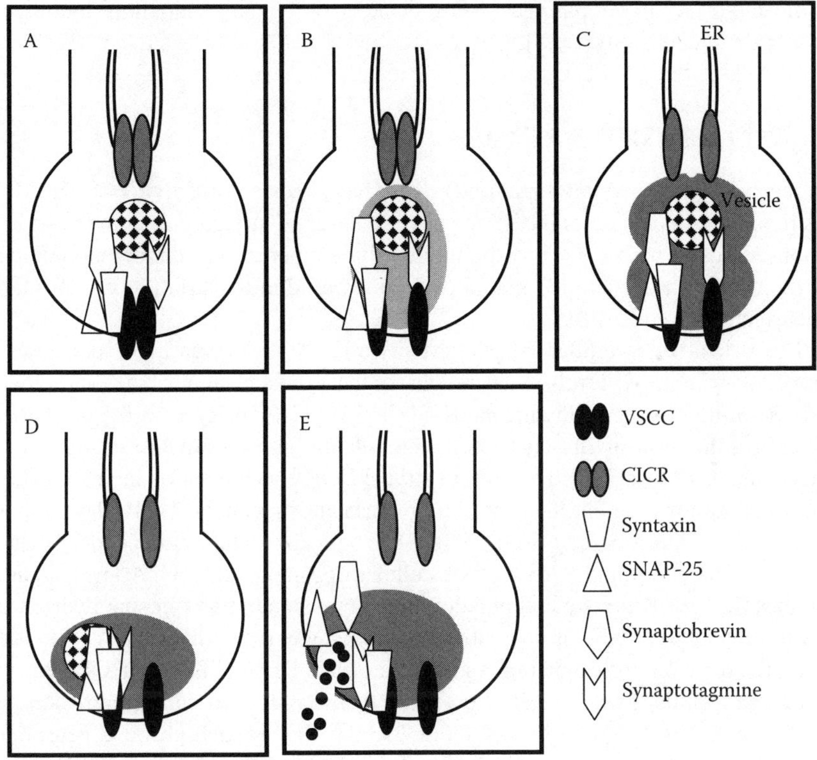


**FIGURE 6.1** Proposed hypothesis regarding N-type VSCC-related exocytosis. Interaction of synprint site of N-type VSCC with syntaxin and SNAP-25 has a biphasic $Ca^{2+}$ dependence with the maximal binding at a range of 10 to 30 $\mu M$, near the threshold for neurotransmitter release. This interaction is regulated by PKC and CaMK-II but not PKA. The synprint site of N-type VSCC binds to synaptotagmin in a $Ca^{2+}$-independent manner but does not bind to synaptobrevin. The synprint site of N-type VSCC binds to the dimetric complex of syntaxin/SNAP-25 and the trimetric complex syntaxin/SNAP-25/synaptobrevin in a $Ca^{2+}$-dependent manner, with the strongest binding at 4 to 18 $\mu M$ $Ca^{2+}$. The synprint site of N-type VSCC and synaptotagmin competitively interact with syntaxin; however, the maximum binding of syntaxin to synaptotagmin requires higher concentration of $Ca^{2+}$ in the range from 100 $\mu M$ to 1 m$M$. As the $Ca^{2+}$ concentration increases beyond 30 $\mu M$, the interaction of syntaxin with the synprint site of N-type VSCC is weakened and the interaction with synaptotagmin is strengthened.

lower affinity for binding to both syntaxin and SNAP-25 than the BI isoform. Although the binding of rbA isoform to both synaptotagmin, in a $Ca^{2+}$-dependent manner (with maximal binding at 10 to 30 $\mu M$), and SNAP-25, in a $Ca^{2+}$-independent manner, has been clearly demonstrated by *in vitro* experiments, the binding to syntaxin is not evident [17]. The interaction between synaptobrevin, which does not bind to synaptotagmin [18], and P-type VSCC has not been clarified yet; however, the cleavage of synaptobrevin was found to decrease $K^+$-evoked dopamine release, but did not affect the dopamine release evoked by the $Ca^{2+}$ ionophore ionomycin, in rat brain synaptosomes [18].

## 6.3 VSCC-SENSITIVE RELEASE

The mechanisms of striatal dopamine release have been studied in detail. The TTX-sensitive, $Ca^{2+}$-dependent, and $K^+$-sensitive releases of striatal dopamine have been demonstrated by numerous investigations. Studies performed under microdialysis conditions have demonstrated that dopamine released in the striatum was primarily of neuronal origin [19–23].

The striatal extracellular DOPA level is affected by increases or decreases in extracellular $Ca^{2+}$ or $K^+$ levels in the absence of the aromatic amino acid decarboxylase inhibitor, *m*-hydroxybenzylhydrazine (NSD-1015) [23]. However, in the presence of NSD-1015, the level of striatal DOPA accumulation is insensitive to changes in the extracellular $Ca^{2+}$ level [23]. The level of striatal DOPA accumulation increased during $K^+$-evoked stimulation, but the sensitivity of striatal extracellular DOPA levels to $K^+$-evoked stimulation was reduced by NSD-1015 in a concentration-dependent manner [23]. Thus, the elevation of striatal extracellular dopamine and DOPA levels induced by either $Ca^{2+}$- or $K^+$-evoked stimulation may not depend upon tyrosine hydroxylase activity. In addition, there are two sources of striatal extracellular DOPA: (1) DOPA released in response to stimulation and (2) the metabolic basal flow of DOPA because striatal extracellular DOPA levels are greater than 1 n*M*, even during perfusion with a $Ca^{2+}$-free perfusion medium containing 40 m*M* $Mg^{2+}$ (which prevents neurotransmitter release from nerve terminals [19,23–25]). Therefore, basal striatal DOPA release is of neuronal origin because the extracellular DOPA level is TTX sensitive, $Ca^{2+}$ dependent, and $K^+$ sensitive, similar to striatal dopamine [23].

It has been well established that a selective N-type VSCC inhibitor, ω-conotoxin GVIA, and a P-type VSCC inhibitor, ω-agatoxin IVA, are irreversible toxic VSCC inhibitors [10]. The effects of more than 3 μ*M* of these VSCC inhibitors on basal striatal extracellular levels of dopamine are irreversible; however, the effects of less than 3 μ*M* of these VSCC inhibitors are reversible [23]. Furthermore, the inhibitory effects of reversible concentrations of VSCC inhibitors (i.e., lower than 3 μ*M*) were concentration dependent, whereas from 3 μ*M* (reversible concentration) to 10 μ*M* (irreversible concentration), the inhibitory effects of VSCC inhibitors reached a plateau. Thus, the effects of lower than 3 μ*M* of VSCC inhibitors on basal striatal dopamine release are probably due to their VSCC antagonistic activity rather than their irreversible toxic effects. The rates at which ω-conotoxin GVIA and ω-agatoxin IVA diffused from the dialysis probe are 1.0 and 0.7% of that of the amounts of corresponding VSCC inhibitors that are perfused for 120 min, respectively. Thus, the estimated concentrations of ω-conotoxin GVIA and ω-agatoxin IVA that cause reversible inhibition of VSCCs are lower than 30 and 20 n*M*, respectively.

The selective N-type VSCC inhibitor, ω-conotoxin GVIA, and the selective P-type VSCC inhibitor, ω-agatoxin IVA, reduced basal striatal releases of dopamine and DOPA in a concentration-dependent manner. ω-Conotoxin GVIA predominantly inhibits striatal basal releases of dopamine ($IC_{50}$ = 0.5 n*M*) and DOPA ($IC_{50}$ = 9.6 n*M*),

**TABLE 6.1**
**Effects of VSCC Inhibitors on Striatal Releases of Dopamine and DOPA**

| | Basal Release | | $Ca^{2+}$-Evoked Release | | $K^+$-Evoked Release | |
|---|---|---|---|---|---|---|
| | **Dopamine** | **DOPA** | **Dopamine** | **DOPA** | **Dopamine** | **DOPA** |
| ω-Conotoxin GVIA ($IC_{50}$) | 0.5 n*M* | 9.6 n*M* | 0.4 n*M* | 10.5 n*M* | >30 n*M* | >30 n*M* |
| ω-Agatoxin IVA ($IC_{50}$) | >20 n*M* | >20 n*M* | — | — | 2.7 n*M* | 0.2 n*M* |

*Note:* The $IC_{50}$ values for ω-conotoxin GVIA and ω-agatoxin IVA of the basal, $Ca^{2+}$-, and $K^+$-evoked releases of striatal dopamine and DOPA were analyzed by using logistic concentration–response curves.

*Source:* Okada, M. et al., Effects of $Ca^{2+}$ channel antagonists on striatal dopamine and DOPA release, studied by *in vivo* microdialysis, *Br. J. Pharmacol.*, 123, 805, 1998. With permission.

whereas basal striatal releases of dopamine and DOPA are reduced by ω-agatoxin IVA weakly; however, the $IC_{50}$ of ω-agatoxin IVA is higher than the reversible inhibition concentration ranges (>20 n*M*)(Table 6.1) [23].

An increase in $Ca^{2+}$ level from 1.2 to 3.4 m*M* in the perfusion medium for 120 min ($Ca^{2+}$-evoked stimulation) increased the levels of striatal dopamine and DOPA releases to 170 and 150%, respectively. The $Ca^{2+}$-evoked striatal releases of dopamine ($IC_{50}$ = 0.4 n*M*) and DOPA ($IC_{50}$ = 10.5 n*M*) were reduced by ω-conotoxin GVIA selectively, whereas $Ca^{2+}$-evoked striatal releases of neither dopamine nor DOPA were affected by ω-agatoxin IVA (Table 6.1) [23].

An increase in $K^+$ level from 2.7 to 50 m*M* in the perfusion medium for 120 min ($K^+$-evoked stimulation) increased the levels of striatal dopamine and DOPA releases to 499 and 653%, respectively. The $K^+$-evoked release of striatal dopamine ($IC_{50}$ = 2.7 n*M*) and DOPA ($IC_{50}$ = 0.2 n*M*) was reduced by ω-agatoxin IVA predominantly; however, although ω-conotoxin GVIA weakly reduced $K^+$-evoked releases, the $IC_{50}$ of these two inhibitors was higher than the reversible inhibition concentration range (>30 n*M*) (Table 6.1) [23].

The question has been raised as to which of the VSCC subtypes are predominantly associated with neurotransmission. In the cerebellum, 21% of the GABAergic Purkinje cells have been shown to be ω-conotoxin GVIA sensitive and 73% to be ω-agatoxin IVA sensitive [26]. In the CA1 region of the hippocampus, which is a glutamatergic neuron-rich area, the rates of occurrence of ω-conotoxin GVIA and ω-agatoxin IVA-sensitive neurons were shown to be 12 and 46%, respectively [26].

These previous demonstrations suggest the existence of two types of releasing mechanisms for striatal dopamine and DOPA, an N-type VSCC-sensitive basal and $Ca^{2+}$-evoked release mechanism and a P-type VSCC-sensitive $K^+$-evoked release mechanism. Furthermore, during the resting stage, the striatal dopamine release was shown to be more sensitive than the striatal DOPA release to N-type VSCC because the $IC_{50}$ value of ω-conotoxin GVIA for striatal dopamine release was lower than that for striatal DOPA release. In contrast to the resting stage, during the depolarization stage, striatal DOPA release was shown to be more sensitive than striatal dopamine release to P-type VSCC because the $IC_{50}$ value of ω-agatoxin IVA for striatal DOPA release was lower than that for striatal dopamine release. Thus, in the striatum, P-type VSCC-sensitive DOPA release is more dominant than N-type VSCC-sensitive release during the depolarization stage, similar to hippocampal glutamatergic neurotransmission.

## 6.4 SNARE-SENSITIVE RELEASE

The botulinum toxins (BoNTs) have seven different proteins composed of two disulfide-linked polypeptide chains [27,28]. They bind specifically to the presynaptic membrane via the heavy chain, while the light chain enters the cytosol of the neurons where it displays a zinc-endopeptidase activity directed at proteins of the neuroexocytosis apparatus [27,28]. BoNT serotypes BoNT/B, BoNT/D, BoNT/F, and BoNT/G cleave, specifically at different single peptide bonds, synaptobrevin, a component of small synaptic vesicles [27,28]. In contrast, the other BoNTs catalyze the hydrolysis of proteins of the presynaptic membrane. Both BoNT/A and BoNT/E cleave SNAP-25 at different sites located within the carboxyl terminus, whereas the specific target of BoNT/C is syntaxin [27,28]. To reach the cytosol, BoNTs, inside vesicles in presynaptic terminals, must necessarily cross the hydrophobic barrier of the vesicle membrane [27,28]. The first step is the binding of the heavy chains of various BoNTs to their respective receptors localized on the surface of the presynaptic membrane. Productive binding is followed by the second step: the internalization of the BoNT-receptor complex vesicles. The acceptors of the different BoNTs may address the respective BoNT-receptor complexes to distinct vesicle subpopulations (step 2) [27,28]. The acidification of the vesicle lumen by a proton-pumping ATPase leads to a conformational change of BoNT. In its acid conformation, BoNT inserts into the lipid bilayer and the light chain translocates across the vesicle membrane (step 3) [27]. The light chain, freed in the cytosol by reduction of the interchain disulfide bond, can display its enzymatic activity on the components of the 20S multisubunit vesicle docking/fusion complex (step 4) [27]. SNARE complexes probably would have turned over many times during the 48 h of BoNTs treatment [27]. Based on these lines of evidence, to clarify the mechanisms of exocytosis using *in vivo* microdialysis, the target brain area should be pretreated with microinjection of a sub-microliter medium containing BoNT, which consists of both light and heavy chains, before insertion of the microdialysis probe. Indeed, in our previous study, the microinjection of BoNT/A, BoNT/B, and BoNT/C reduced hippocampal monoamine release dose- (ranging from 0.01 ng/0.3 μL to 10 ng/0.3 μL) and time (ranging from 18 to 48 h) dependently; however, neither microinjection of light nor heavy chain of BoNT affected monoamine release [29–31].

Cleavage of syntaxin and synaptobrevin by microinjection of respective BoNT/C and BoNT/B decreases basal releases of dopamine and DOPA in a BoNT dose-dependent manner (Table 6.2). The inhibitory effect of BoNT/C on basal releases of dopamine and DOPA is larger than that of BoNT/B (Table 6.2). The microinjection

**TABLE 6.2**
**Effects of BoNTs on Striatal Releases of Dopamine and DOPA**

| | Basal Release | | $Ca^{2+}$-Evoked Release | | $K^+$-Evoked Release | |
|---|---|---|---|---|---|---|
| | Dopamine | DOPA | Dopamine | DOPA | Dopamine | DOPA |
| BoNT/B ($IC_{50}$) | >10 ng | >10 ng | — | — | 0.96 ng | 0.51 ng |
| BoNT/C ($IC_{50}$) | 1.2 ng | 4.1 ng | 0.7 ng | 3.8 ng | >10 ng | >10 ng (0.3 $\mu L^{-1}$) |

*Note:* The $IC_{50}$ values for BoNT/B and BoNT/C of the basal, $Ca^{2+}$-, and $K^+$-evoked releases of striatal dopamine and DOPA were analyzed by using logistic concentration–response curves.

of BoNT/C reduces $Ca^{2+}$-evoked releases of dopamine and DOPA, whereas cleavage of synaptobrevin does not affect basal releases (Table 6.2). Both BoNT/C and BoNT/B reduce $K^+$-evoked releases of dopamine and DOPA dose-dependently (Table 6.2). In contrast to basal release, the inhibitory effect of BoNT/C on $K^+$-evoked release is weaker than that of BoNT/B.

## 6.5 INTERACTION BETWEEN BoNTs AND VSCC INHIBITORS ON RELEASES OF DOPAMINE AND DOPA

To clarify the exocytosis mechanisms of DOPA and glutamate, the interactions between VSCC inhibitors and BoNTs on basal, $Ca^{2+}$-, and $K^+$-evoked releases of dopamine and DOPA were determined. Under the cleavage of syntaxin by microinjection with BoNT/C, the inhibitory effects of ω-conotoxin GVIA on basal, $Ca^{2+}$-, and $K^+$-evoked releases of dopamine and DOPA are abolished (Figure 6.2), whereas ω-agatoxin IVA decreased both basal and $K^+$-evoked releases of dopamine and DOPA (Figure 6.2). Under the cleavage of synaptobrevin by BoNT/B, the inhibitory effects of IVA on basal and $K^+$-evoked releases of dopamine and DOPA are abolished (Figure 6.2), whereas ω-conotoxin GVIA decreased both basal, $Ca^{2+}$-, and $K^+$-evoked releases of both (Figure 6.2).

Both tSNARE syntaxin and vSNARE synaptobrevin are shown to be key players in the $Ca^{2+}$-dependent exocytosis processes [6,12]. The basal, $Ca^{2+}$-, and $K^+$-evoked release of striatal dopamine is regulated by activities of both, VSCCs and SNAREs, using *in vivo* microdialysis. Similar to dopamine, striatal DOPA release is also regulated by VSCCs and SNAREs. The $Ca^{2+}$-evoked releases of dopamine and DOPA are regulated by N-type VSCC and syntaxin selectively but not by P-type VSCC or synaptobrevin. Contrary to $Ca^{2+}$-evoked releases, basal releases of dopamine and DOPA are regulated by N-type VSCC and syntaxin predominantly and by P-type VSCC and synaptobrevin weakly. In addition, $K^+$-evoked releases of dopamine and DOPA are regulated by P-type VSCC and synaptobrevin predominantly, and by N-type VSCC and syntaxin weakly. The experiments on interactions between BoNTs

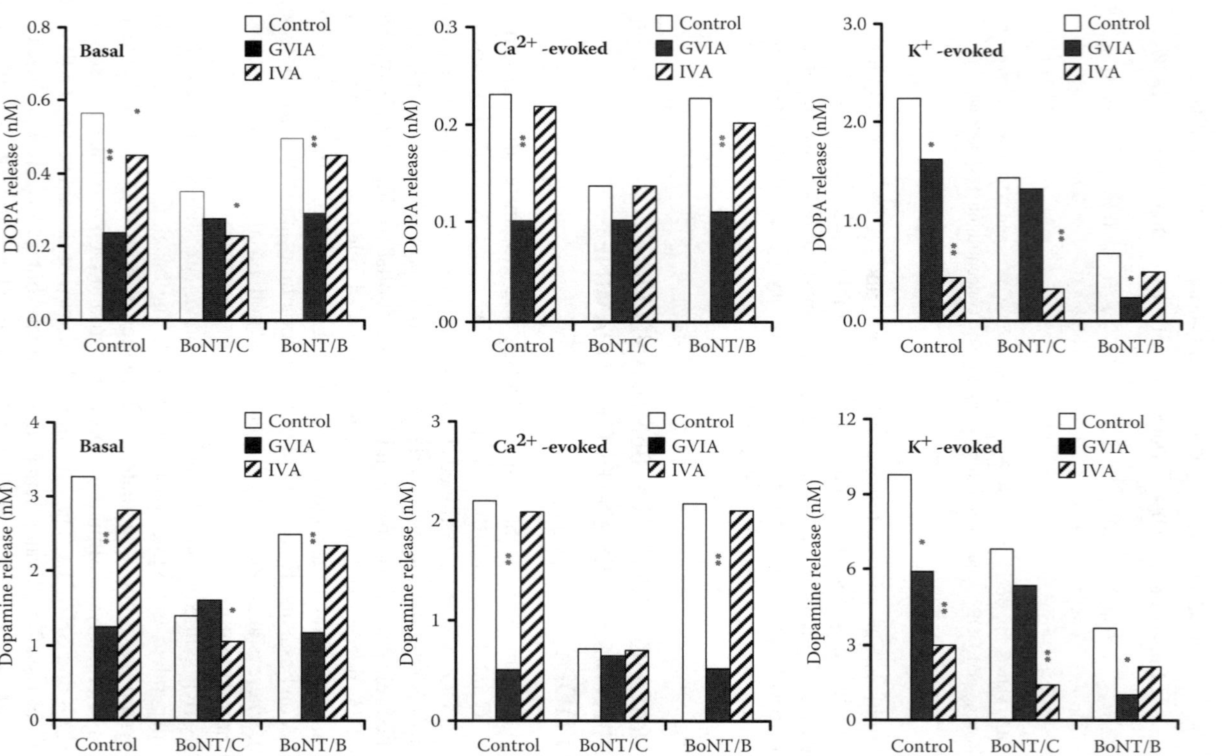


**FIGURE 6.2** Interaction between BoNTs and VSCC inhibitors on basal, $Ca^{2+}$-, and $K^{+}$-evoked releases of striatal dopamine and DOPA. The effects of VSCC inhibitors, a selective N-type VSCC inhibitor, ω-conotoxin GVIA (GVIA) and a selective P-type VSCC inhibitor, ω-agatoxin IVA (IVA), on basal,

and VSCC inhibitors indicate the functional complexes whose components are, at least partially, syntaxin, synaptobrevin, and N-type and P-type VSCCs in exocytosis mechanism, namely, syntaxin with N-type VSCC (syntaxin/N-type) and synaptobrevin with P-type VSCC (synaptobrevin/P-type). The syntaxin/N-type complex is the major regulatory component for both basal and $Ca^{2+}$-evoked releases of striatal dopamine and DOPA; synaptobrevin/P-type is also the major pathway for the $K^{+}$-evoked releases of both.

Several electrophysiological experiments have demonstrated that P-type VSCC is present in high density at central synapses and that glutamatergic neurotransmission primarily requires P-type VSCC, with N-type VSCC playing a secondary role [26,32]. However, the basal striatal DOPA release requires the syntaxin/N-type pathway predominantly, similar to monoamine release [23,29–31,33]. After docking between plasma and vesicle membranes, SNAREs bind together in a parallel, four-helix bundle, with one helix each contributed by synaptobrevin and syntaxin and two contributed by SNAP-25 [34]. The four-helix bundle integrates neurotransmitter exocytosis and possibly the fusion process but does not affect the leave kinetics of the synaptic response [35]. Furthermore, three SNARE complexes between syntaxin, synaptobrevin, and SNAP-25 cooperate to mediate fusion of a single vesicle with plasma membrane and contribute part of a protein and lipidic fusion pore [36]. However, basal and $K^{+}$-evoked releases of DOPA are predominantly inhibited by BoNT/C and BoNT/B, respectively. These contradictory demonstrations indicate that synaptobrevin plays an important role not only in exocytosis but also in synaptic vesicle recycling [37].

## 6.6 CONCLUSION

The striatal extracellular level of DOPA is constituted by two components: DOPA that is released from presynaptic terminals and the metabolic basal flow of DOPA. Both dopamine and DOPA releases in striatum are produced by an exocytosis mechanism, which is composed of two functional complexes, syntaxin/N-type and synaptobrevin/P-type. The activities of these two functional complexes are the result of neuronal excitability. The basal releases of dopamine and DOPA are regulated predominantly by syntaxin/N-type and weakly by synaptobrevin/P-type. In contrast to this, $K^{+}$-evoked releases of dopamine and DOPA are regulated by synaptobrevin/P-type predominantly and by syntaxin/N-type weakly, whereas $Ca^{2+}$-evoked releases are regulated selectively by syntaxin/N-type but not by synaptobrevin/P-type. The dopamine release in response to syntaxin/N-type regulation is more sensitive than that of DOPA, whereas the DOPA release in response to synaptobrevin/P-type is more sensitive than that of dopamine.

**FIGURE 6.2 (Continued)** $Ca^{2+}$-, and $K^{+}$-evoked releases of striatal dopamine and DOPA, under the cleavage of syntaxin by 0.3 ng/0.3 μL BoNT/C or with synaptobrevin by 0.3 ng/0.3 μL BoNT/B. Ordinates indicate the area under curve value of extracellular levels of DOPA (n*M*) and dopamine (n*M*). The effects of VSCC inhibitors on basal, $Ca^{2+}$-, and $K^{+}$-evoked releases under the condition of cleavage with SNAREs were analyzed by two-way analysis of variance (ANOVA) with Tukey's multiple comparison (*$P < .05$, **$P < .01$).

## ACKNOWLEDGMENTS

This study was supported by a grant-in-aid for Scientific Research from the Japanese Ministry of Education, Science, and Culture (13670979, 14570903, and 15659267) and grants from the Hirosaki Research Institute for Neurosciences, the Pharmacopsychiatry Research Foundation, and the Japan Epilepsy Research Foundation.

## REFERENCES

1. Goshima, Y., Kubo, T., and Misu, Y., Transmitter-like release of endogenous 3,4-dihydroxyphenylalanine from rat striatal slices, *J. Neurochem.,* 50, 1725, 1988.
2. Misu, Y. et al., Nicotine releases stereoselectively and $Ca^{2+}$-dependently endogenous 3,4-dihydroxyphenylalanine from rat striatal slices, *Brain Res.,* 520, 334, 1990.
3. Chang, W. Y. and Webster, R. A., Effects of 3-O-methyl dopa on L-DOPA-facilitated synthesis and efflux of dopamine from rat striatal slices, *Br. J. Pharmacol.,* 116, 2637, 1995.
4. Hunter, L.W., Rorie, D.K., and Tyce, G.M., Dihydroxyphenylalanine and dopamine are released from portal vein together with noradrenaline and dihydroxyphenylglycol during nerve stimulation, *J. Neurochem.,* 59, 972, 1992.
5. Berridge, M.J., Neuronal calcium signaling, *Neuron,* 21, 13, 1998.
6. Rettig, J. and Neher, E., Emerging roles of presynaptic proteins in $Ca^{++}$-triggered exocytosis, *Science,* 298, 781, 2002.
7. Sollner, T. et al., A protein assembly-disassembly pathway in vitro that may correspond to sequential steps of synaptic vesicle docking, activation, and fusion, *Cell,* 75, 409, 1993.
8. Sheng, Z.H., Westenbroek, R.E., and Catterall, W.A., Physical link and functional coupling of presynaptic calcium channels and the synaptic vesicle docking/fusion machinery, *J. Bioenerg. Biomembr.,* 30, 335, 1998.
9. Olivera, B.M. et al., Calcium channel diversity and neurotransmitter release: the omega-conotoxins and omega-agatoxins, *Annu. Rev. Biochem.,* 63, 823, 1994.
10. Randall, A. and Tsien, R.W., Pharmacological dissection of multiple types of $Ca^{2+}$ channel currents in rat cerebellar granule neurons, *J. Neurosci.,* 15, 2995, 1995.
11. Kim, D.K. and Catterall, W.A., $Ca^{2+}$-dependent and -independent interactions of the isoforms of the alpha1A subunit of brain $Ca^{2+}$ channels with presynaptic SNARE proteins, *Proc. Natl. Acad. Sci. U.S.A.,* 94, 14782, 1997.
12. Sheng, Z.H. et al., Calcium-dependent interaction of N-type calcium channels with the synaptic core complex, *Nature,* 379, 451, 1996.
13. Turner, K.M., Burgoyne, R.D., and Morgan, A., Protein phosphorylation and the regulation of synaptic membrane traffic, *Trends Neurosci.,* 22, 459, 1999.
14. Yokoyama, C.T., Sheng, Z.H., and Catterall, W.A., Phosphorylation of the synaptic protein interaction site on N-type calcium channels inhibits interactions with SNARE proteins, *J. Neurosci.,* 17, 6929, 1997.
15. Mori, Y. et al., Primary structure and functional expression from complementary DNA of a brain calcium channel, *Nature,* 350, 398, 1991.
16. Starr, T.V., Prystay, W., and Snutch, T.P., Primary structure of a calcium channel that is highly expressed in the rat cerebellum, *Proc. Natl. Acad. Sci. U.S.A.,* 88, 5621, 1991.
17. Rettig, J. et al., Isoform-specific interaction of the alpha1A subunits of brain $Ca^{2+}$ channels with the presynaptic proteins syntaxin and SNAP-25, *Proc. Natl. Acad. Sci. U.S.A.,* 93, 7363, 1996.

18. Schiavo, G. et al., Binding of the synaptic vesicle v-SNARE, synaptotagmin, to the plasma membrane t-SNARE, SNAP-25, can explain docked vesicles at neurotoxin-treated synapses, *Proc. Natl. Acad. Sci. U.S.A.,* 94, 997, 1997.
19. Westerink, B.H., Damsma, G., and de Vries, J. B., Effect of ouabain applied by intrastriatal microdialysis on the in vivo release of dopamine, acetylcholine, and amino acids in the brain of conscious rats, *J. Neurochem.,* 52, 705, 1989.
20. Bergquist, F. et al., Effects of local administration of L-, N-, and P/Q-type calcium channel blockers on spontaneous dopamine release in the striatum and the substantia nigra: a microdialysis study in rat, *J. Neurochem.,* 70, 1532, 1998.
21. Okada, M. et al., Effects of zonisamide on extracellular levels of monoamine and its metabolite, and on $Ca^{2+}$ dependent dopamine release, *Epilepsy Res.,* 13, 113, 1992.
22. Okada, M. et al., Determination of the effects of caffeine and carbamazepine on striatal dopamine release by in vivo microdialysis, *Eur. J. Pharmacol.,* 321, 181, 1997.
23. Okada, M. et al., Effects of $Ca^{2+}$ channel antagonists on striatal dopamine and DOPA release, studied by in vivo microdialysis, *Br. J. Pharmacol.,* 123, 805, 1998.
24. Okada, M. et al., Magnesium ion augmentation of inhibitory effects of adenosine on dopamine release in the rat striatum, *Psychiat. Clin. Neuros.,* 50, 147, 1996.
25. Okada, M. and Kaneko, S., Pharmacological interactions between magnesium ion and adenosine on monoaminergic system in the central nervous system, *Magnesium Res.,* 11, 289, 1998.
26. Takahashi, T. and Momiyama, A., Different types of calcium channels mediate central synaptic transmission, *Nature,* 366, 156, 1993.
27. Humeau, Y. et al., How botulinum and tetanus neurotoxins block neurotransmitter release, *Biochimie,* 82, 427, 2000.
28. Schiavo, G., Rossetto, O., and Montecucco, C., Clostridial neurotoxins as tools to investigate the molecular events of neurotransmitter release, *Semin. Cell Biol.,* 5, 221, 1994.
29. Murakami, T. et al., Determination of effects of antiepileptic drugs on SNAREs-mediated hippocampal monoamine release using in vivo microdialysis, *Br. J. Pharmacol.,* 134, 507, 2001.
30. Okada, M. et al., Adenosine receptor subtypes modulate two major functional pathways for hippocampal serotonin release, *J. Neurosci.,* 21, 628, 2001.
31. Okada, M. et al., Exocytosis mechanism as a new targeting site for mechanisms of action of antiepileptic drugs, *Life Sci.,* 72, 465, 2002.
32. Wheeler, D.B., Randall, A., and Tsien, R.W., Roles of N-type and Q-type $Ca^{2+}$ channels in supporting hippocampal synaptic transmission, *Science,* 264, 107, 1994.
33. Kawata, Y. et al., Pharmacological discrimination between effects of carbamazepine on hippocampal basal, $Ca^{2+}$- and $K^{+}$-evoked serotonin release, *Br. J. Pharmacol.,* 133, 557, 2001.
34. Chen, Y.A. et al., SNARE complex formation is triggered by $Ca^{2+}$ and drives membrane fusion, *Cell,* 97, 165, 1999.
35. Finley, M.F. et al., The core membrane fusion complex governs the probability of synaptic vesicle fusion but not transmitter release kinetics, *J. Neurosci.,* 22, 1266, 2002.
36. Hua, Y. and Scheller, R.H., Three SNARE complexes cooperate to mediate membrane fusion, *Proc. Natl. Acad. Sci. U.S.A.,* 98, 8065, 2001.
37. Maycox, P.R. et al., Clathrin-coated vesicles in nervous tissue are involved primarily in synaptic vesicle recycling, *J. Cell Biol.,* 118, 1379, 1992.

# 7 Release of DOPA from Dog Peripheral Tissues

*Gertrude M. Tyce, Larry W. Hunter, Eric W. Kristensen, Stewart L. Chritton, and Duane K. Rorie*

## CONTENTS



## 7.1 INTRODUCTION

Multiple lines of evidence indicate that L-3,4-dihydroxyphenylalanine (DOPA) has the role of a neurotransmitter in the central nervous system [1]. Herein we describe experiments demonstrating that DOPA is released in the peripheral sympathetic nervous system in response to classical neuronal stimuli in dogs. Specifically, $Ca^{2+}$-dependent DOPA release has been observed from the adrenal medulla [2], from sympathetic ganglia [3] and at noradrenergic nerve endings in veins and arteries [4,5].

The adrenal medulla releases the hormones epinephrine (E) and norepinephrine (NE) in response to cholinergic stimulation. Neuropeptide Y (NPY) and methionine-enkephalin (MET-ENK) are also released from the adrenal medulla in dogs [6] and may act as neuromodulatory substances in this tissue [6]. In addition to these compounds, effluxes of DOPA were demonstrated from the adrenal medulla, and these were compared to the effluxes of the accepted hormones and the neuromodulatory substances. The effluxes of dopamine (DA), the precursor of NE, the effluxes of the catecholamine metabolites 3,4-dihydroxyphenylacetic acid (DOPAC) and 3,4-dihroxyphenylglycol (DHPG), and the effluxes of *O*-methylated metabolites metanephrine (MN), normetanephrine (NMN), and 3-methoxy-4-hydroxyphenylglycol (MHPG) were also measured.

In sympathetic ganglia, the second tissue studied, cholinergic neurons that originate in the spinal cord synapse with noradrenergic neurons that project into peripheral tissues. It is known that in rats, a number of small intensely fluorescent (SIF) cells exist within some sympathetic ganglia [7]. These SIF cells are of interest because they may be interneurons that contain and release DA in response to muscarinic activation [8], in contrast to the postganglionic neurons that release NE in response to nicotinic activation [6]. Early findings of relatively high concentrations of DOPA in dog sympathetic ganglia suggested that DOPA might be released with DA, possibly from the SIF cells, in sympathetic ganglia. The release of DOPA was measured together with the release of DA and NE in response to a number of different stimuli.

In the third tissue studied, the release of DOPA was demonstrated at the terminals of noradrenergic neurons at neuroeffector junctions in blood vessels. In isolated, superfused strips of veins and arteries, overflows of NE, DA, and DOPA were measured under basal conditions, and releases evoked by field-stimulation were quantified. The effects of a series of pharmacologic agents that intervene with the synthesis, metabolism, and disposition of catecholamines are described.

## 7.2 RELEASE OF DOPA FROM DOG ADRENAL GLAND

Adrenal glands removed from mongrel dogs were retrogradely perfused with warmed oxygenated Krebs–Ringer solution [2,6]. During a 70-min stabilization period, levels of E, NE, DA, DOPA, MET-ENK, and NPY declined exponentially (Figure 7.1). In contrast, the levels of the metabolites DHPG, DOPAC, and MHPG tended to increase during the first 40 min of stabilization and then to decrease slowly. The levels of MN and NMN declined slowly throughout. The relative abundance of compounds in perfusates collected between 60 and 70 min of stabilization was E >> NE > DHPG > MN > NMN > DA > DOPAC > MHPG >> DOPA >> MET-ENK >> NPY (Figure 7.1). Removal of $Ca^{2+}$ from the perfusate did not affect basal efflux of any of the compounds measured [2].

When the tissues were stimulated for 2 min with carbachol at 30 $\mu M$ after the stabilization period, the effluxes of E, NE, DA, DOPA, MET-ENK, and NPY rose markedly, but the levels of metabolites did not change significantly (Figure 7.2). The time course of releases of the catecholamines, DOPA, and the neuropeptides were similar. The evoked releases of E, NE, DA, and DOPA were reduced significantly in a $Ca^{2+}$-free perfusate (data not shown).

## 7.3 DOPA RELEASE FROM SYMPATHETIC GANGLIA OF DOGS

The most abundant compound measured in dog sympathetic ganglia was NE (121.7 pmol/mg tissue). An unexpected finding was that the next most abundant compound was DOPA (21.1 pmol/mg). DA, E, and DOPAC were present in the ganglia in similar amounts (12.5, 11.1, and 8.4 pmol/mg, respectively). DHPG was present in the lowest concentration (2.7 pmol/mg) [3]. The concentrations of the compounds measured were essentially the same in ganglia from different lumbar segments.

The average overflows of DOPA, NE, E, and metabolites from superfused ganglia that had not been stimulated are shown in Figure 7.3. There were similar declines

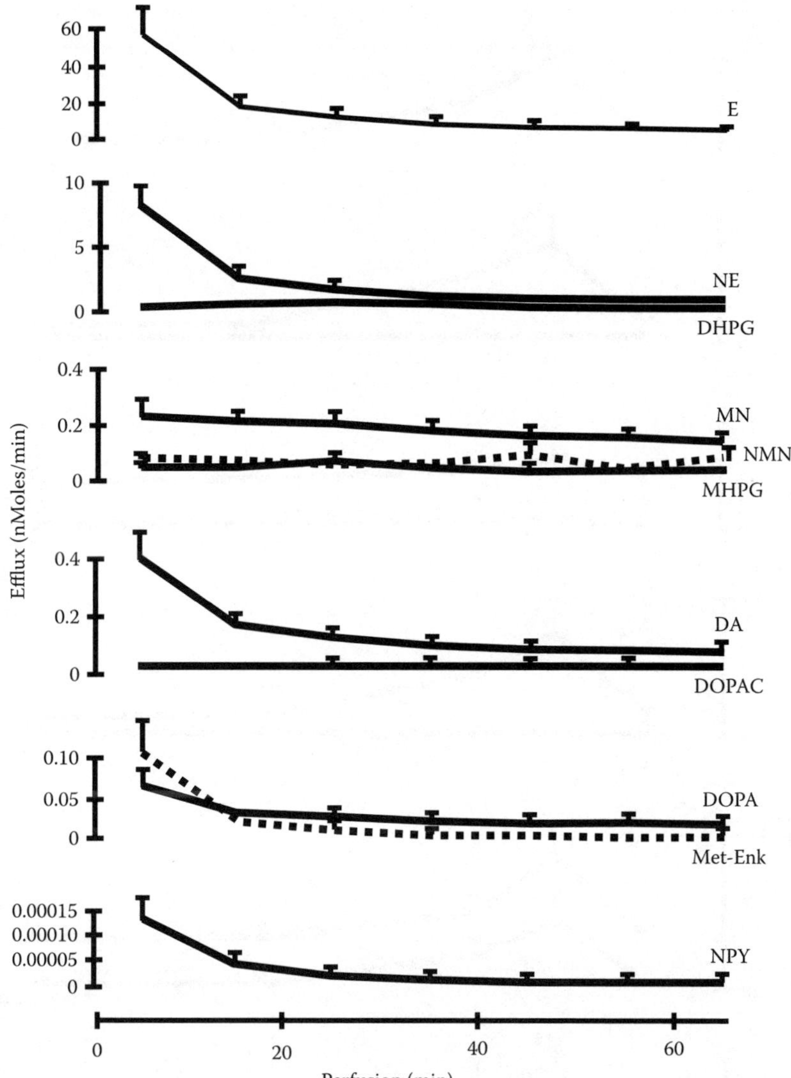


**FIGURE 7.1** Efflux of catecholamines, metabolites, DOPA, and neuropeptides during the initial stabilization period from five perfusions of dog adrenal glands. (From Chritton, S.L. et al., Adrenomedullary secretion of DOPA, catecholamines, catechol metabolites and neuropeptides, *J. Neurochem.*, 69, 2413, 1997. With permission.)

in the first 30 min for DOPA and DA, and these were more rapid than the declines for NE and E. The overflows of DHPG and DOPAC increased in the first 30 min of superfusion and then declined at rates similar to the rates of decline for NE and E.

The evoked releases of compounds in response to 80-m*M* $K^+$ and to 100-μ*M* amphetamine are shown in Figure 7.4. An 80-m*M* $K^+$ stimulus evoked an increased

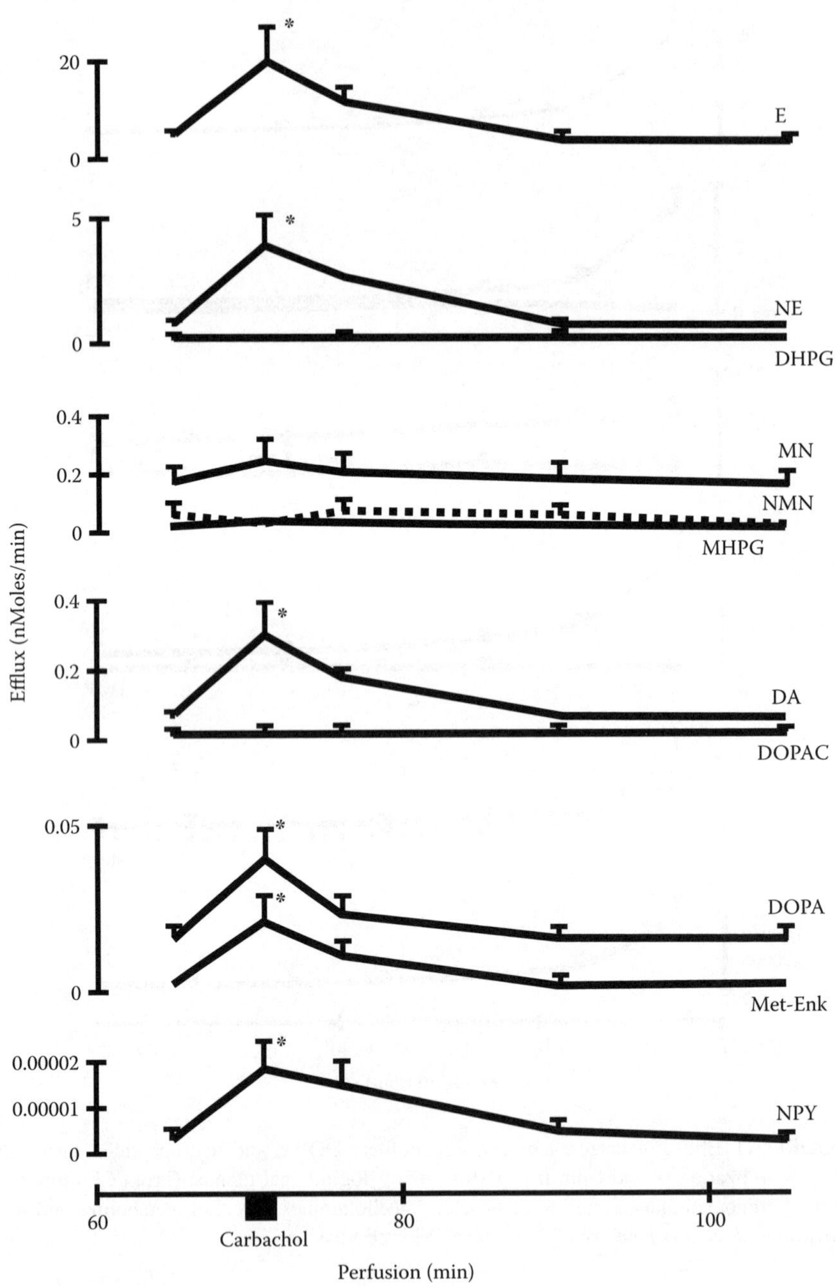


**FIGURE 7.2** Efflux of catecholamines, metabolites, DOPA, and peptides from seven perfused dog adrenal glands (nine for the peptides). *$P < .05$, compared with the preceding basal efflux (sign test). (From Chritton, S.L. et al., Adrenomedullary secretion of DOPA, catecholamines, catechol metabolites and neuropeptides, *J. Neurochem.*, 69, 2413, 1997. With permission.)

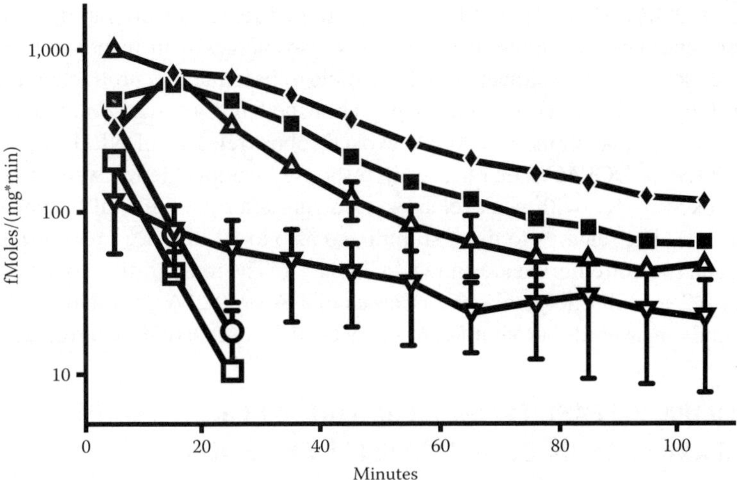


**FIGURE 7.3** Average overflow profiles of DOPA (open circle), DA (open square), DOPAC (filled square), NE (upward triangle), E (downward triangle), and DHPG (diamond) during superfusions of five dog sympathetic ganglia when no stimulus was introduced. (From Kristensen, E.W. et al., Precursors and metabolites of norepinephrine in sympathetic ganglia of the dog, *J. Neurochem.,* 54, 1782, 1990. With permission.)

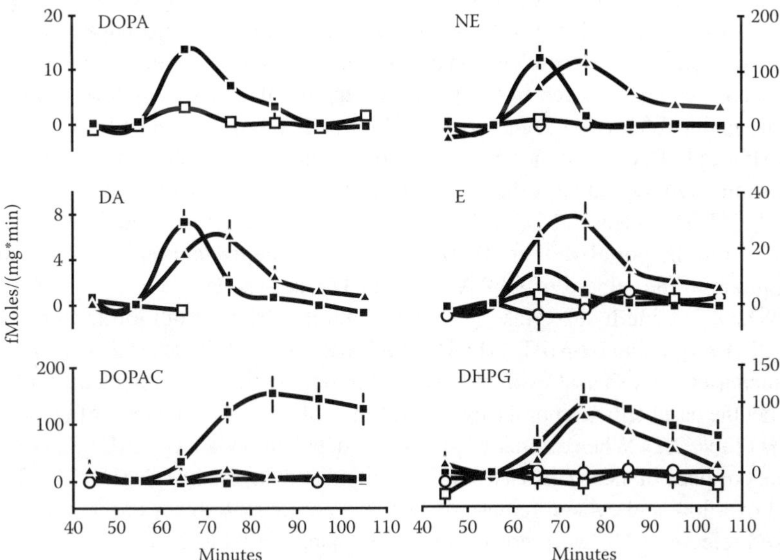


**FIGURE 7.4** Average net evoked releases from superfused dog sympathetic ganglia in response to the specified stimuli for DOPA, DA, DOPAC, NE, E, and DHPG. Data are means ± SE of five experiments. Control (open circle); 80-m*M* $K^+$ and no $Ca^{2+}$ (open square); 100-μ*M* amphetamine (triangle); and 80-m*M* $K^+$ (filled square). (From Kristensen, E.W. et al., Precursors and metabolites of norepinephrine in sympathetic ganglia of the dog, *J. Neurochem.*, 54, 1782, 1990. With permission.)

release of DOPA, DA, NE, and E. These increased releases were rapid, concurrent, and transient. The overflows of metabolites also increased in response to high $K^+$ but these increases were slower, and the decline to baseline was protracted. Increases in overflow of all the compounds were abolished if $Ca^{2+}$ was omitted from the superfusate. Amphetamine at 100 μ*M* evoked robust releases of NE, DA, and E but not of DOPA or DOPAC. The releases in response to amphetamine were slower than those evoked by $K^+$, with peak releases occurring after the removal of the stimulus (Figure 7.4). The releases to these stimuli and also to 40-m*M* $K^+$, to 50-μ*M* amphetamine, and to oxotremorine are shown in Table 7.1. The muscarinic agonist oxotremorine at 0.2 m*M* did not evoke the releases of DA or DOPA or of any of the other compounds measured, but significant releases of E and DHPG occurred at 1 m*M*.

## 7.4 DOPA RELEASE FROM NORADRENERGIC TERMINALS IN BLOOD VESSELS OF DOGS

When helical strips of saphenous veins, portal veins, pulmonary arteries, and mesenteric arteries were superfused with Krebs–Ringer solution, DOPA overflowed from the vessels under basal conditions without nerve stimulation. NE was also detectable in these washout superfusates. The amounts of DOPA in basal superfusates were about 0.2 to 0.3 times those of NE [5]. DA could not be detected in superfusates collected under basal conditions. Predictably, the amounts of NE increased in superfusates in all four vessels during electrical stimulation (ES), and the increases were frequency and $Ca^{2+}$-dependent (Table 7.2). DOPA was also released into superfusates in a frequency- and $Ca^{2+}$ -dependent manner (Table 7.3). The maximum releases of NE in response to ES occurred in the period when stimulation was applied (Table 7.2), but the major releases of DOPA occurred immediately following ES (Table 7.3).

Although DA could not be detected in superfusates collected under basal conditions, DA was always detected during stimulation at 10 Hz and occasionally at 2 Hz [4]. The release of DA followed the same time course as that of NE. Release of DA from the portal vein at 10 Hz was also $Ca^{2+}$-dependent and was of a similar magnitude to the release of DOPA (data for DA not shown).

When tyrosine hydroxylase (TH) was inhibited by the addition to perfusate of α-methyl-*p*-tyrosine (α-MPT, 20 μ*M*), the basal efflux of DOPA was abolished, but the increases above basal evoked by ES were not significantly affected (Table 7.3). Neither the basal effluxes nor the evoked releases of NE were affected by the inhibition of TH (Table 7.2). When aromatic L-amino acid decarboxylase (AAAD) was inhibited by the presence in the superfusates of 3-hydroxybenzylhydrazine (NSD-1015, 20 μ*M*), basal overflow and evoked release of DOPA were increased, but basal overflow and evoked release of NE were not significantly changed (Table 7.2).

In the portal vein, when neuronal and extraneuronal uptakes were blocked by desmethylimipramine (DMI, 1 μ*M*) and corticosterone (CORT, 1 μ*M*), respectively, NEδ-evoked releases, but not basal overflow, were increased (Table 7.2). Neither basal overflow nor the evoked release of DOPA was affected (Table 7.3). The addition of yohimbine (YOH, 0.1 μ*M*) further increased evoked releases but not basal overflows of NE and DOPA (Table 7.2 and Table 7.3).

**TABLE 7.1**
**Effects of Various Stimuli on Net Overflow (pmol/mg) of Each Catechol from Superfused Dog Sympathetic Ganglia**

| | Control | 50 μ*M* Amphetamine | 100 μ*M* Amphetamine | 40-m*M* K$^+$ | 80-m*M* K$^+$ | 80-m*M* K$^+$ no Ca$^{2+}$ | 0.2-m*M* Oxotremorine | 1.0-m*M* Oxotremorine |
|---|---|---|---|---|---|---|---|---|
| DOPA | ND | ND | ND | 0.06 ± 0.04 | 0.22 ± 0.03[a] | 0.05 | −0.05 ± 0.05 | 0.03 ± 0.01 |
| | | | | (2) | (5) | (1) | (2) | (7) |
| DA | ND | 0.06 ± 0.03 | 0.13 ± 0.02[a] | 0.03 ± 0.03 | 0.09 ± 0.02[a] | 0.00 | −0.04 ± 0.02 | ND |
| | | (5) | (5) | (3) | (5) | (1) | (2) | |
| DOPAC | −0.01 ± 0.35 | 0.00 ± 0.26 | 0.30 ± 0.37 | 1.45 ± 0.63 | 5.77 ± 1.14[a] | 0.00 ± 0.20 | 0.37 ± 0.33 | 0.00 ± 0.47 |
| | (6) | (5) | (5) | (7) | (6) | (4) | (4) | (8) |
| NE | 0.00 ± 0.20 | 2.13 ± 0.37[a] | 3.13 ± 0.46[a] | 0.77 ± 0.33 | 1.33 ± 0.27[a] | −0.09 ± 0.15 | −0.26 ± 0.16 | 0.40 ± 0.17 |
| | (6) | (5) | (5) | (7) | (6) | (4) | (4) | (8) |
| E | 0.01 ± 0.03 | 0.84 ± 0.34[a] | 0.81 ± 0.17[a] | 0.35 ± 0.16 | 0.16 ± 0.13 | 0.09 ± 0.09 | 0.03 | 0.22 ± 0.09[a] |
| | (4) | (4) | (5) | (5) | (3) | (3) | (1) | (7) |
| DHPG | 0.03 ± 0.55 | 0.28 ± 0.67 | 2.08 ± 0.34[a] | 1.00 ± 0.79 | 3.56 ± 1.13[a] | −0.41 ± 0.58 | 0.37 ± 0.94 | 1.89 ± 0.23[a] |
| | (6) | (5) | (5) | (7) | (6) | (4) | (4) | (8) |

[a] $P < .05$ vs. control (Student's *t*-test).

*Note:* The method for calculation of net evoked overflow is given in Reference 3.

*Source:* From Kristensen, E.W. et al., Precursors and metabolites of norepinephrine in sympathetic ganglia of the dog, *J. Neurochem.*, 54, 1782, 1990. With permission.

**TABLE 7.2**
**NE Overflow (fmol/mg) by Period in Dog Portal Vein Superfused and Stimulated *In Vitro***

| Treatment | 1 (Basal) | 2 (2 Hz) | 3 (Post 2 Hz) | 4 (Basal) | 5 (10 Hz) | 6 (Post 10 Hz) | 7 (Post 10 Hz) |
|---|---|---|---|---|---|---|---|
| Control, no drugs | 101.3 ± 27.6 | 933.2 ± 130.9[a] | 135.5 ± 24.2 | 34.9 ± 11.1[a] | 5181.2 ± 505.2[a] | 342.2 ± 38.5[a] | 35.9 ± 9.3 |
| $Ca^{2+}$-free | 88.3 ± 23.5 | 71.1 ± 20.1[b] | 56.2 ± 16.9[b] | 26.7 ± 9.2[a] | 32.2 ± 12.3[b] | 18.3 ± 8.2[b] | 8.4 ± 4.9[b] |
| α-MPT | 86.3 ± 14.5 | 752.3 ± 107.5[a] | 100.0 ± 8.9 | 28.7 ± 7.1[a] | 4,017.8 ± 511.6[a] | 250.3 ± 26.0[a] | 27.3 ± 4.1 |
| NSD-1015 | 174.3 ± 41.5 | 917.2 ± 102.1[a] | 169.2 ± 40.6 | 89.8 ± 20.6[b] | 4,482.4 ± 568.6[a] | 476.5 ± 54.9[a] | 95.7 ± 12.6[b] |
| CORT, DMI | 160.9 ± 9.6 | 2,200.8 ± 318.7[a,b] | 876.7 ± 124.7[a,b] | 94.6 ± 6.0[a,b] | 8,127.1 ± 923.8[a,b] | 2,173.6 ± 282.4[a,b] | 462.1 ± 52.7[a,b] |
| CORT, DMI, YOH | 205.2 ± 44.1 | 6,733.8 ± 1,195.7[a–c] | 2,283.4 ± 508.2[a–c] | 107.7 ± 24.3[b] | 16,492.3 ± 3,351.9[a–c] | 4,270.3 ± 627.3[a–c] | 477.5 ± 55.6[a,b] |

*Note:* Values represent the mean ± SE of 4–6 determinations and are adjusted to 100 mg of tissue wet weight. NE was separated on Sep-Pak C-18 cartridges as described in Reference 4. Significant differences, $P < .05$.

[a] Significantly different from the preceding basal value.
[b] Significantly different from control, no drugs.
[c] Significant difference between results with CORT, DMI, YOH, and those with CORT, DMI.

*Source:* From Hunter, L.W. et al., *J. Neurochem.*, 59, 977, 1992. With permission.

**TABLE 7.3**
**DOPA Overflow (fmol/mg) by Period in Dog Portal Vein Superfused and Stimulated *In Vitro***

| Treatment | 1 (Basal) | 2 (2 Hz) | 3 (Post 2 Hz) | 4 (Basal) | 5 (10 Hz) | 6 (Post 10 Hz) | 7 (Post 10 Hz) |
|---|---|---|---|---|---|---|---|
| Control, no drugs | 13.1 ± 4.1 | 14.1 ± 6.5 | 23.6 ± 3.6 | 16.8 ± 6.1 | 36.3 ± 6.8 | 63.4 ± 7.8[a] | 46.0 ± 7.8[a] |
| $Ca^{2+}$-free | 13.9 ± 4.2 | 16.0 ± 2.9 | 10.8 ± 4.3[b] | 13.9 ± 4.0 | 14.7 ± 3.6[b] | 9.3 ± 3.8[b] | 8.4 ± 2.8[b] |
| α-MPT | ND[b] | 7.8 ± 4.8 | ND[b] | ND[b] | 27.3 ± 13.0 | ND[b] | ND[b] |
| NSD-1015 | 123.3 ± 27.6[b] | 178.6 ± 28.6[b] | 286.6 ± 45.0[a,b] | 116.1 ± 32.4[b] | 348.5 ± 63.7[a,b] | 520.1 ± 55.9[a,b] | 422.3 ± 109.3[a,b] |
| CORT, DMI | 20.4 ± 6.6 | 18.0 ± 6.1 | 25.2 ± 6.6 | 20.5 ± 7.1 | 39.5 ± 8.7 | 64.6 ± 9.8[a] | 45.7 ± 8.8[a] |
| CORT, DMI, YOH | 20.9 ± 1.8 | 44.5 ± 2.4[a-c] | 52.5 ± 3.5[a-c] | 24.8 ± 2.9 | 71.9 ± 7.0[a-c] | 101.2 ± 11.5[a-c] | 52.7 ± 5.8[a] |

*Note:* Values represent the mean ± SE of 4–6 determinations and are adjusted to 100 mg of tissue wet weight. DOPA was separated on Sep-Pak C-18 cartridges as described in Reference 4. Significant differences, $P < .05$.

[a] Significantly different from the preceding basal value.
[b] Significantly different from control, no drugs.
[c] Significant difference between results with CORT, DMI, YOH, and those with CORT, DMI. ND, not detected.

*Source:* From Hunter, L.W. et al., *J. Neurochem.*, 59, 975, 1992. With permission.

In control veins, the total overflow plus the evoked releases of DOPA were high relative to the tissue content and added up to more than one half of that in the tissue at the end of the experiment [4]. In contrast, total overflows plus evoked releases of NE were only about one tenth of tissue content [4].

## 7.5 CONCLUSION

Our studies showed that DOPA was released from three sites in the peripheral nervous system. At all three sites, the release was $Ca^{2+}$-dependent. DOPA release was frequency-dependent from the noradrenergic terminals in the blood vessels, and in the sympathetic ganglia, the release was dependent on the concentration of $K^+$. A difference between the three sites was the time course of the DOPA release. In the adrenal gland and in the sympathetic ganglia, the release of DOPA was coincident with the release of the accepted neurotransmitters, but in the blood vessels, DOPA release was delayed in comparison with the release of NE. The washout of DOPA in the adrenal gland was coincident with that of the accepted hormones and neuromodulatory peptides but not with the catecholamine metabolites. In the ganglia, the washout of DOPA was coincident with that of DA thought to be a neurotransmitter in the SIF cells. These similarities of washout suggest a similar cellular or subcellular localization of DOPA with the neurotransmitters.

Thus, in these peripheral tissues, DOPA fits some of the criteria to be a neurotransmitter, i.e., presence, synthesis, metabolism, and physiologic release [1]. Data are not yet available concerning physiologic response and antagonism [1], the other criteria indicative of a neurotransmitter role. Studies addressing these possibilities must await the ready availability of stable DOPA antagonists, e.g., DOPA cyclohexyl ester [9]. A neuromodulatory role for DOPA has been demonstrated in rat vas deferens [10]. In isolated superfused preparations of this tissue, DOPA in high concentrations (30 $\mu M$, similar to those in plasma of patients taking levodopa [11]) was shown to increase the basal but not the evoked releases of NE [10]. It is of interest that basal overflow, but not evoked release of NE, was increased in our NSD-1015 experiments in samples in which DOPA effluxes and releases were approximately tenfold higher than in the controls (Table 7.2 and Table 7.3). However, it is possible that NE concentrations were high because NSD-1015, a hydrazine, had a hitherto unrecognized property to inhibit monoamine oxidase as well as AAAD [4]. In support of this possibility is the fact that concentrations of DHPG in perfusates were greatly reduced in the presence of NSD-1015 [4].

The $Ca^{2+}$-dependency of DOPA release suggests an exocytotic release from vesicles, and the similar washouts of DOPA with DA in the ganglia and with E, NE, and DA in the adrenal gland suggest similar cellular localizations. However, DOPA is synthesized in the neuroplasm in a reaction catalyzed by TH. This enzyme is regulated in part by the neuroplasmic $Ca^{2+}$ concentration. The influx of $Ca^{2+}$ that occurs with nerve depolarization would be expected to increase the activity of TH and thus increase DOPA synthesis. However, in rat striatal slices, enhancement of DOPA overflow and increases in TH activity elicited by high $K^+$ depolarization were not temporally related [12]. In bovine adrenal medulla, the activation of cells with acetylcholine resulted in releases of catecholamines and increases in TH activity that occurred concurrently [13].

Apart from a possible role as a neurotransmitter, DOPA is a precursor of DA and NE. It has been calculated [4] that in the portal vein, about 80% of the DOPA synthesized in neurons is used as a substrate for synthesis of DA and NE, about 8% overflows from the tissue, and what remains in the tissue represents about 14%.

It is not known whether DOPA is localized in different cells from the classical neurotransmitters or in different cellular compartments of the same cell. It is possible that in the sympathetic ganglia both DOPA and DA might be localized in the SIF cells because the washout profiles of the two compounds were so similar. However, we were not able to evoke release of either DOPA or DA by muscarinic stimulation, in contrast to the findings previously reported in rats [8]. These differences might be attributable to differences between species.

Localization within vesicles is necessary for exocytotic release but is not a prerequisite for consideration for a neurotransmitter role for DOPA. Nitric oxide, a compound with multiple neuromodulatory roles, is produced in the neuroplasm and is not stored in vesicles [14]. However, it is not clear how DOPA would leave the nerve cell if not by an exocytotic release, although it is possible that DOPA could leave by a reversal of the amino acid uptake system.

DOPA is present in plasma of many species, including man [15]. The concentrations of DOPA in plasma greatly exceed the concentrations of NE, E, or DA. It is suggested [16,17] that much, but not all, of this DOPA originates from releases from the peripheral sympathetic nervous system similar to the releases described herein.

## ACKNOWLEDGMENTS

These studies were supported in part by National Institutes of Health Grants NS 17858, HL 23217, GM 41797, and HL 07269 and by the Mayo Foundation.

## REFERENCES

1. Misu, Y., Kitahama, K., and Goshima, Y., L-3,4-Dihroxyphenylalanine as a neurotransmitter candidate in the central nervous system, *Pharmacol. Ther.,* 97, 117, 2003.
2. Chritton, S.L. et al., Adrenomedullary secretion of DOPA, catecholamines, catechol metabolites and neuropeptides, *J. Neurochem.,* 69, 2413, 1997.
3. Kristensen, E.W. et al., Precursors and metabolites of norepinephrine in sympathetic ganglia of the dog, *J. Neurochem.,* 54, 1782, 1990.
4. Hunter, L.W., Rorie, D.K., and Tyce, G.M., Dihroxyphenylalanine and dopamine are released from portal vein together with norepinephrine and dihroxyphenylglycol during nerve stimulation, *J. Neurochem.,* 59, 972, 1992.
5. Tyce, G.M. et al., Effluxes of 3,4-dihroxyphenylalanine, 3,4-dihroxyphenylglycol and norepinephrine from four blood vessels during nerve stimulation, *J. Neurochem.,* 84, 833, 1995.
6. Chritton, S.L. et al., Nicotinic- and muscarinic-evoked release of canine adrenal catecholamines and peptides, *Am. J. Physiol.,* 260, R589, 1991.

7. Bell, C. and McLachlan, E.M., Dopaminergic neurons in sympathetic ganglia of the dog, *Proc. R. Soc. Lond. B. Bio.,* 215, 175, 1982.
8. Lutold, B.E., Karoum, F., and Neff, N.E., Activation of rat sympathetic ganglia SIF cell dopamine metabolism by muscarinic agonists, *Eur. J. Pharmacol.,* 54, 21, 1979.
9. Arai, N. et al., DOPA cyclohexyl ester, a competitive DOPA antagonist, protects glutamate release and resultant delayed neuron death by transient ischemia in hippocampus CA1 of conscious rats, *Neurosci. Lett.,* 299, 213, 2001.
10. Dayan, L. and Finburg, J.P.M., L-DOPA increases noradrenaline turnover in central and peripheral nervous systems, *Neuropharmacology,* 45, 524, 2003.
11. Dousa, M.K. et al., L-DOPA biotransformation: correlations of dosage, erythrocyte catechol *O*-methyltransferase and platelet SULT1A3 activities with metabolic pathways in Parkinsonian patients, *J. Neural Transm.,*110, 899, 2003.
12. Goshima, Y., Kubo, T., and Misu,Y., Transmitter-like release of endogenous 3,4-dihroxyphenylalanine from rat striatal slices, *J. Neurochem.* 50, 1725, 1988.
13. Tsutsui, M. et al., Correlation of activation of $Ca^{2+}$/calmodulin-dependent protein kinase 11 with catecholamine secretion and tyrosine hydroxylase activation in cultured bovine adrenal medullary cells, *Mol. Pharmacol.,* 46, 1041, 1994.
14. Snyder, S.H. and Bredt, D.S., Nitric oxide as a neuronal messenger, *Trends Pharmacol. Sci.,*12, 125, 1991.
15. Dousa, M.K., and Tyce, G.M., Free and conjugated plasma catecholamines, DOPA, and 3-O-methyldopa in humans and in various animal species, *Proc. Soc. Exp. Biol. Med.,* 188, 427, 1988.
16. Goldstein, D.S., Eisenhofer, G., and Kopin, I.J., Sources and significance of plasma levels of catechols and their metabolites in humans, *J. Pharmacol. Exp. Ther.,* 306, 800, 2003.
17. Eisenhofer, G., Smolich, J.J., and Esler, M.D., Increased cardiac production of dihydroxyphenylalanine during sympathetic stimulation in anaesthetized dogs, *Neurochem. Int.* 21, 37, 1992.

# Part IV

## Pharmacology

# 8 Presynaptic Responses to Levodopa, Suggesting the Existence of DOPA Recognition Sites

*Yoshio Goshima and Yoshimi Misu*

**CONTENTS**

## 8.1 INTRODUCTION

Since the 1950s [1,2], exogenously applied L-3,4-dihydroxyphenylalanine (levodopa) has been traditionally thought to be an inert amino acid that alleviates the symptoms of Parkinson's disease via its conversion to dopamine by the enzyme aromatic L-amino acid decarboxylase (AADC) [3,4]. In contrast, since 1986 [5–7], we have proposed that endogenous L-3,4-dihydroxyphenylalanine (DOPA) is a neurotransmitter and/or neuromodulator in the central nervous system (CNS), in addition to being a precursor of dopamine [8–10].

Recent evidence suggests that DOPA fulfills several criteria such as biosynthesis, metabolism, active transport, existence, physiological release, competitive antagonism, and physiological or pharmacological responses including interactions with the other neurotransmitter systems [11,12], which must be satisfied before a compound is accepted as a neurotransmitter [13].

The impulse-evoked neurotransmitter release is not only sensitive to an $Na^+$ channel blocker, tetrodotoxin (TTX), but also depends on extracellular $Ca^{2+}$ [14,15]. It has been well documented that the release of neurotransmitter is regulated by various presynaptic receptors [16–27]. In the course of study on presynaptic regulation of endogenous catecholamine release [28,29], we found that electrical field stimulation evoked neurotransmitter-like release of DOPA from rat striatal slices [7]. If DOPA is a neurotransmitter, levodopa or DOPA itself should produce pre- and/or postsynaptic responses through acting on specific recognition sites for DOPA. In hypothalamic slices, nanomolar levodopa facilitates, depending on concentration, the release of noradrenaline evoked by electrical field stimulation. This facilitatory action of levodopa is even seen in the presence or absence of an inhibitor of AADC, and antagonized by (–)-propranolol but not by the (+)-isomer [5]. This action is also antagonized by $\beta_1$- and $\beta_2$-selective adrenoceptor antagonists [30]. Thus, levodopa facilitates noradrenaline release via activation of presynaptic $\beta_1$- and $\beta_2$-adrenoceptors. However, this action of levodopa on the release of noradrenaline has turned out to be due to an indirect action on β-adrenoceptors. In fact, this facilitation is antagonized by a β-adrenoceptor antagonist in a noncompetitive manner [31]. In addition, picomolar levodopa potentiates β-agonist-induced facilitation of the noradrenaline release [32]. In contrast, micromolar levodopa inhibits the noradrenaline release via presynaptic dopamine $D_2$ receptors in the presence of AADC inhibitors [5]. The levodopa-induced inhibition of noradrenaline release is antagonized stereoselectively by dopamine $D_2$ and/or $D_3$ antagonist sulpiride. These biphasic responses to nanomolar and micromolar levodopa are independent of dopamine agonists. Nanomolar dopamine and apomorphine inhibit the noradrenaline release via presynaptic dopamine $D_2$ receptors, whereas micromolar dopamine and apomorphine facilitate it via presynaptic β-adrenoceptors [33]. Some supporting evidence for a presynaptic effect of levodopa *in vivo* comes from studies with positron emission tomography in striata of anesthetized rhesus monkeys [34,35]. By the measure of the levodopa-induced facilitation of noradrenaline release, structure–activity relationships for levodopa were studied in rat hypothalamic slices [31]. We found that DOPA methyl ester (DOPA ME) is a potent competitive antagonist for levodopa in this continuously superfused *in vitro* system. On the other hand, L-*threo*-dihydroxyphenylserine (L-*threo*-DOPS) is an agonist similar to levodopa [36]. DOPA ME, however, is a prodrug for levodopa and is readily hydrolyzed. Then, we explored to find a more potent competitive antagonist by measure of depressor responses to levodopa microinjected into depressor sites of the nucleus tractus solitarii (NTS), the gate of baroreflex neurotransmission and central regulation of arterial blood pressure in the lower brainstem [9–12]. It was evident that DOPA cyclohexyl ester (DOPA CHE) is the most potent and relatively stable competitive antagonist among DOPA ester compounds [37,38]. Competitive antagonism suggests the existence of DOPA recognition sites. In this chapter, we also survey that by uptake and binding studies, DOPA recognition sites appear to differ from DOPA transport sites, dopamine $D_1$- and $D_2$-receptors, $\alpha_2$- and $\beta$-adrenoceptors, $\gamma$-aminobutyric acid (GABA) receptors, and ionotropic glutamate receptors.

## 8.2 EFFECTS OF LEVODOPA ON CATECHOLAMINE RELEASE

It is generally accepted that the effects of levodopa are mediated through its conversion to dopamine by AADC [39–48]. This conversion initiates the resultant phenomena such as accumulation of striatal dopamine in rats [43] and humans [42], displacement of [$^3$H]-dopamine [39], impulse-evoked release of dopamine [40,45] from *in vitro* and *in vivo* rat brain, and increases in the content of 3-methoxytyramine, a dopamine metabolite formed in the synaptic cleft, in the rat striatum [43]. In general, these actions should be blocked by the presence of AADC inhibitors and mimicked by dopamine agonists [41]. The idea that levodopa is an inert amino acid precursor was challenged by measuring endogenous noradrenaline and dopamine released from rat brain slices using high-performance liquid chromatography with electrochemical detection [5,6,30–32].

In superfused rat hypothalamic slices, the effects of levodopa on the spontaneous and evoked release of noradrenaline and dopamine were comparatively studied in the absence and presence of an AADC inhibitor [5].

Under intact AADC activity, the dose–response relationship for the effects of levodopa (0.01 to 10 $\mu M$) on the stimulation-evoked release of noradrenaline shows a biphasic pattern, facilitation at 0.1 $\mu M$, and no effect at 1 and 10 $\mu M$. However, no effects on the spontaneous release of noradrenaline are observed. Levodopa, on the other hand, produces dose-dependent increases in the spontaneous release of dopamine. There is a tendency for levodopa at 0.1 $\mu M$ to increase the evoked release of dopamine and, at 10 $\mu M$, this increase is significant. Only the highest concentration of levodopa (10 $\mu M$) increases the tissue content of dopamine, but it produces no effect on the tissue content of noradrenaline.

In the presence of 10-$\mu M$ *p*-bromobenzyloxyamine (NSD-1055), a central AADC inhibitor, levodopa-induced increases in the spontaneous release and the tissue content of dopamine are prevented. As shown in Figure 8.1, levodopa at 0.1 $\mu M$ again produces facilitation of the evoked release of noradrenaline and reveals significant facilitation of the release of dopamine. In contrast, levodopa at 1 $\mu M$ reveals inhibition of the evoked

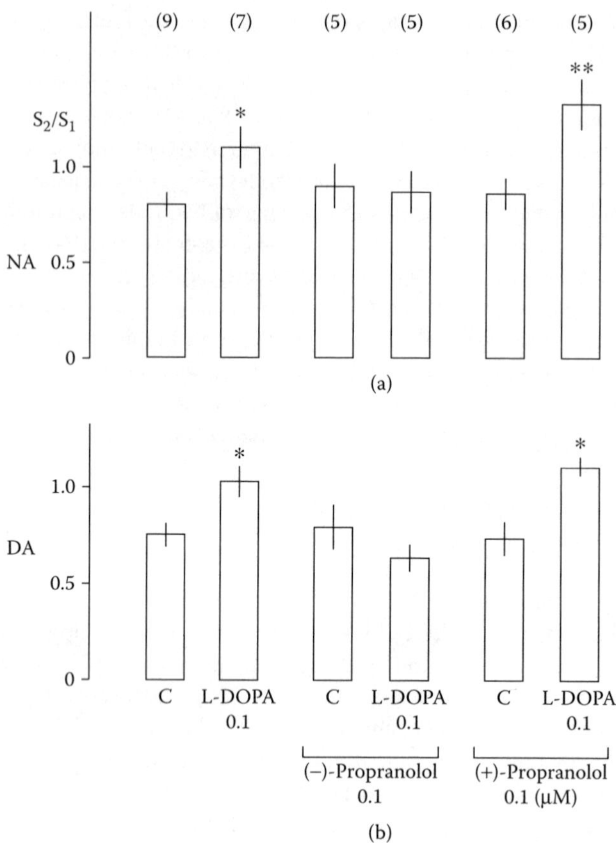


**FIGURE 8.1** Stereoselective antagonism by propranolol against levodopa-induced facilitation of the evoked release of endogenous noradrenaline (NA) and dopamine (DA) from rat hypothalamic slices in the presence of cocaine and NSD-1055. Cocaine (20 μ*M*) and NSD-1055 (10 μ*M*) were added at the start of superfusion. Electrical field stimulation by biphasic impulses of 5 Hz, 2 ms, and 30 mA for 3 min was done twice at 60 ($S_1$) and 90 ($S_2$) min after the start of superfusion. Ordinates show ratios of amounts of noradrenaline (a) and dopamine (b) released during the $S_2$ and $S_1$ periods of stimulation, $S_2/S_1$. Levodopa at 0.1 μ*M* was added to the medium 15 min before $S_2$. (−)- Propandol or (+)-propranolol at 0.1 μ*M* was added from the start of superfusion throughout the experiments. Each column represents the mean and vertical line SE of number of estimations shown in parentheses. Statistical significance: $^*P < .05$ and $^{**}P < .01$, compared to corresponding control (C) (Student's *t*-test). (From Goshima, Y., Kubo, T., and Misu, Y., Biphasic actions of L-DOPA on the release of endogenous noradrenaline and dopamine from rat hypothalamic slices, *Br. J. Pharmacol.*, 89, 229, 1986. With permission.)

release of both noradrenaline and dopamine, without increasing the spontaneous release [5]. Levodopa (0.1 μ*M*)-induced facilitation of the evoked release of noradrenaline and dopamine is antagonized by pretreatment with (−)-propranolol at 0.1 μ*M*, but no effects on the spontaneous release are observed. Significant dissociation is seen between the actions of the (−)- and (+)-isomers of propranolol, thereby indicating stereoselective antagonism by propranolol against levodopa-induced facilitation of the evoked release of noradrenaline and dopamine (Figure 8.1). In contrast, levodopa (1 μ*M*)-induced inhibition of the evoked release of noradrenaline and dopamine is antagonized by pretreatment with (–)-sulpiride at 1 n*M*, without an effect on the spontaneous release. Again, significant dissociation is seen between the actions of the (–)- and (+)-isomers of sulpiride. These results demonstrate that the biphasic actions of levodopa are mediated via the activation of stereoselective presynaptic facilitatory β-adrenoceptors and inhibitory dopamine receptors. The presynaptic nature of the effects of levodopa is further supported by the findings that the impulse-evoked release of noradrenaline and dopamine is $Ca^{2+}$ dependent and TTX sensitive [28,49]. Both types of presynaptic receptors appear to exist on the noradrenergic neuron terminals [28,33,50] and on the dopaminergic neuron terminals [49,51] in the rat hypothalamus, respectively. These biphasic regulatory actions of levodopa are the first to provide evidence for the actions due to levodopa in its own right but not due to its bioconversion to dopamine by AADC. Facilitation of the evoked release of noradrenaline induced by levodopa at 0.1 μ*M* under intact activity of AADC is neither blocked by AADC inhibition nor mimicked by the same concentration of dopamine and apomorphine [33]. Inhibition of the evoked release of noradrenaline induced by levodopa at 1 μ*M* under inhibition of AADC is, for example, different from inhibition of cardiac acceleration in response to sympathetic nerve stimulation elicited by dopamine converted from levodopa acting on presynaptic dopamine receptors [52]. This type of inhibition is blocked by AADC inhibition and mimicked by dopamine. In addition, the biphasic regulatory actions of levodopa are not consistent with the generally accepted idea that the actions of levodopa are manifested through its conversion to dopamine [39–44]. Levodopa, however, elicits dose-dependent increases in the spontaneous release of dopamine, and the highest concentration of levodopa (10 μ*M*) increases the evoked release and tissue content of dopamine under intact activity of AADC. These increases are probably due to the conversion of levodopa to dopamine and reflect intact AADC activity in the slices and may explain the finding that high concentrations of levodopa induce displacement of [$^3$H]-dopamine in rat brain slices [40]. The conversion of levodopa to dopamine and the probable resultant increase in the amounts of dopamine and noradrenaline available for impulse-evoked release may interfere with the manifestation of the inhibitory actions of levodopa, at a micromolar concentration, on the release of both catecholamines in the absence of NSD-1055. In fact, levodopa at 1 μ*M* fails to inhibit the release of noradrenaline, but inhibits the release under the inhibition of AADC [5]. Furthermore, in general, the impulse-evoked release of a neurotransmitter cannot be exactly estimated under the conditions accompanying modifications of the spontaneous release of the neurotransmitter. This idea may also reflect the negative finding concerning the facilitatory effect of levodopa (0.1 μ*M*) on the impulse-evoked release of dopamine in the absence of NSD-1055.

As shown in Table 8.1, similar effects of levodopa on the spontaneous release, the impulse-evoked release, and the tissue content of dopamine are seen in striatal slices as those in hypothalamic slices [6]. Effects of levodopa on the impulse-evoked release of dopamine were also biphasic in striatal slices. Nanomolar levodopa (30 n*M*) facilitates the release of dopamine evoked by electrical field stimulation via presynaptic β-adrenoceptors in the absence and presence of an AADC inhibitor, whereas micromolar levodopa (1 μ*M*) inhibits it under inhibition of AADC via

**TABLE 8.1**
**Effects of Levodopa on Spontaneous and Impulse-Evoked Release of Endogenous Dopamine and the Tissue Content after Superfusion Experiments in the Absence (A) and Presence (B) of Cocaine and NSD-1055 in Rat Striatal Slices**

| Group | Levodopa (μ*M*) | n | $Sp_2/Sp_1$ Ratio | $S_2/S_1$ Ratio | Tissue Content (pmol/mg wet wt) |
|---|---|---|---|---|---|
| A | Control | 8 | 0.66 ± 0.10 | 0.65 ± 0.03 | 31.4 ± 1.8 |
| | 0.03 | 7 | 0.75 ± 0.13 | 0.81 ± 0.07[a] | 38.0 ± 5.9 |
| | 0.1 | 6 | 1.10 ± 0.06[b] | 0.62 ± 0.05 | 29.7 ± 2.3 |
| | 1 | 5 | 4.33 ± 2.00[a] | 0.70 ± 0.12 | 32.1 ± 4.1 |
| | 10 | 5 | 6.08 ± 1.44[b] | 1.37 ± 0.08[b] | 36.8 ± 4.3 |
| | 100 | 5 | 31.5 ± 4.2[b] | 11.3 ± 1.25[b] | 61.9 ± 9.7[b] |
| B | Control | 16 | 0.79 ± 0.04 | 0.73 ± 0.07 | 19.0 ± 1.0 |
| | 0.01 | 9 | 0.73 ± 0.06 | 0.86 ± 0.10 | 18.4 ± 1.1 |
| | 0.03 | 8 | 0.69 ± 0.02 | 1.01 ± 0.09[a] | 19.9 ± 1.4 |
| | 0.1 | 14 | 0.74 ± 0.03 | 0.74 ± 0.06 | 18.3 ± 1.1 |
| | 1 | 9 | 0.76 ± 0.03 | 0.47 ± 0.04[a] | 19.0 ± 1.9 |

*Note:* In A, 0.3-mm slices were superfused with Krebs medium. Electrical field stimulation (5 Hz, 2 ms, 30 mA, 3 min) was done twice 60 ($S_1$) and 90 ($S_2$) min after the start of superfusion. $Sp_2/Sp_1$ and $S_2/S_1$ show the ratio of the spontaneous and evoked release of dopamine immediately before, during, and after the $S_2$ and $S_1$ periods of stimulation, respectively. Levodopa was applied 15 min before $S_2$. In B, the experiments were identical to A, except that 0.7-mm slices were pretreated with 20 μ*M* of cocaine at the start of superfusion and with 10 μ*M* of NSD-1055 15 min before $S_1$. Data are the mean ± SE of n estimations.

[a] *P* < .05.

[b] *P* < .01, compared to corresponding control.

*Source:* From Misu, Y., Goshima, Y., and Kubo, T., Biphasic actions of L-DOPA on the release of endogenous dopamine via presynaptic receptors in rat striatal slices, *Neurosci. Lett.*, 72, 194, 1986. With permission.

presynaptic dopamine $D_2$ receptors. The facilitatory action of levodopa on the evoked release of dopamine is also seen in the striatal slices isolated from mice pretreated with 1-methyl-4-phenyl-1,2,3,6-tetrahydropyridine (MPTP), an animal model for Parkinson's disease [53,54]. In this model, levodopa at 30 *M* facilitates the evoked release of dopamine without modifying the striatal tissue content of dopamine. The minimum concentration of levodopa to induce accumulation of tissue dopamine is 100 n*M*, thereby suggesting that the facilitation of the evoked release of dopamine is the primary action of levodopa in the striatum isolated from the MPTP-treated Parkinson's model mice. In this regard, it is noteworthy that levodopa inhibits the evoked release of dopamine under the inhibition of AADC [5,6], because the AADC activity in the postmortem brains from patients who suffered from Parkinson's disease has been shown to be lowered compared to the control [42]. Evidence for *in vivo* presynaptic effect of levodopa itself in nonhuman primates has been obtained in studies with positron emission tomography [34,35] (see Chapter 9).].

## 8.3 LEVODOPA-INDUCED FACILITATION OF IMPULSE-EVOKED RELEASE OF NORADRENALINE ANTAGONIZED NONCOMPETITIVELY BY (–)-PROPRANOLOL

The fact that levodopa facilitated the impulse-evoked release of noradrenaline and dopamine via presynaptic β-adrenoceptor led us to investigate the mode of antagonism against the effect of levodopa on the release of noradrenaline. This experiment was performed in the further presence of (–)-sulpiride, so that the inhibition of the evoked release of noradrenaline via presynaptic dopamine $D_2$ receptors by the moderate micromolar concentrations of levodopa was minimized [5,31]. Levodopa at 1 n*M* to 1 m*M* facilitates the evoked release of noradrenaline in a concentration-dependent manner without modifying the spontaneous release. (–)-Propranolol (1, 10, and 100 n*M*) reduces the maximal effect of levodopa in a concentration-dependent manner to 54.8, 31.9, and 10.8% of the original maximum, respectively, without rightward shift of the concentration–release curve (Figure 8.2). It is therefore evident that (–)-propranolol antagonizes noncompetitively the levodopa-induced facilitation of the evoked release of noradrenaline. This finding indicates that the recognition site for levodopa relevant to its action differs from presynaptic β-adrenoceptors. This idea is consistent with the finding that levodopa fails to displace the specific binding of [$^3$H]-dihydroalprenolol in membrane preparations from rat brain [31,55] (Table 8.2).

## 8.4 LEVODOPA POTENTIATES PRESYNAPTIC Β- AND $A_2$- BUT NOT ANGIOTENSIN II RECEPTORS TO REGULATE THE RELEASE OF NORADRENALINE FROM RAT HYPOTHALAMIC SLICES

Although levodopa produces facilitatory and inhibitory actions via presynaptic β- and dopamine $D_2$ receptors, respectively, levodopa hardly acts directly on these catecholaminergic receptors (Table 8.2). We therefore sought to determine whether levodopa modulates the functions of these presynaptic receptors. A noneffective concentration of levodopa at 10 pM stereoselectively facilitates isoproterenol-induced increase in the evoked release of noradrenaline [32]. This facilitation is antagonized by DOPA ME completely to the level of increase induced by isoproterenol alone, thereby indicating that levodopa potentiates facilitatory action of isoproterenol on the noradrenaline release. As adrenaline probably acts as an endogenous agonist for presynaptic β-adrenoceptor in rat hypothalamus [50,56], it seems likely that levodopa alone can facilitate the release of noradrenaline or dopamine through modifying the actions of adrenaline released in response to nerve stimulation [19,29]. Picomolar levodopa also potentiates activity of presynaptic inhibitory $\alpha_2$-adrenoceptors, but not those of facilitatory angiotensin II receptors [57]. Thus, the selective analogy is not seen between these presynaptic facilitatory

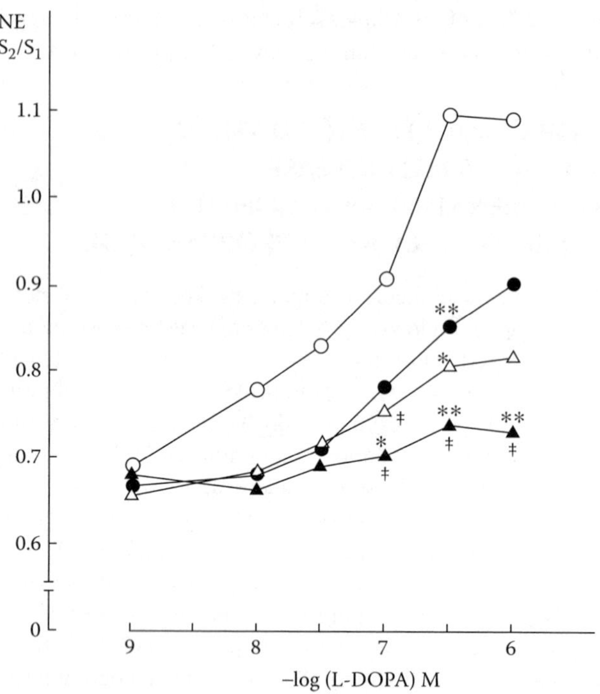


**FIGURE 8.2** Noncompetitive antagonism by (–)-propranolol against levodopa-induced facilitation of the evoked release of noradrenaline (NE) from rat hypothalamic slices in the presence of cocaine (20 μ*M*), NSD-1015 (20 μ*M*), and (–)-sulpiride (1 n*M*). Electrical field stimulation by biphasic pulses of 2 Hz, 2 ms, and 25 V for 3 min was performed twice, 60 ($S_1$) and 90 ($S_2$) min after the start of superfusion. Cocaine, (–)-sulpiride, and NSD-1015 was applied, respectively, 60 min, 60 min, and 20 min before $S_1$. Pretreatment with (–)-propranolol was initiated 10 min before $S_1$ and continued throughout the experiments. Levodopa was applied 15 min before $S_2$. Figure shows concentration–release curves for levodopa in the absence (○, control) and presence of (–)-propranolol at 1 n*M* (●), 10 n*M* (△), and 100 n*M* (▲). Ordinate shows the ratio of the amount of noradrenaline released during the $S_2$ and $S_1$ periods of stimulation, $S_2/S_1$. Abscissa shows –log (L-DOPA) concentrations. Data shown are means of four to seven estimations. The SE bars are omitted for clarity. **P* < .05, ***P* < .01 (Student's *t*-test), and ‡*P* < .05 (Dunnett's test), compared to corresponding control. (From Goshima, Y., Nakamura, S., and Misu, Y., L-Dihydroxyphenylalanine methyl ester is a potent competitive antagonist of the L-dihydroxyphenylalanine-induced facilitation of the evoked release of endogenous norepinephrine from rat hypothalamic slices, *J. Pharmacol. Exp. Ther.*, 258, 466, 1991. With permission.)

angiotensin II receptors and β-adrenoceptors. Levodopa appears to be a mother compound not only as a precursor for catecholamines but also as a facilitator for a catecholaminergic presynaptic process that is commonly involved in signal transduction pathways mediated via inhibitory dopamine $D_2$ receptors, inhibitory $\alpha_2$-adrenoceptors, and facilitatory β-adrenoceptors. It remains to be determined, however, how levodopa signaling modulates the functions of these presynaptic receptors.

**TABLE 8.2**
**Effects of Levodopa-Related Compounds (1 m*M*) on Specific Binding of Tritiated Catecholaminergic Receptor Ligands**

| Drugs | [$^3$H]Dihydroalprenolol | [$^3$H]SCH23390 | [$^3$H]Spiperone | [$^3$H]Rauwolscine |
|---|---|---|---|---|
| Dopamine | n.t. | 0.3 ± 0.4 (2.68 μ*M*) | 12 ± 2 (2.4 μ*M*) | n.t. |
| Levodopa | No effect[a] | 90 ± 3 | 98 ± 16 | n.t. |
| DOPA ME | No effect[a] | 73 ± 6 (3.22 m*M*) | 61 ± 10 (1.93 m*M*) | (Ki = 40 μ*M*) |
| DOPA CHE | n.t. | 43 ± 2 (0.60 m*M*) | 68 ± 6 (1.86 m*M*) | n.t. |
| DOPA CPE | n.t. | 54 ± 9 (1.12 m*M*) | 65 ± 7 (1.95 m*M*) | n.t. |
| DOPA CPDME | n.t. | 20 ± 2 (0.31 m*M*) | 37 ± 11 (0.73 m*M*) | n.t. |

*Note:* Data represent the mean ± SE expressed as a percentage of the control binding (n = 3). The concentrations shown in parentheses are $IC_{50}$ values, and n.t. stands for not tested.

[a] Levodopa and DOPA ME tested did not affect the binding of [$^3$H]dihydroalprenolol.

*Source:* From Goshima, Y., Nakamura, S., and Misu, Y., L-Dihydroxyphenylalanine methyl ester is a potent competitive antagonist of the L-dihydroxyphenylalanine-induced facilitation of the evoked release of endogenous norepinephrine from rat hypothalamic slices, *J. Pharmacol. Exp. Ther.*, 258, 466, 1991; Furukawa, N. et al., L-DOPA cyclohexyl ester is a novel potent and relatively stable competitive antagonist against L-DOPA among several L-DOPA ester compounds, *Jpn. J. Pharmacol.*, 82, 40, 2000. With permission.

## 8.5 EFFECTS OF LEVODOPA ON ACETYLCHOLINE RELEASE

The effects of levodopa on the release of neurotransmitters other than catecholamines also have been observed in various experimental systems [58–61]. The effects of levodopa on peripheral cholinergic neurotransmission were investigated with intracellular recordings from submucous plexus neurons of the guinea pig cecum [60]. Levodopa at 30 n*M* augments the amplitude of fast excitatory postsynaptic potentials (EPSPs), but does not affect depolarization elicited by puff application of acetylcholine (ACh). The augmenting effect of levodopa on the fast EPSPs is blocked by DOPA ME. On the other hand, the fast EPSPs are depressed by 10-μ*M* levodopa, but are transiently augmented after rinsing the drug. DOPA ME does not affect the inhibitory action of levodopa on the fast EPSPs, but antagonizes the potentiation unmasked following rinsing levodopa. Consistently, levodopa at 30 nM facilitates the transient increase in the intracellular $Ca^{2+}$ concentrations evoked by the somatic action potential ($\Delta[Ca^{2+}]$AP), but at 10 μ*M* inhibits the increase in $\Delta[Ca^{2+}]AP$ [60]. This finding suggests that levodopa at low concentrations enhances the transient increase in $\Delta[Ca^{2+}]AP$, increasing the ACh release, but at high concentrations diminishes $\Delta[Ca^{2+}]AP$, inhibiting the peripheral cholinergic neurotransmission.

## 8.6 SPONTANEOUS RELEASE OF ACh INHIBITED BY LEVODOPA FROM THE STRIATUM OF EXPERIMENTAL PARKINSON'S MODEL RATS

The striatal cholinergic system has been implicated in the pathophysiology of movement disorders such as Parkinson's disease, but the cellular mechanisms underlying cholinergic neuronal function are still unknown [62]. We studied the effects of levodopa on the basal release of ACh from the striatum of experimental Parkinson's disease model rats. In rats given nigrostriatal dopaminergic hemilesion with 6-hydroxydopamine (6-OHDA) injected into the unilateral medial forebrain bundle, local perfusion of levodopa (10 to 100 n*M*) under intact AADC activity inhibits the basal release of ACh during ipsilateral striatal microdialysis, compared with sham-operated rats [59] (Figure 8.3). The effect of levodopa is stereoselective, as dextrodopa produces no effect. The inhibition elicited by levodopa is independent of dopamine converted from levodopa, because dopamine (100 n*M*) administered at the same dose to the highest dose of levodopa elicits no inhibition. The levodopa-induced inhibition of the basal release of ACh is not antagonized by sulpiride, a dopamine $D_2/D_3$ antagonist, whereas quinpirole, a selective dopamine $D_2$ agonist, and 7-hydroxy-*N*,*N*-di-*n*-propyl-2-aminotetralin (7-OHDPAT), a selective dopamine $D_3$ agonist, inhibit the basal release of ACh in a completely sulpiride-sensitive manner [63,64] (Figure 8.4). In addition, a selective dopamine $D_1$ agonist reversely increases the basal release of ACh in a manner sensitive to a dopamine $D_1$ antagonist [63].

## 8.7 VESICULAR GLUTAMATE RELEASED BY LEVODOPA FROM *IN VITRO* STRIATA

Levodopa causes the release of endogenous glutamate from striatal slices with an $ED_{50}$ value of 140 μ*M* [58] in a partially TTX-sensitive and $Ca^{2+}$-dependent manner, which suggests, at least in part, vesicular release of glutamate (Figure 8.5 and Figure 8.6). This release is elicited by levodopa itself, because it is also observed under inhibition of AADC with 3-hydroxybenzylhydrazine (NSD-1015) and is antagonized by DOPA ME in a competitive manner. This effect of levodopa is independent of dopamine, because dopamine (300 μ*M*) elicits no glutamate release. Micromolar levodopa also releases glutamate from cultured striatal neurons in the absence of NSD-1015 [65]. Although intracellular signal transduction mechanisms for some excitatory effects of levodopa are not clear at present, it is highly possible that levodopa-induced glutamate release is involved in neuronal degeneration elicited by levodopa itself.

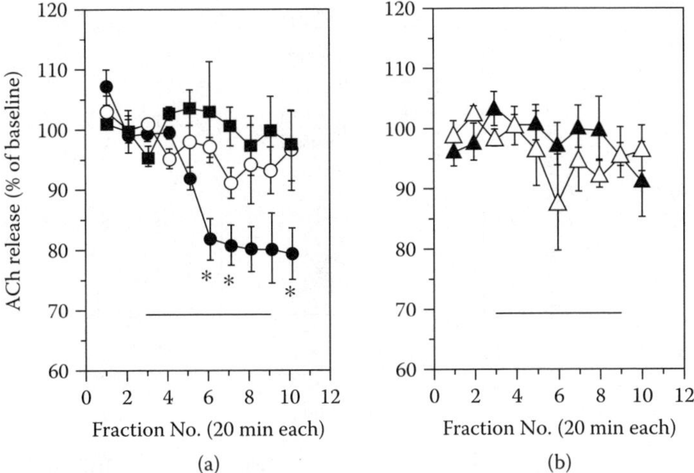


**FIGURE 8.3** Time course of the effects of levodopa and dextrodopa (a) and dopamine (b) on basal ACh release from the striatum of sham-operated and 6-OHDA-lesioned rats. Ringer solution containing eserine sulfate (10 μ*M*) and ascorbate (0.28 m*M*) was preperfused for 120 min, and then perfusates were collected every 20 min. The mean absolute ACh release in four fractions was taken as control (100%). Test drugs, 10-n*M* levodopa (○ in shame-operated, ● in lesioned), 10-n*M* dextrodopa (■ in lesioned), and 100-n*M* dopamine (△ in sham-operated, ▲ in lesioned) were locally infused into the striatum via a probe for further 120 min (6 fractions), as indicated with a horizontal bar in the figure. Statistical significance: $*P < .05$, compared to the corresponding sham-operated (Student's *t*-test). Data are the mean ± SE from 3–6 separate experiments. (From Ueda, H. et al., L-DOPA inhibits spontaneous acetylcholine release from the striatum of experimental Parkinson's model rats, *Brain Res.*, 698, 213, 1995. With permission.)

Primarily cultured striatal neurons are protected by ascorbic acid, an antioxidant, against neuron death elicited by both levodopa and dextrodopa in 3 d in culture and by dopamine in 3 and 10 d in culture [66]. These findings suggest that degradation products generated by autoxidation or enzymatic oxidation of levodopa and/or dopamine into reactive free radicals, DOPA quinones, and/or dopamine quinones appear to be involved in this type of neuron death (see Chapter 2, Chapter 18, and Chapter 19). Furthermore, levodopa elicits both stereoselective and antioxidant-insensitive neurotoxicity in 10 d in culture. The application of TTX, $Ca^{2+}$ omission, and $Mg^{2+}$ addition protect against this kind of neuronal death [65]. The antagonist of the *N*-methyl-D-aspartate (NMDA) receptor ion channel domain (+)-5-methyl-10,11-dihydro-5*H*-dibenzol[a,d]cyclohepten-5,10-imine maleate (MK-801) [66,67] and non-NMDA antagonists [66,68] antagonize neuronal death that results from glutamate release [65]. It appears to be mediated by vesicular release of glutamate [58]. Glutamate can increase $Ca^{2+}$ influx via its receptors [67,68], which, forming the complex of $Ca^{2+}$ with calmodulin, can activate neuronal nitric oxide (NO) synthase to produce neurotoxic NO against the cell and mitochondrial membranes, then leading to neuron death

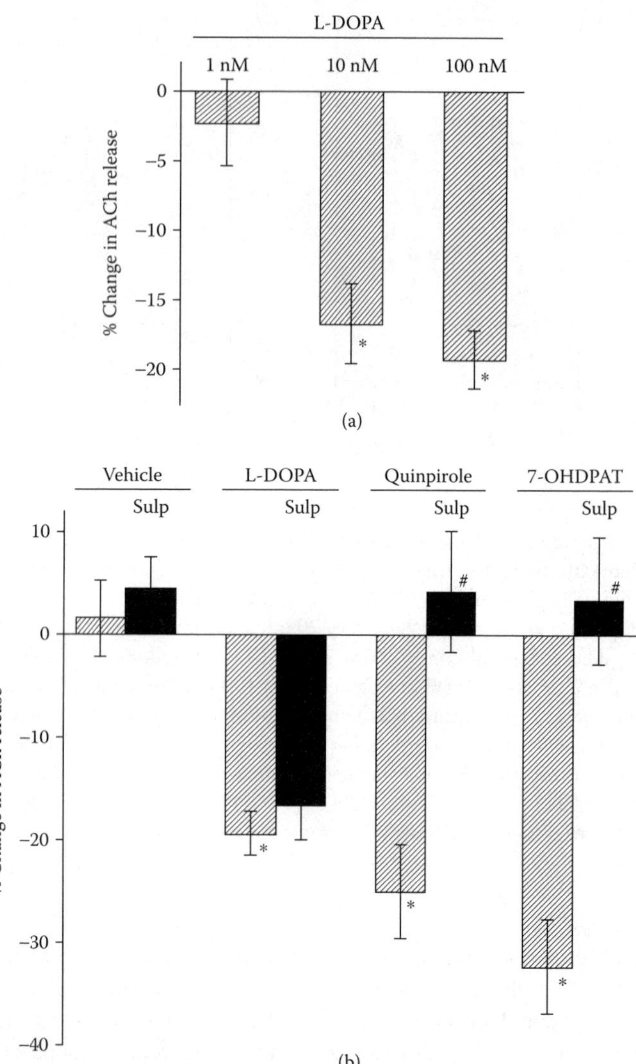


**FIGURE 8.4** Concentration-dependent effects of levodopa (L-DOPA) on basal ACh release from the striatum of 6-OHDA-lesioned rats and lack of antagonism by (–)-sulpiride (Sulp). (a) The average of ACh release per fraction (20 min) from six fractions after the beginning of levodopa perfusion, as percentage change in ACh release. Average of absolute ACh release in four fractions was taken as control (100%) in each experiment. Data are the mean ± SE from 7–10 separate experiments. (b) Effects of (–)-sulpiride on inhibition of ACh release induced by levodopa (0.1 μ*M*), quinpirole (1 μ*M*), and 7-OHDPAT (1 μ*M*). (–)-Sulpiride at 1 μ*M* was infused 60 min prior to the challenge of agonists including levodopa. Data are from 4–10 separate experiments. $*P < .05$, compared with vehicle and $\#P < .05$, compared with each agonist alone (Student's *t*-test). (From Ueda, H. et al., L-DOPA inhibits spontaneous acetylcholine release from the striatum of experimental Parkinson's model rats, *Brain Res.*, 698, 213, 1995. With permission.)

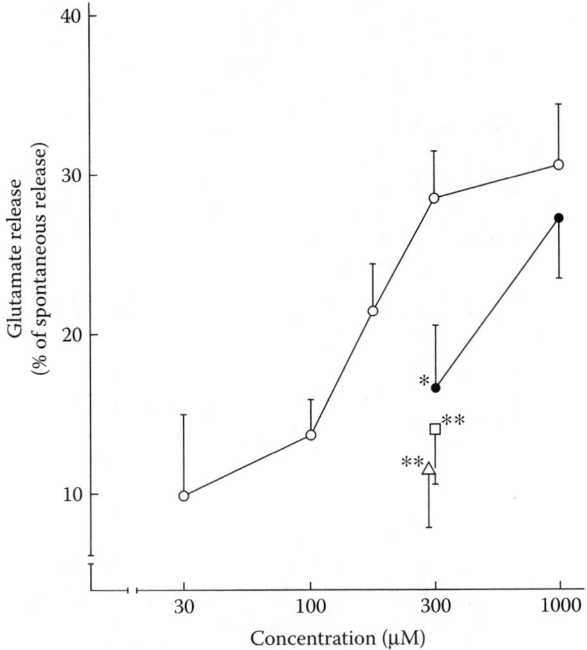


**FIGURE 8.5** Competitive antagonism by DOPA ME (●) against release of endogenous glutamate induced by levodopa (○) from rat superfused striatal slices. Levodopa, dextrodopa (□), and dopamine (△) were applied for 8 min, 72 min after the start of superfusion. Pretreatment with DOPA ME was performed 20 min before levodopa application. Ordinate shows amount of levodopa-induced release of glutamate. Each value represents percentage of absolute value of the spontaneous release estimated in a sample immediately before levodopa application. Values are mean ± SE of at least five determinations. $*P < .05$ and $**P < .01$, compared with levodopa at 300 μ*M* alone (unpaired Student's *t*-test). (From Goshima, Y. et al., L-DOPA induces $Ca^{2+}$-dependent and tetrodotoxin-sensitive release of endogenous glutamate from rat striatal slices, *Brain Res.*, 617, 167, 1993. With permission.)

[69]. The further details on the glutamate release and their pathophysiological relevance will be discussed in Chapters 18 and 20.

## 8.8 STRUCTURE–ACTIVITY RELATIONSHIPS FOR LEVODOPA

If recognition sites for levodopa exist, we can expect to find agonists related to and competitive antagonists for levodopa. Comparative activities of levodopa analogs were determined by using the facilitation of the evoked release of noradrenaline via presynaptic β-adrenoceptors in hypothalamic slices under essentially complete inhibition of AADC. NSD-1015 (20 μ*M*) is expected to inhibit the activity of AADC by 99.6% in a noncompetitive manner [30], and has been utilized as a central AADC inhibitor *in vivo* as well [70].

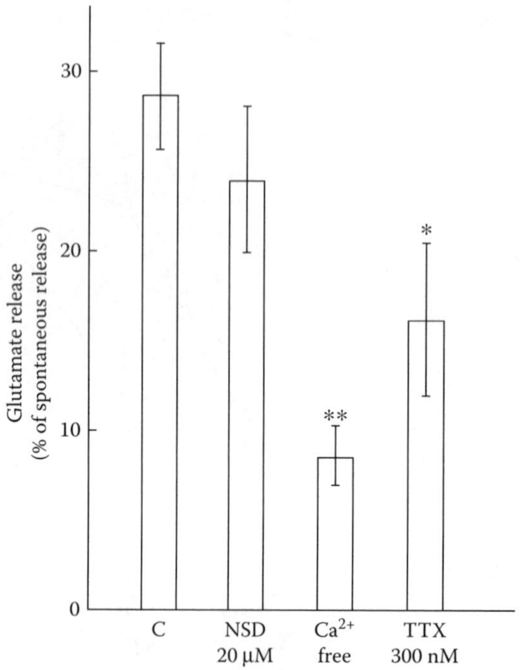


**FIGURE 8.6** Effects of NSD-1015, TTX, and $Ca^{2+}$ deprivation on levodopa (300 μ*M*)-induced glutamate release from rat superfused striatal slices. Pretreatment with NSD-1015 (20 μ*M*), TTX (300 n*M*), or $Ca^{2+}$ deprivation was performed 20 min before levodopa application. Data shown are mean ± SE from at least five estimations. $*P < .05$ and $**P < .01$, compared with levodopa alone (C) (unpaired Student's *t*-test). Other details are as in Figure 8.5. (From Goshima, Y. et al., L-DOPA induces $Ca^{2+}$-dependent and tetrodotoxin-sensitive release of endogenous glutamate from rat striatal slices, *Brain Res.*, 617, 167, 1993. With permission.)

### 8.8.1 First Screening for Compounds Exhibiting Levodopa-Like Action

Levodopa, but not dextrodopa, facilitates the impulse-evoked release of noradrenaline, indicating that the activity of levodopa is stereoselective in nature, in common with many neurotransmitters [31]. In the first screening process, the activities of levodopa analogs were determined alone on the evoked release of noradrenaline. As summarized in Figure 8.7, it is clear that L-phenylalanine, which lacks hydroxy groups at carbon 3 and 4 positions of the aromatic ring of levodopa, produces no facilitation. The effectiveness of levodopa is abolished by substitution for the hydroxy group of carbon 3 with a methoxy group, as in the case of 3-*O*-methyldopa, a metabolite by catechol-O-methyltransferase (COMT) [71], and for the hydroxy groups of carbon 3 and/or 4 with such phosphorylated ester as DOPA phosphate [72]. Substitution for the amino group of the side chain of α-methyl-dopa with a hydrazino group also produces no facilitation, as in the case of carbidopa. 3,4-Dihydroxyphenylacetic acid (DOPAC), which lacks the amino group of the side chain, also produces no facilitation. Thus, the agonistic activity of levodopa

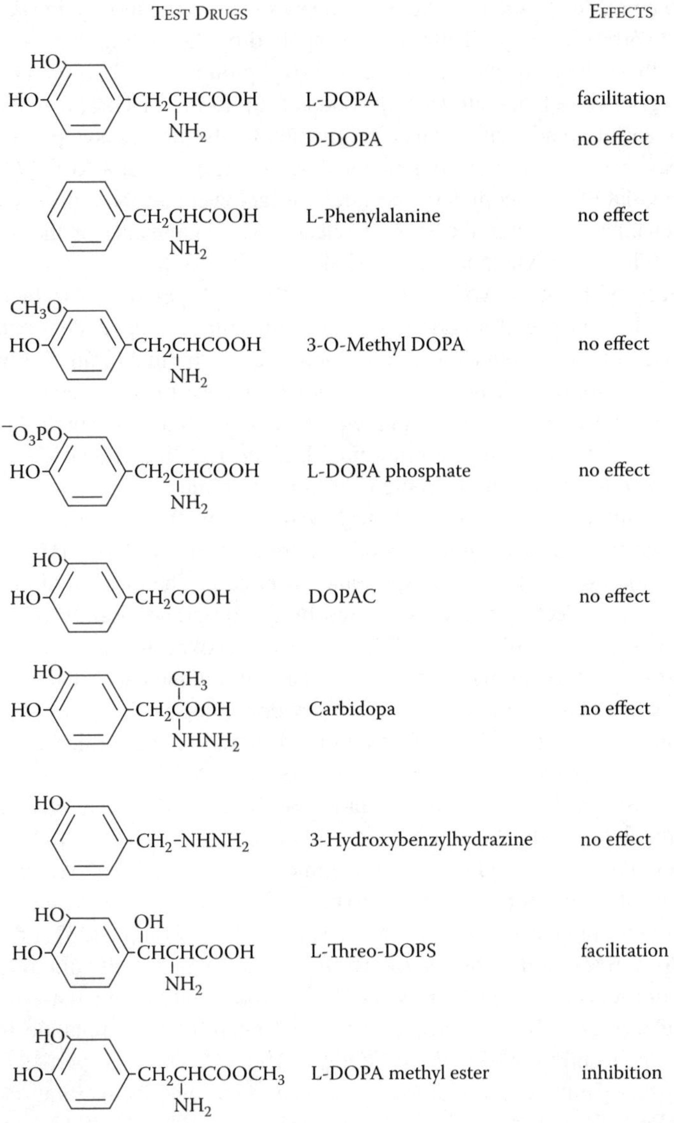

| Test Drugs | | Effects |
|---|---|---|
| HO, HO–$C_6H_3$–$CH_2CH(NH_2)COOH$ | L-DOPA | facilitation |
| | D-DOPA | no effect |
| $C_6H_5$–$CH_2CH(NH_2)COOH$ | L-Phenylalanine | no effect |
| $CH_3O$, HO–$C_6H_3$–$CH_2CH(NH_2)COOH$ | 3-O-Methyl DOPA | no effect |
| $^{-}O_3PO$, HO–$C_6H_3$–$CH_2CH(NH_2)COOH$ | L-DOPA phosphate | no effect |
| HO, HO–$C_6H_3$–$CH_2COOH$ | DOPAC | no effect |
| HO, HO–$C_6H_3$–$CH_2C(CH_3)(NHNH_2)COOH$ | Carbidopa | no effect |
| HO–$C_6H_4$–$CH_2$-$NHNH_2$ | 3-Hydroxybenzylhydrazine | no effect |
| HO, HO–$C_6H_3$–$CH(OH)CH(NH_2)COOH$ | L-Threo-DOPS | facilitation |
| HO, HO–$C_6H_3$–$CH_2CH(NH_2)COOCH_3$ | L-DOPA methyl ester | inhibition |

**FIGURE 8.7** The chemical structures of levodopa (L-DOPA) analogs and summarized effects of the analogs alone on the impulse-evoked release of noradrenaline from rat hypothalamic slices in the presence of cocaine and NSD-1015 in comparison with the facilitation via activation of presynaptic β-adrenoceptors induced by levodopa at 1–100 n*M*. Electrical field stimulation by biphasic pulses of 2 Hz, 2 ms, and 25 V for 3 min was done twice at 60 ($S_1$) and 90 ($S_2$) min after the start of superfusion. Cocaine (20 μ*M*) and NSD-1015 (20 μ*M*) were applied 60 min and 20 min before $S_1$. The analogs were applied 15 min before $S_2$. The effects were evaluated by the ratio of evoked noradrenaline release during $S_2$ and $S_1$ periods of stimulation.

appears to be related to such structural features as its catechol moiety, in addition to the amino and carboxyl groups. This idea is supported by the finding that L-*threo*-DOPS, which has these three constituents and a hydroxyl group at the β position of levodopa, elicits an agonistic activity similar to levodopa [36]. L-*threo*-DOPS, a synthetic amino acid precursor of noradrenaline, has been thought to alleviate the symptoms of Parkinson's disease by its conversion to noradrenaline via the action of AADC [73,74]. However, L-*threo*-DOPS, at picomolar concentrations far lower than those to elicit conversion to noradrenaline, facilitates the evoked release of noradrenaline in the absence and presence of the AADC inhibitor in hypothalamic slices [36].

In slices with intact AADC, L-*threo*-DOPS at 1 p*M* to 1 n*M* facilitated the impulse-evoked release of noradrenaline in a concentration-dependent manner without increases in the spontaneous release and elicited the maximum facilitation at 1 n*M*. At 10 n*M* to 10 μ*M*, however, the evoked release of noradrenaline tended to decrease gradually from the maximum with tendency of increases in the spontaneous release. Only the highest concentration of L-*threo*-DOPS at 100 μ*M* increases the spontaneous release and tissue content of noradrenaline.

Under inhibition of AADC with NSD-1015 (20 μ*M*), the concentration-dependent facilitation of the evoked release of noradrenaline at 1 p*M* to 1 n*M* was similarly observed without increases in the spontaneous release. The maximum facilitation at 1 n*M* was stereoselective and antagonized by (–)-propranolol at 10 n*M*. At 10 n*M* to 1 μ*M* of L-*threo*-DOPA, no facilitation of the evoked release was seen, which differed from the maximum facilitation at 1 n*M*, without increases in the spontaneous release. Involvement of presynaptic inhibitory dopamine $D_2$ receptors [5] and inhibitory $\alpha_2$-adrenoceptors [57] in regulation of noradrenaline release induced by levodopa is also evident in the case of L-*threo*-DOPS. (–)-Sulpiride at 1 n*M* and yohimbine at 10 n*M* restored no facilitation of the evoked release of noradrenaline induced by L-*threo*-DOPS at 100 n*M* completely to the level of the maximum facilitation. Presynaptic inhibitory dopamine $D_2$ receptors and $\alpha_2$-adrenoceptors probably existing on noradrenaline nerve terminals in rat hypothalamic slices are tonically functioning, because (–)-sulpiride (10 to 100 n*M*) alone [33] and yohimbine (10 n*M* to 1 μ*M*) alone [28] facilitate the evoked release of noradrenaline in a concentration-dependent manner. NSD-1015 partially inhibits increase in the spontaneous release and blocks increase in tissue content of noradrenaline induced by the highest concentration of L-*threo*-DOPA at 100 μ*M*. These properties of L-*threo*-DOPS are very similar to those of levodopa. The stereoselective facilitation of evoked noradrenaline release induced by levodopa [31] and L-*threo*-DOPS [36] is an important finding, as discussed in the following text.

Muraki et al. have tested the effects of various catecholamines, sympathomimetics, and related compounds for their ability to potentiate the voltage-dependent calcium current (Ica) evoked in single cells isolated from the taenia of the guinea pig's cecum [75]. They found that levodopa, dopamine, isoproterenol, adrenaline, and noradrenaline show almost equal potency to potentiate Ica. However, the racemic mixtures of the optical isomers of isoproterenol, adrenalin, and noradrenaline, and (+)-isoproterenol, are equipotent with the (–)-isomers of these drugs. Based on the structure–activity relationship, they term the receptor activated by catecholamines to increase Ica a *C-receptor* in view of its sensitivity to catechol. This property of

*C-receptor* appears to differ from that of recognition sites for levodopa and L-*threo*-DOPS, because levodopa [31] and L-*threo*-DOPS [36] itself shows stereoselective activity on the impulse-evoked release of noradrenaline from superfused rat hypothalamic slices.

### 8.8.2 L-*Threo*-DOPS as Levodopa Agonist

In addition to the similarities between levodopa and L-*threo*-DOPS discussed in Subsection 8.8.1, the facilitation of evoked noradrenaline release induced by picomolar concentrations of L-*threo*-DOPA from hypothalamic slices [36] is competitively antagonized by DOPA ME with a $pA_2$ value of 13.6 in a qualitative manner similar to levodopa [31]. On the other hand, it is noncompetitively antagonized by (−)-propranolol in a manner similar to levodopa. These findings suggest that this artificial amino acid is the agonist acting by itself on DOPA recognition sites, in addition to its role as a precursor of noradrenaline [73,74].

Indeed, hypotension and bradycardia in response to microinjection of nanomolar levodopa (10 to 100 ng) [76] into depressor sites of the NTS are mimicked by the lower nanomolar doses of L-*threo*-DOPS (0.3 to 10 ng) [77]. These responses to L-*threo*-DOPS are dose dependent, stereoselective, and DOPA-ME sensitive, but NSD-1015 insensitive. In addition, similar depressor and bradycardic responses to L-*threo*-DOPS are observed in depressor sites of the caudal ventrolateral medulla (CVLM) [78] in accordance with the case of levodopa [79]. The CVLM receives excitatory baroreflex inputs from the NTS and projects inhibitory GABA-containing neurons to the rostral ventrolateral medulla (RVLM) [9–12], the exit of the baroreflex neurotransmission to the thoracic spinal cord, participating in central regulation of arterial blood pressure in the lower brainstem. However, in pressor sites of the RVLM, wide dose ranges of microinjected L-*threo*-DOPS failed to elicit the expected hypertension and tachycardia [78] in contrast with levodopa [80]. Competitive antagonism produced by DOPA ME against responses to L-*threo*-DOPS in hypothalamic slices [36] and the tonic function of endogenously released DOPA suggested by responses to DOPA ME alone in the NTS, the CVLM, and the RVLM (shown in Section 8.10) also suggest the existence of an L-*threo*-DOPS-like endogenous substance in the CNS [78]. It appears likely, however, that there is an L-*threo*-DOPS-responsive subtype of DOPA recognition sites in the hypothalamus [36], the NTS [77], and the CVLM [78], whereas an L-*threo*-DOPS-unresponsive subtype of the DOPA recognition sites exists in the RVLM [78].

### 8.8.3 Second Screening for Antagonistic Activities of Levodopa Analogs

In the second screening process for antagonistic activities of levodopa analogs such as L-phenylalanine, L-leucine, carbidopa, DOPA phosphate, and DOPA ME, the last analog was a potent competitive antagonist against levodopa in continuously superfused hypothalamic slices [31]. For screening for antagonism by levodopa-related analogs against levodopa-induced facilitation of the evoked release of noradrenaline, (−)-sulpiride was added to minimize the inhibition of the release of noradrenaline via presynaptic dopamine $D_2$ receptors by the moderate micromolar concentrations of levodopa in the presence of an AADC inhibitor [5]. Under this condition, levodopa at 1 n*M* to 1 μ*M* facilitates the evoked release of noradrenaline in a concentration-dependent manner. Carbidopa (0.1 to 10 n*M*) antagonizes the levodopa-induced facilitation and roughly shifts the concentration–facilitation curve for levodopa to the right. This suggests that carbidopa, a peripheral AADC inhibitor, is a competitive antagonist. This analog, however, seems to be not as suitable, because the concentration dependency for this antagonism is not as clear. L-DOPA phosphate (0.1 n*M*) antagonizes the levodopa (100 and 300 n*M*)-induced facilitation, and this analog, at 0.01 and 0.1 n*M*, tends to produce a roughly parallel shift of the curve for levodopa to the right. Again, L-DOPA phosphate cannot be a useful antagonist, because the concentration dependency at 0.01 and 0.1 n*M* was not so clear, and this analog, at 1 n*M*, produces no antagonism, seeming to elicit a complex triphasic response. Possible metabolites and/or unknown actions of this analog itself might prevent an antagonistic action against levodopa. Neither L-phenylalanine nor L-leucine, both of which are large neutral amino acids, at 1 μ*M* antagonizes the facilitatory action of levodopa. On the other hand, the concentration–facilitation curve for levodopa is progressively displaced to the right in the presence of increasing concentrations of DOPA ME, a carboxylic acid ester of levodopa, and the maximal effect of levodopa is equipotent to that seen in control slices (Figure 8.8). No reduction from the original maximum is seen. The Schild plots give a straight line with a slope of 1.00. The $pA_2$ extrapolated from the Schild plots is 8.9 [31].

### 8.8.4 Third Screening for Potent Competitive Antagonists Using Depressor Responses to Levodopa Microinjected into Depressor Sites of NTS in Anesthetized Rats

DOPA ME is the first compound that we found as a competitive antagonist for levodopa. Its usefulness as an antagonist, however, is limited to experimental conditions under which it is supplied continuously in brain slices [31,32,36, 57,58], or its effect can be exerted within several minutes after a single microinjection into baroreflex neurotransmission centers in the lower brainstem [76–81,88,89]. This compound has been originally developed as a prodrug for levodopa and is readily hydrolyzed [82–84]. Indeed, no antagonism but even potentiation for some responses to levodopa is seen in the case of intracerebroventricular application of DOPA ME [85]. Searching for a stable and potent competitive antagonist, we have thus continued to synthesize and screen the effectiveness of a series of DOPA ester compounds with chemically bulky structures against esterases that cause the conversion of DOPA ME to levodopa. Three novel compounds, DOPA CHE, DOPA cyclopentyl ester (DOPA CPE), and DOPA cyclopentyldimethyl ester (DOPA CPDME), were obtained (Figure 8.9). Screening of DOPA antagonists by monitoring the impulse-evoked release of noradrenaline from superfused rat hypothalamic slices has turned out to be a time-consuming work. Thus, we studied the antagonistic activities of these candidates against depressor responses of anesthetized rats to levodopa microinjected into the NTS [37,38]. DOPA CHE, DOPA CPE, and DOPA CPDME at 1 μg microinjected into depressor sites of the NTS elicit or tend to elicit more marked antagonism against depressor

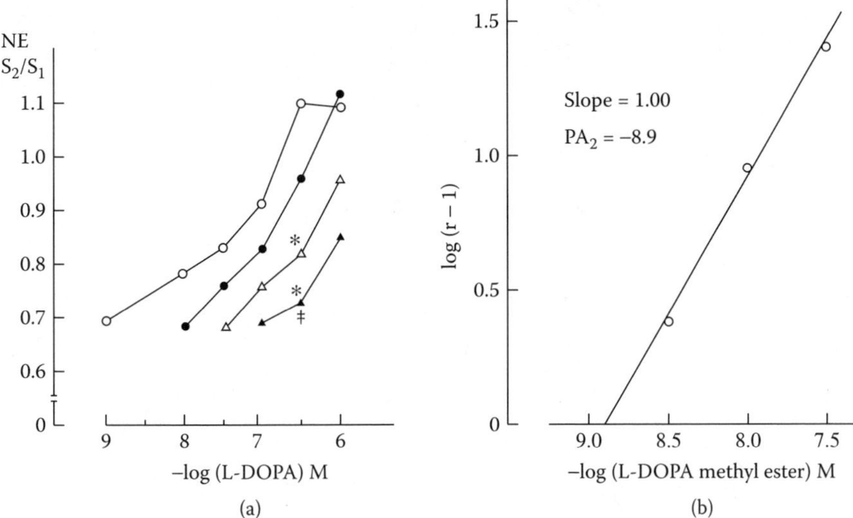


**FIGURE 8.8** Competitive antagonism by DOPA ME (L-DOPA methyl ester) against facilitation of the evoked release of noradrenaline (NE) induced by levodopa (L-DOPA) from rat hypothalamic slices in the presence of cocaine (20 μ*M*), NSD-1015 (20 μ*M*), and (–)-sulpiride (1 n*M*). Pretreatment with DOPA ME was initiated 10 min before $S_1$ and continued throughout the experiments. (a) Concentration–release curves for levodopa in the absence (○, control) and presence of DOPA ME at 3 n*M* (●), 10 n*M* (△), and 30 n*M* (▲). *$P < .01$, (Student's *t*-test) and ‡$P < .01$ (Dunnett's test), compared to corresponding control. Other details are as in Figure 8.2. (b) Schild plots for the antagonism in (a). The regression of log [dose–ratio (r)–1] on log [DOPA ME] yields a straight line with a slope = 1.00 and a $pA_2$ of 8.9. (From Goshima, Y., Nakamura, S., and Misu, Y., L-Dihydroxyphenylalanine methyl ester is a potent competitive antagonist of the L-dihydroxyphenylalanine-induced facilitation of the evoked release of endogenous norepinephrine from rat hypothalamic slices, *J. Pharmacol. Exp. Ther.*, 258, 466, 1991. With permission.)

responses to 60-ng levodopa, compared to DOPA ME [38]. At 100 ng, DOPA CHE elicits the most potent antagonism. DOPA ME (300 ng) and DOPA CHE (30 ng) elicit a competitive type of antagonism (Figure 8.10) [37] in accordance with findings on the parameters of facilitation of evoked noradrenaline release via presynaptic β-adrenoceptors from hypothalamic slices [5,31]. We further explored to find DOPA ester compounds with a prolonged antagonistic activity, compared to DOPA ME. At 1 μg, duration of the antagonistic activity of the single microinjection of DOPA CHE is approximately at least three times longer than that of DOPA ME (Figure 8.11). Being consistent with this finding, the conversion from DOPA CHE at 1 μ*M* perfused via a probe to extracellular DOPA, monitored in perfusate samples obtained during microdialysis of the nucleus accumbens, is the lowest among these ester compounds and less than half of that from DOPA ME (Figure 8.12). These findings suggest that DOPA CHE is the most potent and relatively stable competitive antagonist among these DOPA ester compounds for exogenously applied levodopa and endogenously released DOPA.

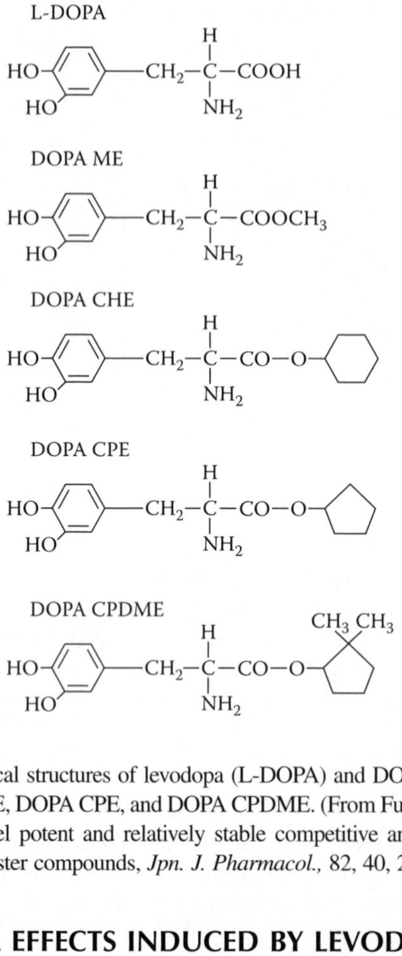

L-DOPA

DOPA ME

DOPA CHE

DOPA CPE

DOPA CPDME

**FIGURE 8.9** The chemical structures of levodopa (L-DOPA) and DOPA ester compounds such as DOPA ME, DOPA CHE, DOPA CPE, and DOPA CPDME. (From Furukawa, N. et al., L-DOPA cyclohexyl ester is a novel potent and relatively stable competitive antagonist against L-DOPA among several L-DOPA ester compounds, *Jpn. J. Pharmacol.,* 82, 40, 2000. With permission.)

## 8.9 BEHAVIORAL EFFECTS INDUCED BY LEVODOPA IN CONSCIOUS RATS ANTAGONIZED BY SYSTEMIC ADMINISTRATION OF DOPA CHE

In conscious rats pretreated with NSD-1015 (100 mg/kg, i.p.), levodopa (100 mg/kg, i.p.) elicits marked locomotor activity, and the noneffective dose of levodopa (30 mg/kg) potentiates locomotor activity elicited by quinpirole (0.01 to 1 mg/kg, s.c.), a dopamine $D_2$ agonist [86]. We recently confirmed these findings and further found that the systemic administration of DOPA CHE (40 to 100 mg/kg, i.p.) suppresses partially, in a dose-dependent manner, the locomotor activity elicited by levodopa (100 mg/kg) [87]. In addition, DOPA CHE (100 mg/kg) completely antagonizes the licking behavior induced by levodopa (100 mg/kg). Furthermore, a noneffective dose of levodopa (20 mg/kg) surely potentiates locomotor activity induced by quinpirole (0.3 mg/kg). Importantly, the systemic administration of a low dose of DOPA CHE (10 mg/kg) antagonizes this levodopa-induced potentiation to the level of locomotor activity elicited by quinpirole alone (see Chapter 15). DOPA CHE can be used as a tool to antagonize responses to levodopa and released DOPA in experimental systems *in vivo*.

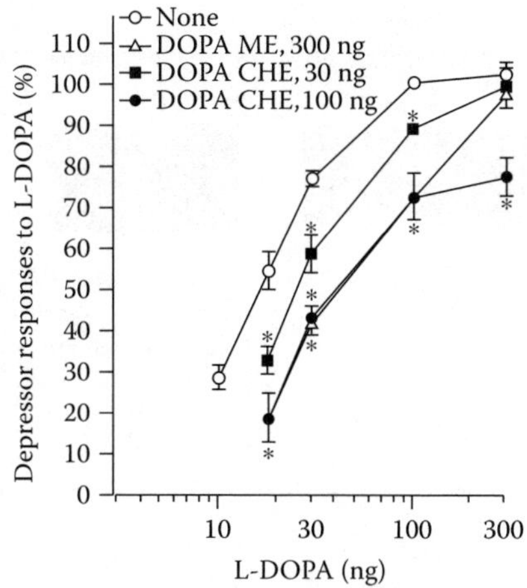


**FIGURE 8.10** DOPA CHE is a potent competitive antagonist in comparison with DOPA ME against depressor responses to levodopa (L-DOPA) microinjected into depressor sites of the NTS in anesthetized rats. The effects were determined 1 min after microinjection of antagonists. Ordinate shows depressor responses to levodopa as percentage of a control and the depressor response of each rat to levodopa at 100 ng before microinjection of antagonists. Each value shows the mean ± SE (n = 5–75). The control Δ blood pressor was –43 ± 1 mm Hg (n = 75). **P* < .05 vs. none (Dunn's multiple comparison test). (From Misu, Y. et al., *Jpn. J. Pharmacol.*, 7, 307, 1997. With permission.)

## 8.10 ANTAGONISM BY DOPA ANTAGONISTS AGAINST RESPONSES TO ENDOGENOUSLY RELEASED DOPA

DOPA ME and DOPA CHE can antagonize responses to endogenously released DOPA. In the NTS [81], CVLM [79], and RVLM [80] of the lower brainstem, the tonic function of DOPA recognition sites has been proven by the antagonism of DOPA ME against basally released DOPA. The bilateral microinjection of DOPA ME alone (1 to 2 μg × 2) elicits hypertension and tachycardia in the depressor NTS [81] and depressor CVLM [79], and hypotension and bradycardia in pressor RVLM [80]. All of these responses to the competitive antagonist alone, which suggest the physiological activation of depressor and pressor sites elicited by basally released DOPA, are abolished following inhibition of the biosynthesis of DOPA induced by the prior injection of α-methyl-*p*-tyrosine (200 mg/kg, i.p.) (see Chapter 11). In addition, in depressor sites of the NTS, the bilateral microinjection of DOPA ME (1 μg × 2) antagonizes reflex bradycardia following DOPA release elicited by baroreceptor activation in response to infusion of phenylephrine (i.v.) [81]. DOPA ME (1 μg) microinjected unilaterally antagonizes hypotension and bradycardia elicited ipsilaterally by either DOPA released following the electrical stimulation of the aortic depressor nerve (ADN) [81] or exogenously microinjected levodopa [76] in a similar manner. In depressor sites of the CVLM, the unilateral microinjection of

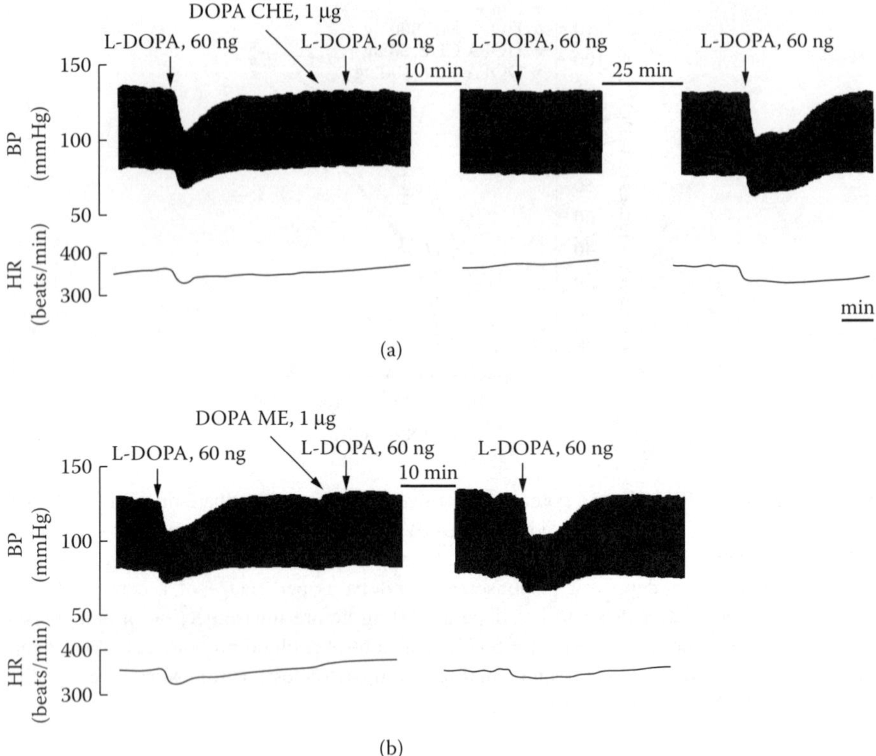


**FIGURE 8.11** A typical trace of long-lasting antagonism by 1 μg DOPA CHE (a), compared to DOPA ME (b), against depressor and bradycardic responses of anesthetized rats to 60-ng levodopa (L-DOPA) microinjected into depressor sites of the NTS. DOPA antagonists were microinjected 1 min prior to levodopa. (From Furukawa, N. et al., L-DOPA cyclohexyl ester is a novel potent and relatively stable competitive antagonist against L-DOPA among several L-DOPA ester compounds, *Jpn. J. Pharmacol.*, 82, 40, 2000. With permission.)

DOPA ME (1 μg) partially inhibits depressor responses by approximately 60% but not bradycardic responses to DOPA released following ipsilateral stimulation of the ADN [88]. In pressor sites of the RVLM, DOPA ME (1.5 × 2 and 3.0 × 2 μg) microinjected bilaterally antagonizes pressor and tachycardiac responses to DOPA released by electrical stimulation of the unilateral posterior hypothalamic nucleus in a dose-dependent manner [89]. Further details are discussed in Chapter 11.

Furthermore, DOPA released by 10-min four-vessel occlusion during striatal microdialysis of conscious rats appears to cause release of glutamate and resultant mild delayed neuron death in the striatum elicited by ischemia [90]. Intrastriatal perfusion of nanomolar DOPA CHE (10 to 100 n*M*) via microdialysis probes antagonizes release of glutamate and protects against delayed neuron death by ischemia. In contrast with the striatum, the hippocampal CA1 region is highly vulnerable to brain ischemia [67,91]. Ischemia of 5-min duration elicits a slight degree of glutamate release

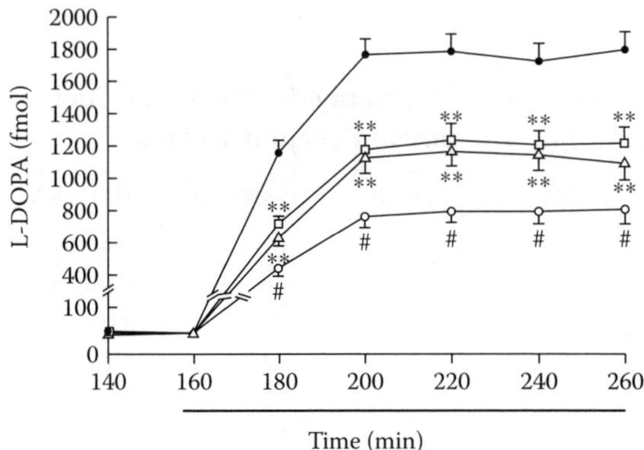


**FIGURE 8.12** Time courses of conversion from DOPA ester compounds to extracellular DOPA during microdialysis of the nucleus accumbens in anesthetized rats. Ringer solution was perfused at a rate of 2 ml/min. Perfusates were collected every 20 min. DOPA ME (●), DOPA CPE (□), DOPA CPDME (△), and DOPA CHE (○) at 1 μ*M* (n = 6) were perfused via a probe at a horizontal bar 160 min after the start of perfusion throughout the experiments. Ordinate shows the mean ± SE of the absolute value of DOPA. ***P* < .005 and #*P* < .0001, compared to the corresponding DOPA ME (Dunn's multiple comparison test). (From Furukawa, N. et al., L-DOPA cyclohexyl ester is a novel potent and relatively stable competitive antagonist against L-DOPA among several L-DOPA ester compounds, *Jpn. J. Pharmacol.,* 82, 40, 2000. With permission.)

without detectable amounts of DOPA in perfusate samples during the hippocampal microdialysis and mild delayed neuron death [92]. These events elicited by 5-min ischemia are abolished by intrahippocampal perfusion of nanomolar DOPA CHE (100 n*M*), probably via antagonism against DOPA released. Although the release of DOPA is below assay sensitivity in this design, DOPA release that is less than the measurable extracellular level (0.4 n*M*) [90] appears to cause release of glutamate and resultant delayed neuron death elicited by this ischemia. Extremely low concentrations of levodopa (3 to 10 p*M*) exert responses that potentiate activities of presynaptic facilitatory β-adrenoceptors [32] and inhibitory $\alpha_2$-adrenoceptors [57] (see Chapter 20).

## 8.11 INTERACTIONS OF LEVODOPA WITH RECEPTORS FOR OTHER NEUROTRANSMITTER SYSTEMS AND WITH DOPA TRANSPORT SITES

DOPA ester compounds and levodopa do not displace or hardly displace the selective binding of tritiated dopamine $D_1$ and $D_2$ receptor ligands or tritiated $\alpha_2$-adrenoceptor and β-adrenoceptor ligands in rat brain membrane preparations (Table 8.2) [31,38]. In addition, DOPA ME fails to displace the selective binding of tritiated GABA [93]. Furthermore, DOPA ME antagonizes depressor and bradycardic responses to levodopa microinjected into depressor sites of the NTS [76] and the CVLM [79]

**TABLE 8.3**
**Effects of Levodopa-Related Compounds (1 m*M*) on Specific Binding of Tritiated Ionotropic Glutamate Receptor Ligands**

| Compounds | [³H]-AMPA | [³H]-Kainate | [³H]-MK-801 | [³H]-DCKA | [³H]-CGP39653 |
|---|---|---|---|---|---|
| Levodopa | 22 ± 3 | 83 ± 6 | 93 ± 1 | 77 ± 2 | 88 ± 5 |
| DOPA CHE | 90 ± 4 | 84 ± 5 | 50 ± 2 | 98 ±13 | 96 ± 3 |
| DOPA CPE | 102 ± 6 | 98 ± 34 | 37 ± 1 | 135 ± 15 | 85 ± 4 |
| DOPA CPDME | 101 ± 11 | 78 ± 27 | 43 ± 2 | 126 ± 10 | 97 ± 16 |
| DOPA ME | 113 ± 8 | 78 ± 27 | 43 ± 2 | 126 ±10 | 97 ±16 |
| L-threo-DOPS | 91 ± 8 | 115 ± 3 | 91 ± 6 | 52 ± 9 | 66 ±13 |
| α-Methyl-DOPA | 94 ± 6 | 103 ± 8 | 86 ± 5 | 83 ± 9 | 99 ±13 |
| α-Methyl-*p*-tyrosine | 103 ± 4 | 106 ± 5 | 81 ± 2 | 102 ±10 | 94 ± 2 |
| Carbidopa | 94 ± 7 | 103 ± 9 | 92 ± 11 | 31 ±11 | 95 ± 5 |
| DOPA phosphates | 91 ± 2 | 80 ± 1 | 105 ± 3 | 50 ± 5 | 55 ± 11 |
| 3-*O*-Methyl-DOPA | 96 ± 8 | 88 ± 8 | 100 ± 5 | 88 ± 3 | 79 ±12 |
| Kynurenic acid | 88 ± 5 | 51 ± 9 | 101 ± 2 | 1± 2 | 34 ± 2 |
| Glutamate | 21 ± 9 | 1 ± 1 | n.t. | n.t. | 1 ± 1 |
| Kainate | 4 ± 3 | 0 ± 0 | n.t. | n.t. | n.t. |
| CNQX | 0 ± 0 | 2 ± 2 | n.t. | n.t. | n.t. |
| MK-801 | n.t. | n.t. | 0 ± 0 | n.t. | n.t. |
| Glycine | n.t. | n.t. | n.t. | 0 ± 0 | n.t. |

*Note:* Data represent the mean ± SE expressed as a percentage of the control binding (n = 3), and n.t. is not tested.

*Source:* From Miyamae, T. et al., Some interactions of L-DOPA and its related compounds with glutamate receptors, *Life Sci.*, 64, 1045, 1999; Furukawa, N. et al., L-DOPA cyclohexyl ester is a novel potent and relatively stable competitive antagonist against L-DOPA among several L-DOPA ester compounds, *Jpn. J. Pharmacol.,* 82, 40, 2000. With permission.

and pressor and tachycardic responses to levodopa microinjected into pressor sites of the RVLM [80]. In contrast, DOPA ME does not antagonize glutamate-induced cardiovascular responses similar to levodopa in the NTS [76], the CVLM [79], and the RVLM [80], suggesting that DOPA recognition sites differ from glutamate receptors.

In Table 8.3 [38,94], we further explored whether levodopa and DOPA ester compounds interact with ionotropic glutamate receptors. We used tritiated DL-α-amino-3-hydroxy-5-methyl-4-isoxazole propionic acid (AMPA) and kainate to label

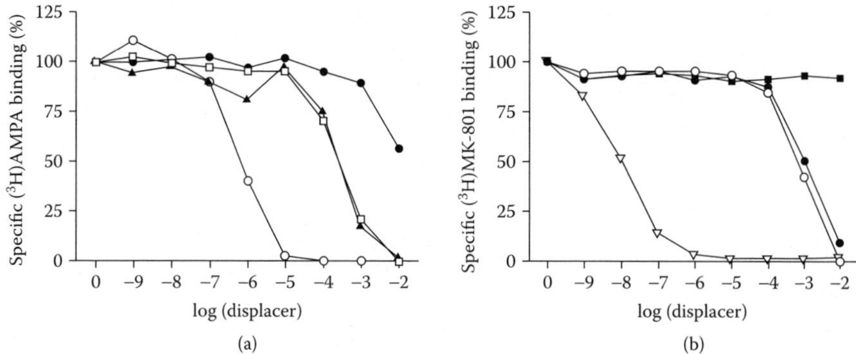


**FIGURE 8.13** Effects of levodopa-related compounds (1 m*M*) on specific binding of [$^3$H]-AMPA (a) and [$^3$H]-MK-801 (b) in rat brain membrane preparations. (a) Levodopa (□), levodopa + ascorbic acid (200 μ*M*) (▲), DOPA ME (●), and CNQX (○). (b) DOPA ME (●), levodopa (■), DOPA CHE (○), and MK-801 (▽). (From Miyamae, T. et al., Some interactions of L-DOPA and its related compounds with glutamate receptors, *Life Sci.*, 64, 1045, 1999. With permission.)

AMPA receptors and kainate receptors, respectively. We also used tritiated MK-801, 5,7-dichloro-kynurenic acid (DCKA), and DL-(*E*)-2-amino-4-propyl-5-phospho-3-pentenoic acid (CGP39653) to label the ion channel domain of NMDA receptors, NMDA glycine site, and NMDA binding site, respectively. Among these specific binding sites, levodopa acts only on AMPA receptors with low affinity. $IC_{50}$ value of levodopa against tritiated AMPA binding is 260 μ*M*, being higher compared with 18 μ*M* of kainate, 2.8 μ*M* of glutamate, and 0.6 μ*M* of 6-cyano-7-quinoxaline-2,3-dione (CNQX) (Figure 8.13A). DOPA ME and DOPA CHE elicit no modifications (Figure 8.13A and Table 8.3). In contrast, DOPA ester compounds act only on the ion channel domain of NMDA receptors with $IC_{50}$ values in the millimolar range. Respective $IC_{50}$ value of DOPA ME and DOPA CHE against tritiated MK-801 binding is 1 and 0.68 m*M*, being far higher compared with 10 n*M* of cold MK-801 (Figure 8.13B). Levodopa produces no modification. These findings suggest that DOPAergic agonist and competitive antagonists do not interact on ionotropic glutamate receptors.

In an attempt to find specific receptors for DOPA, we explored whether levodopa can induce current responses in *Xenopus laevis* oocytes injected with poly $A^+$RNA extracted from rat brain [94]. Levodopa induced inward current responses in the oocytes held at –70 mV. The $EC_{50}$ of levodopa to induce the responses, however, was around 2 m*M*. Levodopa currents at 1 m*M* are inwardly rectifying with the reversal potential of around –30 mV in current voltage relationships and are abolished by substitution of $Na^+$ and $K^+$ with tetramethylammonium$^+$ but not modified by $Cl^-$ with isothionate. These electrophysiological properties suggest that levodopa opens nonselective cation channels. Kainate (100 μ*M*) elicits inward currents far larger than levodopa. These currents are reduced by 68% in the presence of levodopa (10 m*M*) (Figure 8.14). Both levodopa (3 m*M*) currents and kainate (100 μ*M*)

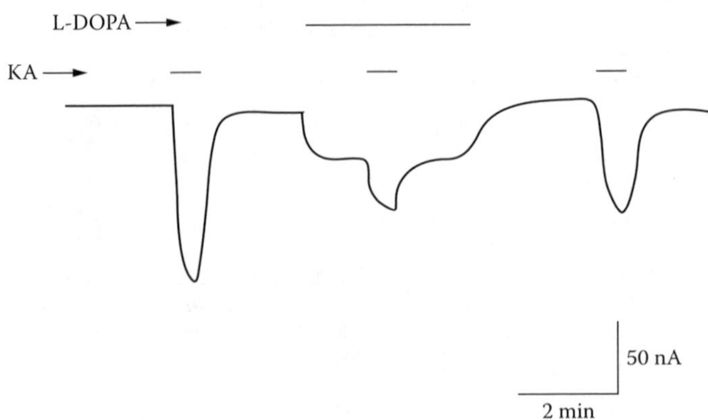


**FIGURE 8.14** A typical trace for summation study on inward currents induced by kainate (KA) at 100 μ*M* and levodopa (L-DOPA) at 10 m*M* in *Xenopus laevis* oocytes injected with rat forebrain poly $A^+$ RNA and held at –70 mV. Kainate and levodopa were applied at each horizontal bar. (From Miyamae, T. et al., Some interactions of L-DOPA and its related compounds with glutamate receptors, *Life Sci.*, 64, 1045, 1999. With permission.)

currents are abolished by kynurenic acid at 1 m*M* or CNQX at 100 μ*M*. These results suggest that levodopa elicits some responses via non-NMDA receptors. This idea is further supported in oocytes coinjected with AMPA receptor subunits GluR1-4 cRNAs. Levodopa indeed induced concentration-dependent inward currents, showing activation of GluR1-4 channel subunits expressed, although the current amplitude is far smaller even at the highest concentration, compared to kainite. Levodopa thus acts as a partial agonist for AMPA receptors. These findings are consistent with the binding experiments demonstrating that levodopa inhibits specific binding of [$^3$H]-AMPA with $IC_{50}$ of 260 μ*M* (Figure 8.13) [94]. This activity is not blocked by ascorbic acid, an antioxidant. Therefore, it is likely that levodopa itself interacts with AMPA receptors, but not through conversion to some oxidized derivatives sharing the properties of AMPA agonists such as trihydroxyphenylalanine [95–99]. Although the physiological relevance of DOPA as a weak AMPA agonist is unknown, these findings may well explain previous observations that levodopa at millimolar orders induces a weak excitatory current in frog and rat spinal cord [100], rat hippocampal neurons, and some neurotoxicity in chick embryo retina [96]. On the other hand, levodopa does not appear to interact with metabotropic glutamate receptors (mGluR) at least in *Xenopus laevis* oocytes, because levodopa (100 μ*M*) produces no response in the *Xenopus laevis* oocytes expressing mGluR1 receptor [94].

The other potential targets for these DOPA ester compounds may include transporters for levodopa. The transport of levodopa across the neuronal plasma membrane is cocaine insensitive, which is common for large neutral amino acids [31]. We also found that there exists an $Na^+$-dependent active transport system of levodopa [101,102]. DOPA CHE, however, does not inhibit the uptake of labeled levodopa into oocytes (Figure 8.15) [101].

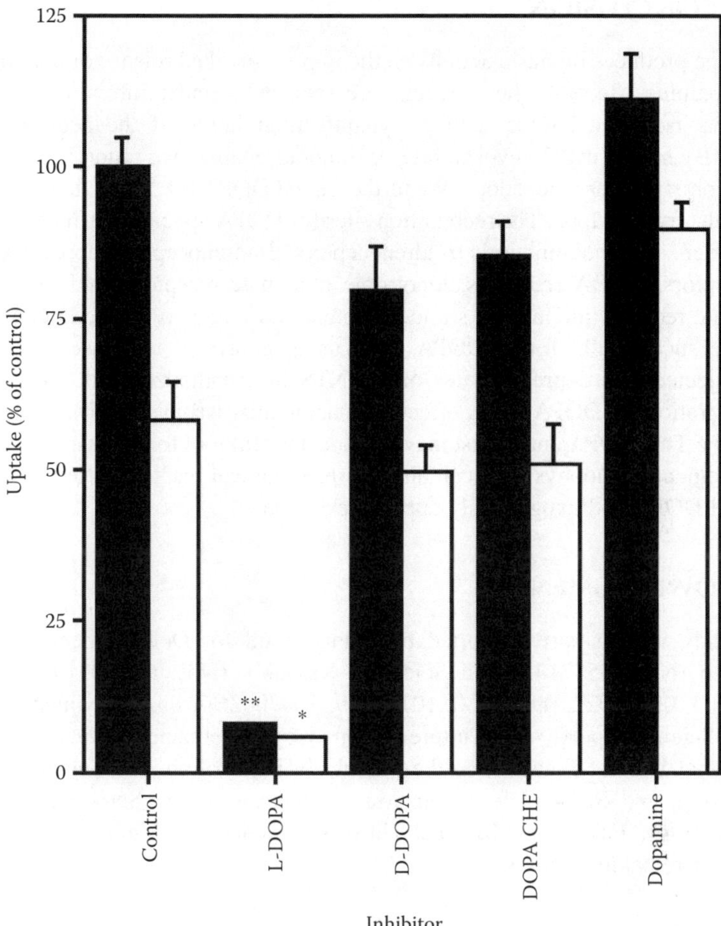


**FIGURE 8.15** Inhibition study of labeled levodopa (L-DOPA) uptake by levodopa-related compounds. [$^{14}$C]-Levodopa (30 μ*M*) uptake into *Xenopus laevis* oocytes injected with rabbit intestinal epithelium poly A$^+$ RNA was performed with 1-m*M* nonradioactive compounds in the presence (filled columns) or absence (open columns) of Na$^+$. Levodopa uptake is expressed as a percentage of the control uptake without inhibitor in the presence of Na$^+$. Data are mean ± SE (n = 6 to 10). *$P < .05$ and **$P < .01$, vs. corresponding control uptake in the absence or presence of Na$^+$. (From Ishii, H. et al., Involvement of rBAT in Na$^+$-dependent and -independent transport of the neurotransmitter candidate L-DOPA in *Xenopus laevis* oocytes injected with rabbit small intestinal epithelium poly A$^+$ RNA, *Biochim. Biophys. Acta,* 1466, 61, 2000. With permission.)

DOPAergic agonist and competitive antagonists should act at the same sites, which appear to differ from DOPA transport sites, catecholamine receptors, GABA receptors, and ionotropic glutamate receptors. To support the identification of DOPA as a neurotransmitter, it is essential that DOPA recognition sites exist on which DOPA or levodopa acts to elicit physiological or pharmacological responses.

## 8.12 CONCLUSION

Levodopa produces biphasic actions on the impulse-evoked release of noradrenaline and dopamine. Because these actions are seen even under inhibition of AADC, levodopa itself can induce such presynaptic modulation of the neurotransmitter release. By measuring the evoked release of noradrenaline, we found L-*threo*-DOPS as an agonist similar to levodopa. We further found DOPA ME as the first competitive antagonist for levodopa. The recognition sites for DOPA appear to differ from DOPA transporters, catecholaminergic $\alpha$-adrenoceptors, $\beta$-adrenoceptors, dopamine $D_1$ and $D_2$ receptors, GABA receptors, ionotropic glutamate receptors, and metabotropic glutamate receptor mGluR1. As a more potent and relatively stable DOPA antagonist, we successfully found DOPA CHE using depressor responses to levodopa microinjected into depressor sites of the NTS in anesthetized rats. The systemic administration of DOPA CHE effectively antagonizes some *in vivo* responses to levodopa. This DOPA antagonist may provide a useful tool for assessment of *in vivo* physiological, pathophysiological, and pharmacological responses to endogenously released DOPA and exogenously applied levodopa.

## ACKNOWLEDGMENTS

This study was in part supported by grants-in-aid for Developmental Scientific Research (No. 06557143) and Scientific Research (Nos. 61480119, 03454146, 07407003, 09877022, 09280280, 10176229, 10470026) from the Ministry of Education, Science, Sports, and Culture, Japan. We also obtained grants from Mitsui Life Social Welfare Foundation, the Mitsubishi Foundation, the Uehara Memorial Foundation, and SRF, all in Japan. Many thanks to Sanae Sato, Department of Pharmacology, Fukushima Medical University School of Medicine, Fukushima, Japan, for remaking figures.

## REFERENCES

1. Carlsson, A., Lindqvist, M., and Magnusson, T., 3,4-Dihydroxyphenylalanine and 5-hydroxytryptophan as reserpine antagonists, *Nature,* 180, 1200, 1957.
2. Ehringer, H. and Hornykiewicz, O., Verteilung von Noradrenaline und Dopamine (3-Hydroxytyramin) im Gehirn des Menschen und ihr Verhalten bei Erkrankungen des extrapyramidalen Systems, *Klin. Wochenschr.,* 38, 1236, 1960.
3. Bartholini, G. et al., Increase of cerebral catecholamines by 3,4-dihydroxyphenylalanine after inhibition of peripheral decarboxylase, *Nature,* 215, 852, 1967.
4. Hefti, F. and Melamed, E., L-DOPA's mechanism of action in Parkinson's disease, *Trends Neurosci.,* 3, 229, 1980.
5. Goshima, Y., Kubo, T., and Misu, Y., Biphasic actions of L-DOPA on the release of endogenous noradrenaline and dopamine from rat hypothalamic slices, *Br. J. Pharmacol.,* 89, 229, 1986.
6. Misu, Y., Goshima, Y., and Kubo, T., Biphasic actions of L-DOPA on the release of endogenous dopamine via presynaptic receptors in rat striatal slices, *Neurosci. Lett.,* 72, 194, 1986.

7. Goshima, Y., Kubo, T., and Misu, Y., Transmitter-like release of endogenous 3,4-dihydroxyphenylalanine from rat striatal slices, *J. Neurochem.,* 50, 1725, 1988.
8. Misu, Y. and Goshima, Y., Is L-dopa an endogenous neurotransmitter?, *Trends Pharmacol. Sci.,* 14, 119, 1993.
9. Misu, Y., Ueda, H., and Goshima, Y., Neurotransmitter-like actions of L-DOPA, *Adv. Pharmacol.,* 32, 427, 1995.
10. Misu, Y. et al., Neurobiology of L-DOPAergic systems, *Prog. Neurobiol.,* 49, 415, 1996.
11. Misu, Y., Goshima, Y., and Miyamae, T., Is DOPA a neurotransmitter?, *Trends Pharmacol. Sci.,* 23, 262, 2002.
12. Misu, Y., Kitahama, K., and Goshima, Y., L-3,4-Dihydroxyphenylalanine as a neurotransmitter candidate in the central nervous system, *Pharmacol. Ther.,* 97, 117, 2003.
13. Hoffman, B.B. and Taylor, P., Neurotransmission: the autonomic and somatic motor nervous systems, in *Goodman & Gilman's The Pharmacological Basis of Therapeutics,* Hardman, J.G., Limbird, L.E., and Gilman, A.G., Eds., McGraw-Hill, New York, 2001, 115.
14. Katz, B. and Miledi, R., Tetrodotoxin resistant electric activity in presynaptic terminals, *J. Physiol.,* 203, 459, 1969.
15. Augustine, G.J., How does calcium trigger neurotransmitter release?, *Curr. Opin. Neurobiol.,* 11, 320, 2001.
16. Langer, S.Z., 25 years since the discovery of presynaptic receptors: present knowledge and future perspectives, *Trends Pharmacol. Sci.,* 18, 95, 1997.
17. Starke, K., Presynaptic autoreceptors in the third decade: focus on alpha2-adrenoceptors, *J. Neurochem.,* 78, 685, 2001.
18. Chesselet, M., Presynaptic regulation of neurotransmitter release in the brain: facts and hypothesis, *Neuroscience,* 12. 347, 1984.
19. Misu, Y. and Kubo, T., Presynaptic β-adrenoceptors, *Med. Res. Rev.,* 6, 197, 1986.
20. Wall, P.D., Do nerve impulses penetrate terminal arborizations? A pre-presynaptic control mechanism, *Trends Neurosci.,* 18, 99, 1995.
21. MacDermott, A.B., Role L.W., and Siegelbaum S.A., Presynaptic ionotropic receptors and the control of transmitter release, *Annu. Rev. Neurosci.,* 22, 443, 1999.
22. Schoepp, D.D., Unveiling the functions of presynaptic metabotropic glutamate receptors in the central nervous system, *J. Pharmacol. Exp. Ther.,* 299, 12, 2001.
23. Vitten, H. and Isaacson, J.S., Synaptic transmission: exciting times for presynaptic receptors, *Curr. Biol.,* 11, R695, 2001.
24. Kalsner, S., Autoreceptors do not regulate routinely neurotransmitter release: focus on adrenergic systems, *J. Neurochem.,* 78, 676, 2001.
25. Raiteri, M., Presynaptic autoreceptors, *J. Neurochem.,* 78, 673, 2001.
26. Boehm, S. and Kubista, H., Fine tuning of sympathetic transmitter release via ionotropic and metabotropic presynaptic receptors, *Pharmacol. Rev.,* 54, 43, 2002.
27. Kamiya, H., Kainate receptor-dependent presynaptic modulation and plasticity, *Neurosci. Res.,* 42, 1, 2002.
28. Ueda, H., Goshima, Y., and Misu Y., Presynaptic mediation by $\alpha_2$-, $\beta_1$- and $\beta_2$-adrenoceptors of endogenous noradrenaline and dopamine release from slices of rat hypothalamus, *Life Sci.,* 33, 371, 1983.
29. Goshima, Y., Kubo, T., and Misu, Y., Autoregulation of endogenous epinephrine release via presynaptic adrenoceptors in the rat hypothalamic slice, *J. Pharmacol. Exp. Ther.,* 235, 248, 1985.

30. Goshima, Y., Nakamura, S., and Misu, Y., L-DOPA facilitates the release of endogenous norepinephrine and dopamine via presynaptic $\beta_1$- and $\beta_2$-adrenoceptors under essentially complete inhibition of L-aromatic amino acid decarboxylase in rat hypothalamic slices, *Jpn. J. Pharmacol.,* 53, 47, 1990.
31. Goshima, Y., Nakamura, S., and Misu, Y., L-Dihydroxyphenylalanine methyl ester is a potent competitive antagonist of the L-dihydroxyphenylalanine-induced facilitation of the evoked release of endogenous norepinephrine from rat hypothalamic slices, *J. Pharmacol. Exp. Ther.,* 258, 466, 1991.
32. Goshima, Y. et al., Picomolar concentrations of L-DOPA stereoselectively potentiate activities of presynaptic β-adrenoceptors to facilitate the release of endogenous noradrenaline from rat hypothalamic slices, *Neurosci. Lett.,* 129, 214, 1991.
33. Misu Y. et al., Presynaptic inhibitory dopamine receptors on noradrenergic nerve terminals: analysis of biphasic actions of dopamine and apomorphine on the release of endogenous norepinephrine in rat hypothalamic slices, *J. Pharmacol. Exp. Ther.,* 235, 771–777, 1985.
34. Tedroff, J. et al., L-DOPA modulates striatal dopaminergic function in vivo: evidence from PET investigations in nonhuman primates, *Synapse,* 25, 56, 1997.
35. Torstenson, R. et al., Effect of apomorphine infusion on dopamine synthesis rate relates to dopaminergic tone, *Neuropharmacology,* 37, 989, 1998.
36. Yue, J.-L. et al., L-Dopa-like regulatory actions of L-threo-3,4-dihydroxyphenylserine on the release of endogenous noradrenaline via presynaptic receptors in rat hypothalamic slices, *J. Pharm. Pharmacol.,* 44, 990, 1992.
37. Misu, Y. et al., L-DOPA cyclohexyl ester is a novel stable and potent competitive antagonist against L-DOPA, as compared to L-DOPA methyl ester, *Jpn. J. Pharmacol.,* 7, 307, 1997.
38. Furukawa, N. et al., L-DOPA cyclohexyl ester is a novel potent and relatively stable competitive antagonist against L-DOPA among several L-DOPA ester compounds, *Jpn. J. Pharmacol.,* 82, 40, 2000.
39. Ng, K.Y. et al., Dopamine: stimulation-induced release from central neurons, *Science,* 172, 487, 1971.
40. Ng, K.Y., Colburn, R.W., and Kopin, I.J., Effects of L-DOPA on efflux of cerebral monoamines from synaptosomes, *Nature,* 230, 331, 1971.
41. Bunney, B.S., Aghajanian, G.K., and Roth, R.H., Comparison of effects of L-DOPA, amphetamine and apomorphine on firing rate of rat dopaminergic neurones, *Nature New Biol.,* 245, 123, 1973.
42. Lloyd K.G., Davidson L., and Hornykiewicz O., The neurochemistry of Parkinson's disease. Effect of L-DOPA therapy, *J. Pharmacol. Exp. Ther.,* 195, 453, 1975.
43. Ponzio, F. et al., Does acute L-DOPA increase active release of dopamine from dopaminergic neurons?, *Brain Res.,* 273, 45, 1983.
44. Melamed E. et al., Suppression of L-dopa-induced circling in rats with nigral lesions by blockade of central dopa-decarboxylase: implications for mechanism of action of L-dopa in parkinsonism, *Neurology,* 34, 1566, 1984.
45. Koshimura, K. et al., L-DOPA administration enhances exocytotic dopamine release in vivo in the rat striatum, *Life Sci.,* 51, 747, 1992.
46. Neff, N.H., Wemlinger, T.A., and Hadjiconstantinou, M., SCH 23390 enhances exogenous L-DOPA decarboxylation in nigrostriatal neurons, *J. Neural Transm.,* 107, 429, 2000.
47. Lopez, A. et al., Mechanisms of the effects of exogenous levodopa on the dopamine-denervated striatum, *Neuroscience,* 103, 639, 2001.
48. Shen, K.Z. et al., Dopamine receptor supersensitivity in rat subthalamus after 6-hydroxydopamine lesions, *Eur. J. Neurosci.,* 18, 2967, 2003.

49. Ueda, H., Goshima, Y., and Misu, Y., Presynaptic $\alpha_2$- and dopamine-receptor-mediated inhibitory mechanisms and dopamine nerve terminals in the rat hypothalamus, *Neurosci. Lett.*, 40, 157, 1983.
50. Ueda, H. et al., Involvement of epinephrine in the presynaptic beta adrenoceptor mechanism of norepinephrine release from rat hypothalamic slices, *J. Pharmacol. Exp. Ther.*, 232, 507, 1985.
51. Ueda, H. et al., Adrenaline involvement in the presynaptic β-adrenoceptor-mediated mechanism of dopamine release from slices of the rat hypothalamus, *Life Sci.*, 34, 1087, 1984.
52. Lokhandwala, M.F. and Buckley, J.P., The effect of L-DOPA on peripheral sympathetic nerve function: role of presynaptic dopamine receptors, *J. Pharmacol. Exp. Ther.*, 204, 362, 1978.
53. Arai, N. et al., Evaluation of a 1-methyl-4-phenyl-1,2,3,6-tetrahydropyridine (MPTP)-treated C57 black mouse model for parkinsonism, *Brain Res.*, 515, 57, 1990.
54. Goshima, Y. et al., Nanomolar L-DOPA facilitates release of dopamine via presynaptic β-adrenoceptors: comparative studies on the actions in striatal slices from control and 1-methyl-4-phenyl-1,2,3,6-tetrahydropyridine (MPTP)-treated C57 black mice, an animal model for Parkinson's disease, *Jpn. J. Pharmacol.*, 55, 93, 1991.
55. Bylund, D.B. and Snyder, S.H., Beta adrenergic receptor binding in membrane preparations from mammalian brain, *Mol. Pharmacol.*, 12, 568, 1976.
56. Starke, K., Gothert, M., and Kilbinger, H., Modulation of neurotransmitter release by presynaptic autoreceptors, *Physiol. Rev.*, 69, 864, 1989.
57. Sato, K. et al., L-Dopa potentiates presynaptic inhibitory $\alpha_2$-adrenoceptor- but not facilitatory angiotensin II receptor-mediated modulation of noradrenaline release from rat hypothalamic slices, *Jpn. J. Pharmacol.*, 62, 119, 1993.
58. Goshima, Y. et al., L-DOPA induces $Ca^{2+}$-dependent and tetrodotoxin-sensitive release of endogenous glutamate from rat striatal slices, *Brain Res.*, 617, 167, 1993.
59. Ueda, H. et al., L-DOPA inhibits spontaneous acetylcholine release from the striatum of experimental Parkinson's model rats, *Brain Res.*, 698, 213, 1995.
60. Hirai, K., Katayama, Y., and Misu, Y., L-DOPA induces concentration-dependent facilitation and inhibition of presynaptic acetylcholine release in the guinea-pig submucous plexus, *Brain Res.*, 718, 105, 1996.
61. Akbar, M. et al., Inhibition by L-3,4-dihydroxyphenylalanine of hippocampal CA1 neurons with facilitation of noradrenaline and γ-aminobutyric acid release, *Eur. J. Pharmacol.*, 414, 197, 2001.
62. Calabresi, P. et al., Acetylcholine-mediated modulation of striatal function, *Trends Neurosci.*, 23, 120, 2000.
63. Sato, K. et al., Supersensitization of intrastriatal dopamine receptors involved in opposite regulation of acetylcholine release in Parkinson's model rats, *Neurosci. Lett.*, 173, 59, 1994.
64. Sato, K. et al., 6-OHDA-induced lesion of the nigrostriatal dopaminergic neurons potentiates the inhibitory effect of 7-OHDPAT, a selective $D_3$ agonist, on acetylcholine release during striatal microdialysis in conscious rats, *Brain Res.*, 655, 233, 1994.
65. Maeda, T. et al., L-DOPA neurotoxicity is mediated by glutamate release in cultured rat striatal neurons, *Brain Res.*, 771, 159, 1997.
66. Cheng, N. et al., Differential neurotoxicity induced by L-DOPA and dopamine in cultured striatal neurons, *Brain Res.*, 743, 278, 1996.
67. Choi, D.W., Glutamate neurotoxicity and diseases of the nervous system, *Neuron*, 1, 623, 1988.

68. Sheardown, M.J. et al., 2,3-Dihydroxy-6-nitro-7-sulfamoyl benzo(F) quinoxaline: a neuroprotectant for cerebral ischemia, *Science,* 247, 571, 1990.
69. Dawson, T.M., Dawson, V.L., and Snyder, S.H., Nitric oxide as a mediator of neurotoxicity, *NIDA Res. Monogr.,* 136, 258, 1993.
70. Carlsson, A., Functional significance of drug-induced changes in brain monoamine levels, in *Progress in Brain Research,* Vol. 8, Himwich, H.E. and Himwich, W.A., Eds., Elsevier, Amsterdam, 1964, 9.
71. Bertrand, A. and Weil-Fugazza, J., Sympathectomy does not modify the levels of dopa or dopamine in the rat dorsal root ganglion, *Brain Res.,* 681, 201, 1995.
72. McLane, J., Osber, M., and Pawelek, J.M., Phosphorylated isomers of L-DOPA stimulate MSH binding capacity and responsiveness to MSH in cultured melanoma cells, *Biochem. Biophys. Res. Commun.,* 145, 719, 1987.
73. Bartholini, J. et al., The stereoisomers of 3,4-dihydroxyphenylserine as precursors of norepinephrine, *J. Pharmacol. Exp. Ther.,* 193, 523, 1975.
74. Narabayashi, H. et al., DL-threo-3,4-Dihydroxyphenylserine for freezing symptom in parkinsonism, *Adv. Neurol.,* 40, 497, 1984.
75. Muraki, K. et al., Receptor for catecholamines responding to catechol which potentiates voltage-dependent calcium current in single cell from guinea-pig taenia caeci, *Br. J. Pharmacol.,* 111, 1154, 1994.
76. Kubo, T. et al., Evidence for L-DOPA systems responsible for cardiovascular control in the nucleus tractus solitarii of the rat, *Neurosci. Lett.,* 140, 153, 1992.
77. Miyamae, T. et al., Depressor action of L-*threo*-dihydroxyphenylserine in the rat nucleus tractus solitari, *Eur. J. Pharmacol.,* 300, 105, 1996.
78. Furukawa, N. et al., An L-DOPA-like depressor action of L-*threo*-dihydroxyphenylserine in the rat caudal ventrolateral medulla, *Life Sci.* 61, 1177, 1997.
79. Yue, J.-L., Goshima, Y., and Misu, Y., Transmitter-like L-3,4-dihydroxyphenylalanine tonically functions to mediate vasodepressor control in the caudal ventrolateral medulla of rats, *Neurosci. Lett.,* 159, 103, 1993.
80. Yue, J.-L. et al., Evidence for L-DOPA relevant to modulation of sympathetic activity in the rostral ventrolateral medulla of rats, *Brain Res.,* 629, 310, 1993.
81. Yue, J.-L. et al., Baroreceptor-aortic nerve-mediated release of endogenous L-3,4-dihydroxyphenylalanine and its tonic function in the nucleus tractus solitarii of rats, *Neuroscience,* 62, 145, 1994.
82. Hanson, L.C. and Utley, J.D., Biochemical and behavioral effects of L-DOPA methyl ester in cats treated with reserpine, *Psychopharmacologia,* 8, 140, 1965.
83. Cooper, D.R. et al., L-Dopa esters as potential prodrugs: behavioural activity in experimental models of Parkinson's disease, *J. Pharm. Pharmacol.,* 39, 627, 1987.
84. Cooper, D.R. et al., L-Dopa esters as a potential prodrugs: effect on brain concentration of dopamine metabolites in reserpinized mice, *J. Pharm. Pharmacol.,* 39, 809, 1987.
85. Nakamura, S. et al., Endogenously released DOPA is probably relevant to nicotine-induced increases in locomotor activities of rats, *Jpn. J. Pharmacol.,* 62, 107, 1993.
86. Nakamura, S. et al., Non-effective dose of exogenously applied L-DOPA itself stereoselectively potentiates postsynaptic $D_2$ receptor-mediated locomotor activities of conscious rats, *Neurosci. Lett.,* 170, 22, 1994.
87. Matsushita, N., Misu, Y., and Goshima, Y., Antagonism by DOPA cyclohexyl ester against behavioral effects of L-DOPA in rats, *J. Pharmacol. Sci.,* 91 (Suppl. I), 202P, 2003.
88. Miyamae, T. et al., L-DOPAergic components in the caudal ventrolateral medulla in baroreflex neurotransmission, *Neuroscience,* 92, 137, 1999.

89. Nishihama, M. et al., An L-DOPAergic relay from the posterior hypothalamic nucleus to the rostral ventrolateral medulla and its cardiovascular function in anesthetized rats, *Neuroscience,* 92, 123, 1999.
90. Furukawa, N. et al., Endogenously released DOPA is a causal factor for glutamate release and resultant delayed neuronal cell death by transient ischemia in rat striata, *J. Neurochem.,* 76, 815, 2001.
91. Obrenovitch, T.P. and Richards, D.A. Extracellular neurotransmitter changes in cerebral ischaemia, *Cerebrovasc. Brain Metab. Rev.,* 7, 1, 1995.
92. Arai, N. et al., DOPA cyclohexyl ester, a competitive DOPA antagonist, protects glutamate release and resultant delayed neuron death by transient ischemia in hippocampus CA1 of conscious rats, *Neurosci. Lett.,* 299, 213, 2001.
93. Honjo, K. et al., GABA may function tonically via $GABA_A$ receptors to inhibit hypotension and bradycardia by L-DOPA microinjected into depressor sites of the nucleus tractus solitarii in anesthetized rats, *Neurosci. Lett.,* 261, 93, 1999.
94. Miyamae, T. et al., Some interactions of L-DOPA and its related compounds with glutamate receptors, *Life Sci.,* 64, 1045, 1999.
95. Aizenman, E. et al., A 3,4-dihydroxyphenylalanine oxidation product is a non-*N*-methyl-D-aspartate glutamatergic agonist in rat cortical neurons, *Neurosci. Lett.,* 116, 168, 1990.
96. Olney, J.W. et al., Excitotoxicity of L-DOPA and 6-OH-DOPA: implications for Parkinson's and Huntington's diseases, *Exp. Neurol.,* 108, 269, 1990.
97. Cha, J.-H.J. et al., Trihydroxyphenylalanine (6-hydroxy-DOPA) displaces [$^3$H]AMPA binding in rat striatum, *Neurosci. Lett.,* 132, 55, 1991.
98. Newcomer, T.A., Rosenberg, P.A., and Aizenman, E., TOPA quinone, a kainite-like agonist and excitotoxin, is generated by a catecholaminergic cell line, *J. Neurosci.,* 15, 3172, 1995.
99. Prado, B. et al., Toxic effects of L-DOPA on mesencephalic cell cultures: protection with antioxidants, *Brain Res.,* 682, 133, 1995.
100. Biscoe, T.J. et al., Structure-activity relations of excitatory amino acids on frog and rat spinal neurons, *Br. J. Pharmacol.,* 58, 373, 1976.
101. Ishii, H. et al., Involvement of rBAT in $Na^+$-dependent and -independent transport of the neurotransmitter candidate L-DOPA in *Xenopus laevis* oocytes injected with rabbit small intestinal epithelium poly $A^+$ RNA, *Biochim. Biophys. Acta,* 1466, 61, 2000.
102. Sugaya, Y. et al., Autoradiographic studies using L-[$^{14}$C]DOPA and L-[$^3$H]DOPA reveal regional $Na^+$-dependent uptake of the neurotransmitter candidate L-DOPA in the CNS, *Neuroscience,* 104, 1, 2000.

# 9 The Neuroregulatory Properties of Levodopa: Evidence and Potential Role in the Treatment of Parkinson's Disease

*Joakim Tedroff*

## CONTENTS



## 9.1 INTRODUCTION

Parkinson's disease is a progressive neurodegenerative disorder clinically characterized by tremor, rigidity, bradykinesia, and postural instability as a consequence of dopamine depletion in the striatum arising from degeneration of dopaminergic neurons in the substantia nigra pars compacta of the midbrain. The mainstay of the treatment in Parkinson's disease is L-3,4-dihydroxyphenylalanine (L-DOPA, levodopa), the precursor of dopamine. Exogenous levodopa therapy dramatically improves the akinesia and bradykinesia that characterize Parkinson's disease [1].

However, most Parkinson's disease patients experience fluctuations in their motor response, including shortened response duration and dyskinesia, after several years of therapy [2–3]. Whether the motor response fluctuation is the consequence of disease progression, long-term levodopa therapy, or both is not clear [4–5].

The pathophysiology underlying these pharmacodynamic changes that accompany chronic levodopa treatment is not known. The prevailing view points to postsynaptic mechanisms as a major cause of the on–off fluctuations [6]. However, little is known about the nature of such presumed postsynaptic mechanisms. Various theories, ranging from changes in dopamine receptor affinity to changes occurring in the basal ganglia downstream from the dopaminergic system, have been put forward.

The pharmacodynamics of levodopa therapy undergoes marked changes with the disease progression as well as with the duration of levodopa therapy. An increase in magnitude of the response to levodopa during long-term therapy has been noted in several studies in Parkinson's disease patients [4,7,8]. Likewise, the latency to onset of response and to peak response shortens during chronic therapy with levodopa [4,7,9,10]. The seemingly shortened latent period before the response may also mean that the response is becoming more dramatic with more rapid onset and greater magnitude during chronic levodopa treatment. Despite these evident disadvantages, levodopa still remains the most effective therapy for Parkinson's disease and is used in an overwhelming majority of patients. In due course, all patients with Parkinson's disease will require levodopa for their symptomatic treatment. In this respect levodopa is clearly more effective than dopamine agonists (predominantly D2 and D3 receptor agonists). However, once motor complications develop, manipulation of levodopa doses and adjunctive antiparkinson medications are of limited help in reducing the problems. For these reasons there is tremendous interest among clinical researchers and the pharmaceutical industry to find ways to prevent the development of dyskinesia and motor fluctuations as well as to develop better ways to treat them when they do emerge.

## 9.2 NEUROMODULATION BY EXOGENOUS LEVODOPA: POTENTIAL IMPLICATIONS FOR CHANGES IN LEVODOPA PHARMACODYNAMICS WITH PROGRESSION OF PARKINSON'S DISEASE

Traditionally, levodopa is believed to reverse the motor symptoms of Parkinson's disease following its conversion by aromatic amino acid DOPA decarboxylase (AADC) to dopamine. More recently, there has been a new wave of interest in the mechanism of action of levodopa, and it has been suggested that endogenous L-DOPA may be a neurotransmitter or neuromodulator in its own right [11,12]. Because levodopa still remains the mainstay of treatment in Parkinson's disease, the concept of a neuromodulatory role of L-DOPA has potentially important therapeutic implications for the treatment of Parkinson's disease. During recent years, a considerable amount of new information has emerged supporting an underlying role for L-DOPA *per se* in the pharmacodynamic changes occurring with chronic levodopa treatment in Parkinson's disease. In support of this concept are the findings that acute or chronic administration of levodopa affects dopamine receptor status and AADC activity, the type and severity of the effects depending on the duration of the treatment [13]. There is evidence of an increase in ligand binding to striatal D1 and D2 receptors, as confirmed *ex vivo* [14], and to D2 receptors, as shown *in vivo* by means of positron emission tomography (PET), following acute levodopa administration in rats [15]. At present, little is known about direct L-DOPA effects on D2 dopamine receptors although this issue is of interest in the context of Parkinson's disease. Also, the concept of a neurotransmitter role for L-DOPA is supported by its ability to facilitate the release of dopamine and noradrenaline via presynaptic β-adrenoceptors *in vitro* [16] and to stimulate glutamate and γ-aminobutyric acid (GABA) release from rat striatal slices under AADC inhibition [17–19]. L-DOPA also releases glutamate from the striatum of conscious rats [20] and has been shown to inhibit the basal release of acetylcholine in a model of Parkinson's disease [21]. Although a recognition site or receptor for L-DOPA has not been identified, L-DOPA methyl ester competitively antagonizes the facilitatory effect of L-DOPA on noradrenaline release in rat hypothalamic slices [22].

These aforementioned properties of L-DOPA clearly point toward an important role of L-DOPA *per se* in the therapeutic effects in Parkinson's disease. However, owing to the complexity of the cortico-basal ganglia system and the multitude of neurotransmitters involved in this adaptable system, as well as the lack of specific orally active L-DOPA antagonists, the relative impact of such effects in Parkinson's disease cannot be elucidated.

### 9.2.1 Locomotor Deficits Induced by Nigrostriatal Lesions Not Reversed by L-DOPA but D2 Agonist Efficacy Modulated

To study the effects of L-DOPA, a large number of studies have utilized the centrally acting AADC inhibitor, 3-hydroxybenzylhydrazine (NSD-1015). It was recently demonstrated that NSD-1015 inhibits the reversal of motor deficits produced by exogenous levodopa administration to the 1-methyl-4-phenyl-1,2,3,6-tetrahydropyridine (MPTP)-treated primate model of Parkinson's disease [23], suggesting that L-DOPA itself does not produce locomotor activity. The picture in the rodent is less clear as both increases and decreases in levodopa-induced motor behavior have been reported following NSD-1015 treatment [24–28]. Perhaps the most convincing evidence for a postsynaptic modulatory effect of exogenous levodopa is the effect on locomotor activity in experimental animals. Increased accumulation of endogenous L-DOPA, produced by blockade of central AADC activity using NSD-1015, was reported to potentiate the locomotor activity produced by the D2 receptor agonist quinpirole in normal rats [28]. Furthermore, a subthreshold dose of levodopa potentiated quinpirole-induced locomotor activity following NSD-1015 treatment of both normal and 6-hydroxydopamine (6-OHDA)-lesioned rats [27]. NSD-1015 blocks central AADC activity; it may also act as a monoamine oxidase inhibitor and, thus, maintain striatal dopamine concentrations by reducing dopamine metabolism [23].

Such data suggest a role for endogenous L-DOPA both in the initiation of motor activity and in the actions of dopamine agonist drugs, again explaining why levodopa may be more effective in treating Parkinson's disease than dopamine agonists.

### 9.2.2 L-DOPA *Per Se* Modulating Presynaptic Dopaminergic Regulation

Dopamine-containing fibers arising from the substantia nigra and ventral tegmental area, with projection to the striatum and cortical/limbic areas, are characterized by an extraordinary plasticity. The compensatory activity is based on various homeostatic mechanisms that guarantee the functional stability of the neuronal system. Of major importance for the homeostatic control of dopaminergic systems are the dopamine autoreceptors regulating release and synthesis of dopamine [29]. These autoreceptors appear to be of D2 receptor type. This transsynaptic feedback control is influenced by medication that affects the dopaminergic systems such as exogenous levodopa and dopamine agonists.

Using PET and L-[$^{11}$C]DOPA a direct measure of the presynaptic utilization of exogenous levodopa can be calculated *in vivo* [30]. The method can be used to calculate a unidirectional rate constant (i.e., Ki) for the biotransformation of L-[$^{11}$C]DOPA to [$^{11}$C]dopamine in striatal and cortical brain regions. In monkeys and humans pharmacological challenges using dopaminomimetics produce measurable changes in L-[$^{11}$C]DOPA Ki, suggesting that the presynaptic turnover of exogenous levodopa is influenced by dopaminergic tone [31]. Hence, the method offers an opportunity to study the functional tone of the presynaptic dopaminergic system *in vivo*. A consistent finding from pharmacological challenge studies is that the magnitude of the changes in dopaminergic tone induced by various pharmacological perturbations is highly baseline tone dependent [12,32]. A robust finding from such studies is that pharmacological doses of exogenous levodopa and apomorphine induce opposing effects on L-[$^{11}$C]DOPA Ki. Exogenous levodopa was demonstrated to increase L-[$^{11}$C]DOPA Ki in a baseline state-dependent manner, and similar findings were reported when monkeys were pretreated with 6R-tetrahydrobiopterin (6R-BH4), a cofactor promoting tyrosine hydroxylation, suggesting a similar role for endogenously derived L-DOPA [33]. To support this notion the researchers also showed that the effect was further augmented when 6R-BH4 was coadministered with tyrosine, the precursor amino acid. On the other hand, pharmacological doses of apomorphine, a mixed D1 and D2 receptor agonist, decreased this rate constant, also in a state-dependent manner. Thus, it was demonstrated that a dopamine agonist reduced the presynaptic turnover of exogenous levodopa, which is in agreement with the notion of a presynaptic autoreceptor-mediated inhibitory control of dopaminergic activity. In this respect, the enhancing effects of exogenous levodopa or endogenously derived L-DOPA on L-[$^{11}$C]DOPA Ki are more difficult to explain. It is plausible that such a state-dependent versatile effect represents a fundamental homeostasis-regulating mechanism for dopaminergic function *in vivo* and that this system of dopaminergic control may undergo changes as a consequence of degeneration such as occurs in Parkinson's disease.

### 9.2.3 L-DOPA Effects in Parkinson's Disease: Lessons from PET

The progressive loss of dopamine leading to the symptoms of Parkinson's disease can be compensated to a certain extent at the presynaptic and/or postsynaptic level. Several mechanisms are set in motion to compensate for such a dopamine deficiency

[34]. Among them, increased dopamine turnover probably occurs even at very early stages of the disease [35]. In the early stages of the disease autoreceptors are functional responding with down regulation in response to dopamine agonist challenge [36,37]. Using a [$^{11}$C]raclopride displacement technique, it could be demonstrated that this also accounts for dopamine release following a standardized acute levodopa challenge in Parkinson's disease patients [38]. In this study, it was shown that the magnitude of dopamine release following levodopa challenge increases with disease duration and severity and is most pronounced in the putamen, the striatal region with the most severe dopaminergic nerve cell loss.

As for L-[$^{11}$C]DOPA turnover, marked modifications occur with the progression of Parkinson's disease. In early and uncomplicated Parkinson's disease, the net effect of therapeutic apomorphine and levodopa is to decrease the L-[$^{11}$C]DOPA Ki, an effect that is most pronounced in the putamen, and, particularly, in its most dorsal parts [35,36]. In patients with long-standing and advanced Parkinson's disease, therapeutic apomorphine does not affect L-[$^{11}$C]DOPA Ki in any striatal subregion. A particularly pertinent finding is that a therapeutic levodopa challenge, which decreases striatal L-[$^{11}$C]DOPA Ki in mildly ill Parkinson's disease patients, paradoxically increases L-[$^{11}$C]DOPA Ki in patients with advanced disease and in those who suffer from pronounced levodopa-induced motor fluctuations [39]. In Figure 9.1,

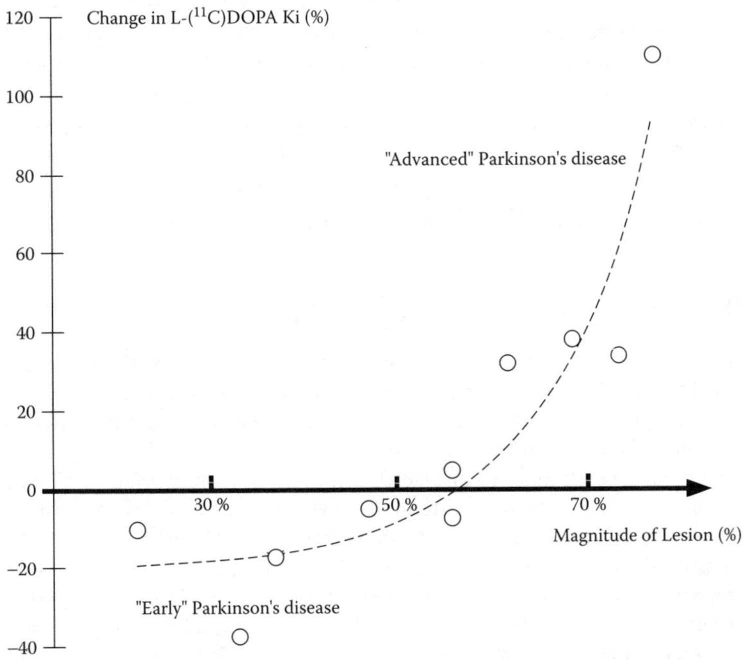


**FIGURE 9.1** Effects on the L-[$^{11}$C]DOPA Ki, as measured by PET, of a standardized therapeutic levodopa challenge in ten patients with Parkinson's disease. The net effect of the change in this rate constant in the putamen is given and related to the severity of dopaminergic lesion for each individual patient (i.e., baseline value as compared to a sample of age-matched healthy volunteers).

the results from these studies are presented. The graph demonstrates the net effect on L-[$^{11}$C]DOPA Ki in the putamen following a standardized therapeutic levodopa challenge in patients with Parkinson's disease of various severities. The net levodopa-induced change in this rate constant is related to the degree of lesion with reference to an age-matched healthy population. As seen from this figure, in patients with mild Parkinson's disease and a lesser degree of lesion, a therapeutic levodopa administration results in a decreased putaminal L-[$^{11}$C]DOPA Ki, whereas with increasing lesion severity, this rate constant starts to increase, to be markedly augmented in the most severely disabled patients, suggesting that the loss of dopamine feedback control with increasing lesion severity results in a predominance for the reinforcing effects of L-DOPA on residual presynaptic nerve terminals.

## 9.3 CONCLUSION

There is by now a substantial bulk of evidence supporting the concept suggesting that L-DOPA may be a neurotransmitter or neuromodulator in its own right. This concept clearly has important therapeutic implications because levodopa is a widely used mainstay treatment for Parkinson's disease. The pharmacological evidence supporting the concept suggest that L-DOPA has reinforcing properties on a number of dopamine-mediated functions, as well as effects on other important neurotransmitter systems. The data also suggest that L-DOPA has properties to modulate presynaptic dopaminergic function. In Parkinson's disease, it is suggested that the reinforcing properties of L-DOPA start to dominate as the disease becomes more advanced. This theory is well in line with results from pharmacodynamic studies suggesting that the levodopa response becomes more dramatic, with more rapid onset and greater magnitude, with the progression of Parkinson's disease. Orally active specific L-DOPA antagonists are therefore expected to decrease the magnitude of motor response fluctuations in patients with advanced Parkinson's disease.

## REFERENCES

1. Cotzias, G.C., Van Woert, M.H., and Schiffer, L.M., Aromatic amino acids and modification of parkinsonism, *N. Engl. J. Med.,* 276, 374, 1967.
2. Marsden, C.D. and Parkes, J.D., "On-off" effects in patients with Parkinson's disease on chronic levodopa therapy, *Lancet,* 1, 292, 1976.
3. Marsden, C.D., Parkinson's disease, *J. Neurol. Neurosurg. Psychiatry,* 57, 672, 1994.
4. Nutt, J.G. et al., Effect of long-term therapy on the pharmacodynamics of levodopa: relation to on–off phenomenon, *Arch. Neurol.,* 49, 1123, 1992.
5. Kostic, V.S. et al., The effect of stage of Parkinson's disease at the onset of levodopa therapy on development of motor complications, *Eur. J. Neurol.,* 9, 9, 2002.
6. Chase, T.N., Konitsiotis, S., and Oh, J.D., Striatal molecular mechanisms and motor dysfunction in Parkinson's disease, *Adv. Neurol.,* 86, 355, 2001.
7. Colosimo, C. et al., Motor response to acute dopaminergic challenge with apomorphine and levodopa in Parkinson's disease: implications for the pathogenesis of the on-off phenomenon, *J. Neurol. Neurosurg. Psychiatry,* 61, 634, 1996.
8. Nutt, J.G. et al., Evolution of the response to levodopa during the first 4 years of therapy, *Ann. Neurol.,* 51, 686, 2002.

9. Contin, M. et al., Longitudinal monitoring of the levodopa concentration–effect relationship in Parkinson's disease, *Neurology,* 44, 1287, 1994.
10. Sohn, Y.H. et al., Levodopa peak response time reflects severity of dopamine neuron loss in Parkinson's disease, *Neurology,* 44, 755, 1994.
11. Misu, Y. and Goshima, Y., Is L-dopa an endogenous neurotransmitter?, *Trends Pharmacol. Sci.,* 14, 119, 1993.
12. Tedroff, J. et al., L-DOPA modulates striatal dopaminergic function in vivo: evidence from PET investigations in nonhuman primates, *Synapse,* 25, 56, 1997.
13. Opacka-Juffry, J. and Brooks, D.J., L-Dihydroxyphenylalanine and its decarboxylase: new ideas on their neuroregulatory roles, *Movement Disord.,* 10, 241, 1995.
14. Murata, M. and Kanazawa, I., Repeated L-DOPA administration reduces the ability of dopamine storage, *Neurosci. Res.,* 16, 15, 1993.
15. Hume, S.P. et al., Effect of L-DOPA and 6-hydroxydopamine lesioning on [$^{11}$C]raclopride binding in rat striatum, quantified using PET, *Synapse,* 21, 45, 1995.
16. Goshima, Y., Nakamura, S., and Misu, Y., L-DOPA facilitates the release of endogenous norepinephrine and dopamine via presynaptic $b_1$- and $b_2$-adrenoceptors under essentially complete inhibition of L-aromatic amino acid decarboxylase in rat hypothalamic slices, *Jpn. J. Pharmacol.,* 53, 47, 1990.
17. Aceves, J. et al., L-DOPA stimulates the release of [$^3$H] gamma-aminobutyric acid in the basal ganglia of 6-OHDA lesioned rats, *Neurosci. Lett.,* 121, 223, 1991.
18. Goshima, Y. et al., L-DOPA induces $Ca^{2+}$-dependent and tetrodotoxin-sensitive release of endogenous glutamate from rat striatal slices, *Brain Res.,* 617, 167, 1993.
19. Misu, Y. et al., DOPA causes glutamate release and delayed neuron death by brain ischemia in rats, *Neurotoxicol. Teratol.,* 24, 629, 2002.
20. Misu, Y. et al., Nanomolar L-dopa induces $Ca^{2+}$-dependent and tetrodotoxin-sensitive release of glutamate from striata of conscious rat, *J. Neurochem.,* 65, 154, 1995
21. Ueda, H. et al., L-DOPA inhibits spontaneous acetylcholine release from the striatum of experimental Parkinson's model rats, *Brain Res.,* 698, 213, 1995.
22. Goshima, Y., Nakamura, S., and Misu, Y., L-Dihydroxyphenylalanine methyl ester is a potent competitive antagonist of the L-dihydroxyphenylalanine-induced facilitation of the evoked release of endogenous norepinephrine from rat hypothalamic slices, *J. Pharmacol. Exp. Ther.,* 258, 466, 1991.
23. Treseder, S.A., Jackson, M., and Jenner, P., The effects of central aromatic amino acid DOPA decarboxylase inhibition on the motor actions of L-DOPA and dopamine agonists in MPTP-treated primates, *Br. J. Pharmacol.,* 129, 1355, 2000.
24. Goodale, G. and Moore, K. A., Comparison of the effects of decarboxylase inhibitors on L-DOPA-induced circling behaviour and the conversion of DOPA to dopamine in the brain, *Life Sci.,* 19, 701, 1976.
25. Melamed, E. et al., Suppression of L-dopa-induced circling in rats with nigral lesions by blockade of central dopa decarboxylase, *Neurology,* 34, 1566, 1984.
26. Nakazoto, T. and Akiyama, A., Effect of exogenous L-DOPA on behavior in the rat: an in vivo voltammetric study, *Brain Res.,* 490, 332, 1989.
27. Nakamura, S. et al., Non-effective dose of exogenously applied L-DOPA itself stereoselectively potentiates postsynaptic $D_2$-receptor-mediated locomotor activities of conscious rats, *Neurosci. Lett.,* 170, 22, 1994.
28. Yue, J.-L. et al., Endogenously released L-DOPA itself tonically functions to potentiate $D_2$ receptor-mediated locomotor activities of conscious rats, *Neurosci. Lett.,* 170, 107, 1994.
29. Wolf, M.E. and Roth, R.H., Autoreceptor regulation of dopamine synthesis, *Ann. N. Y. Acad. Sci.,* 604, 323, 1990.

30. Tedroff, J. et al., Estimation of regional cerebral utilization of [$^{11}$C ]-L-3,4-dihydroxyphenylalanine (DOPA) in the primate by positron emission tomography, *Acta Neurol. Scand.,* 85,166, 1992.
31. Tedroff, J. et al., Functional positron emission tomographic studies of striatal dopaminergic activity. Changes induced by drugs and nigrostriatal degeneration, *Adv. Neurol.,* 69, 443, 1995.
32. Torstenson, R. et al., Effect of apomorphine infusion on dopamine synthesis rate relates to dopaminergic tone, *Neuropharmacology,* 37, 989, 1998.
33. Watanabe, Y. et al., Elevation of $^{11}$C -dopamine turnover in vivo by peripheral administration of 6R-tetrahydrobiopterin in monkey striatum, in *Pteridine and related biogenic amines in neuropsychiatry, pediatrics and immunology,* Bleu, ?.?. et al., Eds., Lakeshore Publishing Company, City name, 1991, 353.
34. Calne, D.B. and Zigmond, M.J., Compensatory mechanisms in degenerative neurologic diseases. Insights from parkinsonism, *Arch. Neurol.,* 48, 361, 1991.
35. Tedroff, J. et al., Regulation of dopaminergic activity in early Parkinson's disease, *Ann. Neurol.,* 46, 359, 1999.
36. Ekesbo, A. et al., Dopamine autoreceptor function is lost in advanced Parkinson's disease, *Neurology,* 52,120, 1999.
37. Maeda, T. et al., Loss of regulation by presynaptic dopamine D2 receptors of exogenous L-DOPA-derived dopamine release in the dopaminergic denervated striatum, *Brain Res.* 817, 185, 1999.
38. Tedroff, J. et al., Levodopa-induced changes in synaptic dopamine in patients with Parkinson's disease measured by [$^{11}$C]raclopride displacement and PET, *Neurology,* 46, 1430, 1996.
39. Torstenson, R. et al., Differential effects of levodopa on dopaminergic function in early and advanced Parkinson's disease, *Ann. Neurol.,* 41, 334, 1997.

# 10 General Survey of Blood Pressure Regulation in Lower Brainstem and Responses to Levodopa and Glutamate

*Takao Kubo*

## CONTENTS



## 10.1 INTRODUCTION

The lower brainstem cardiovascular regulatory mechanism has been identified by studies conducted during the last two decades. Lower brainstem nuclei involved in the tonic and baroreflex blood pressure control include the nucleus tractus solitarius (NTS), caudal ventrolateral medulla (CVLM), and rostral ventrolateral medulla (RVLM). Primary baroreceptor afferent fibers synapse on second-order neurons located within the NTS. The NTS in turn sends an excitatory projection to the CVLM that then sends an inhibitory projection to bulbospinal sympatho-excitatory neurons in the RVLM. The bulbospinal neurons in the RVLM have monosynaptic projections to sympathetic preganglionic cells in the spinal cord and are thought to subserve a sympathoexcitatory function. This chapter will discuss the roles of these medullary regions in arterial blood pressure regulation, especially baroreflex regulation and the roles of glutamate and endogenous L-3,4-dihydroxyphenylalanine (DOPA) in the reflex pathways. Most of the references cited here refer to studies conducted in the rat. The reader is referred to several excellent, relevant reviews [1–5].

## 10.2 ARTERIAL BLOOD PRESSURE REGULATION MECHANISM IN THE LOWER BRAINSTEM

Baroreceptor afferent fibers of the carotid sinus and aortic nerves consist of myelinated and unmyelinated fiber groups; the cell bodies of the nerves are located in the nodose ganglion and the petrosal ganglion, respectively. It has been established that these buffer nerves terminate in the intermediate to caudal portion of the NTS near the obex [6]. There is no difference in the topographical distribution of terminals of myelinated and unmyelinated fibers. The NTS is also the first central synapse for the chemoreceptor afferents. The afferents from the carotid body terminate in a midline area around the calamus scriptorius in the commissural subnucleus of the NTS.

The NTS neuron, as the second-order barosensitive neuron, receives not only inputs from both carotid sinus and aortic nerves but also inputs from diverse sources, including other vagal afferents and the defense area in the hypothalamus [6,7]. In addition, some barosensitive NTS neurons are interneurons innervated by local short axons.

The CVLM was first identified as a vasodepressor area. In the rat, the CVLM is located in an area ventral to the nucleus ambiguus and dorsal to the medial aspect of the lateral reticular nucleus. The CVLM contains neurons that have a monosynaptic projection from the NTS and that are integral to baroreceptor-mediated control of sympathetic vasomotor outflow and arterial pressure [8–10]. Further, it has been suggested that there are two functionally distinct divisions in the CVLM [3,11]. One is the rostral CVLM, which is involved in the control of sympathetic outflow and in mediating the baroreceptor reflex, and the other is the caudal CVLM, which is also involved in the control of sympathetic outflow but is not involved in the baroreceptor reflex. In support of this idea, an anatomical study demonstrated that the rate of RVLM-projecting CVLM neurons that express Fos in response to baroreceptor activation is greater in the rostral CVLM than in the caudal CVLM [12]. The RVLM-projecting CVLM neurons are indicated to use $\gamma$-aminobutyric acid (GABA) as an inhibitory neurotransmitter [13,14].

The RVLM was first identified as a vasopressor area. In the rat, the RVLM is a subdivision of the nucleus paragigantocellularis lateralis and surrounds and includes the C1 adrenaline neurons [15]. The RVLM plays an essential role in the maintenance of tonic and reflex control of arterial blood pressure. The RVLM contains non-C1 neurons and C1 adrenaline neurons. The non-C1 neurons have intrinsic pacemaker activity [16] and project directly to the sympathetic preganglionic neurons in the intermediolateral cell column (IML) of the thoracic spinal cord to maintain sympathetic tone. The C1 adrenaline neuron also projects to the IML but its contribution to the generation of sympathetic vasomotor tone is unclear. When injected into the RVLM, an immunotoxin produced by conjugating the ribosomal toxin saporin to an anti-dopamin-β-hydroxylase antibody (anti-DBH-SAP), caused the loss of 84% of C1 adrenaline neurons while sparing non-C1 neurons in the RVLM [17,18]. The loss of bulbospinal adrenergic neurons by the anti-DBH-SAP treatment in the RVLM did not alter the ability of RVLM to maintain sympathetic nerve activity and arterial blood pressure at rest but this loss reduced the sympathoexcitatory and pressor responses evoked by RVLM stimulation [18]. Thus, it appears that the C1 adrenaline neuron is critical for the full expression of the sympathoexcitatory responses generated by the RVLM.

As stated earlier, the RVLM receives tonic GABAergic inputs from the CVLM [14,15,19]. A part of these inputs is involved in mediation of the baroreceptor reflex and another part is not. The former CVLM GABA neurons may be present in the rostral CVLM and the latter in the caudal CVLM. There exist $GABA_A$ and $GABA_B$ receptors in the rat RVLM. Many pharmacological studies have demonstrated that $GABA_A$ receptors are involved in the tonic and baroreflex inhibition of RVLM pressor neurons [20,21]. The physiological role of $GABA_B$ receptors in the RVLM is still unclear.

Another vasopressor area called the caudal pressor area (CPA) was found in the caudalmost ventrolateral medulla. Chemical stimulation or inhibition of the CPA caused a marked increase or fall of arterial blood pressure [22,23], which depended on the activity of neurons in the RVLM [24]. Another vasodepressor area called midline medullary depressor area was found in the caudal medullary raphe. Chemical stimulation of this area caused depressor responses accompanied by inhibition of the discharge of RVLM neurons and, thus, inhibition of sympathetic nerve discharge [25]. In rabbits, the inhibition of RVLM neural activity induced by caudal midline medulla stimulation has been demonstrated to be mediated by GABAergic inhibition of neurons in the RVLM [26]. The exact physiological roles of both pressor and depressor areas remain to be established.

## 10.3 ARTERIAL BLOOD PRESSURE REGULATION ROLE OF GLUTAMATE IN THE LOWER BRAINSTEM

Ever since the description of Talman et al. [27] that L-glutamate is the neurotransmitter of baroreceptor afferent nerve fibers, many lines of evidence supporting it have been obtained. For example, blockade of glutamate receptors with kynurenate in the NTS attenuated or abolished transmission between the baroreceptor afferents and second-order NTS neurons [28,29]. Activation of the baroreflex by intravenous infusion of phenylephrine increased glutamate in dialysates perfused in the rat NTS [30,31]. Ablation of the afferent cell bodies in the nodose ganglion reduced the high-affinity uptake of L-glutamate in the NTS [32]. After injection of horseradish peroxidase into the nodose ganglion, electron microscopic analysis revealed an enrichment of glutamate immunoreactivity in boutons identified as vagus nerve sensory afferents in the NTS [33].

Glutamate receptors have been divided into two categories; ionotropic glutamate receptors and metabotropic glutamate receptors (mGluR). The ionotropic glutamate receptor is divided into non-*N*-methyl-D-aspartate (NMDA)

(DL-α-amino-3-hydoxy-5-methyl-4-isoxazole propionic acid [AMPA] and kainate) receptors and NMDA receptors. Both non-NMDA and NMDA receptors are involved in baroreflexes or baroreceptor-evoked discharge of NTS neurons [28,29,34,35]. Anatomical studies have demonstrated the presence of NMDA receptor immunoreactivity in dendrites, axons, and axon terminals in the medial NTS of the rat [36], suggesting that NMDA receptors are located not only postsynaptically on their target neurons but also presynaptically on vagal afferents in the NTS. These findings implicate NMDA receptors in the autoregulation of the presynaptic release and in the postsynaptic responses to glutamate at the level of the first central synapse in the NTS.

At least eight mGluR subtypes have been identified and they can be categorized into group I (mGluR1 and mGluR5), group II (mGluR2 and mGluR3), and group III (mGluR4, mGluR6, mGluR7, and mGluR8) on the basis of similarities of agonist pharmacology, primary sequence, and G-protein-effector coupling. In rats, mRNA for all eight mGluR subtypes is detected in the NTS [37] and in the afferent-fiber cell bodies in the nodose ganglia [38], suggesting that mGluRs may be located pre- and postsynaptically in the rat NTS. Studies using selective mGluR agonists and antagonists have provided evidence for functional mGluRs in the modulation of cardiovascular responses in the rat NTS [39]. From electrophysiological studies recording postsynaptic currents evoked by stimulation of primary sensory afferents, synaptic transmission in the rat NTS is suggested to be depressed by presynaptic mGluRs [38], while *in vivo* microdialysis experiments reveal facilitatory mGluRs regulating glutamate release in the rat NTS [41]. The group I mGluRs are suggested to contribute to tonic cardiovascular control and baroreflex regulation in the NTS [42–44]. Finally, L-glutamate microinjected into the NTS acted at both ionotropic glutamate receptors and mGluRs, and blockade of both classes of glutamate receptors is required to completely block the cardiovascular responses to microinjection of L-glutamate in the NTS [40].

Glutamate is also indicated to be the neurotransmitter in the pathway from the NTS to the CVLM for the expression of baroreceptor reflex. CVLM microinjection of the NMDA receptor antagonist 2-amino-5-phosphonovaleric acid (AP5) or the non-NMDA receptor antagonist 6-cyano-7-nitroquinoxaline-2,3-dione (CNQX) partially inhibited the depressor response and the inhibition of RVLM unit activity in response to the electrical stimulation of aortic nerve, and mixtures of AP5 and CNQX abolished both responses to aortic nerve stimulation [44–46]. These findings indicate that both NMDA and non-NMDA receptors are involved in mediating the baroreflex responses in the CVLM. On the other hand, AP5 but not CNQX injected into the CVLM increased the basal blood pressure [45]. Thus, it appears that only NMDA receptors are involved in maintaining basal arterial blood pressure in the CVLM.

Glutamate in the RVLM is suggested to be an excitatory neurotransmitter involved in cardiovascular responses. Using double-labeling technique, the RVLM is indicated to receive projections of glutamate-immunoreactive neurons from various nuclei in the limbic system, hypothalamus, midbrain, pons, and medulla oblongata [47]. Microinjection of non-NMDA and NMDA agonists into the rat RVLM caused a pressor response and kynurenate injected into the RVLM blocked the responses [48,49], indicating that both NMDA and non-NMDA receptors in the RVLM may participate in cardiovascular regulation.

Neurons of the rat RVLM are endowed not only with ionotropic glutamate receptors but also with mGluRs [52]. RVLM microinjections of a selective mGluR agonist increased arterial blood pressure and this pressor response was not inhibited by NMDA and non-NMDA receptor antagonists [53], suggesting that mGluRs in the RVLM may participate in cardiovascular regulation. From studies using group I, II, and III mGluR agonists and antagonists in rat RVLM, it is suggested that all of these three subtypes of mGluR participate in the cardiovascular responses induced by L-glutamate and that group I mGluR may play an important role in the maintenance of arterial blood pressure [54].

The RVLM receives monosynaptic projections from the commissural subnucleus of the NTS [50], the first central synapse for the chemoreceptor afferents. Microinjections of the NMDA receptor antagonist AP5 into the rat RVLM inhibited the carotid chemoreceptor stimulation-induced pressor response, whereas this response was not affected by injection of the non-NMDA receptor antagonist CNQX [48,49]. AP5 inhibited the pressor response, to aspartate but not to glutamate in the rat RVLM. Chemoreceptor stimulation caused a release of aspartate but not of glutamate in the rat RVLM [51]. These observations are consistent with the hypothesis that aspartate in the rat RVLM is involved in mediating the chemoreceptor reflex.

Many studies have demonstrated that sympathoexcitatory projections from the RVLM to the IML use glutamate as an excitatory transmitter. For example, electron microscopic analysis revealed that, following injections into the RVLM, anterogradely transported *Phaseolus vulgaris*-leucoagglutinin was colocalized with glutamate immunoreactivity in axon terminals in the IML. Blockade of spinal glutamate receptors reduced basal arterial blood pressure, resting sympathetic nerve activity, and the pressor response obtained by stimulation of the RVLM [55].

It is still unclear whether adrenaline itself in the C1 bulbospinal adrenaline neuron is involved in mediating the pressor response obtained by RVLM stimulation. A histochemical study has demonstrated that all C1 bulbospinal neurons contain phosphate-activated glutaminase, a glutamate synthesizing enzyme [56], and the vesicular glutamate transporter BNPI/VGLUT2 mRNA, a marker of glutamatergic neurons [57]. Thus, it is probable that glutamate released from C1 bulbospinal adrenaline neuron terminals may be involved in mediating the action obtained by RVLM stimulation.

## 10.4 ARTERIAL BLOOD PRESSURE REGULATION ROLE OF DOPA IN LOWER BRAINSTEM

Although DOPA has been believed to be only a precursor for dopamine, the lines of evidence accumulated by recent studies suggest that DOPA itself fulfills the criteria for neurotransmitters and/or neuromodulators, including synthesis, metabolism, active transport, presence, physiological release, competitive antagonism, and responses [5,58].

In the NTS, pharmacological studies have suggested that DOPA is a neurotransmitter of the primary baroreceptor afferents [59]. Exogenously injected levodopa caused a dose-dependent decrease in arterial blood pressure in the absence and presence of the central aromatic L-amino acid decarboxylase inhibitor 3-hydroxybenzylhydrazine (NSD-1015). Microinjection of DOPA methyl ester (DOPA ME) into the rat NTS antagonized the hypotension and bradycardia elicited by either the microinjection of levodopa into the same site or electrical stimulation of the aortic nerve [59,60]. Furthermore, baroreceptor activation by intravenous infusion of phenylephrine elicited DOPA release during microdialysis [59]. Electrical stimulation of the aortic nerve elicited tetrodotoxin-sensitive DOPA release during microdialysis of the NTS. Histochemical studies also support the transmitter role of DOPA in the primary baroreceptor afferents. Tyrosine hydroxylase and DOPA immmunoreactivity has been found in the rat nodose ganglion. Following denervation of the aortic nerve distal to the ganglion, decreases in tyrosine hydroxylase and DOPA immunoreactivity were detected in the cell bodies in the nodose ganglion and in the centrally projecting terminals in the dorsal vagal complex [59].

In connection with glutamate, prior microinjection of kynurenate into the rat NTS markedly reduced depressor and bradycardic responses to levodopa [61]. However, receptor-binding studies indicated that DOPA recognition sites are different from ionotropic glutamate receptors [62]. Thus, it has been proposed that DOPA is a neurotransmitter of the primary aortic nerve and that glutamate is a neurotransmitter of secondary neurons or interneurons in the neuronal microcircuits of the NTS [5].

DOPA is also suggested to be a neurotransmitter for carrying baroreflex information in the CVLM [63]. Levodopa injected into the CVLM produced a dose-dependent depressor response in the absence and presence of NSD-1015. Depressor responses induced by levodopa CVLM injection or aortic nerve stimulation were antagonized by DOPA ME [63,64]. During microdialysis of the CVLM, baroreceptor activation elicited selective DOPA release. Electrical stimulation of the aortic nerve elicited α-methyl-*p*-tyrosine (α-MPT)-sensitive DOPA release during microdialysis of the CVLM. Local inhibition of the biosynthesis of DOPA with the tyrosine hydroxylase inhibitor α-MPT infused into the CVLM abolished DOPA release and reduced the hypotension evoked by aortic nerve stimulation.

In the RVLM, it is suggested that DOPA is relevant to the modulation of sympathetic activity [65]. The spontaneous DOPA release in the RVLM was partially $Ca^{2+}$ dependent and tetrodotoxin sensitive and was reduced by α-MPT. Levodopa injected into the RVLM produced a dose-dependent pressor response in the absence and presence of NSD-1015, whereas DOPA ME similarly injected produced a prolonged hypotension. In addition, it is suggested that there exists a DOPAergic pressor relay from the posterior hypothalamic nucleus (PHN) to the RVLM. The electrical stimulation of the PHN elicited selective DOPA release with hypertension and tachycardia, and these responses to PHN stimulation were inhibited by tetrodotoxin in the RVLM [66]. In the rat PHN, some cells display markedly DOPA-positive immunoreactivity with very weakly dopamine-positive immunoreactivity [67]. The rat PHN neurons send descending projections to the RVLM [68].

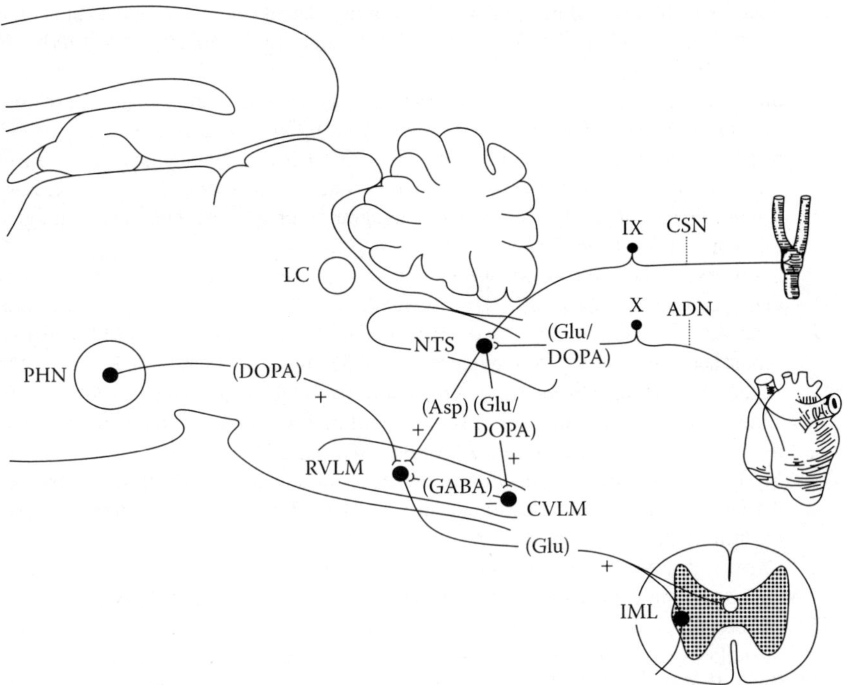


**FIGURE 10.1** Baroreflex pathways, central regulatory mechanisms of arterial blood pressure, and probable glutamate (Glu)- and DOPA-mediated relays in the lower brainstem of rats. Abbreviations: ADN, aortic depressor nerve; Asp, aspartate; CSN, carotid sinus nerve; CVLM, caudal ventrolateral medulla; IML, intermediolateral cell column of the thoracic spinal cord; LC, locus coeruleus; NTS, nucleus tractus solitarius; PHN, posterior hypothalamic nucleus; RVLM, rostral ventrolateral medulla. (Adapted from Misu, Y. et al., Neurobiology of L-DOPAergic systems, *Prog. Neurobiol.*, 49, 415, 1996. With permission.)

## 10.5 CONCLUSION

The lower brainstem nuclei, especially the NTS, the CVLM, and the RVLM play essential roles in baroreflex neurotransmission and the central regulation of arterial blood pressure (Figure 10.1). The excitatory amino acid glutamate is suggested to be a neurotransmitter of the arterial baroreceptor reflex in the NTS and the CVLM. DOPA is also suggested to be a neurotransmitter of the baroreflex in the brainstem nuclei. The exact roles of both of these amino acid transmitters in brainstem cardiovascular regulation remain to be clarified.

## REFERENCES


1. Sun, M.-K., Pharmacology of reticulospinal vasomotor neurons in cardiovascular regulation, *Pharmacol. Rev.,* 48, 465, 1996.
2. Aicher, S.A. et al., Anatomical substrates for baroreflex sympathoinhibition in the rat, *Brain Res. Bull.,* 51, 107, 2000.

3. Sved, A.F., Ito, S., and Madden, C.J., Baroreflex dependent and independent roles of the caudal ventrolateral medulla in cardiovascular regulation, *Brain Res. Bull.,* 51, 129, 2000.
4. Sapru, H.N., Glutamate circuits in selected medullo-spinal areas regulating cardiovascular function, *Clin. Exp. Pharmacol. Physiol.,* 29, 491, 2002.
5. Misu, Y., Kitahama, K., and Goshima, Y., L-3,4-Dihydroxyphenylalanine as a neurotransmitter candidate in the central nervous system, *Pharmacol. Ther.,* 97, 117, 2003.
6. Jordan, D. and Spyer, K.M., Brainstem integration of cardiovascular and pulmonary afferent activity, *Prog. Brain Res.,* 67, 295, 1986.
7. Hilton, S.M., The defense-arousal system and its relevance for circulatory and respiratory control, *J. Exp. Biol.,* 100, 159, 1982.
8. Agarwal, S.K., Gelsema, A.J., and Calaresu, F.R., Inhibition of rostral VLM by baroreceptor activation is relayed through caudal VLM, *Am. J. Physiol.,* 258, R1271, 1990.
9. Aicher, S.A. et al., Monosynaptic projections from the nucleus tractus solitarii to C1 adrenergic neurons in the rostral ventrolateral medulla: comparison with input from the caudal ventrolateral medulla, *J. Comp. Neurol.,* 373, 62, 1996.
10. Yu, D. and Gordon, F.J., Anatomical evidence for a bi-neuronal pathway connecting the nucleus tractus solitarius to caudal ventrolateral medulla to rostral ventrolateral medulla in the rat, *Neurosci. Lett.,* 205, 21, 1996.
11. Cravo, S.L., Morrison, S.F., and Reis, D.J., Differentiation of two cardiovascular regions within caudal ventrolateral medulla, *Am. J. Physiol.,* 261, R985, 1991.
12. Badoer, E. et al., Localization of barosensitive neurons in the caudal ventrolateral medulla which project to the rostral ventrolateral medulla, *Brain Res.,* 657, 258, 1994.
13. Chan, R.K.W. and Sawchenko, P.E., Organization and transmitter specificity of medullary neurons activated by sustained hypertension: implications for understanding baroreceptor reflex circuitry, *J. Neurosci.,* 18, 371, 1998.
14. Minson, J.B. et al., c-fos identifies GABA-synthesizing barosensitive neurons in caudal ventrolateral medulla, *Neuroreport,* 8, 3015, 1997.
15. Ross, C.A. et al., Tonic vasomotor control by the rostral ventrolateral medulla: effect of electrical or chemical stimulation of the area containing C1 adrenaline neurons on arterial pressure, heart rate, and plasma catecholamines and vasopressin, *J. Neurosci.,* 4, 474, 1984.
16. Sun, M.-K. et al., Reticulospinal pacemaker neurons of the rostral ventrolateral medulla with putative sympathoexcitatory function: an intracellular study in vitro, *Brain Res.,* 442, 229, 1988.
17. Madden, C.J. et al., Lesions of the C1 catecholaminergic neurons of the ventrolateral medulla in rats using anti-DβH-saporin, *Am. J. Physiol.,* 277, R1063, 1999.
18. Schreihofer, A.M., Stornetta, R.L., and Guyenet, P.G., Regulation of sympathetic tone and arterial pressure by rostral ventrolateral medulla after depletion of C1 cells in rat, *J. Physiol. (Lond.),* 529, 221, 2000.
19. Willette, R.N. et al., Vasopressor and depressor areas in the rat medulla: identification by microinjection of L-glutamate, *Neuropharmacology,* 22, 1071, 1983.
20. Amano, M. and Kubo, T., Involvement of both $GABA_A$ and $GABA_B$ receptors in tonic inhibitory control of blood pressure in the rostral ventrolateral medulla of the rat, *Naunyn-Schmiedebs. Arch. Pharmacol.,* 348, 146, 1993.
21. Li, Y.-W. and Guyenet, P.G., Neuronal inhibition by a $GABA_B$ receptor agonist in the rostral ventrolateral medulla of the rat, *Am. J. Physiol.,* 268, R428, 1995.
22. Possas, O.S. et al., A fall in arterial blood pressure produced by inhibition of the caudalmost ventrolateral medulla: the caudal pressor area, *J. Auton. Nerv. Syst.,* 49, 235, 1994.

23. Natarajan, M. and Morrison, S.F., Sympathoexcitatory CVLM neurons mediate responses to caudal pressor area stimulation, *Am. J. Physiol.*, 279, R264, 2000.
24. Sun, W. and Panneton, W.M., The caudal pressor area of the rat: its precise location and projections to the ventrolateral medulla, *Am. J. Physiol.*, 283, R768, 2002.
25. Verberne, A.J., Sartor, D.M., and Berke, A., Midline medullary depressor responses are mediated by inhibition of RVLM sympathoexcitatory neurons in rats, *Am. J. Physiol.*, 276, R1054, 1999.
26. Coleman, M.J. and Dampney, R.A.L., Sympathoinhibition evoked from caudal midline medulla is mediated by GABA receptors in rostral VLM, *Am. J. Physiol.*, 274, R318, 1998.
27. Talman, W.T., Perrone, M.H., and Reis, D.J., Evidence for L-glutamate as the neurotransmitter of baroreceptor afferent nerve fibers, *Science*, 209, 813, 1980.
28. Kubo, T. and Kihara, M., Unilateral blockade of excitatory amino acid receptors in the nucleus tractus solitarii produces an inhibition of baroreflexes in rats, *Naunyn-Schmiedebs Arch. Pharmacol.*, 343, 317, 1991.
29. Ohta, H. and Talman, W.T., Both NMDA and non-NMDA receptors in the NTS participate in the baroreceptor reflex in rats, *Am. J. Physiol.*, 267, R1065, 1994.
30. Ohta, H., Li, X., and Talman, W.T., Release of glutamate in the nucleus tractus solitarii in response to baroreflex activation in rats, *Neuroscience*, 74, 29, 1996.
31. Lawrence, A.J. and Jarrott, B., L-glutamate as a neurotransmitter at baroreceptor afferents: evidence from in vivo microdialysis, *Neuroscience*, 58, 585, 1994.
32. Lewis, S.J. et al., Reduced glutamate binding in rat dorsal vagal complex after nodose ganglionectomy, *Brain Res. Bull.*, 21, 913, 1988.
33. Sykes, R.M., Spyer, K.M., and Izzo, P.N., Demonstration of glutamate immunoreactivity in vagal sensory afferents in the nucleus tractus solitarius of the rat, *Brain Res.*, 762, 1, 1997.
34. Zhang, J. and Mifflin, S.W., Influences of excitatory amino acid receptor agonists on nucleus of the solitary tract neurons receiving aortic depressor nerve inputs, *J. Pharmacol. Exp. Ther.*, 282, 639, 1997.
35. Dean, C. et al., Modulation of arterial baroreflexes by antisense oligodeoxynucleotides to NMDAR1 receptors in the nucleus tractus solitarius, *J. Auton. Nerv. Syst.*, 74, 109, 1998.
36. Aicher, S.A., Sharma, S., and Pickel, V.M., N-methyl-D-aspartate receptors are present in vagal afferents and their dendritic targets in the nucleus tractus solitarius, *Neuroscience*, 91, 119, 1999.
37. Hoang, C.J. and Hay, M., Expression of metabotropic glutamate receptors in nodose ganglia and the nucleus of the solitary tract, *Am. J. Physiol.*, 281, H457, 2001.
38. Chen, C.Y. et al., Synaptic transmission in nucleus tractus solitarius is depressed by Group II and III but not Group I presynaptic metabotropic glutamate receptors in rats, *J. Physiol. (Lond.)*, 538, 773, 2002.
39. Jones, N.M. et al., Type I and II metabotropic glutamate receptors mediate depressor and bradycardic actions in the nucleus tract of anaesthetized rats, *Eur. J. Pharmacol.*, 380, 129, 1999.
40. Foley, C.M. et al., Glutamate in the nucleus of the solitary tract activates both ionotropic and metabotropic glutamate receptors, *Am. J. Physiol.*, 275, R1858, 1998.
41. Jones, N.M., Monn, J.A., and Beart, P.M., Type I and II metabotropic glutamate receptors regulate the outflow of [$^{3}$H]D-aspartate and [$^{14}$C]gamma-aminobutyric acid in rat solitary nucleus, *Eur. J. Pharmacol.*, 353, 43, 1998.
42. Lin, L.H. and Talman, W.T., Colocalization of GluR 1 and neuronal nitric oxide synthase in rat nucleus tractus solitarii neurons, *Neuroscience*, 106, 801, 2001.

43. Matsumura, K. et al., Subtypes of metabotropic glutamate receptors in the nucleus of the solitary tract of rats, *Brain Res.,* 842, 461, 1999.
44. Kubo, T., Kihara, M., and Misu, Y., Ipsilateral but not contralateral blockade of excitatory amino acid receptors in the caudal ventrolateral medulla inhibits aortic baroreceptor reflex in rats, *Naunyn-Schmiedebs Arch. Pharmacol.,* 343, 46, 1991.
45. Jung, R., Bruce, E.N., and Katona, P.G., Cardiorespiratory responses to glutamatergic antagonists in the caudal ventrolateral medulla of rats, *Brain Res.,* 564, 286, 1991.
46. Miyawaki, T. et al., Role of AMPA/kainate receptors in transmission of the sympathetic baroreflex in rat CVLM, *Am. J. Physiol.,* 272, R800, 1997.
47. Takayama, K. and Miura, M., Difference in distribution of glutamate-immunoreactive neurons projecting into the subretrofacial nucleus in the rostral ventrolateral medulla of SHR and WKY: a double-labeling study, *Brain Res.,* 570, 259, 1992.
48. Amano, M., Asari, T., and Kubo, T., Excitatory amino acid receptors in the rostral ventrolateral medulla mediate hypertension induced by carotid body chemoreceptor stimulation, *Naunyn-Schmiedebs Arch. Pharmacol.,* 349, 549, 1994.
49. Sun, M.-K. and Reis, D.J., NMDA receptor mediated sympathetic chemoreflex excitation of RVL-spinal vasomotor neurons in rats, *J. Physiol. (Lond.),* 482, 53, 1995.
50. Aicher, S.A. et al., Monosynaptic projections from the nucleus tractus solitarii to C1 adrenergic neurons in the rostral ventrolateral medulla: comparison with input from the caudal ventrolateral medulla, *J. Comp. Neurol.,* 373, 62, 1996.
51. Kubo, T. et al., Evidence for the involvement of endogenous aspartate in the mediation of carotid chemoreceptor reflexes in the rostral ventrolateral medulla of the rat, *Neurosci. Lett.,* 232, 103, 1997.
52. Brailoiu, G.C., Dun, S.L., and Dun, N.J., Glutamate receptor subunit immunoreactivity in neurons of the rat rostral ventrolateral medulla, *Auton. Neurosci.,* 98, 55, 2002.
53. Tsuchihashi, T. and Averill, D.B., Metabotropic glutamate receptors in the ventrolateral medulla of rats, *Hypertension,* 21, 739, 1993.
54. Tsuchihashi, T. et al., Metabotropic glutamate receptor subtypes involved in cardiovascular regulation in the rostral ventrolateral medulla of rats, *Brain Res. Bull.,* 52, 279, 2000.
55. Chalmers, J. et al., Amino acid neurotransmitters in the central control of blood pressure and in experimental hypertension, *J. Hypertens.,* 10 (Suppl. 7), S27, 1992.
56. Minson, J.B. et al., Glutamate in spinally projecting neurons of the rostral ventral medulla, *Brain Res.,* 555, 326, 1991.
57. Stornetta, R.L. et al., Vesicular glutamate transporter DNP/VGLUT2 is expressed by both C1 adrenergic and nonaminergic presympathetic vasomotor neurons of the rat medulla, *J. Comp. Neurol.,* 444, 207, 2002.
58. Misu, Y. et al., Neurobiology of L-DOPAergic systems, *Prog. Neurobiol.*, 49, 415, 1996.
59. Yue, J.-L. et al., Baroreceptor-aortic nerve-mediated release of endogenous L-3,4-dihydroxyphenylalanine and its tonic function in the nucleus tractus solitarii of rats, *Neuroscience,* 62,145, 1994.
60. Kubo, T. et al., Evidence for L-DOPA systems responsible for cardiovascular control in the nucleus tractus solitarii of the rat, *Neurosci. Lett.,* 140, 153, 1992.
61. Yamanashi, K. et al., Involvement of nitric oxide production via kynurenic acid-sensitive glutamate receptors in DOPA-induced depressor responses in the nucleus tractus solitarii of anesthetized rats, *Neurosci. Res.,* 43, 231, 2002.
62. Miyamae, T. et al., Some interactions of L-DOPA and its related compounds with glutamate receptors, *Life Sci.,* 64, 1045, 1999.

63. Miyamae, T. et al., L-DOPAergic components in the caudal ventrolateral medulla in baroreflex neurotransmission, *Neuroscience* 92, 137, 1999.
64. Yue, J.-L., Goshima, Y., and Misu, Y., Transmitter-like L-3,4-dihydroxyphenylalanine tonically functions to mediate vasodepressor control in the caudal ventrolateral medulla of rats, *Neurosci. Lett.,* 159, 103, 1993.
65. Yue, J.-L. et al., Evidence for L-DOPA relevant to modulation of sympathetic activity in the rostral ventrolateral medulla of rats, *Brain Res.,* 629, 310, 1993.
66. Nishihama, M. et al., An L-DOPAergic relay from the posterior hypothalamic nucleus to the rostral ventrolateral medulla and its cardiovascular function in anesthetized rats, *Neuroscience,* 92, 123, 1999.
67. Tison, F. et al., Immunohistochemistry of endogenous L-DOPA in the rat posterior hypothalamus, *Histochemistry,* 93, 655, 1990.
68. Vertes, R.P. and Crane, A.M., Descending projections of the posterior nucleus of the hypothalamus: *Phaseolus vulgaris* leucoagglutinin analysis in the rat, *J. Comp. Neurol.,* 354, 607, 1996.

Baroreflex is the principal neuronal mechanism by which the cardiovascular system is regulated under a negative feedback control. Figure 11.1 represents the survey of these events [8–11,24–30] focused mainly on cardiovascular responses to exogenously microinjected levodopa and DOPA methyl ester (DOPA ME), a competitive antagonist for DOPA [31], and endogenously released DOPA in the depressor NTS, depressor CVLM, and pressor RVLM. Arterial baroreceptors are located both in the aortic arch and the carotid sinus. The primary sensory baroreceptor afferents, the aortic depressor nerve (ADN) and carotid sinus nerve (CSN), terminate in depressor sites of the NTS. Glutamate, a representative of excitatory amino acids, has been regarded as the most probable neurotransmitter of the primary baroreceptor afferents [24], the second-order interneurons, and the following microcircuit neurons within depressor sites of the NTS [29]. Glutamate is also believed to be neurotransmitter of the direct excitatory pathway from the NTS to depressor sites of the CVLM [32–34], and the excitatory pathway from the RVLM to the thoracic intermediolateral cell column (IML) [35] among many neurotransmitter candidates. The neurotransmitter of the inhibitory neurons projecting from the CVLM to the RVLM is accepted generally to be γ-aminobutyric acid (GABA) [27,36,37].

In the case of DOPA, during microdialysis of the NTS [14], CVLM [19], and RVLM [22] in anesthetized rats, basal DOPA release is markedly inhibited by α-methyl-*p*-tyrosine (α-MPT, 200 mg/kg, i.p.), an inhibitor of tyrosine hydroxylase (TH), a rate-limiting enzyme for biosynthesis of catecholamines. Basal DOPA release is in part inhibited by application of tetrodotoxin (TTX) and deprivation of extracellular $Ca^{2+}$ performed via microdialysis probes (see Chapter 5). DOPA appears to be released at least partially via the spontaneous activity of some neurons. High $K^+$-induced depolarization releases DOPA in a $Ca^{2+}$-dependent manner in these three areas. $Ca^{2+}$ ions appear to be primarily involved in the process of DOPA release rather than that of activation of TH [38].

In the NTS, intermittent strong electrical stimulation of the ADN releases DOPA with hypotension and bradycardia [14]. The evoked DOPA release is suppressed by TTX. In addition, baroreceptor stimulation by infusion of phenylephrine (i.v.) elicits DOPA release and reflex bradycardia, both of which are suppressed by bilateral sinoaortic denervation. These findings suggest that DOPA is a neurotransmitter of the primary baroreceptor afferents to carry baroreflex information to the NTS. Neurons showing immunocytochemically TH-positive, DOPA-positive, AADC-negative, and dopamine-negative reactivity exist in some regions of the CNS, including the NTS [14,39,40], which may contain DOPA as an end product [9,11]. Furthermore, unilateral denervation of the ADN peripheral to the ganglion nodosum containing the cell bodies decreases immunocytochemical reactivity of TH and DOPA but not that of dopamine and dopamine-β-hydroxylase in the ipsilateral medial and intermediate subdivisions of the NTS and the dorsal motor nucleus of the vagus (DMV) [14]. Decreases in TH-immunoreactivity and DOPA-immunoreactivity are also observed in the ipsilateral ganglion nodosum. The ADN appears to utilize DOPA but not dopamine or noradrenaline as a neurotransmitter, and the long axons of the ADN having a monosynaptic property appear to terminate directly in depressor sites or indirectly in some secondary neurons in synapse with depressor sites within the neuronal microcircuits of the NTS.

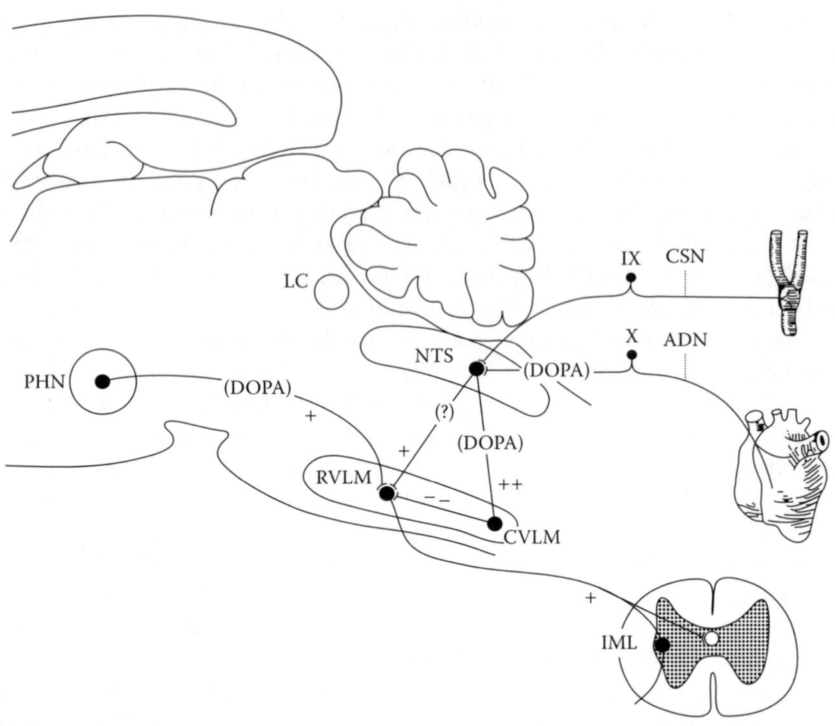


**FIGURE 11.1** Hypothesized DOPA-mediated neuronal pathways for baroreflex and central regulation of arterial blood pressure in the lower brainstem of anesthetized rats. Cardiovascular responses to exogenous microinjection of levodopa and a competitive antagonist DOPA ME and to endogenously released DOPA in depressor sites of the NTS and CVLM and in pressor sites of the RVLM are summarized. The unilateral microinjection of nanomolar levodopa produces hypotension and bradycardia in the NTS and CVLM and hypertension and tachycardia in the RVLM. These responses are neither inhibited by a central AADC inhibitor nor mimicked by the same low-dose ranges of dopamine, noradrenaline, and adrenaline, but antagonized by the ipsilateral microinjection of DOPA ME in these three areas. The bilateral microinjection of DOPA ME alone elicits hypertension and tachycardia in the depressor NTS and CVLM and hypotension and bradycardia in the pressor RVLM. These responses are suppressed following inhibition of DOPA biosynthesis by the prior i.p. injection of α-MPT, a TH inhibitor. These findings suggest the tonic function of endogenously released DOPA for resting integration of arterial blood pressure in these three centers. Electrical stimulation of the ADN releases TTX-sensitive DOPA in the NTS and α-MPT-sensitive DOPA in the CVLM and elicits hypotension and bradycardia. The hypotension and bradycardia are antagonized by the ipsilateral microinjection of DOPA ME in the NTS in a manner similar to exogenously applied levodopa. This hypotension is also antagonized by the ipsilateral microinjection of DOPA ME in the CVLM. Phenylephrine-induced baroreceptor activation elicits DOPA release in the NTS and reflex bradycardia, both of which are abolished by bilateral sinoaortic denervation. Reflex bradycardia is also antagonized by the bilateral microinjection of DOPA ME in the NTS. These findings suggest a tonic function of DOPA for baroreflex integration in the NTS. Phenylephrine-induced baroreceptor activation also elicits DOPA release in the CVLM, which is suppressed by acute lesion of the ipsilateral NTS. Unilateral electrical stimulation of the PHN releases DOPA in the ipsilateral RVLM and elicits

In contrast with the NTS, no immunocytochemical evidence for existence of neurons having DOPA as an end product has been shown in the CVLM and the RVLM [9,11]. Notwithstanding, during microdialysis of the CVLM, intermittent intensive electrical stimulation of the ADN releases DOPA selectively, which is suppressed by inhibition of DOPA biosynthesis with α-MPT locally infused into the CVLM [21]. Newly synthesized DOPA appears to be released by ADN stimulation. No increases in extracellular levels of dopamine, noradrenaline, and adrenaline are seen. In addition, baroreceptor activation by phenylephrine also releases DOPA selectively, which is suppressed by acute electrolytic lesion of the ipsilateral NTS. No increases in extracellular levels of dopamine, noradrenaline, and adrenaline are observed. Furthermore, electrolytic lesion of the unilateral NTS decreases selectively the tissue content of DOPA by 45% without decreases in that of dopamine, noradrenaline, and adrenaline in the ipsilateral CVLM. These findings suggest that NTS projects a DOPAergic depressor relay with a monosynaptic property directly to depressor sites or indirectly to some neurons near depressor sites at least in the ipsilateral CVLM [21]. A DOPAergic baroreceptor–ADN–NTS–CVLM depressor relay appears to exist for carrying baroreflex information to the CVLM [14,21]. From findings obtained in the CVLM, we can safely say that the sensitivity of immunocytochemical analysis to identify DOPAergic neurons appears to be lower compared with biochemical approaches in combination with denervation experiments. This is also the case in the RVLM.

The RVLM receives excitatory inputs from the various regions in the CNS including the NTS, hypothalamic paraventricular nucleus, and lateral hypothalamic area [8,9,25,26,30]. In the posterior hypothalamic nucleus (PHN) (A11 cell group) some cells display TH-positive, weakly AADC-positive [41], and markedly DOPA-positive immunocytochemical reactivity with very weakly dopamine-positive immunocytochemical reactivity [42]. The rat PHN neurons send descending projections to the RVLM [43]. Intermittent strong electrical stimulation of the unilateral PHN releases DOPA selectively during microdialysis of the ipsilateral RVLM with hypertension and tachycardia [23]. The evoked DOPA release is abolished by TTX. No increases in extracellular levels of dopamine, noradrenaline, and adrenaline are seen. In addition, electrolytic lesion of the bilateral PHN decreases selectively the tissue content of DOPA by 50% without decreases in that of dopamine, noradrenaline, and adrenaline in the unilateral RVLM. These findings suggest that PHN projects a DOPAergic pressor relay having a monosynaptic property directly to pressor sites or indirectly to some neurons near pressor sites in the RVLM in a cross-innervating manner.

The evidence for competitive antagonism is one of several important criteria for establishment of a compound as a neurotransmitter. DOPA ME is the first competitive

**FIGURE 11.1 (Continued)** hypertension and tachycardia. TTX perfused into the RVLM suppresses DOPA release and partially inhibits hypertension with a slight inhibition of tachycardia. This hypertension and tachycardia are in part antagonized by the bilateral microinjection of DOPA ME in the RVLM. These findings suggest the existence of a DOPA-mediated baroreceptor–ADN–NTS depressor relay, a DOPA-mediated baroreceptor–ADN–NTS–CVLM depressor relay, and a DOPA-mediated PHN–RVLM pressor relay in the lower brainstem of anesthetized rats. Other details are shown in Figure 5.2. (Modified from Misu, Y., Kitahama, K., and Goshima, Y., L-3,4-Dihydroxyphenylalanine as a neurotransmitter candidate in the central nervous system, *Pharmacol. Ther.,* 97, 117, 2003. With permission.)

antagonist to be identified for levodopa [31,44]. Among DOPA ester compounds, DOPA cyclohexyl ester (DOPA CHE) is the most potent and relatively stable competitive antagonist for levodopa [45,46]. These two antagonists antagonize responses to both exogenously applied levodopa [13,19,22,31,44–48] and endogenously released DOPA [14,19,21–23,49,50]. Competitive antagonism suggests the existence of DOPA recognition sites.

DOPA CHE does not inhibit the uptake of labeled levodopa into *Xenopus laevis* oocytes [51]. Levodopa up to 1 m$M$ does not displace the selective binding of tritiated dopamine $D_1$ and $D_2$ receptor ligands in rat brain membrane preparations [46]. DOPA ester compounds containing DOPA ME and DOPA CHE inhibit slightly the binding of these ligands with $IC_{50}$ values of millimolar range, which are far higher than those of dopamine (2 to 3 $\mu M$) [46]. DOPA ME fails to displace the selective binding of a tritiated β-adrenoceptor ligand, whereas this antagonist inhibits slightly the selective binding of tritiated rauwolscine, an $\alpha_2$-adrenoceptor ligand [31]. The affinity of DOPA ME for the binding site of rauwolscine, however, is negligible (Ki = 40 $\mu M$), compared with that (Ki = 2 to 20 n$M$) of typical $\alpha_2$-adrenoceptor agonists and antagonists [52]. In addition, DOPA ME inhibits hardly the selective binding of tritiated GABA [16]. Furthermore, we did the binding assay for ionotropic glutamate receptors with available tritiated ligands to label the ion channel domain of *N*-methyl-D-aspartate (NMDA) receptors, NMDA binding site, NMDA glycine site, DL-α-amino-3-hydroxy-5-methyl-4-isoxazol propionic acid (AMPA) receptors, and kainate receptors, respectively [46,53]. Among these binding sites, DOPA ester compounds act only on NMDA ion channel domain with $IC_{50}$ values in the millimolar range, whereas levodopa acts only on AMPA receptors with low affinity. DOPAergic agonist and competitive antagonists should act on the same sites, which appear to differ from DOPA transport sites, catecholamine receptors, GABA receptors, and ionotropic glutamate receptors. In addition, the prior microinjection of DOPA ME fails to antagonize cardiovascular responses to glutamate in the NTS [13], CVLM [19], RVLM [22], and to NMDA in the CVLM [21]. To support the identification of DOPA as a neurotransmitter, it is essential that there exist DOPA recognition sites, on which DOPA or levodopa acts to elicit physiological or pharmacological responses.

In this chapter, we mainly survey cardiovascular responses to exogenously microinjected levodopa and DOPA ME, and physiologically released DOPA in the NTS, CVLM, and RVLM in the lower brainstem.

## 11.2 RESPONSES TO LEVODOPA MICROINJECTED INTO DEPRESSOR SITES OF NTS AND CVLM, AND PRESSOR SITES OF RVLM IN ANESTHETIZED RATS

Male Wistar rats anesthetized with urethane (1.2 g/kg, i.p.) were usually used. Figure 11.2A to Figure 11.2D show representative traces of responses to levodopa (L-DOPA) microinjected into depressor sites of the NTS in the presence of 3-hydroxybenzylhydrazine (NSD-1015), a central AADC inhibitor [13]. When delivered at the depressor sites, the medial area of the NTS, identified with the levo type of glutamate (monosodium salt) at 30 ng, levodopa at 10 to 100 ng elicits dose-dependent decreases in arterial blood pressure (BP) and heart rate (HR)

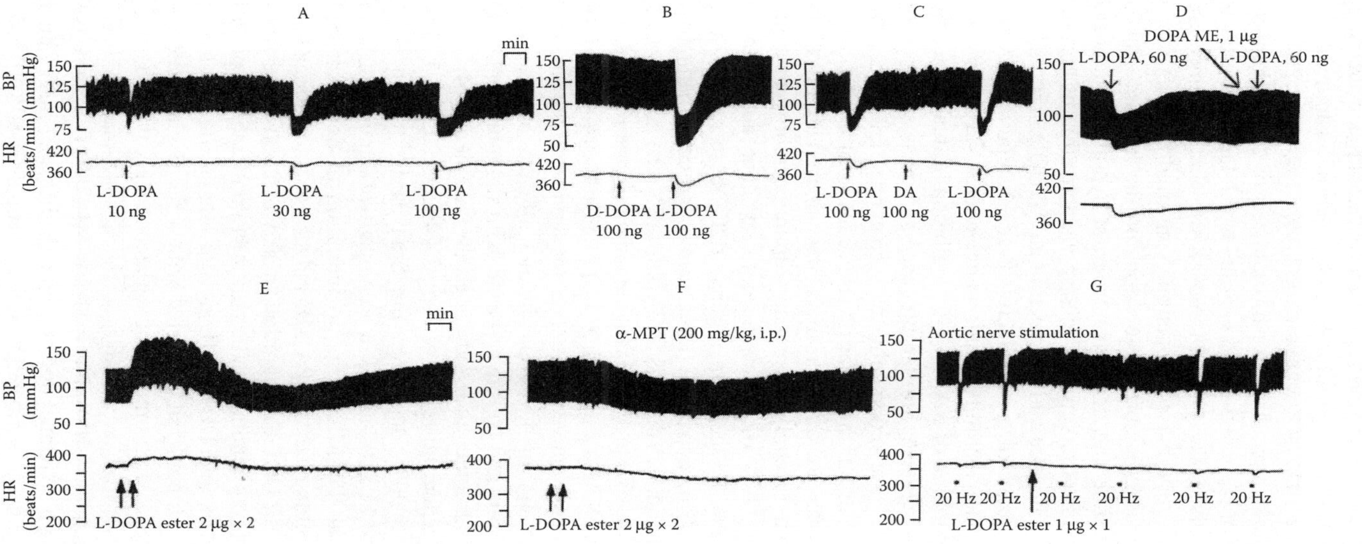


**FIGURE 11.2** Typical traces of responses to levodopa (L-DOPA) and a competitive antagonist DOPA ME (L-DOPA ester) microinjected into depressor sites of the NTS and to DOPA released by electrical stimulation of the ADN (aortic nerve), and antagonism induced by DOPA ME in anesthetized rats. Hypotension and bradycardia elicited by the unilateral microinjection of levodopa are (A) dose dependent, (B) stereoselective, (C) not mimicked by dopamine (DA) at the same highest dose to levodopa, (D) antagonized by ipsilateral microinjection of DOPA ME, are also seen under inhibition of AADC, and are postsynaptic in nature. The bilateral microinjection of DOPA ME alone elicits hypertension and tachycardia (E), which are suppressed by the prior i.p. injection of β-MPT, an inhibitor of TH, involved in the biosynthesis of DOPA from tyrosine (F). These findings suggest that basally released DOPA, acting on its recognition sites, functions tonically to activate depressor sites of the NTS. A further important finding is that the unilateral microinjection of DOPA ME antagonizes hypotension and bradycardia elicited by either the ipsilateral electrical ADN stimulation (G) or exogenous microinjection of levodopa (D) in a similar manner. (Modified from Misu, Y., Kitahama, K., and Goshima, Y., L-3,4-Dihydroxyphenylalanine as a neurotransmitter candidate in the central nervous system, *Pharmacol. Ther.*, 97, 117, 2003. With permission.)

(Figure 11.2A). The same responses are observed in the absence of NSD-1015 (15, 16, 18, 45, 46), suggesting the action of levodopa by itself but not bioconversion to dopamine. These responses to levodopa, similar to many neurotransmitters, are stereoselective, because dextrodopa (D-DOPA) elicits no effect (Figure 11.2B).

DOPA is, as a matter of course, a precursor for biosynthesis of catecholamines and α-methyl DOPA, a prodrug of α-methyl-noradrenaline, which has been used as a centrally acting antihypertensive drug [12]. Moreover, the first evidence for responses to levodopa in its own right but not dopamine converted from levodopa is the presynaptic biphasic regulatory actions of levodopa on the impulse-evoked release of endogenous noradrenaline and dopamine from rat hypothalamic slices [5]. Nanomolar levodopa facilitates noradrenaline release via activation of presynaptic β-adrenoceptors both in the absence and presence of an AADC inhibitor. This facilitation is antagonized by propranolol, a nonselective β-adrenoceptor antagonist [12]. In contrast, micromolar levodopa inhibits noradrenaline release via presynaptic dopamine $D_2$ receptors under inhibition of AADC. This inhibition is antagonized by sulpiride, a dopamine $D_2/D_3$ antagonist. Similar presynaptic biphasic effects of levodopa on the impulse-evoked release of dopamine are seen in rat striatal slices [6].

In the NTS, however, depressor and bradycardic responses to 50-ng levodopa microinjected into the depressor sites are seen in a similar manner even after destruction of noradrenergic neurons in the medulla oblongata with intracerebroventricular injection of 6-hydroxydopamine [13]. These responses to levodopa appear to be postsynaptic in nature. In addition, as shown in Figure 11.2C, these responses to levodopa are not mimicked by 100-ng dopamine (DA), corresponding to the highest dose of levodopa used. Noradrenaline at 100 ng is also confirmed to be noneffective. We confirmed that the minimum effective doses of dopamine, noradrenaline, α-methyl-noradrenaline, and adrenaline are around 1 μg [14,54]. Microinjection of isoproterenol at 0.45 to 28 μg, a nonselective β-adrenoceptor agonist, elicits no effect [55]. Furthermore, responses to 100-ng levodopa are not affected by the prior microinjection of 100-ng propranolol and 100-ng sulpiride, with slight but negligible inhibition induced by 100-ng yohimbine, an $\alpha_2$-adrenoceptor antagonist [14].

On the other hand, the prior microinjection of 1-μg DOPA ME 1 min before levodopa antagonizes, by 70 to 90%, depressor and bradycardic responses to levodopa at 60 to 100 ng (Figure 11.2D) [13,46] in a competitive manner [45]. An important finding is that 1-μg DOPA ME elicits no inhibition of depressor and bradycardic responses to 30-ng glutamate. DOPA-ME-induced antagonism against levodopa appears to be selective for DOPA recognition sites, which may differ from ionotropic glutamate receptors. From binding studies in rat brain membrane preparations, it is evident that levodopa and DOPA ME do not interact on ionotropic glutamate receptors [53]. Among binding sites labeled with tritiated glutamatergic ligands, levodopa displaces only specific binding of tritiated AMPA with the $IC_{50}$ of 260 μ*M*, whereas DOPA ME does only that of tritiated (+)-5-methyl-10,11-dihydro-5*H*-dibenzol[a,d]cyclohepten-5,10-imine maleate (MK-801), a selective antagonist of the NMDA receptor ion channel domain, with the $IC_{50}$ of 1 m*M*, which is far higher, by five orders of magnitude, compared with that of cold MK-801 (10 n*M*). Furthermore, DOPA ME up to 1 m*M* fails to displace specific binding of tritiated ligands to label MNDA binding site and NMDA glycine site. DOPA ME, however, has some limitation as the levodopa antagonist. The unilateral microinjection of DOPA ME alone elicits slight decreases in BP and HR [13], which are probably due to bioconversion to DOPA. DOPA ME is a representative prodrug for DOPA and is readily hydrolyzed to DOPA [56]. Decreases in BP and HR elicited by levodopa are restored within 10 min after pretreatment with DOPA ME [46].

Levodopa at 10 to 100 ng microinjected into depressor sites of the CVLM elicits decreases in BP and HR in a manner similar to the NTS [19]. These responses are dose dependent in the absence and presence of NSD-1015, stereoselective, postsynaptic, and markedly antagonized by 1-μg DOPA ME without modifications of decreases in BP and HR induced by 100-ng glutamate. In addition, the prior microinjection of 1-μg DOPA ME fails to antagonize hypotension and bradycardia induced by 0.3-ng NMDA [21], confirming selective antagonism via DOPA recognition sites but not nonselective antagonism via NMDA receptors. In the case of CVLM, however, dopamine and noradrenaline at 100 ng elicit slight but far smaller decreases in BP and HR, compared with 100-ng levodopa [19].

Figure 11.3A and Figure 11.3B show summarized data of pressor and tachycardiac responses to levodopa at 30 to 300 ng microinjected into pressor sites of the RVLM [22]. These responses are dose dependent in the absence and presence of NSD-1015 (Figure 11.3A), stereoselective, postsynaptic, and markedly antagonized by 1.5-μg DOPA ME without modifications of increases in BP and HR elicited by glutamate (Glu) at 300 ng (Figure 11.3B). Microinjection of dopamine, noradrenaline, and adrenaline at 300 ng elicits no effect.

There is a difference of the used doses of levodopa between the depressor NTS or CVLM (10 to 100 ng) and the pressor RVLM (30 to 300 ng) [13,19,22]. This difference might reflect the possibility that the total tonicity of the excitatory pathways from various regions of the CNS to the RVLM [8,9,25,26,30] is less than the tonicity of the predominant excitatory baroreceptors–primary baroreceptor afferents–NTS–CVLM pathway.

## 11.3 TONIC FUNCTION OF BASALLY RELEASED DOPA VIA ACTIVATION OF DOPA RECOGNITION SITES FOR RESTING INTEGRATION OF ARTERIAL BP IN NTS, CVLM, AND RVLM IN ANESTHETIZED RATS

It is evident that basally released DOPA activates DOPA recognition sites and functions tonically to integrate resting arterial BP in the depressor NTS [14]. The microinjection of DOPA ME (L-DOPA ester) alone at 1 and 2 μg (Figure 11.2E) into the bilateral depressor sites of the NTS in the absence of NSD-1015 elicits initial increases followed by decreases in BP and HR in a dose-dependent manner. Furthermore, pretreatment with α-MPT (200 mg/kg, i.p., 80 min prior to microinjection of DOPA ME) suppresses initial increases in BP and HR and unmasks initial decreases in BP and HR (Figure 11.2F). α-MPT markedly decreases basal DOPA release during microdialysis of the NTS. DOPA ME appears to produce hypertension and tachycardia following antagonism against newly synthesized and basally released DOPA to activate the depressor sites, which function tonically to integrate

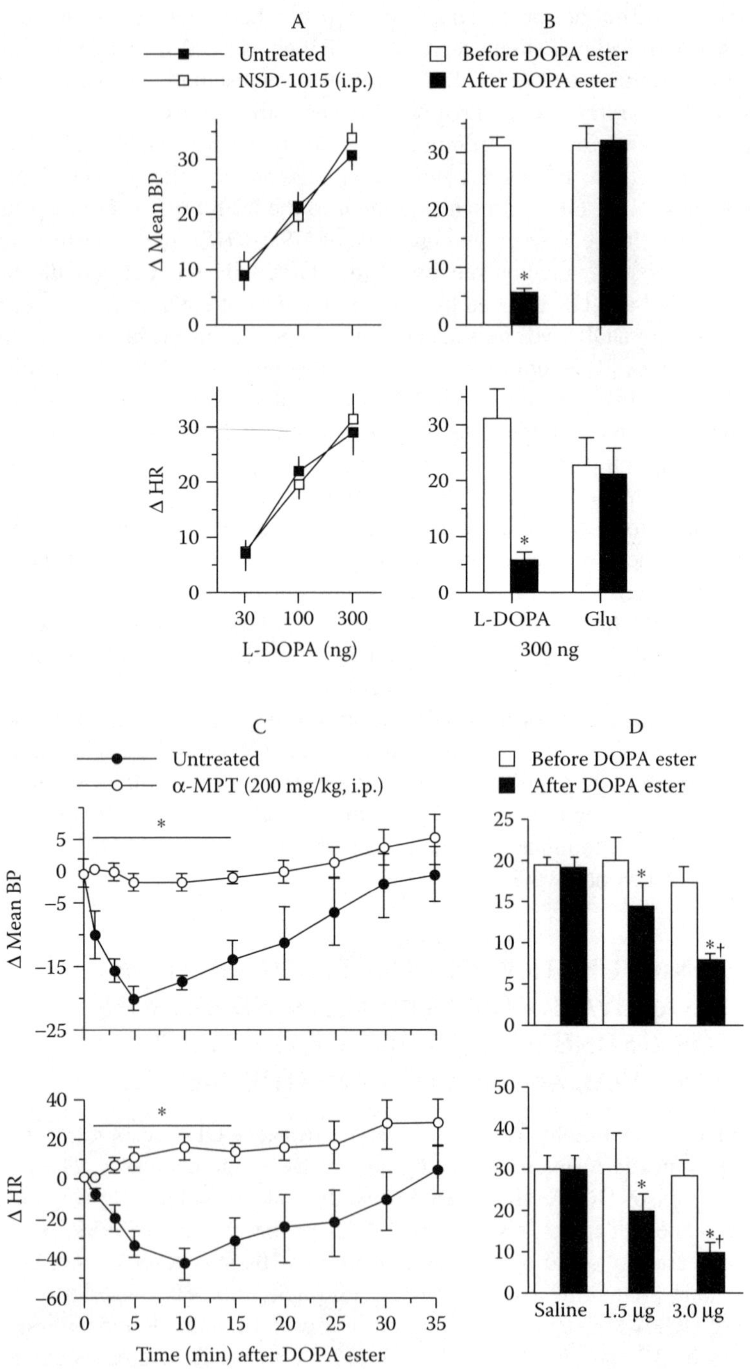


**FIGURE 11.3** Summarized data of responses to microinjection of levodopa (L-DOPA) and DOPA ME (DOPA ester) into pressor sites of the RVLM and responses to DOPA released

resting arterial BP in the NTS. Decreases in BP and HR elicited by DOPA ME are probably due to bioconversion to DOPA.

It is also evident in the CVLM that newly synthesized and basally released DOPA activates DOPA recognition sites located in the depressor sites and functions tonically to integrate resting arterial BP [19]. The microinjection of DOPA ME alone at 1 μg into the bilateral depressor sites of the CVLM and pretreatment with α-MPT produce findings similar to those obtained in the NTS.

It is further evident in the RVLM that newly synthesized and basally released DOPA activates DOPA recognition sites located in the pressor sites and functions tonically to integrate resting arterial BP [22]. Microinjection of DOPA ME alone at 2 μg into the bilateral pressor sites produces decreases in BP and HR, which are suppressed by pretreatment with α-MPT (Figure 11.3C). DOPA ME appears to produce hypotension and bradycardia following antagonism against basally released DOPA to activate the pressor sites.

## 11.4 DOPA ME ANTAGONIZED HYPOTENSION AND BRADYCARDIA INDUCED BY EITHER ELECTRICAL STIMULATION OF ADN OR EXOGENOUSLY MICROINJECTED LEVODOPA IN NTS

If DOPA is a neurotransmitter of the primary baroreceptor afferent ADN terminating in depressor sites of the NTS, electrical stimulation of the ADN elicits DOPA release accompanied by hypotension and bradycardia, which should be antagonized by

**FIGURE 11.3 (Continued)** by electrical stimulation of the PHN and antagonism induced by DOPA ME in anesthetized rats. Dose–response curves for pressor and tachycardiac responses of untreated (■) and NSD-1015-treated (□) rats to the unilateral microinjection of levodopa into pressor sites of the RVLM (A). Effects of DOPA ME, determined 1 min after the unilateral microinjection at 1.5 μg, on responses of NSD-1015-treated rats to levodopa and glutamate (Glu) (B). Effects of the bilateral microinjection of 2-μg DOPA ME alone on mean BP and HR in untreated (●) and α-MPT-treated (○) rats (C). α-MPT was i.p. injected 80 min before microinjection. Dose-dependent antagonism by the bilateral microinjection of 1.5 and 3.0-μg DOPA ME against pressor and tachycardiac responses to electrical stimulation of the unilateral electrical stimulation of the PHN (33 Hz, 0.2 mA, 0.1 msec, and 10 sec every 3 to 4 min) (D). Effects of DOPA ME and saline were determined by the peak inhibition of pressor and tachycardiac responses to the first or second PHN stimulation after microinjection. Values are mean ± SE (n = 5 to 6 in A to C, and n = 9 to 18 in D). $^*P < .01$ vs. the value before DOPA ME in B and vs. untreated in C (Student's *t*-test). $^*P < .05$ vs. before microinjection (paired *t*-test) and $^\dagger P < .05$ vs. saline (one-way ANOVA/Bonferroni's test) in D. (Modified from Yue, J.-L. et al., Evidence for L-DOPA relevant to modulation of sympathetic activity in the rostral ventrolateral medulla of rats, *Brain Res.*, 629, 310, 1993, and Nishihama, M. et al., An L-DOPAergic relay from the posterior hypothalamic nucleus to the rostral ventrolateral medulla and its cardiovascular function in anesthetized rats, *Neuroscience*, 92, 123, 1999. With permission.)

DOPA ME. Indeed, during microdialysis of the NTS, strong and intermittent electrical stimulation of the ADN (100 Hz, 8 V, 0.1 msec, for 4 min, and four such episodes over a 20-min interval) releases DOPA in a repetitive, constant, and TTX-sensitive manner with hypotension and bradycardia. Mild and brief stimulation of the ADN (20 Hz, 3 V, 0.1 msec, for 10 sec) also elicits decreases in BP and HR in a repetitive and constant manner (Figure 11.2G). It is an important finding that DOPA ME at 1 μg microinjected into the unilateral depressor sites antagonizes hypotension and bradycardia elicited ipsilaterally by either DOPA released following the ADN stimulation [14] or exogenously microinjected levodopa [13,45,46] in a similar manner (Figure 11.2D and Figure 11.2G). It is highly probable that DOPA is a neurotransmitter of the primary sensory baroreceptor afferent ADN terminating in depressor sites of the NTS.

## 11.5 DOPA ME ANTAGONIZED BAROREFLEX BRADYCARDIA ELICITED BY PHENYLEPHRINE IN NTS

If DOPA is a neurotransmitter of the primary baroreceptor afferents, baroreceptor activation by phenylephrine produces DOPA release in the NTS and reflex bradycardia. Indeed, intensive and sustained infusion of phenylephrine (50 μg/kg/min for 20 min, i.v.) elicits DOPA release during microdialysis of the NTS and baroreflex bradycardia, both of which are abolished by bilateral sinoaortic denervation without modification of hypertension [14] (see Figure 5.3 in Chapter 5). Furthermore, the microinjection of DOPA ME at 1 μg into the bilateral depressor sites antagonizes reflex bradycardia induced by mild and brief infusion of phenylephrine (15 μg/10 μl/min, infused for 10 sec, i.v.). DOPA released by baroreceptor stimulation appears to function for baroreflex integration of arterial BP via activation of DOPA recognition sites within depressor sites of the NTS.

## 11.6 FURTHER EVIDENCE IN CVLM

During microdialysis of the unilateral CVLM, strong and intermittent electrical stimulation of the ipsilateral ADN (20 Hz, 3 V, 0.3 msec, for 3 min and six such episodes over a 30-min interval) elicits repetitively and constantly selective DOPA release, hypotension, and bradycardia [21]. No release of dopamine, 3,4-dihydroxyphenylacetic acid, a metabolite of dopamine, noradrenaline, or adrenaline is observed. Local inhibition of DOPA biosynthesis with 30-μ*M* α-MPT infused into the ipsilateral CVLM abolishes this DOPA release, but partially inhibits (approximately by 60%) hypotension elicited by ADN stimulation. These findings suggest that a DOPAergic signal is carried at least from the unilateral ADN to the ipsilateral CVLM, whereas catecholamines are not involved in this neurotransmission. α-MPT, however, hardly inhibits bradycardia in response to ADN stimulation.

We further confirmed effects of DOPA ME and $CoCl_2$, a nontoxic and mainly presynaptic $Ca^{2+}$ channel blocker for neurotransmission [57,58], microinjected into the depressor sites, on depressor and bradycardic responses to mild and brief stimulation of the ispilateral ADN (20 Hz, 3 V, 0.1 msec, for 10 sec). The release of putative neurotransmitters other than DOPA in response to ADN stimulation is also to be generally inhibited by $CoCl_2$. DOPA ME at 1 μg and $CoCl_2$ at 119 ng partially inhibit (approximately by 60%) hypotension in a manner similar to α-MPT. Thus, the remaining DOPA-ME- and $CoCl_2$-insensitive component of hypotension might be due to signal transmission from the unilateral medial NTS to the contralateral depressor sites of the CVLM [59].

In contrast, bradycardic responses to ADN stimulation are hardly inhibited by the ipsilateral application of α-MPT, and even by DOPA ME and $CoCl_2$ into the CVLM. These findings suggest that the main pathways of the preganglionic efferents of baroreceptor cardiovagal reflexes, which originate from the nucleus ambiguus and DMV [60–62], do not pass through depressor sites of the CVLM. On the other hand, DOPA ME markedly antagonizes bradycardic responses to levodopa microinjected into the depressor sites, which appears to be mainly due to a decrease in the tonicity of the cardiac sympathetic nerve. This type of decrease in response to ADN stimulation appears to be a minor component.

## 11.7 FURTHER EVIDENCE IN RVLM

During microdialysis of the unilateral RVLM, intense electrical stimulation of the ipsilateral PHN (50 Hz, 0.3 mA, 0,1 msec, twice for 5 min at an interval of 5 min) selectively releases DOPA in a repetitive and constant manner accompanied by hypertension and tachycardia [23]. However, no dopamine, noradrenaline, or adrenaline is released. TTX at 1 μ*M* perfused into the ipsilateral RVLM inhibits markedly evoked DOPA release (see Figure 5.5 in Chapter 5), but partially inhibits hypertension with slight but significant inhibition of tachycardia. These findings suggest that the unilateral PHN projects a DOPAergic pressor relay at least to the ipsilateral RVLM, whereas no catecholamines are involved in this neurotransmission. TTX-induced partial inhibition of hypertension and tachycardia in response to PHN stimulation, however, is very weak, compared with the suppression of evoked DOPA release, which is in part explained by signal transmission from the unilateral PHN to contralateral RVLM [63]. In accordance with this idea, as shown in Figure 11.3D, DOPA ME at 1.5 and 3.0 μg microinjected into the bilateral pressor sites dose-dependently antagonizes pressor and tachycardiac responses to mild and transient stimulation of the unilateral PHN (33 Hz, 0.2 mA, 0.1 msec, for 10 sec) [23]. This antagonism, however, is still incomplete, suggesting that the remaining responses are probably due to further intact neuronal activity of the direct pathway from the PHN to the IML of the thoracic spinal cord [64].

TTX perfused into the unilateral RVLM elicits no changes of resting mean BP and HR, which are sufficiently maintained by the neuronal activity of the pathway through the contralateral RVLM [63] and of the direct pathway from the PHN to the IML [64]. The sympathetic preganglionic neurons in the IML further receive direct projections from the various regions in the CNS such as the paraventricular nucleus, the A5 noradrenergic cell group, and the caudal raphe nuclei [65].

## 11.8 ALTERED DOPAERGIC SYSTEMS IN NTS, CVLM, AND RVLM OF SPONTANEOUSLY HYPERTENSIVE RATS

Lesions of the NTS and sinoaortic denervation produce transient neurogenic hypertension [66,67]. The spontaneously hypertensive rat (SHR) has been frequently used as a genetic model for essential hypertension. Involvement of the CNS in the development and maintenance of hypertension has been suggested. For example, sensitivity of baroreceptors in stroke-prone SHR [68] and inhibitory cardiopulmonary reflex in young and adult SHR [69] decrease, compared with age-matched Wistar Kyoto rats (WKY). In SHR, a pressor response is enhanced in the NTS [70], whereas a depressor response is less active in the RVLM [36]. Microinjection of TTX into the bilateral CVLM produces marked hypertension in adult WKY up to a BP level similar to that in age-matched SHR, whereas no such tonic effect of TTX is seen in SHR [71]. Alterations of the catecholaminergic systems in the brainstem appear to be involved in the development and maintenance of hypertension in SHR [72–74]. AADC but not TH activity decreases in the lower brainstem of young and adult SHR [72], whereas TH activity increases in the NTS of Wistar rats with elevated BP at 1 week after sinoaortic denervation [67].

In contrast, no evidence has been shown concerning alterations in the basal release of neurotransmitter candidates and the tonic neuronal activity to release basal neurotransmitter candidates during microdialysis of the NTS, CVLM, and RVLM of SHR in comparison with WKY. Our idea is that DOPA is a neurotransmitter of the primary baroreceptor afferents, the depressor relay from the NTS to the CVLM, and the pressor relay from the PHN to the RVLM, to play a role in central regulation of arterial BP in the lower brainstem. Thus, we attempted to clarify whether or not basal DOPA release, TTX-sensitive tonic neuronal activity to release basal DOPA, TH and AADC activity, and postsynaptic responses to the microinjection of levodopa are altered in the NTS [15], CVLM [20], and RVLM [15] in adult SHR and age-matched WKY.

TH and AADC activity was assayed in the tissues dissected from the lower brainstem. The caudal dorsomedial medulla region including the bilateral NTS was dissected out between the levels 1.5 mm rostral and 1.0 mm caudal to the obex. We dissected the bilateral CVLM regions between the levels 1.0 mm rostral and 0.5 mm caudal to the pyramidal decussation and the bilateral RVLM regions between the levels 0 and 2 mm caudal to the trapezoid bodies [9].

The resting mean BP and HR are always elevated in SHR (137 ± 3 mmHg and 388 ± 5 beats/min, n = 14, $P < .05$, respectively, Student's *t*-test), compared with WKY (88 ± 3 mmHg and 358 ± 5 beats/min, n = 14) in microinjection experiments for the NTS and RVLM [15]. Using HPLC-ECD, basal DOPA release is consistently detectable during microdialysis of the NTS, CVLM, and RVLM in the two strains. Basal DOPA release is partially inhibited by 1-μ*M* TTX [15,20], as in the case of normotensive Wistar rats [14,19,22], except the case of the CVLM in SHR [20] (Figure 11.4). TTX-sensitive tonic neuronal activity to release basal DOPA was calculated as the total absolute value of basal DOPA release before TTX perfusion minus that after TTX perfusion.

### 11.8.1 NTS

As shown in Figure 11.4, the absolute value of basal DOPA release is consistently lower in the depressor NTS of adult SHR, compared with age-matched WKY [15]. Furthermore, 1-μ*M* TTX reduces gradually and partially basal DOPA release to the same levels in the two strains 2 h after perfusion, which means that TTX reduces it by different amounts between SHR and WKY. TTX-sensitive tonic neuronal activity to release basal DOPA is lower in the depressor NTS of SHR, compared with WKY. These findings appear to be parallel to the observation that the sensitivity of baroreceptors in stroke-prone SHR is diminished, compared with WKY [68], DOPA is indeed a neurotransmitter of the primary baroreceptor afferents. Decreases in TTX-sensitive tonic neuronal activity to release basal DOPA and decreases in its absolute amounts in the NTS of SHR appear to be further due to the disturbance of the process of DOPA release itself, but not secondarily due to decrease in biosynthesis or increase in bioconversion of DOPA. TH activity to synthesize DOPA from tyrosine inversely increases in the caudal dorsomedial medulla region including the bilateral NTS of SHR, compared with WKY, whereas no difference of AADC activity to convert DOPA to dopamine is seen between SHR and WKY [15].

The increase in TH activity in the NTS of SHR, compared with WKY, appears to be a reflection of hypertension in SHR. This finding is not consistent with the observation that no difference of TH activity is seen in the whole lower brainstem between young and adult SHR and WKY [72]. It is important, however, to measure regional TH activity at least in the region including the NTS or more directly in the NTS itself. Indeed, our finding is consistent with increased TH activity in the NTS of Wistar rats, the BP of which was elevated at 1 week after bilateral sinoaortic denervation [67]. Alterations in content and turnover of catecholamines in the brainstem are shown in the development and maintenance of hypertension in SHR [72–74]. However, presynaptic and postsynaptic components of the DOPAergic system proposed in the NTS differ from catecholaminergic components, as shown in normotensive Wistar rats [13,14].

By microinjection into the depressor sites, levodopa at 10 to 300 ng elicits dose-dependent hypotension and bradycardia in SHR and WKY, which is consistent with those in normotensive Wistar rats [13] (Figure 11.2). The significant difference is that at higher dose ranges (100 and 300 ng), depressor but not bradycardic responses of SHR to levodopa are more marked, compared with WKY. This is in a paradoxical direction for the maintenance of hypertension in SHR. One explanation is that a depressor response, in general, can reach a greater depressor level when control BP is higher as in the case of SHR, compared with WKY. In addition, an increase in sensitivity of postsynaptic DOPA recognition sites might be responsible for greater hypotension following a functional denervation effect of the primary baroreceptor afferents because the tonic neuronal activity to release basal DOPA is decreased in the NTS of SHR, compared with WKY.

### 11.8.2 CVLM

The absolute value of basal DOPA release is consistently lower in the depressor CVLM of SHR than that in WKY (Figure 11.4) [20], as in the case of NTS [15].

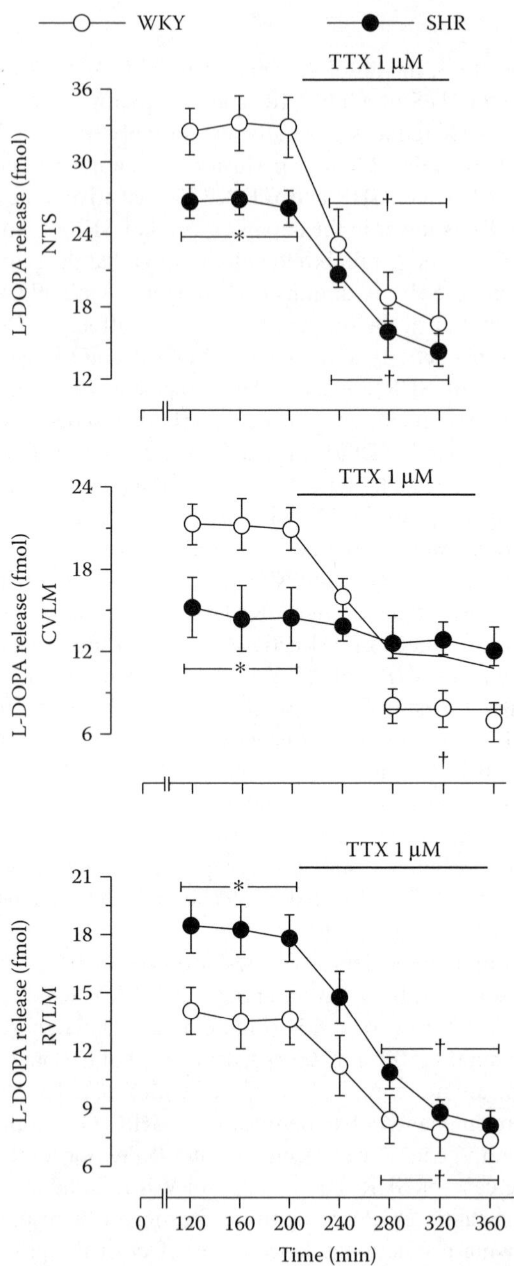


**FIGURE 11.4** Time courses of basal DOPA (L-DOPA) release and its TTX-sensitivity during microdialysis of the NTS, CVLM, and RVLM in adult SHR (●) and age-matched WKY (○). Male 15- to 16-week-old rats of the two strains were anesthetized with urethane (1.2 g/kg, i.p.). Normal Ringer solution was perfused at a rate of 1 μl/min, and 1 μ*M* TTX was perfused at upper horizontal bars 200 min after the start of perfusion. Perfusates were collected successively

This decrease in basal DOPA release in SHR in comparison with WKY is not secondarily due to decrease in biosynthesis or increase in conversion of DOPA, because no differences of TH and AADC activity are seen between the bilateral CVLM areas dissected from SHR and WKY [20]. The TTX-sensitive tonic neuronal activity to release basal DOPA is seen in the CVLM of WKY, which is consistent with the finding in Wistar rats [19], whereas it is completely lost in the CVLM of SHR [20]. This is consistent with the finding that microinjection of TTX into depressor sites of the bilateral CVLM produces marked hypertension in WKY, whereas this type of tonic influence is not seen in SHR [71]. Thus, the loss of TTX-sensitive neuronal activity to release basal DOPA in the CVLM of SHR appears to play an important role for the maintenance of hypertension.

By microinjection, levodopa at 10 to 300 ng elicits dose-dependent hypotension and bradycardia in depressor sites of the CVLM of the two strains, being consistent with normotensive Wistar rats [19]. Depressor but not bradycardic responses of SHR to higher dose ranges of levodopa (100 and 300 ng) are greater, compared with WKY. This is in a paradoxical direction for the maintenance of hypertension in SHR. One explanation is that a depressor response can reach a greater depressor level when control BP is higher as in the case of SHR, compared with WKY, as in the case of the NTS. In addition, the tonic neuronal activity to release basal DOPA is completely lost in the CVLM of SHR. Thus, in comparison with the NTS of SHR, an increase in sensitivity of postsynaptic DOPA recognition sites in the CVLM of SHR following a functional denervation effect of a DOPAergic NTS–CVLM depressor relay appears to play a significant role in eliciting greater hypotension in SHR, compared with WKY.

### 11.8.3 RVLM

The absolute value of basal DOPA release is consistently higher during microdialysis of the pressor RVLM of SHR compared with WKY (Figure 11.4) [15]. The TTX-sensitive tonic neuronal activity to release basal DOPA is also enhanced in SHR, compared with WKY. These alterations might be related to dysfunction of the DOPAergic baroreceptors–primary baroreceptor afferents–NTS–CVLM depressor relay in SHR, compared with WKY. These alterations appear to be not secondarily due to increase in biosynthesis of DOPA from tyrosine, because no difference of TH activity is seen in the bilateral RVLM areas between the two strains. In contrast, AADC activity to convert DOPA to dopamine decreases in the RVLM region of SHR, compared to WKY, being consistent with the observation in the whole lower brainstem of SHR [72].

**FIGURE 11.4 (Continued)** every 40 min. Ordinates show the absolute values of basal DOPA release. Pair number of estimation for WKY and SHR was 8 in the NTS, 7 in CVLM, and 7 in RVLM. Each value represents mean ± S.E. $^*P < .05$ vs. WKY and $^\dagger P < .05$ vs. the value immediately before TTX perfusion (Student's *t*-test in the NTS and RVLM and Bonferroni's *t*-test in the CVLM). Note that there is no significance in the CVLM of SHR. (From Misu, Y. et al., Neurobiology of L-DOPAergic systems, *Prog. Neurobiol.*, 49, 415, 1996. With permission.)

By microinjection, levodopa at 10 to 600 ng elicits dose-dependent hypertension and tachycardia in the pressor sites of both strains, being consistent with normotensive Wistar rats [22]. Pressor but not tachycardiac responses of SHR to levodopa are slightly but significantly greater than those of WKY, especially at lower dose ranges (10 to 30 ng) without an increase in the maximal responses at higher dose ranges (300 and 600 ng). The greater response might be mainly related to an increase in the affinity of DOPA recognition sites within the pressor sites of SHR rather than an increase in the number of DOPA recognition sites. It appears likely that alterations of presynaptic and postsynaptic components of the DOPAergic system in the RVLM of SHR play a major role to maintain hypertension.

### 11.8.4 Concluding Remarks on SHR

In SHR, compared with WKY, basal DOPA release decreases in the depressor NTS and CVLM, and increases in the pressor RVLM, and TTX-sensitive tonic neuronal activity to release basal DOPA is decreased in the NTS, lost in the CVLM, and increased in the RVLM. In SHR, AADC activity to convert DOPA to dopamine decreases, and sensitivity of pressor responses to levodopa increases in the RVLM. All of these presynaptic and postsynaptic alterations appear to be involved in the maintenance of hypertension in SHR. DOPA appears to be a neurotransmitter of baroreflex neurotransmission pathways in the lower brainstem and to play a role in central regulation of arterial BP.

## 11.9 Interactions between GABA and DOPA in NTS

GABA is a representative inhibitory neurotransmitter in the CNS. GABA-containing neurons have been found in the NTS [27,75]. Evidence exists to support an inhibitory role of GABAergic interneurons in the modulation of baroreflex function within the NTS [76–79]. A small proportion of GABAergic neurons in the baroreceptor afferent zone of the NTS shows c-fos immunoreactivity colabeled for glutamic acid decarboxylase mRNA in response to phenylephrine-induced hypertension [27].

Postsynaptic depressor and bradycardic responses to electrical ADN stimulation are attenuated by muscimol, a $GABA_A$ agonist, and nipecotic acid, a GABA uptake inhibitor, but are potentiated by bicuculline, a $GABA_A$ antagonist, microinjected into the depressor sites [78]. GABA appears to function tonically to inhibit baroreflex inputs via $GABA_A$ receptors. Thus, if DOPA is a neurotransmitter of the ADN, depressor and bradycardic responses to exogenously microinjected levodopa should be inhibited by GABAergic agonists and facilitated by bicuculline. At first, we tried to clarify this idea. In addition, we tried to clarify whether or not antagonism by DOPA ME against basally released DOPA to activate the depressor sites [14] can modulate pressor and tachycardiac responses to GABA.

Furthermore, GABA appears to regulate tonically the basal release of established neurotransmitters primarily via activation of inhibitory $GABA_A$ receptors in the rat CNS. Muscimol inhibits the basal release of acetylcholine in the striatum [80],

5-hydroxytryptamine in the dorsal raphe nucleus [81], and noradrenaline in the median preoptic nucleus area [82], whereas bicuculline alone facilitates the basal release of acetylcholine [80], 5-hydroxytryptamine [81], and noradrenaline [82]. Inhibitory $GABA_A$ receptors in the prefrontal cortex appear to function tonically to regulate dopamine release in the dorsolateral striatum [83]. Thus, if it is the case that DOPA is a neurotransmitter of the ADN terminating in the depressor sites, GABA ought to tonically inhibit basal DOPA release during microdialysis of the NTS. We tried to clarify whether or not muscimol perfused via probes inhibits basal DOPA release in a bicuculline-sensitive manner and whether or not a higher dose of bicuculline alone facilitates it.

### 11.9.1 GABA Tonically Inhibits Depressor and Bradycardic Responses to Levodopa in Depressor Sites of NTS

Mean BP and HR of male Wistar rats (250 to 350 g) anesthetized with urethane are 82 ± 1 mmHg and 364 ± 7 beats/min (n = 58) in this series of experiments [16]. The microinjection of levodopa at 10 to 60 ng into the unilateral depressor sites elicits dose-dependent decreases in BP and HR in the absence of NSD-1015. Control responses to 30-ng levodopa are shown in Figure 11.5A. In contrast, GABA at 3 to 300 ng elicits dose-dependent increases in BP and HR. The maximum ΔBP and ΔHR elicited by the highest dose of GABA (300 ng) are 18 ± 3 mmHg and 17 ± 3 beats/min (n = 4). Nipecotic acid at 100 ng alone also increases BP and HR (Figure 11.5C). Bicuculline at 10 ng alone elicits marked decreases in BP and HR (Figure 11.5D). These findings suggest a tonic function of endogenously released GABA to inhibit baroreflex inputs in the depressor sites via activation of postsynaptic $GABA_A$ receptors, being consistent with a previous report from our laboratory [77]. Inconsistent findings are also shown. The unilateral microinjection of GABA (100 ng) elicits no effect in rats anesthetized with halothane [24]. The bilateral microinjection of nipecotic acid increases BP without altering HR in rats anesthetized with chloralose, whereas the bilateral microinjection of bicuculline slightly decreases BP [76]. These discrepancies appear to be derived from differences in anesthetics, rat strains, or other experimental conditions.

Control responses to 30-ng levodopa are slightly inhibited by pretreatment with the lowest dose of GABA (3 ng) (Figure 11.5B), and significantly inhibited by the moderate doses of GABA at 10 and 30 ng (data not shown), when pressor and tachycardiac responses to GABA alone returned to basal levels. Control responses are also inhibited by 100-ng nipecotic acid (Figure 11.5C). In contrast, control responses are markedly potentiated by pretreatment with 10-ng bicuculline, when depressor responses to the antagonist alone returned to a basal level (Figure 11.5D). Basally released GABA appears to tonically inhibit depressor and bradycardic responses to levodopa via activation of $GABA_A$ receptors. Some physiological receptor–receptor interaction appears to occur between postsynaptic excitatory DOPA recognition sites and inhibitory $GABA_A$ receptors in the depressor sites. The DOPA recognition sites appear to differ from $GABA_A$ receptors, because DOPA ME up to 10 μ*M* fails to displace the specific binding of tritiated GABA and elicits only 9 and 16% displacement at 100 μ*M* and 1 m*M*, respectively, in rat brain membrane preparations [16].

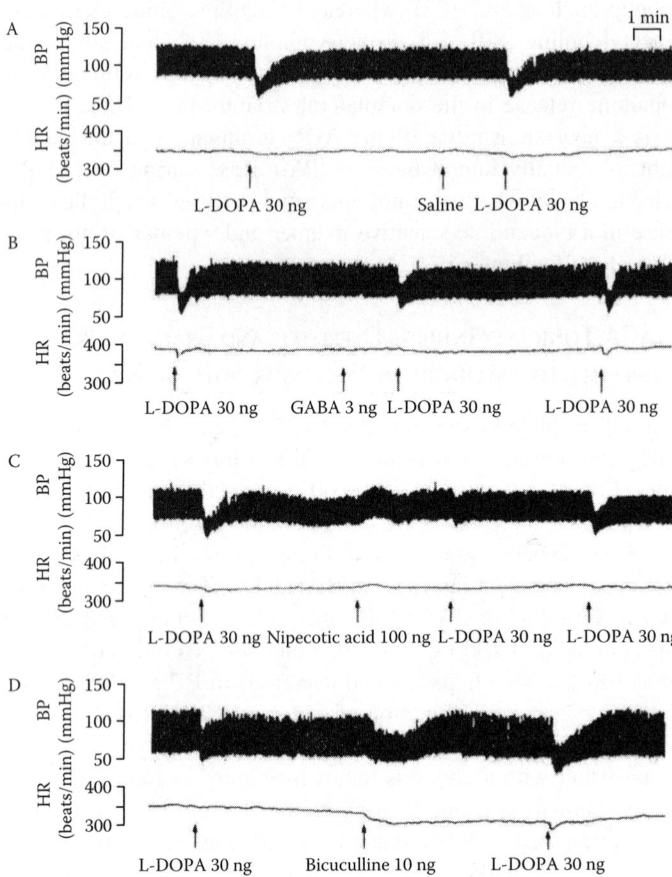


**FIGURE 11.5** Typical traces of effects of pretreatment with saline (A), 3-ng GABA (B), 100-ng nipecotic acid (C), and 10-ng bicuculline (D) on decreases in BP and HR elicited by microinjection of 30-ng levodopa (L-DOPA) into depressor sites of the unilateral NTS in anesthetized rats. Levodopa was challenged when vascular responses to pretreated drugs returned to basal levels. (From Honjo, K. et al., GABA may function tonically via $GABA_A$ receptors to inhibit hypotension and bradycardia by L-DOPA microinjected into depressor sites of the nucleus tractus solitarii in anesthetized rats, *Neurosci. Lett.*, 261, 93, 1999. With permission.)

### 11.9.2 GABA Appearing to Inhibit Tonic Function of Basally Released DOPA to Activate Depressor Sites in NTS

The tonic function of basally released DOPA to activate the depressor sites is evident. The microinjection of DOPA ME alone into the bilateral depressor sites elicits dose-dependent pressor and tachycardiac responses, which are abolished following inhibition of biosynthesis of DOPA by pretreatment with α-MPT accompanied with marked decreases in basal DOPA release [14]. Taken together with this finding, GABA ought to inhibit the tonic function of basally released DOPA to activate the depressor sites. Indeed, pressor and tachycardiac responses to the highest dose of GABA (300 ng) decreases by half at a time when the tonic function of basally released DOPA appeared to be antagonized by pretreatment with the unilateral microinjection of 1-μg DOPA ME. Exogenously applied GABA appears to induce hypertension and tachycardia, at least partially, following inhibition of the tonic function of basally released DOPA to activate postsynaptic depressor sites in the NTS.

GABA-induced postsynaptic inhibition of responses to exogenously microinjected levodopa and basally released DOPA appears to be a factor for the inhibitory roles of GABA in baroreflex neurotransmission in depressor sites of the NTS.

### 11.9.3 GABA Appearing to Tonically Inhibit Basal DOPA Release via Activation of Presynaptic $GABA_A$ Receptors in NTS

During microdialysis of the NTS, basal DOPA release becomes stable 160 min after the start of perfusion and remains constant throughout the experimental periods (Figure 11.6A and Figure 11.6B), being consistent with the previous observation [14]. Muscimol at 10 to 100 μ*M* inhibits basal DOPA release in a dose-dependent manner (Figure 11.6A). The highest dose of muscimol (100 μ*M*) elicits a ceiling phenomenon of approximately 30% inhibition, the degree of which is consistent with a TTX-sensitive and $Ca^{2+}$-dependent component of basal DOPA release under an anesthetized condition [14]. Muscimol at 100 μ*M* appears to completely suppress a neurotransmitter component of basal DOPA release elicited by spontaneous neuronal activity.

As shown in Figure 11.6B, the 30-μ*M*-muscimol-induced inhibition is fully antagonized by a noneffective dose of bicuculline at 10 μ*M*. Importantly, a higher dose of bicuculline at 30 μ*M* alone facilitates basal DOPA release in a time course similar to that of muscimol-induced inhibition. These findings suggest a tonic function of endogenously released GABA to inhibit basal DOPA release evoked by spontaneous neuronal activity via activation of presynaptic inhibitory $GABA_A$ receptors probably located on DOPAergic neurons in the depressor sites. GABA-induced presynaptic inhibition of basal DOPA release appears to be a factor for an inhibitory effect of GABA on baroreflex inputs in depressor sites of the NTS.

### 11.9.4 Conclusive Remarks on Interactions between GABA and DOPA in Depressor Sites of NTS

Presynaptic and postsynaptic interactions between inhibitory GABA system and excitatory DOPA system in the NTS appear to further support the idea that DOPA is a neurotransmitter of the ADN terminating in the depressor sites.

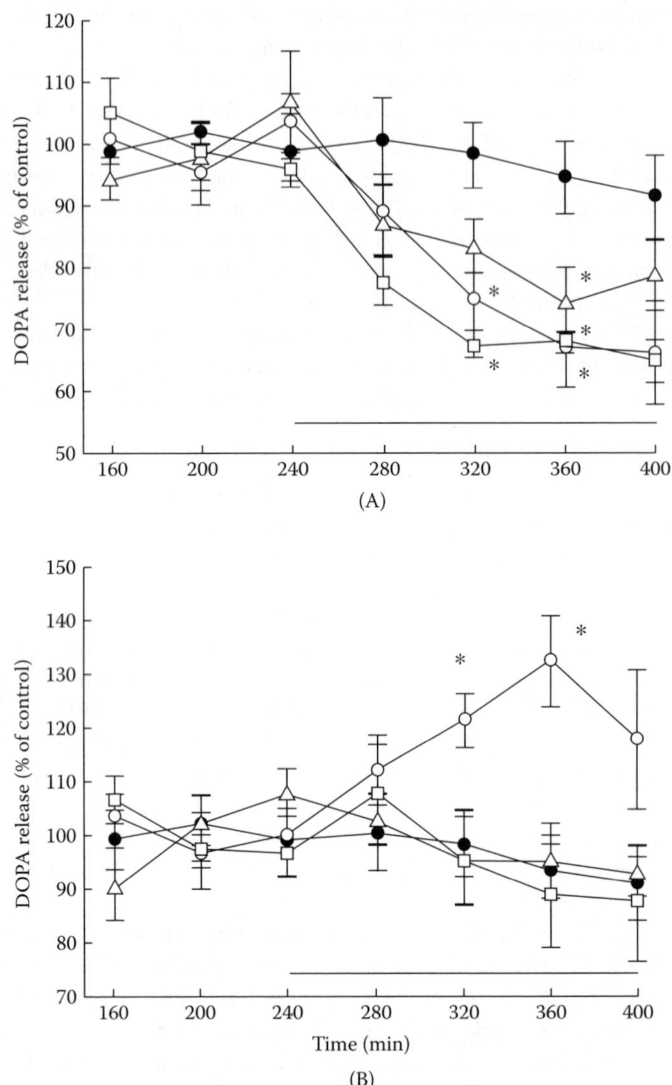


**FIGURE 11.6** Time courses of dose-dependent inhibition of basal DOPA release induced by 10- to 100-μ*M* muscimol (A), antagonism by a noneffective dose of 10-μ*M* bicuculline against 30-μ*M*-muscimol-induced inhibition, and facilitation of basal DOPA release elicited by 30-μ*M* bicuculline alone (B) during microdialysis of the NTS in anesthetized rats. Ringer solution was perfused at a rate of 1.6 ml/min. Dialysates were successively collected every 40 min. Ordinates show basal DOPA release as percentage of control, the mean absolute values of the initial stable three successive samples 160 to 240 min after the start of perfusion. In A, vehicle (●) and muscimol at 10 (△), 30 (○), and 100 (□) μ*M* were perfused via a probe at a horizontal bar 240 min after the start of perfusion and continued throughout the experiments. Respective control (fmol) was 26.4 ± 3.4 for vehicle (n = 9) and 18.3 ± 2.5, 18.2 ± 3.5,

## 11.10 IS THERE A NOVEL PATHWAY CONSISTING OF THE DOPA-CONTAINING PRIMARY BARORECEPTOR AFFERENT ADN AND THE GLUTAMATE-CONTAINING SECONDARY NEURONS IN NEURONAL MICROCIRCUITS OF THE NTS?

We have accumulated evidence to support the idea that DOPA is a highly probable neurotransmitter candidate of the primary baroreceptor afferents terminating in the depressor sites [8–11]. Electrical stimulation of the ADN and baroreceptor activation by phenylephrine elicit DOPA release during microdialysis of the NTS [14]. DOPA ME antagonizes hypotension and bradycardia in response to either electrical stimulation of the ADN [14] or exogenous microinjection of levodopa into the depressor sites [13,45,46], and further antagonizes reflex bradycardia induced by phenylephrine [14]. On the other hand, this antagonist elicits no antagonism against hypotension and bradycardia induced by glutamate [13]. Glutamate has been believed to be the most probable neurotransmitter of the primary baroreceptor afferents [24,28,29] and the second-order and following-order interneurons within the neuronal microcircuits of the depressor sites [29]. Endogenous basal glutamate is measurable during microdialysis of the NTS [84]. However, the evoked release of glutamate over the basal level is apparently not observed following ADN and baroreceptor stimulation, which is consistent with other findings [85]. However, we can expect the existence of some interactions between DOPAergic and glutamatergic systems in the NTS, because relatively high micromolar concentrations of levodopa release glutamate by itself from *in vitro* striatal slices [44] and primary cultured neurons [86] and from *in vivo* striata [49]. DOPA released by four-vessel occlusion appears to be an upstream factor for glutamate release and resultant delayed neuron death in rat striatum and hippocampal CA1 region [49,50].

Microinjection of kynurenate, a broad-spectrum ionotropic glutamate receptor antagonist, into the depressor sites antagonizes depressor and bradycardic responses to electrical ADN stimulation [24,87–89]. Tonic hypertension and tachycardia are also always induced by kynurenate alone. However, this antagonist elicits no

**FIGURE 11.6 (Continued)** and 25.8 ± 3.6 for muscimol at 10 (n = 7), 30 (n = 7), and 100 (n = 5) μ*M*. In B, vehicle (●), bicuculline at 10 (△) and 30 (○) μ*M* alone, and 10-μ*M* bicuculline with 30-μ*M* muscimol (□) were perfused. Respective control (fmol) was 26.4 ± 3.4 for vehicle (n = 9), 14.4 ± 2.2 and 14.2 ± 5.1 for bicuculline at 10 (n = 7) and 30 (n = 6) μ*M* alone, and 20.4 ± 2.6 for 10-μ*M* bicuculline with 30-μ*M* muscimol (n = 10). Data are means ± SE. **P* < .05 vs. vehicle (ANOVA/Dunn's multiple comparison test). (From Goshima, Y. et al., The evidence for tonic GABAergic regulation of basal L-DOPA release via activation of inhibitory $GABA_A$ receptors in the nucleus tractus solitarii in anesthetized rats, *Neurosci. Lett.*, 261, 155, 1999. With permission.)

[24,87,88] or only partial [89,90] antagonism against depressor and bradycardic responses to exogenously microinjected glutamate. It has been proposed that an unknown excitatory amino acid, the responses to which are antagonized by kynurenate, might be a primary neurotransmitter of the ADN [87,88].

Since 1980, the growing evidence has shown that an endothelium-derived relaxing factor found by Furchgott and Zawadzki [91], which was identified later as nitric oxide (NO), is involved in a lot of peripheral [92] and central [93] physiological functions. NO elicits central neurotransmission through activation of soluble guanylate cyclase and increases in cyclic GMP [93]. NO has also been implicated in neurodegenerative processes [94–96]. NO is catalyzed from L-arginine via activation of brain NO synthase (NOS) [97]. In addition, the accumulating evidence has suggested that NO plays an important role in baroreflex neurotransmission in the NTS [98]. Phenylephrine-induced hypertension shows extensive colocalization of fos immunoreactivity and makers for the NO phenotype in the baroreceptor afferent zone in the NTS [27]. The degenerating vagal axon terminals in the NTS following the unilateral denervation of the ganglion nodosum are NOS-immunoreactive, and the major distribution of this immunoreactivity is found in the nondegenerating intrinsic axons [99]. Antisense oligodeoxynucleotide (oligo) to neuronal NOS (nNOS) suppresses immunohistochemical reactivity of nNOS in the NTS [100]. L-Arginine microinjected into the depressor sites elicits hypotension and bradycardia, which are inhibited by 7-nitroindazole, an nNOS inhibitor [101]. The microinjection of nNOS antisense oligo alone into the bilateral depressor sites elicits pressor responses, suggesting the tonic function of nNOS to activate the depressor sites [100].

Furthermore, studies using dual immunoelectron microscopy reveal the colocalization of nNOS and NMDA receptors within fine neuronal processes in rat brain [102]. Several lines of evidence suggest the existence of reciprocal interactions between glutamate and NO in the NTS [103–105]. Unilateral microinjection of glutamate, NMDA, AMPA, and L-arginine into the depressor sites elicits hypotension and bradycardia [103]. These responses to ionotropic glutamate agonists are inhibited by prior administration of 7-nitroindazole, whereas responses to L-arginine are attenuated by prior administration of NMDA and non-NMDA antagonists. During microdialysis of the NTS, the extracellular level of glutamate is markedly increased by the administration of NO solution and is reduced by perfusion with a NOS inhibitor $N^G$-monomethyl-L-arginine (L-NMMA), whereas the extracellular NO level is increased by the administration of NMDA and AMPA [104]. NMDA also elicits increases in the extracellular levels of NO and glutamate with hypotension and bradycardia, all responses of which are inhibited by another NOS inhibitor $N^G$-nitro-L-arginine methyl ester [105].

Based on these findings, we attempted to clarify whether kynurenate attenuates depressor and bradycardic responses to levodopa microinjected into the depressor sites, and whether any levodopa–glutamate–NO and/or levodopa–NO–glutamate cascades are involved in levodopa-induced responses in the NTS [18]. In each group of this series of experiments, the resting mean BP was 85 ± 5 mmHg (n = 5) to 91 ± 8 mmHg (n = 4), and HR was 358 ± 3 beats/min (n = 5) to 415 ± 9 beats/min (n = 9) in urethane-anesthetized rats.

### 11.10.1 Levodopa Appearing to Produce Depressor and Bradycardic Responses via Activation of Kynurenate-Sensitive Ionotropic Glutamate Receptors and nNOS in NTS

Kynurenate at 600 pmol microinjected into the unilateral depressor sites, which blocks depressor and bradycardic responses to ADN stimulation [89], elicits no effect on resting mean BP and HR [18]. Microinjection of kynurenate 2 min prior to levodopa reduces by 75 to 80% depressor and bradycardic responses to 300-pmol levodopa in a reversible manner, suggesting the existence of a major pathway mediated by ionotropic glutamate receptors involved in these levodopa-induced responses.

At first, it has to be mentioned that DOPA recognition sites appear to differ from kynurenate-sensitive ionotropic glutamate receptors. Kynurenate at 1 m*M* markedly displaces the selective binding of tritiated ligands for NMDA binding site, NMDA glycine site, and non-NMDA kainate receptors, whereas levodopa and DOPA ester compounds up to 1 m*M* fail to displace this for these kynurenate-sensitive binding sites in rat brain membrane preparations [46,53]. Thus, how does levodopa induce kynurenate-sensitive depressor and bradycardic responses? It is possible that microinjected levodopa releases endogenous glutamate via DOPA recognition sites as an upstream event prior to activation of kynurenate-sensitive glutamate receptors (Figure 11.7). Levodopa releases glutamate from striatal slices in a DOPA-ME-sensitive manner [44], and DOPA released by four-vessel occlusion triggers to produce glutamate release during microdialysis of striatum and hippocampal CA1

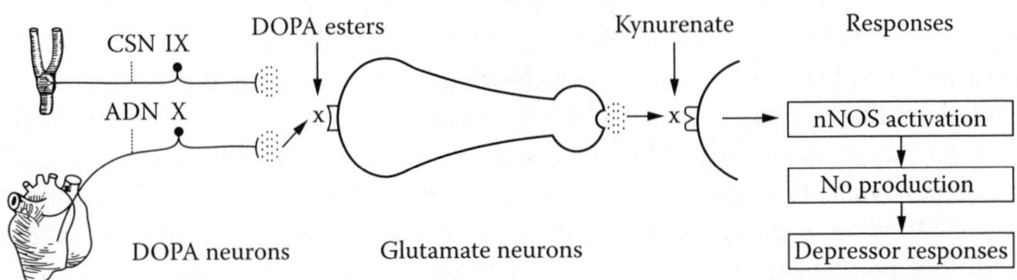


**FIGURE 11.7** A proposed novel pathway consisting of DOPA-containing neurons as the primary baroreceptor afferent ADN and secondary glutamate-containing neurons within neuronal microcircuits in depressor sites of the NTS in anesthetized rats. DOPA, by acting on its recognition sites, might release undetectable but functioning glutamate from second-order neurons. DOPA ester compounds (DOPA esters) antagonize depressor and bradycardic responses to either electrical ADN stimulation or the microinjection of levodopa, but not those to the microinjection of glutamate. Kynurenate elicits actions almost similar to DOPA ester compounds. DOPA recognition sites appear to differ from kynurenate-sensitive ionotropic glutamate receptors. (Modified from Misu, Y., Kitahama, K., and Goshima, Y., L-3,4-Dihydroxyphenylalanine as a neurotransmitter candidate in the central nervous system, *Pharmacol. Ther.,* 97, 117, 2003. With permission.)

region and the resultant delayed neuron death, both of which are protected by DOPA CHE [49,50].

We tried to clarify whether levodopa perfused via probes releases glutamate during microdialysis of the NTS. However, levodopa evokes no glutamate release over the basal release, which is consistent with the observation that electrical ADN or vagal stimulation and baroreceptor stimulation by phenylephrine evokes no glutamate release [84,85] under the same experimental conditions that result in DOPA release [14,84]. Although the exact reasons remain unknown, this is probably due to some small amount of glutamate release evoked from the primary baroreceptor afferents and/or the secondary interneurons, compared with the abundant amount of basal release in extracellular space. In some reports, however, baroreceptor activation by phenylephrine can evoke glutamate release during microdialysis of the NTS [106,107]. Exact reasons for these discrepancies between negative [84,85] and positive [106,107] findings on glutamate release evoked by baroreceptor activation are unknown at present. Thus, we think that via activation of DOPA recognition sites under our experimental conditions, the microinjection of levodopa into the depressor sites might release undetectable but functioning glutamate, which leads to activation of kynurenate-sensitive ionotropic glutamate receptors [18].

As the next step, we explored whether there are interactions between NOS and levodopa. L-NMMA at 100 nmol microinjected elicits slight and brief pressor responses. The maximal ΔBP is 11 ± 6 mmHg (n = 6). L-NMMA at 100 nmol 10 min prior to levodopa inhibits by 85 to 90% depressor and bradycardic responses to 300-pmol levodopa in a stereoselective and reversible manner. L-NMMA is a competitive inhibitor for all three NOS isozymes with low specificity such as nNOS, endothelial NOS, and inducible NOS [108]. It is unlikely that inducible NOS is involved in levodopa-induced responses, because L-NMMA inhibits it in an irreversible manner in this experimental design [108]. In addition, nNOS antisense oligo at 20 pmol elicits no effect on resting mean BP and HR and, importantly, reduces (by 65%) depressor and bradycardic responses to 300-pmol levodopa, challenged 45 min after its prior microinjection. It is consistent with the observation that this nNOS antisense oligo markedly suppresses the nNOS immunoreactivity in the NTS only 45 min after its pretreatment [100]. The nNOS sense and scrambled oligos (20 pmol) produce no effect on resting mean BP and HR and depressor and bradycardic responses to 300-pmol levodopa. These findings indicate that nNOS is involved in levodopa-induced responses in the depressor sites. Furthermore, NO production via activation of nNOS appears to be a downstream event following activation of glutamate receptors because prior microinjection of kynurenate at the same dose (600 pmol) to reduce responses to levodopa fails to affect depressor and bradycardic responses to 30-nmol L-arginine (n = 9) [18].

### 11.10.2 Concluding Remarks on Interactions among Levodopa, Glutamate, and nNOS in Depressor Sites of NTS

Herein, we propose a novel pathway consisting of the DOPA-containing primary baroreceptor afferent ADN and the glutamate-containing secondary neurons within neuronal microcircuits in depressor sites of the NTS (Figure 11.7). Taken together with findings obtained by immunocytochemical and functional approaches with denervation of the ADN, the DOPA-containing long axons of the ADN appear to terminate directly in depressor sites or indirectly in some second-order neurons in synapse with depressor sites within neuronal microcircuits of the NTS to carry baroreflex information [13–15]. In addition to the primary baroreceptor afferents, glutamate is regarded as a probable neurotransmitter of the second-order interneurons in the depressor sites [29]. It appears likely that levodopa releases undetectable but functioning glutamate from the second-order glutamate interneurons within the neuronal microcircuits to activate kynurenate-sensitive ionotropic glutamate receptors. Based on the assumption that glutamate is a second-order neurotransmitter, the undetectable evoked release [18,84,85] might be reasonable. A following cascade of signal transduction appears to be activation of nNOS, and then production of NO, which leads to depressor and bradycardic responses [18]. There are three additional possibilities: (1) DOPA recognition sites are located on or near the terminals of glutamate neurons as the primary baroreceptor afferent ADN; (2) responses to levodopa that remained after application of kynurenate are antagonized by metabotropic glutamate antagonists acting on metabotropic glutamate receptors that might be located at perisynaptic and/or extrasynaptic membranes [90]; (3) some component of DOPA released from the primary baroreceptor afferent ADN directly activates DOPA recognition sites in the depressor sites [7–11].

## 11.11 CONCLUSION

Exogenously microinjected levodopa elicits in its own right cardiovascular responses in the depressor NTS, depressor CVLM, and pressor RVLM in the lower brainstem of anesthetized rats. Endogenously released DOPA appears to be involved in resting and baroreflex integration of arterial BP in the NTS and CVLM, and in resting integration of arterial BP in the RVLM. DOPA appears to be a neurotransmitter of the primary baroreceptor afferent ADN–NTS depressor pathway, the NTS–CVLM depressor pathway, and the PHN–RVLM pressor pathway.

Decreases in basal DOPA release and tonic neuronal activity to release basal DOPA in the depressor NTS and CVLM, increases in these measures in the pressor RVLM, and decrease in AADC activity and increase in sensitivity of pressor responses to levodopa in the RVLM appear to maintain hypertension in SHR.

In the NTS, GABA appears to tonically inhibit basal DOPA release and DOPAergic baroreflex inputs via activation of presynaptic and postsynaptic $GABA_A$ receptors.

Herein, we propose a novel pathway consisting of the DOPA-containing primary baroreceptor afferent ADN and the glutamate-containing secondary neurons within neuronal microcircuits in depressor sites of the NTS. Following cascades appear to be nNOS activation, the production of NO to function, which leads to depressor and bradycardic responses.

## ACKNOWLEDGMENTS

This work was mainly supported by grants-in-aid for Developmental Scientific Research (No. 06557143) and General Scientific Research (No. 61480119, 03454146, 07407003, 10470026) from the Ministry of Education, Science, Sports, and Culture, Japan, and by grants from the Mitsubishi Foundation, the Uehara

Memorial Foundation, Japanese Heart Foundation, Mitsui Life Social Welfare Foundation, and SRF, all in Japan. Many thanks go to Sanae Sato, Department of Pharmacology, Fukushima Medical University School of Medicine, Fukushima 960-1295, Japan, for remaking figures.

## REFERENCES

1. Carlsson, A., Lindqvist, M., and Magnusson, T., 3,4-Dihydroxyphenylalanine and 5-hydroxytryptophan as reserpine antagonists, *Nature,* 180, 1200, 1957.
2. Ehringer, H. and Hornykiewicz, O., Verteilung von Noradrenaline und Dopamine (3-Hydroxytyramin) im Gehirn des Menschen und ihr Verhalten bei Erkrankungen des extrapyramidalen Systems, *Klin. Wochenschr,* 38, 1236, 1960.
3. Bartholini, G. et al., Increase of cerebral catecholamines by 3,4-dihydroxyphenylalanine after inhibition of peripheral decarboxylase, *Nature,* 215, 852, 1967.
4. Hefti, F. and Melamed, E., L-DOPA's mechanism of action in Parkinson's disease, *Trends Neurosci.,* 3, 229, 1980.
5. Goshima, Y., Kubo, T., and Misu, Y., Biphasic actions of L-DOPA on the release of endogenous noradrenaline and dopamine from rat hypothalamic slices, *Br. J. Pharmacol.,* 89, 229, 1986.
6. Misu, Y., Goshima, Y., and Kubo, T., Biphasic actions of L-DOPA on the release of endogenous dopamine via presynaptic receptors in rat striatal slices, *Neurosci. Lett.,* 72, 194, 1986.
7. Misu, Y. and Goshima, Y., Is L-dopa an endogenous neurotransmitter?, *Trends Pharmacol. Sci.,* 14, 119, 1993.
8. Misu, Y., Ueda, H., and Goshima, Y., Neurotransmitter-like actions of L-DOPA, *Adv. Pharmacol.,* 32, 427, 1995.
9. Misu, Y. et al., Neurobiology of L-DOPAergic systems, *Prog. Neurobiol.,* 49, 415, 1996.
10. Misu, Y., Goshima, Y., and Miyamae, T., Is DOPA a neurotransmitter?, *Trends Pharmacol. Sci.,* 23, 262, 2002.
11. Misu, Y., Kitahama, K., and Goshima, Y., L-3,4-Dihydroxyphenylalanine as a neurotransmitter candidate in the central nervous system, *Pharmacol. Ther.,* 97, 117, 2003.
12. Hoffman, B.B. and Taylor, P., Neurotransmission: the autonomic and somatic motor nervous systems, in *Goodman & Gilman's The Pharmacological Basis of Therapeutics,* Hardman, J.G., Limbird, L.E., and Gilman, A.G., Eds., McGraw-Hill, New York, 2001, 115.
13. Kubo, T. et al., Evidence for L-DOPA systems responsible for cardiovascular control in the nucleus tractus solitarii of the rat, *Neurosci. Lett.,* 140, 153, 1992.
14. Yue, J.-L. et al., Baroreceptor-aortic nerve-mediated release of endogenous L-3,4-dihydroxyphenylalanine and its tonic function in the nucleus tractus solitarii of rats, *Neuroscience,* 62, 145, 1994.
15. Yue, J.-L. et al., Altered tonic L-3,4-dihydroxyphenylalanine systems in the nucleus tractus solitarii and the rostral ventrolateral medulla of spontaneously hypertensive rats, *Neuroscience,* 67, 95, 1995.
16. Honjo, K. et al., GABA may function tonically via $GABA_A$ receptors to inhibit hypotension and bradycardia by L-DOPA microinjected into depressor sites of the nucleus tractus solitarii in anesthetized rats, *Neurosci. Lett.,* 261, 93, 1999.
17. Goshima, Y. et al., The evidence for tonic GABAergic regulation of basal L-DOPA release via activation of inhibitory $GABA_A$ receptors in the nucleus tractus solitarii in anesthetized rats, *Neurosci. Lett.,* 261, 155, 1999.

18. Yamanashi, K. et al., Involvement of nitric oxide production via kynurenic acid-sensitive glutamate receptors in DOPA-induced depressor responses in the nucleus tractus solitarii of anesthetized rats, *Neurosci. Res.,* 43, 231, 2002.
19. Yue, J.-L., Goshima, Y., and Misu, Y., Transmitter-like L-3,4-dihydroxyphenylalanine tonically functions to mediate vasodepressor control in the caudal ventrolateral medulla of rats, *Neurosci. Lett.,* 159, 103, 1993.
20. Miyamae, T. et al., Loss of tonic neuronal activity to release L-DOPA in the caudal ventrolateral medulla of spontaneously hypertensive rats, *Neurosci. Lett.,* 198, 37, 1995.
21. Miyamae, T. et al., L-DOPAergic components in the caudal ventrolateral medulla in baroreflex neurotransmission, *Neuroscience,* 92, 137, 1999.
22. Yue, J.-L. et al., Evidence for L-DOPA relevant to modulation of sympathetic activity in the rostral ventrolateral medulla of rats, *Brain Res.,* 629, 310, 1993.
23. Nishihama, M. et al., An L-DOPAergic relay from the posterior hypothalamic nucleus to the rostral ventrolateral medulla and its cardiovascular function in anesthetized rats, *Neuroscience,* 92, 123, 1999.
24. Talman, W.T., Perrone, M.H., and Reis, D.J., Evidence for L-glutamate as the neurotransmitter of baroreceptor afferent nerve fibers, *Science,* 209, 813, 1980.
25. Urbanski, R.W. and Sapru, H.N., Evidence for a sympathoexcitatory pathway from the nucleus tractus solitarii to the ventrolateral medullary pressor area, *J. Auton. Nerv. Syst.,* 23, 161, 1988.
26. Sun, M.-K., Pharmacology of reticulospinal vasomotor neurons in cardiovascular regulation, *Pharmacol. Rev.,* 48, 465, 1996.
27. Chan, R.K.W. and Sawchenko, P.E., Organization and transmitter specificity of medullary neurons activated by sustained hypertension: implications for understanding baroreceptor reflex circuitry, *J. Neurosci.,* 18, 371, 1998.
28. Aicher, S.A. et al., Anatomical substrates for baroreflex sympathoinhibition in the rat, *Brain Res. Bull.,* 51, 107, 2000.
29. Andresen, M.C. et al., Cellular mechanisms of baroreceptor integration at the nucleus tractus solitarius, *Ann. N. Y. Acad. Sci.,* 940, 132, 2001.
30. Sved, A.F. et al., Excitatory inputs to the RVLM in the context of the baroreceptor reflex, *Ann. N.Y. Acad. Sci.,* 940, 247, 2001.
31. Goshima, Y., Nakamura, S., and Misu, Y., L-Dihydroxyphenylalanine methyl ester is a potent competitive antagonist of the L-dihydroxyphenylalanine-induced facilitation of the evoked release of endogenous norepinephrine from rat hypothalamic slices, *J. Pharmacol. Exp. Ther.,* 258, 466, 1991.
32. Gordon, F.J., Aortic baroreceptor reflexes are mediated by NMDA receptors in caudal ventrolateral medulla, *Am. J. Physiol.,* 252, R628, 1987.
33. Kubo, T., Kihara, M., and Misu, Y., Ipsilateral but not contralateral blockade of excitatory amino acid receptor in the caudal ventrolateral medulla inhibits aortic baroreceptor reflex in rats, *Naunyn-Schmiedeberg's Arch. Pharmacol.,* 343, 46, 1991.
34. Miyawaki, T. et al., Role of AMPA/kainate receptors in transmission of the sympathetic baroreflex in rat CVLM, *Am. J. Physiol.,* 272, R800, 1997.
35. Morrison, S.F. et al., Rostral ventrolateral medulla: a source of the glutamatergic innervation of the sympathetic intermidiolateral nucleus, *Brain Res.,* 562, 126, 1991.
36. Kubo, T. et al., Cardiovascular effects of L-glutamate and γ-aminobutyric acid injected into the rostral ventrolateral medulla in normotensive and spontaneously hypertensive rats, *Arch. Int. Pharmacodyn.,* 279, 150, 1986.
37. Minson, J.B. et al., c-fos identifies GABA-synthesizing barosensitive neurons in caudal ventrolateral medulla, *NeuroReport,* 8, 3015, 1997.

38. Goshima, Y., Kubo, T., and Misu, Y., Transmitter-like release of endogenous 3,4-dihydroxyphenylalanine from rat striatal slices, *J. Neurochem.,* 50, 1725, 1988.
39. Jaeger, C.B. et al., Aromatic L-amino acid decarboxylase in the rat brain: immunocytochemical localization in neurons of the brain stem, *Neuroscience,* 11, 691, 1984.
40. Tison, F. et al., Endogenous L-DOPA in the rat dorsal vagal complex: an immunocytochemical study by light and electron microscopy, *Brain Res.,* 497, 260, 1989.
41. Skagerberg, G. et al., Studies on dopamine-, tyrosine hydroxylase- and aromatic L-amino acid decarboxylase-containing cells in the rat diencephalon: comparison between formaldehyde-induced histofluorescence and immunofluorescence, *Neuroscience,* 24, 605, 1988.
42. Tison, F. et al., Immunohistochemistry of endogenous L-DOPA in the rat posterior hypothalamus, *Histochemistry,* 93, 655, 1990.
43. Vertes, R.P. and Crane, A.M., Descending projections of the posterior nucleus of the hypothalamus: *Phaseolus vulgaris* leucoagglutinin analysis in the rat, *J. Comp. Neurol.,* 354, 607, 1996.
44. Goshima, Y. et al., L-DOPA induces $Ca^{2+}$-dependent and tetrodotoxin-sensitive release of endogenous glutamate from rat striatal slices, *Brain Res.,* 617, 167, 1993.
45. Misu, Y. et al., L-DOPA cyclohexyl ester is a novel stable and potent competitive antagonist against L-DOPA, as compared to L-DOPA methyl ester, *Jpn. J. Pharmacol.,* 75, 307, 1997.
46. Furukawa, N. et al., L-DOPA cyclohexyl ester is a novel potent and relatively stable competitive antagonist against L-DOPA among several L-DOPA ester compounds, *Jpn. J. Pharmacol.,* 82, 40, 2000.
47. Goshima, Y. et al., Picomolar concentrations of L-DOPA stereoselectively potentiate activities of presynaptic β-adrenoceptors to facilitate the release of endogenous noradrenaline from rat hypothalamic slices, *Neurosci. Lett.,* 129, 214, 1991.
48. Akbar, M. et al., Inhibition by L-3,4-dihydroxyphenylalanine of hippocampal CA1 neurons with facilitation of noradrenaline and γ-aminobutyric acid release, *Eur. J. Pharmacol.,* 414, 197, 2001.
49. Furukawa, N. et al., Endogenously released DOPA is a causal factor for glutamate release and resultant delayed neuronal cell death by transient ischemia in rat striata, *J. Neurochem.,* 76, 815, 2001.
50. Arai, N. et al., DOPA cyclohexyl ester, a competitive DOPA antagonist, protects glutamate release and resultant delayed neuron death by transient ischemia in hippocampus CA1 in conscious rats, *Neurosci. Lett.,* 299, 213, 2001.
51. Ishii, H. et al., Involvement of rBAT in $Na^+$-dependent and -independent transport of the neurotransmitter candidate L-DOPA in *Xenopus laevis* oocytes injected with rabbit small intestinal epithelium poly $A^+$ RNA, *Biochim. Biophys. Acta,* 1466, 61, 2000.
52. Perry, B. and U'Prichard, D.C., [$^3$H]-Rauwolscine (α-yohimbine): a specific antagonist radioligand for brain $\alpha_2$-adrenergic receptors, *Eur. J. Pharmacol.,* 76, 461, 1981.
53. Miyamae, T. et al., Some interactions of L-DOPA and its related compounds with glutamate receptors, *Life Sci.,* 64, 1045, 1999.
54. Kubo, T. and Misu, Y., Pharmacological characterization of the α-adrenoceptors responsible for a decrease of blood pressure in the nucleus tractus solitarii of the rat, *Naunyn-Schmiedeberg's Arch. Pharmacol.,* 317, 120, 1981.
55. Zandberg, P., De Jong, W., and De Wied, D., Effect of catecholamine-receptor stimulating agents on blood pressure after local application in the nucleus tractus solitarii of the medulla oblongata, *Eur. J. Pharmacol.,* 55, 43, 1979.
56. Cooper, D.R. et al., L-DOPA esters as potential prodrugs; behavioural activity in experimental models of Parkinson's disease, *J. Pharm. Pharmacol.,* 39, 627, 1987.

57. Cechetto, D.F. and Chen, S.J., Hypothalamic and cortical sympathetic responses relay in the medulla of the rat, *Am. J. Physiol.,* 263, R544, 1992.
58. Giancola, S.B., Roder, S., and Ciriello, J., Contribution of caudal ventrolateral medulla to the cardiovascular responses elicited by activation of bed nucleus of the stria terminalis, *Brain Res.,* 606, 162, 1993.
59. Ross, C.A., Ruggiero, D.A., and Reis, D.J., Projections from the nucleus tractus solitarii to the rostral ventrolateral medulla, *J. Comp. Neurol.,* 242, 511, 1985.
60. Nosaka, S., Yasunaga, K., and Tamai, S., Vagal cardiac preganglionic neurons: distribution, cell types, and reflex discharges, *Am. J. Physiol.,* 243, R92, 1982.
61. Izzo, P.N., Deuchars, J., and Spyer, K.M., Localization of cardiac vagal preganglionic motoneurons in the rat: immunocytochemical evidence of synaptic inputs containing 5-hydroxytryptamine, *J. Comp. Neurol.,* 327, 572, 1993.
62. Standish, A. et al., Central neuronal circuit innervating the rat heart defined by transneuronal transport of pseudorabies virus, *J. Neurosci.,* 15, 1998, 1995.
63. Schwanzel-Fukuda, M., Morrell, J.I., and Pfaff, D.W., Localization of forebrain neurons which project directly to the medulla and spinal cord of the rat by retrograde tracing with wheat germ agglutinin, *J. Comp. Neurol.,* 226, 1, 1984.
64. Saper, C.B. et al., Direct hypothalamo-autonomic connection, *Brain Res.,* 117, 305, 1976.
65. Dampney, R.A., Functional organization of central pathways regulating the cardiovascular system, *Physiol. Rev.,* 74, 323, 1994.
66. Doba, N. and Reis, D.J., Role of central and peripheral adrenergic mechanisms in neurogenic hypertension produced by brainstem lesions in rat, *Circulation Res.,* 34, 293, 1974.
67. Chalmers, J.P., Petty, M.A., and Reid, J.L., Participation of adrenergic and noradrenergic neurons in central connections of arterial baroreceptor reflexes in the rat, *Circ. Res.,* 45, 516, 1979.
68. Luft, F.C. et al., Baroreceptor reflex effect on sympathetic nerve activity in stroke-prone spontaneously hypertensive rats, *J. Auton. Nerv. Syst.,* 17, 199, 1986.
69. Verberne, A.J., Young, N.A., and Louis, W.J., Impairment of inhibitory cardiopulmonary vagal reflexes in spontaneously hypertensive rats, *J. Auton. Nerv. Syst.,* 23, 63, 1988.
70. Catelli, J.M. and Sved, A.F., Enhanced pressor response to GABA in the nucleus tractus solitarii of the spontaneously hypertensive rat, *Eur. J. Pharmacol.,* 151, 243, 1988.
71. Smith, J.K. and Barron, K.W., Cardiovascular effects of L-glutamate and tetrodotoxin microinjected into the rostral and caudal ventrolateral medulla in normotensive and spontaneously hypertensive rats, *Brain Res.,* 506, 1, 1990.
72. Yamori, Y., Lovenberg, W., and Sjoerdsma A., Norepinephrine metabolism in brainstem of spontaneously hypertensive rats, *Science,* 170, 544, 1970.
73. Versteeg, D.H.G. et al., Catecholamine content of individual brain regions of spontaneously hypertensive rats (SH-rats), *Brain Res.,* 112, 429, 1976.
74. Patel, K.P., Kline, R.L., and Mercer, P.F., Noradrenergic mechanisms in the brain and peripheral organs of normotensive and spontaneously hypertensive rats at various ages, *Hypertension,* 3, 682, 1981.
75. Izzo, P.N., Sykes, R.M., and Spyer, K.M., γ-Aminobutyric acid immunoreactive structures in the nucleus tractus solitarii: a light and electron microscopic study, *Brain Res.,* 591, 69, 1992.
76. Catelli, J.M., Giakas, W.J., and Sved, A.F., GABAergic mechanisms in nucleus tractus solitarius alter blood pressure and vasopressin release, *Brain Res.,* 403, 279, 1987.

77. Kubo, T. and Kihara, M., Evidence for the presence of GABAergic and glycine-like systems responsible for cardiovascular control in the nucleus tractus solitarii of the rat, *Neurosci. Lett.,* 74, 331, 1987.
78. Kubo, T. and Kihara, M., Evidence for γ-aminobutyric acid receptor-mediated modulation of the aortic baroreceptor reflex in the nucleus tractus solitarii of the rat, *Neurosci. Lett.,* 89, 156, 1988.
79. Sved, A.F. and Tsukamoto, K., Tonic stimulation of $GABA_B$ receptors in the nucleus tractus solitarius modulates the baroreceptor reflex, *Brain Res.,* 592, 37, 1992.
80. Anderson, J.J. et al., $GABA_A$ and $GABA_B$ receptors differentially regulate striatal acetylcholine release in vivo, *Neurosci. Lett.,* 160, 126, 1993.
81. Tao, R., Ma, Z., and Auerbach, S.B., Differential regulation of 5-hydroxytryptamine release by $GABA_A$ and $GABA_B$ receptors in midbrain raphe nuclei and forebrain of rats, *Br. J. Pharmacol.,* 119, 1375, 1996.
82. Sakamaki, K. et al., GABAergic modulation of noradrenaline release in the median preoptic nucleus in the rat, *Neurosci. Lett.,* 342, 77, 2003.
83. Matsumoto, M. et al., Involvement of $GABA_A$ receptors in the regulation of the prefrontal cortex on dopamine release in the rat dorsolateral striatum, *Eur. J. Pharmacol.,* 482, 177, 2003.
84. Misu, Y. et al., Endogenous L-DOPA but not glutamate or GABA is released by aortic nerve and baroreceptor stimulation in the nucleus tractus solitarii (NTS) of rats, *Jpn. J. Pharmacol.,* 64 (Suppl. I), 342P, 1994.
85. Sved, A.F. and Curtis, J.T., Amino acid neurotransmitters in nucleus tractus solitarius: an in vivo microdialysis study, *J. Neurochem.,* 61, 2089, 1993.
86. Maeda, T. et al., L-DOPA neurotoxicity is mediated by glutamate release in cultured rat striatal neurons, *Brain Res.,* 771, 159, 1997.
87. Leone, C. and Gordon, F.J., Is L-glutamate a neurotransmitter of baroreceptor information in the nucleus of the tractus solitarius?, *J. Pharmacol. Exp. Ther.,* 250, 953, 1989.
88. Pawloski-Dahm, C. and Gordon, F.J., Evidence for a kynurenate-insensitve glutamate receptor in nucleus tractus solitarii, *Am. J. Physiol.,* 262, H1611, 1992.
89. Kubo, T. and Kihara, M., Unilateral blockade of excitatory amino acid receptors in the nucleus tractus solitarii produces an inhibition of baroreflexes in rats, *Naunyn-Schmiedeberg's Arch. Pharmacol.,* 343, 317, 1991.
90. Foley, C.M. et al., Cardiovascular response to group I metabotropic glutamate receptor activation in NTS, *Am. J. Physiol.,* 276, R1469, 1999.
91. Furchgott, R.F. and Zawadzki, J.V., The obligatory role of endothelial cells in the relaxation by acetylcholine, *Nature,* 288, 373, 1980.
92. Moncada, S. and Higgs, A., The L-arginine-nitric oxide pathway, *N. Engl. J. Med.,* 329, 2002, 1993.
93. Garthwaite, J. and Boulton, C.L., Nitric oxide signaling in the central nervous system, *Annu. Rev. Physiol.,* 57, 683, 1995.
94. Lipton, S.A. et al., A redox-based mechanism for the neuroprotective and neurodestructive effects of nitric oxide and related nitroso compounds, *Nature,* 364, 626, 1993.
95. Dawson, V.L. et al., Mechanisms of nitric oxide-mediated neurotoxicity in primary brain cultures, *J. Neurosci.,* 13, 2651, 1993.
96. Misu, Y. et al., DOPA causes glutamate release and delayed neuron death by brain ischemia in rats, *Neurotoxicol. Teratol.,* 24, 629, 2002.
97. Wu, G. and Morris, S.M., Jr., Arginine metabolism: nitric oxide and beyond, *Biochem. J.,* 336, 1, 1998.

98. Talman, W.T., Nitrooxidergic transmission in the nucleus tractus solitarii, *Ann. N.Y. Acad. Sci.,* 835, 225, 1997.
99. Lin, L.-H. et al., Direct evidence for nitric oxide synthase in vagal afferents to the nucleus tractus solitarii, *Neuroscience,* 84, 549, 1998.
100. Maeda, M. et al., Injection of antisense oligos to nNOS into nucleus tractus solitarii increases blood pressure, *Neuroreport,* 10, 1957, 1999.
101. Lin, H.C. et al., Nitric oxide signaling pathway mediates the L-arginine-induced cardiovascular effects in the nucleus tractus solitarii of rats, *Life Sci.,* 65, 2439, 1999.
102. Aoki, C. et al., NMDA-R1 subunit of the cerebral cortex co-localizes with neuronal nitric oxide synthase at pre- and postsynaptic sites and in spine, *Brain Res.,* 750, 25, 1997.
103. Lin, H.C., Wan, F.J., and Tseng, C.J., Modulation of cardiovascular effects produced by nitric oxide and ionotropic glutamate receptor interaction in the nucleus tractus solitarii of rats, *Neuropharmacology,* 38, 935, 1999.
104. Lin, H.C. et al., Reciprocal regulation of nitric oxide and glutamate in the nucleus tractus solitariii of rats, *Eur. J. Pharmacol.,* 407, 83, 2000.
105. Matsuo, I. et al., Glutamate release via NO production evoked by NMDA in the NTS enhances hypotension and bradycardia in vivo, *Am. J. Physiol. Regul. Integr. Comp. Physiol.,* 280, R1285, 2001.
106. Lawrence, A.J. and Jarrott, B., L-Glutamate as a neurotransmitter at baroreceptor afferents: evidence from in vivo microdialysis, *Neuroscience,* 58, 585, 1994.
107. Ohta, H., Li, X., and Talman, W.T., Release of glutamate in the nucleus tractus solitarii in response to baroreflex activation in rats, *Neuroscience,* 74, 29, 1996.
108. Reif, D.W. and McCreedy, S.A., N-nitro-L-arginine and N-monomethyl-L-arginine exhibit a different pattern of inactivation toward the three nitric oxide synthases, *Arch. Biochem. Biophys.,* 320, 170, 1995.

# 12 Glutamatergic and Nitroxidergic Neurotransmission in the Nucleus Tractus Solitarii

*William T. Talman and Li-Hsien Lin*

## CONTENTS



## 12.1 INTRODUCTION

As the primary site of termination of cardiovascular and visceral afferent fibers of the vagus and glossopharyngeal nerves, the nucleus tractus solitarii (NTS) [1–3] plays a critical role in the regulation of blood pressure and peripheral blood flow. Direct stimulation of the NTS, or the cardiovascular afferent fibers that terminate within it, leads to marked changes in arterial blood pressure and regional blood flow [4,5]. Additionally, lesions within the NTS lead to acute hypertension both in humans [6] and in experimental animals [7,8].

A considerable amount of evidence has accumulated in support of the hypothesis that glutamate (GLU) is a neurotransmitter released from cardiovascular afferent terminals in the NTS [9], but GLU is not alone in holding that distinction. Among other putative transmitters purported to participate in the central cardiovascular control by the NTS is nitric oxide (NO) [10]. This chapter will consider some of the studies from our own laboratory, as well as some studies from other laboratories, that support a link between glutamatergic and nitroxidergic neurons in cardiovascular regulation through the NTS.

In fact, a link between GLU and NO in central neurotransmission has been considered for quite some time. Early studies showed that activation of GLU receptors in the brain led to synthesis and release of NO [11,12]. Acting through NO synthase (NOS) and release of NO, GLU might then effect the physiological responses attributed to it through activation of soluble guanylate cyclase (sGC) and increases in cyclic GMP [12,13]. Other studies have shown that the relationship between GLU and NOS is more than casual, in that destruction of GLU receptors eliminates activation of sGC by GLU even though an NO donor could still activate the same neuronal pools of the enzyme by acting "downstream" of the GLU receptor [12]. Influences of GLU receptor activation on NO synthesis were first associated with *N*-methyl-D-aspartate (NMDA) receptor activation, but subsequent studies have shown that kainate, metabotropic (ACPD responsive), and α-amino-3-hydroxy-5-methylisoxozole-proprionic acid (AMPA) receptor agonists also enhance synthesis of NO [14–17]. The relationship between GLU and NO is more complex than simply that of GLU receptor activation leading to NO synthesis. Antagonists of GLU receptors may themselves lead to release of NO [18]; NO acts presynaptically as well as postsynaptically [19,20] to provide a feedback for release of GLU [21], and the two compounds may produce integrated responses in the brain.

They may act together to participate in such critical activities as long-term potentiation in the hippocampus [22] and long-term depression in the cerebellum [23]. Their integrated responses have been better studied at central sites other than the NTS but some studies now suggest that NO release may also be linked with GLU receptor activation in the NTS [24,25]. It should be clear that important insights into integrative and transduction mechanisms, both in the NTS and other central sites, may derive from studying nitroxidergic neuronal transmission in the NTS. Indeed, such studies could shed light not only on GLU and NO but also on other transmitter mechanisms in the NTS in which L-3,4-dihydroxyphenylalanine (DOPA) [26,27], substance P [28–30], acetylcholine [31], and γ-aminobutyric acid (GABA) [32–36], to name but a few, have been implicated in cardiovascular reflex control, with links to NO in the NTS or other central sites [37–39].

With regard to the integration of GLU and NO, this integration may be intracellular in some sites. In such cases, the two could act through a common transduction pathway. At other sites, intercellular processes could account for integration and could involve activation of nitroxidergic neurons by GLU receptor activation. Data from our studies of the NTS support direct and integrative effects of glutamatergic and nitroxidergic neurons on transmission of cardiovascular reflex signals within the NTS.

## 12.2 NEURONAL NITRIC OXIDE SYNTHASE IN THE NTS

We have used histochemical, immunohistochemical, and *in situ* hybridization techniques to demonstrate [40], as have others [41], that neuronal NO synthase (nNOS) is expressed in neurons and terminals in the dorsal vagal complex and NTS [42,43]. Although an estimate of the contribution made by cardiovascular afferents to nNOS labeling in the NTS has varied between studies [44], our own work conclusively shows that nNOS is present in vagal afferents and their terminals in the NTS [40]. Thus, NO may be released from those terminals in the NTS.

We used two approaches in our study of nNOS and vagal afferents [40]. First, we injected the anterograde tracer biotinylated dextran amine (BDA) into the nodose ganglion of anesthetized rats to label and visualize the vagus nerve axon terminals in the NTS. At the light microscopic level, we studied the projections and terminations and compared them with neurons and neuritic processes stained for nNOS immunoreactivity (IR). We found that nNOS-IR was concentrated in the regions of the NTS in which vagal afferent fibers terminated. In the second phase of the studies, we utilized electron microscopic analysis of the NTS after removal of one nodose ganglion in an attempt to determine whether degenerating vagus nerve terminals contained nNOS [40]. Three days after removal of the ganglion, we were able to identify numerous degenerating terminals that contained nNOS-IR. We found that 20 to 30% of the neurons in the nodose ganglia that had been removed were immunopositive for nNOS.

To confirm the endogenous production of nNOS in vagal afferent terminals in the NTS, we utilized *in situ* hybridization techniques in another study [44] in which we removed a nodose ganglion to induce degeneration of vagus nerve terminals in the NTS. We used a cDNA probe provided by David Bredt to confirm the presence of nNOS mRNA in the NTS and compared the staining on the deafferented side with that on the control side and on both sides in control animals in which the nodose ganglion had not been removed. The density of nNOS-IR and mRNA in the NTS was reduced by approximately 15% on the deafferented side. These findings further supported our immunoelectron microscopic studies [40]; but, as with those earlier studies, they also supported the possibility of there being a large contribution of intrinsic neurons and/or nonvagal afferents to the nNOS-IR in the NTS. Our own studies suggest that intrinsic structures contribute substantially to that pool. We identified many immunopositive dendrites, some immunopositive neurons, and some nondegenerating immunopositive axons in the NTS, but we suggest that the large number of labeled dendrites in the NTS favor a contribution by intrinsic neurons, rather than nonvagal afferent axons, to the residual nNOS-IR seen in the NTS after vagus nerve deafferentation. In recent studies [45] with a direct bearing on the link between nNOS in the NTS and baroreflex control, we have found nNOS-IR in central terminals of labeled aortic depressor nerve in the rat.

## 12.3 ANATOMICAL BASIS FOR A LINK BETWEEN GLUTAMATE AND NITRIC OXIDE IN NTS

Many pharmacological studies have suggested a link between GLU and NO in the NTS. Anatomical studies are now beginning to complement pharmacology and further support the link. With developing technology, support from these anatomical studies has strengthened. We have performed a series of immunohistochemical, confocal microscopic studies to address the question. We have reported [46] that GLU-IR and nNOS-IR colocalized in some of the same NTS neurons, whereas other NTS neurons contained only one of them (Figure 12.1). In this study and in all of our subsequent studies of nNOS-IR, we found that distribution of the enzyme was

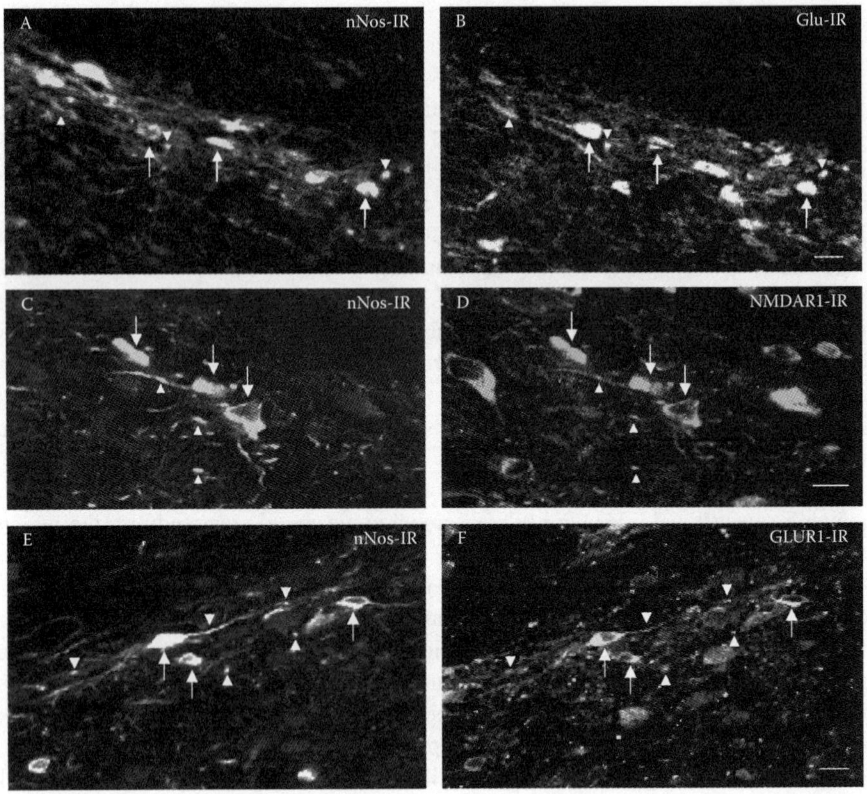


**FIGURE 12.1** Colocalization of nNOS and GLU (A, B), NMDAR1 (C, D), and GLUR1 (E, F) in the NTS. A, B: Confocal images (in gray scale) of nNOS immunoreactivity (nNOS-IR), shown in A, and GLU immunoreactivity (GLU-IR), shown in B, in the dorsolateral subnucleus of the NTS of the same section. Arrows and arrowheads indicate examples of double-labeled neurons and fibers, respectively. C, D: Confocal images of nNOS-IR, shown in C, and NMDAR1-IR, shown in D, in the dorsolateral subnucleus of the NTS of the same section. E, F: Confocal images of nNOS-IR, shown in E, and GLUR1-IR, shown in F, in the dorsolateral subnucleus of the NTS. Arrows and arrowheads indicate examples of double-labeled neurons and fibers, respectively. Scale bar = 10 μm. (Adapted from Talman, W.T. et al., Nitroxidergic influences on cardiovascular control by NTS: a link with glutamate, *Ann. N.Y. Acad. Sci.,* 940, 169, 2000; Lin, L.-H. and Talman, W.T., Colocalization of GluR1 and neuronal nitric oxide synthase in rat nucleus tractus solitarii neurons, *Neuroscience,* 106, 801, 2001. With permission.)

not uniform among NTS subnuclei. We found a similar heterogeneous distribution of neurons that colocalized GLU-IR and nNOS-IR. Double-labeled neurons were most prominent in the central subnucleus, a region that receives afferents from the stomach, mouth, and esophagus [47]. However, neurons in subnuclei that receive cardiovascular afferents also prominently colocalized GLU-IR and nNOS-IR. The two immunolabels were not confined to neurons in the NTS but were also seen in fibers that passed through the NTS, as well as in fibers that lay in close apposition to other fibers that contained one or both markers.

We utilized, in more recent studies, double and triple immunolabeling of nNOS and vesicular GLU transporters (VGLUT), which are acknowledged to provide better labeling of glutamatergic neuronal elements than does GLU itself. We have found VGLUT1, VGLUT2, and VGLUT3 in the nodose ganglion, in vagal afferents in the NTS (unpublished observations), and in subnuclei of the NTS [48,49]. In addition, nNOS-IR was present in VGLUT2-IR-containing fibers and VGLUT3-IR-containing fibers and neurons in the NTS.

Although these studies provided evidence that nNOS was found both within glutamatergic neurons and in neuronal elements making synaptic contact with those neurons in NTS, they did not link nNOS with transduction of glutamatergic signals in the NTS. Thus, we sought to determine the relationship between nNOS and GLU receptors in the NTS. We performed confocal double- and triple-labeling immunohistochemical studies to identify the relationship between specific GLU receptors and nNOS in the NTS. In the first of those studies [50], we found very few neurons that contained nNOS without also containing NMDAR1-IR, but we did find neurons that contained only NMDAR1-IR (Figure 12.1). Neurons that contained both nNOS-IR and NMDAR1-IR were present in cardiovascular regions of the NTS as well as in subnuclei that associated with other physiological functions. In electron microscopic studies, we have identified synapses formed by some nNOS-containing terminals and NMDAR1-IR postsynaptic dendritic membranes (unpublished observations). We have also found terminals that synapse with NMDAR1-IR-containing dendrites but do not contain nNOS-IR themselves.

As predicted, we found that the association between nNOS and GLU neuronal elements was not confined to NMDA receptors (Figure 12.1). Similar to NMDAR1-IR, GLUR1-IR was found in virtually every NTS neuron that contained nNOS-IR.

One of the principal mechanisms in signal transduction effected by NO is the activation of sGC. Thus, it follows that neuronal elements containing NMDA and AMPA receptors should contain sGC if GLU and NO were linked in the signaling by those neurons. Our recent studies also support that link in showing colocalization of nNOS and sGC. We found [52] that many NTS neurons contained both nNOS-IR and sGC-IR and that double-labeled neurons were present in all NTS subnuclei. There were also cells that contained nNOS-IR or sGC-IR alone.

## 12.4 PHARMACOLOGICAL STUDIES LINKING GLU AND NO IN CARDIOVASCULAR CONTROL THROUGH NTS

Some researchers have shown that the activity of NTS neurons is reduced in the presence of a NOS inhibitor and have suggested that the nitroxidergic input to the NTS may tonically influence this effect [53]. We have found that introduction of an NO donor *S*-nitrosocysteine into the NTS leads to depressor and bradycardic responses such as those produced by GLU in anesthetized animals [54]. However, responses to the *S*-nitrosothiols differed from GLU in that the latter produces depressor responses in anesthetized animals and pressor responses in awake animals, whereas *S*-nitrosothiols microinjected into the NTS elicit depressor responses in either animal model [55]. We suggested that responses to *S*-nitrosothiols depended on NO as a component of the molecule but not on release of NO itself from the molecule. That conjecture was supported by the persistence of cardiovascular responses to the *S*-nitrosothiols in the presence of hemoglobin, which would be expected to scavenge NO from the extracellular space and block responses. We also found that the cardiovascular responses to *S*-nitrosocysteine were not replicated by injection of other NO donors or even NO itself. Finding that responses to *S*-nitrosocysteine only occurred with injection of the levo isomer of the molecule, we further conjectured that there could be a receptor to which the NO-containing molecule could bind to elicit the cardiovascular effects we had seen. Our findings were largely replicated by another laboratory [56] in its studies of ventilatory control through the NTS. It remains uncertain whether NO is itself the mediator of the physiological responses or whether it is a component of another active molecule.

However, as with NO, *S*-nitrosothiols act, at least in part, through activation of sGC. Cardiovascular effects elicited by injection of *S*-nitrosothiols into the NTS were blocked by the sGC inhibitor [57] methylene blue [58]. Furthermore, inhibition of sGC also blocked the cardiopulmonary (Bezold–Jarisch) reflex [59]. However, it is less clear what role sGC may play in baroreflex control by the NTS. Some studies that used more selective blockers of guanylate cyclase did not directly address baroreflex transmission in the NTS [25], and our own studies failed to show an effect of sGC blockade on baroreflex responses [60]. The role of NO itself in baroreflex transmission has been somewhat controversial with some studies suggesting that it may contribute [61] and others reporting that NO in the NTS may not contribute to baroreflex modulation of sympathetic nerve activity [62]. If GLU and NO are indeed linked in baroreflex control through the NTS, the latter finding is inconsistent with studies of GLU and the baroreflex, which have found GLU receptors to be critical for full expression of the baroreflex [63]. In a recent study in our lab, we found that inhibition of nNOS by bilateral injection of the nNOS inhibitor AR-R 17477 led to both hypertension and inhibition of the arterial baroreflex (Figure 12.2) [64]. Those findings suggest that NO synthesized by neurons plays an excitatory role in baroreflex transmission through the NTS but, intriguingly, others [65] have shown that NO synthesized by eNOS may mediate inhibition of the baroreflex. It then is possible that inhibition of both nNOS and eNOS in earlier studies led to the removal of both excitatory and inhibitory influences on the reflex and, thus, the failure to define a role for NO in reflex transmission. As with the baroreflex, definition of the role of NO in arterial chemoreflex transmission continues to evolve. Such a role is supported by the presence of NOS in the carotid body and alteration of chemoreflex activity after peripheral administration of an NOS inhibitor [66,67].

Our anatomical studies suggest that NO and GLU may interact both pre- and postsynaptically, a suggestion that is supported by one study, which suggests that the two may even act on the same neurons in NTS [10]. sGC plays an important role in the transduction of signals initiated by NO [68]. Therefore, we first studied whether cardiovascular responses elicited by microinjection of GLU into the NTS

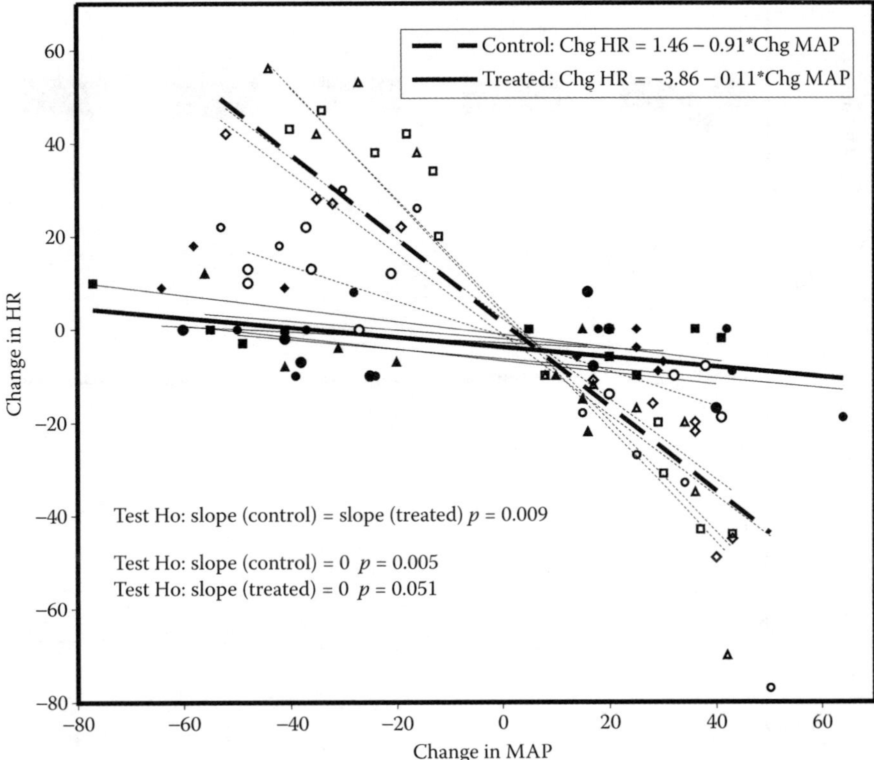


**FIGURE 12.2** Injection of the nNOS inhibitor AR-R 17477 (7.5 nmol) bilaterally into NTS in five rats significantly reduced baroreflex heart rate responses (Change in HR; y-axis) to increases and decreases in mean arterial pressure (Change in MAP; x-axis) when compared with responses prior to injections into NTS. Each set of open symbols represents data points for each animal under control conditions and dashed lines show the regression line for each animal. The bold dashed line is the estimate of the mean regression line under control conditions and represents a regression coefficient of 0.91. Each set of black symbols represents data points for each animal after treatment, and solid lines show the regression line for each animal. The bold solid line is the estimate of the mean regression line after treatment and represents a regression coefficient of 0.11. (From Talman, W.T. and Nitschke Dragon, D., Transmission of arterial baroreflex signals depends on neuronal nitric oxide synthase, *Hypertension,* 43, 820, 2004. With permission.)

were affected by inhibition of the enzyme. We found that injection of the guanylate cyclase inhibitor methylene blue into the NTS significantly attenuated responses to subsequent injection of GLU and agonists of ionotropic GLU receptors at the same site [69]. In subsequent studies that utilized more selective inhibitors of sGC, we found that depressor and bradycardic responses to NMDA were reduced in a dose-dependent manner by prior injection of LY83583. Injection of the still more selective inhibitor 1*H*-[1,2,4]oxadiazolo[4,3,-a]quinoxalin-1-one (ODQ) [70] led to the inhibition of responses to ionotropic (Figure 12.3) agonists, whereas responses to

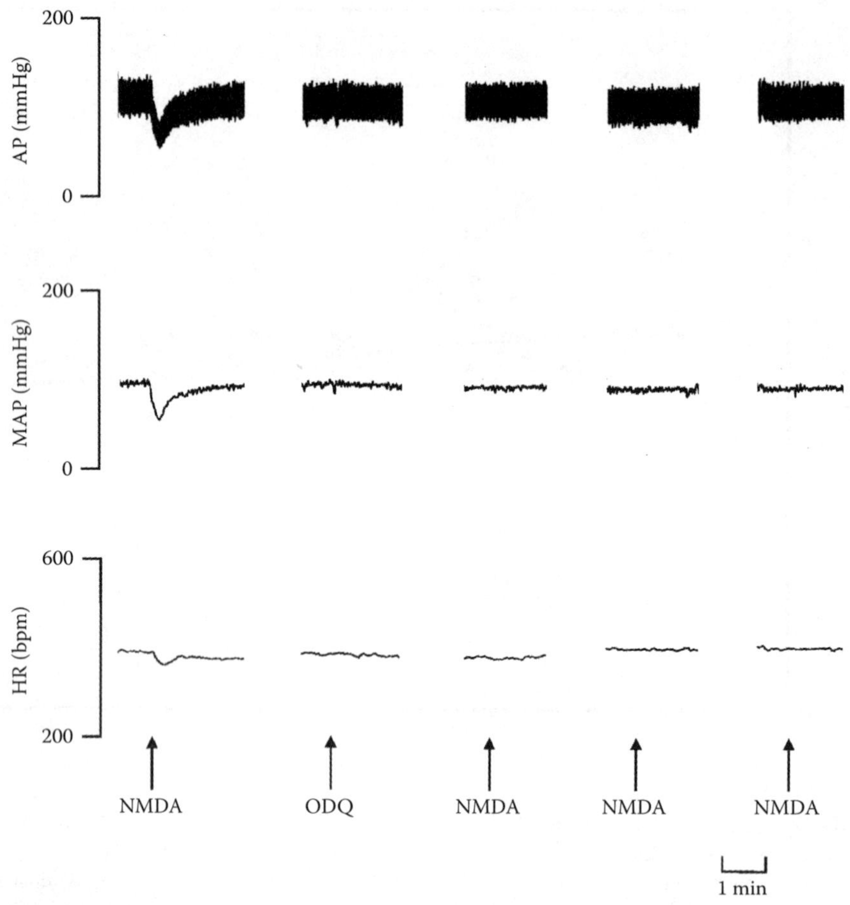


**FIGURE 12.3** Inhibition of sGC by injection of ODQ (200 pmol shown here) into NTS elicits dose-dependent reduction of depressor and bradycardic responses to injection of the ionotropic agonist NMDA (3 pmol) at the same site. Injection of ODQ at the same site in NTS after the second injection of NMDA blocked responses to subsequent injections of NMDA 10, 20, and 30 min later (right columns). Injection of vehicle (not shown) did not affect responses to NMDA. (From Chianca, D.A., Jr. et al., *Am. J. Physiol. Heart Circ. Physiol.*, 286, H1521, 2004. With permission.)

metabotropic agonists were spared, as they had been by methylene blue (Figure 12.4). Our findings with ODQ were consistent with an earlier report that ODQ inhibits activation of GLU receptors [25]. We conjecture that sparing of metabotropic receptors by ODQ accounts for the persistence of baroreflex activity after injection of the inhibitor bilaterally into the NTS [71].

In subsequent studies, we sought to determine whether inhibition of nNOS blocks responses to GLU agonists. As we had predicted, we found that intraperitoneal injection of the nNOS inhibitor 7-nitroindazole (7-NI) significantly attenuated

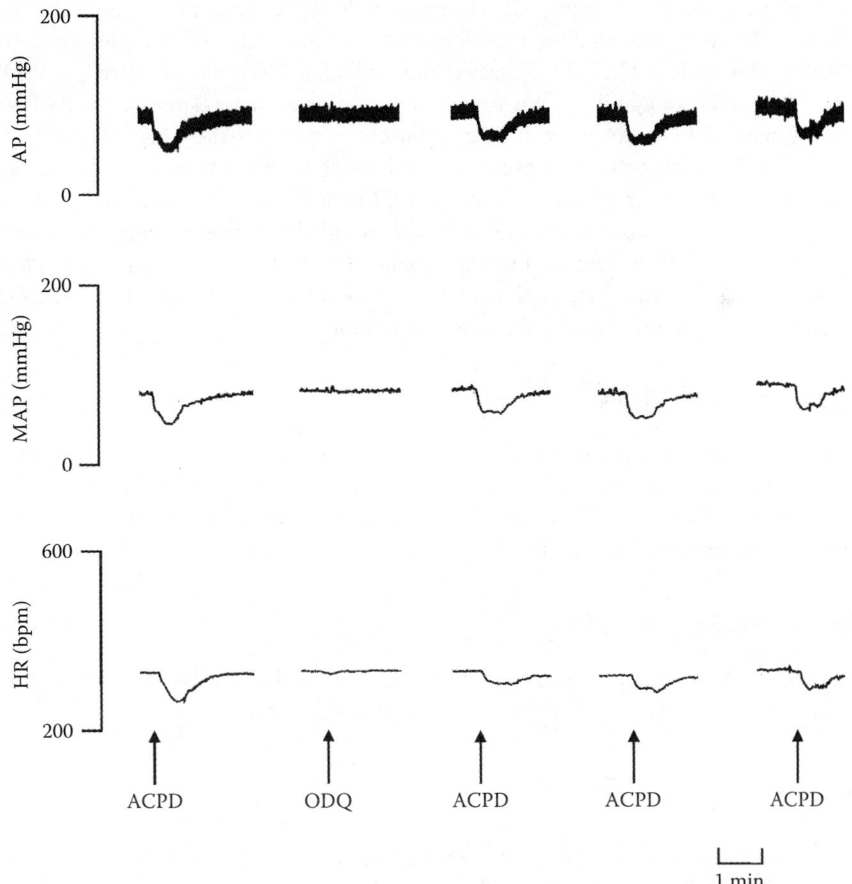


**FIGURE 12.4** Inhibition of sGC by injection of ODQ (200 pmol) into NTS does not affect depressor or bradycardic responses to injection of the metabotropic agonist ACPD (75 pmol) at the same site. ACPD, similar to NMDA, elicited reproducible depressor and bradycardic responses (left column) that were not affected 10, 20, or 30 min after ODQ (right 3 columns, respectively). (From Chianca, D.A., Jr. et al., *Am. J. Physiol. Heart Circ. Physiol.*, 286, H1521, 2004. With permission.)

responses to NMDA microinjected unilaterally into the NTS of anesthetized rats [69]. The attenuation of responses elicited by NMDA was reversed by administration of L-arginine, the precursor for synthesis of NO by nNOS. D-arginine did not alter effects of 7-NI. In more recent studies, we found that 7-NI similarly attenuated responses to injection of AMPA and ACPD into the NTS (unpublished observations).

Although these findings were intriguing and consistent with other published findings [25], we felt that the studies were limited for several reasons. Specifically, intraperitoneal delivery of 7-NI meant that the exact location at which the inhibitor was acting to block GLU agonists could not be known; at the same time, injection of 7-NI into the NTS might militate against the specificity of the inhibitor for nNOS [72]. In recent studies, therefore, we used the aqueous soluble, selective nNOS inhibitor AR-R 17477 [73] injected into the NTS to determine the effects of nNOS blockade on GLU agonists. We found that cardiovascular responses to NMDA, AMPA, and ACPD were in each case significantly attenuated by prior injection of AR-R 17477, as were the responses to L-arginine, the precursor for synthesis of NO [74,75]. In light of the integration between GLU and DOPA as described by Misu et al. [39], we hypothesized that AR-R 17477 would also inhibit responses to DOPA injected into the NTS. Our preliminary studies (unpublished observations) indeed demonstrate significant attenuation of the responses to DOPA injected into NTS when preceded by injection of the nNOS inhibitor.

## 12.5 CONCLUSION

Thus, these studies suggest that there is not only a link between GLU and NO in the transmission and transduction of cardiovascular reflex signals through the NTS, but also that NO may play an integral role in the transmission of signals by other putative transmitters, such as DOPA, in the NTS.

## ACKNOWLEDGMENTS

The work was funded in part by National Institutes of Health R01 HL59593 and a Merit Review by the U.S. Department of Veterans Affairs.

## REFERENCES

1. Panneton, W.M. and Loewy, A.D., Projections of the carotid sinus nerve to the nucleus of the solitary tract in the cat, *Brain Res.,* 191, 239, 1980.
2. Wallach, J.H. and Loewy, A.D., Projections of the aortic nerve to the nucleus tractus solitarius in the rabbit, *Brain Res.,* 188, 247, 1980.
3. Kalia, M. and Mesulam, M.-M., Brain stem projections of sensory and motor components of the vagus complex in the cat: II. laryngeal, tracheobronchial, pulmonary, cardiac, and gastrointestinal branches, *J. Comp. Neurol.,* 193, 467, 1980.
4. Yin, M. et al., Hemodynamic effects elicited by stimulation of the nucleus tractus solitarii, *Hypertension,* 23 (Suppl.), I73, 1994.
5. Colombari, E. et al., Hemodynamic effects of L-glutamate in NTS of conscious rats: a possible role of vascular nitrosyl factors, *Am. J. Physiol.,* 274, H1066, 1998.
6. Montgomery, B.M., The basilar artery hypertensive syndrome, *Arch. Intern. Med.,* 108, 559, 1961.
7. Doba, N. and Reis, D.J., Acute fulminating neurogenic hypertension produced by brainstem lesions in the rat, *Circ. Res.,* 32, 584, 1973.
8. Talman, W.T., Perrone, M.H., and Reis, D.J., Acute hypertension after the local injection of kainic acid into the nucleus tractus solitarii of rats, *Circ. Res.,* 48, 292, 1981.
9. Gordon, F.J. and Talman, W.T., Role of excitatory amino acids and their receptors in bulbospinal control of cardiovascular function, in *Central Neural Mechanisms in Cardiovascular Regulation,* Vol. 2, Kunos, G. and Ciriello, J., Eds., Birkhauser, New York, 1992, p. 209, chap. 7.

10. Tagawa, T. et al., Nitric oxide influences neuronal activity in the nucleus tractus solitarius of rat brainstem slices, *Circ. Res.,* 75, 70, 1994.
11. Garthwaite, J. et al., NMDA receptor activation induces nitric oxide synthesis from arginine in rat brain slices, *Eur. J. Pharmacol.,* 172, 413, 1989.
12. Garthwaite, J. and Garthwaite, G., Cellular origins of cyclic GMP responses to excitatory amino acid receptor agonists in rat cerebellum *in vitro, J. Neurochem.,* 48, 29, 1987.
13. De Vente, J. et al., Immunocytochemistry of cGMP in the cerebellum of the immature, adult, and aged rat: the involvement of nitric oxide. A micropharmacological study, *Eur. J. Neurosci.,* 2, 845, 1990.
14. Garthwaite, J., Southam, E., and Anderton, M., A kainate receptor linked to nitric oxide synthesis from arginine, *J. Neurochem.,* 53, 1952, 1989.
15. Kiedrowski, L., Costa, E., and Wroblewski, J.T., Glutamate receptor agonists stimulate nitric oxide synthase in primary cultures of cerebellar granule cells, *J. Neurochem.,* 58, 335, 1992.
16. Southam, E., East, S.J., and Garthwaite, J., Excitatory amino acid receptors coupled to the nitric oxide/cyclic GMP pathway in rat cerebellum during development, *J. Neurochem.,* 56, 2072, 1991.
17. Okada, D., Two pathways of cyclic GMP production through glutamate receptor-mediated nitric oxide synthesis, *J. Neurochem.,* 59, 1203, 1992.
18. Marin, P. et al., Non-classical glutamate receptors, blocked by both NMDA and non-NMDA antagonists, stimulate nitric oxide production in neurons, *Neuropharmacology,* 32, 29, 1993.
19. Larkman, A.U. and Jack, J.J., Synaptic plasticity: hippocampal LTP, *Curr. Opin. Neurobiol.,* 5, 324, 1995.
20. Hawkins, R.D., Zhuo, M., and Arancio, O., Nitric oxide and carbon monoxide as possible retrograde messengers in hippocampal long-term potentiation, *J. Neurobiol.,* 25, 652, 1994.
21. Segieth, J. et al., Nitric oxide regulates excitatory amino acid release in a biphasic manner in freely moving rats, *Neurosci. Lett.,* 200, 101, 1995.
22. Izumi, Y., Clifford, D.B., and Zorumski, C.F., Inhibition of long-term potentiation by NMDA-mediated nitric oxide release, *Science,* 257, 1273, 1992.
23. Shibuki, K. and Okada, D., Endogenous nitric oxide release required for long-term synaptic depression in the cerebellum, *Nature,* 349, 326, 1991.
24. Di Paola, E.D., Vidal, M.J., and Nisticò, G., L-Glutamate evokes the release of an endothelium-derived relaxing factor-like substance from the rat nucleus tractus solitarius, *J. Cardiovasc. Pharmacol.,* 17 (Suppl. 3), S269, 1991.
25. Lin, H.C., Wan, F.J., and Tseng, C.-J., Modulation of cardiovascular effects produced by nitric oxide and ionotropic glutamate receptor interaction in the nucleus tractus solitarii of rats, *Neuropharmacology,* 38, 935, 1999.
26. Kubo, T. et al., Evidence for L-DOPA systems responsible for cardiovascular control in the nucleus tractus solitarii of the rat, *Neurosci. Lett.,* 140, 153, 1992.
27. Yue, J.-L. et al., Baroreceptor-aortic nerve-mediated release of endogenous L-3,4-dihydroxyphenylalanine and its tonic depressor function in the nucleus tractus solitarii of rats, *Neuroscience,* 62, 145, 1994.
28. Miura, M., Takayama, K., and Okada, J., Study of possible transmitters in the solitary tract nucleus of the cat involved in the carotid sinus baro- and chemoreceptor reflex, *J. Auton. Nerv. Syst.,* 19, 179, 1987.
29. Potts, J.T. and Fuchs, I.E., Naturalistic activation of barosensitive afferents release substance P in the nucleus tractus solitarius of the cat, *Brain Res.,* 893, 155, 2001.

30. Riley, J. et al., Ablation of NK1 receptors in rat nucleus tractus solitarii blocks baroreflexes, *Hypertension,* 40, 823, 2002.
31. Criscione, L., Reis, D.J., and Talman, W.T., Cholinergic mechanisms in the nucleus tractus solitarii and cardiovascular regulation in the rat, *Eur. J. Pharmacol.,* 88, 47, 1983.
32. Bousquet, P. et al., Evidence for a neuromodulatory role of GABA at the first synapse of the baroreceptor reflex pathway. Effects of GABA derivatives injected into the NTS, *Naunyn-Schmiedeberg's Arch. Pharmacol.,* 319, 168, 1982.
33. Catelli, J.M., Giakas, W.J., and Sved, A.F., GABAergic mechanisms in nucleus tractus solitarius alter blood pressure and vasopressin release, *Brain Res.,* 403, 279, 1987.
34. Suzuki, M., Kuramochi, T., and Suga, T., GABA receptor subtypes involved in the neuronal mechanisms of baroreceptor reflex in the nucleus tractus solitarii of rabbits, *J. Auton. Nerv. Syst.,* 43, 27, 1993.
35. Sved, A.F. and Tsukamoto, K., Tonic stimulation of $GABA_B$ receptors in the nucleus tractus solitarius modulates the baroreceptor reflex, *Brain Res.*, 592, 37, 1992.
36. Zhang, J. and Mifflin, S.W., Receptor subtype specific effects of GABA agonists on neurons receiving aortic depressor nerve inputs within the nucleus of the solitary tract, *J. Auton. Nerv. Syst.,* 73, 170, 1998.
37. Spike, R.C., Todd, A.J., and Johnston, H.M., Coexistence of NADPH diaphorase with GABA, glycine, and acetylcholine in rat spinal cord, *J. Comp. Neurol.,* 335, 320, 1993.
38. Krowicki, Z.K. and Hornby, P.J., The inhibitory effect of substance P on gastric motor function in the nucleus raphe obscurus is mediated via nitric oxide in the dorsal vagal complex, *J. Auton. Nerv. Syst.,* 58, 177, 1996.
39. Yamanashi, K. et al., Involvement of nitric oxide production via kynurenic acid-sensitive glutamate receptors in DOPA-induced depressor responses in the nucleus tractus solitarii of anesthetized rats, *Neurosci. Res.,* 43, 231, 2002.
40. Lin, L.-H. et al., Direct evidence for nitric oxide synthase in vagal afferents to the nucleus tractus solitarii, *Neuroscience,* 84, 549, 1998.
41. Ruggiero, D.A. et al., Central and primary visceral afferents to nucleus tractus solitarii may generate nitric oxide as a membrane-permeant neuronal messenger, *J. Comp. Neurol.,* 364, 51, 1996.
42. Kristensson, K. et al., Co-induction of neuronal interferon-gamma and nitric oxide synthase in rat motor neurons after axotomy: a role in nerve repair or death?, *J. Neurocytol.,* 23, 453, 1994.
43. Lawrence, A.J. et al., The distribution of nitric oxide synthase-, adenosine deaminase- and neuropeptide Y-immunoreactivity through the entire rat nucleus tractus solitarius: effect of unilateral nodose ganglionectomy, *J. Chem. Neuroanat.,* 15, 27, 1998.
44. Lin, L.-H. et al., Up-regulation of nitric oxide synthase and its mRNA in vagal motor nuclei following axotomy in rat, *Neurosci. Lett.,* 221, 97, 1997.
45. Lin, L.-H. and Talman, W.T., Ultrastructural evidence for neuronal nitric oxide synthase in aortic depressor nerve afferents in the nucleus tractus solitarii, *Soc. Neurosci. Abst.,* 861.7, 2002.
46. Lin, L.-H., Emson, P.C., and Talman, W.T., Apposition of neuronal elements containing nitric oxide synthase and glutamate in the nucleus tractus solitarii of rat: a confocal microscopic analysis, *Neuroscience,* 96, 341, 2000.
47. Altschuler, S.M. et al., Viscerotopic representation of the upper alimentary tract in the rat: sensory ganglia and nuclei of the solitary and spinal trigeminal tracts, *J. Comp. Neurol.,* 283, 248, 1989.
48. Lin, L.-H. et al., Vesicular glutamate transporter colocalizes with syndrome, neuronal nitric oxide synthase in rat nucleus tractus solitarii, *Neuroscience,* 123, 247, 2004.

49. Lin, L.-H. and Talman, W.T., Vesicular glutamate transporter 3 is colocalized with neuronal nitric oxide synthase in the nucleus tractus solitarii and nodose ganglion, *FASEB J.,* 18, A1262, 2004.
50. Lin, L.-H. and Talman, W.T., *N*-methyl-D-aspartate receptors on neurons that synthesize nitric oxide in rat nucleus tractus solitarii, *Neuroscience,* 100, 581, 2000.
51. Lin, L.-H. and Talman, W.T., Colocalization of GluR1 and neuronal nitric oxide synthase in rat nucleus tractus solitarii neurons, *Neuroscience,* 106, 801, 2001.
52. Lin, L.-H. and Talman, W.T., Soluble guanylate cyclase and neuronal nitric oxide synthase are colocalized in rat nucleus tractus solitarii, *FASEB J.,* 18, A1262, 2004.
53. Ma, S., Abboud, F.M., and Felder, R.B., Effects of L-arginine-derived nitric oxide synthesis on neuronal activity in nucleus tractus solitarius, *Am. J. Physiol. Regul. Integr. Comp. Physiol.,* 268, R487, 1995.
54. Ohta, H. et al., Actions of *S*-nitrosocysteine in the nucleus tractus solitarii are unrelated to release of nitric oxide, *Brain Res.,* 746, 98, 1997.
55. Machado, B.H. and Bonagamba, L.G.H., Microinjection of *S*-nitrosocysteine into the nucleus tractus solitarii of conscious rats decreases arterial pressure but L-glutamate does not, *Eur. J. Pharmacol.,* 221, 179, 1992.
56. Lipton, A.J. et al., *S*-nitrosothiols signal the ventilatory response to hypoxia, *Nature,* 413, 171, 2001.
57. Garthwaite, J., Charles, S.L., and Chess-Williams, R., Endothelium-derived relaxing factor release on activation of NMDA receptors suggests role as intercellular messenger in the brain, *Nature,* 336, 385, 1988.
58. Lewis, S.J. et al., Microinjection of *S*-nitrosocysteine into the nucleus tractus solitarii decreases arterial pressure and heart rate via activation of soluble guanylate cyclase, *Eur. J. Pharmacol.,* 202, 135, 1991.
59. Lewis, S.J. et al., Processing of cardiopulmonary afferent input within the nucleus tractus solitarii involves activation of soluble guanylate cyclase, *Eur. J. Pharmacol.,* 203, 327, 1991.
60. Chianca, D.A., Jr. et al., NMDA receptors in the nucleus tractus solitarii are linked to soluble guanylate cyclase, *Am. J. Physiol. Heart Circ. Physiol.,* 2003.
61. Vargas da Silva, S. et al., Blockers of the L-arginine-nitric oxide-cyclic GMP pathway facilitate baroreceptor resetting, *Hypertension,* 23 (Suppl.), I60, 1994.
62. Zanzinger, J., Czachurski, J., and Seller, H., Effects of nitric oxide on sympathetic baroreflex transmission in the nucleus tractus solitarii and caudal ventrolateral medulla in cats, *Neurosci. Lett.,* 197, 199, 1995.
63. Ohta, H. and Talman, W.T., Both NMDA and non-NMDA receptors in the NTS participate in the baroreceptor reflex in rats, *Am. J. Physiol.,* 267, R1065, 1994.
64. Talman, W.T. and Nitschke Dragon, D., Transmission of arterial baroreflex signals depends on neuronal nitric oxide synthase, *Hypertension,* 43, 820, 2004.
65. Paton, J.F. et al., Adenoviral vector demonstrates that angiotensin II-induced depression of the cardiac baroreflex is mediated by endothelial nitric oxide synthase in the nucleus tractus solitarii of the rat, *J. Physiol. (Lond.),* 531, 2, 2001.
66. Grimes, P.A. et al., Nitric oxide synthase occurs in neurons and nerve fibers of the carotid body, *Adv. Exp. Med. Biol.,* 360, 221, 1994.
67. Trzebski, A. et al., Carotid chemoreceptor activity and heart rate responsiveness to hypoxia after inhibition of nitric oxide synthase, *Adv. Exp. Med. Biol.,* 360, 285, 1994.
68. Moncada, S., The 1991 Ulf von Euler Lecture. The L-arginine: nitric oxide pathway, *Acta Physiol. Scand.,* 145, 201, 1992.
69. Talman, W.T. et al., Nitroxidergic influences on cardiovascular control by NTS: a link with glutamate, *Ann. N.Y. Acad. Sci.,* 940, 169, 2001.

70. Abi-Gerges, N. et al., A comparative study of the effects of three guanylyl cyclase inhibitors on the L-type $Ca^{2+}$ and muscarinic $K^{+}$ currents in frog cardiac myocytes, *Br. J. Pharmacol.,* 121, 1369, 1997.
71. Foley, C.M. et al., Cardiovascular response to group I metabotropic glutamate receptor activation in NTS, *Am. J. Physiol. Regul. Integr. Comp. Physiol.,* 276, R1469, 1999.
72. Zagvazdin, Y. et al., Evidence from its cardiovascular effects that 7-nitroindazole may inhibit endothelial nitric oxide synthase *in vivo, Eur. J. Pharmacol.,* 303, 61, 1996.
73. Johansson, C. et al., The neuronal selective nitric oxide inhibitor AR-R 17477, blocks some effects of phencyclidine, while having no observable behavioural effects when given alone, *Pharmacol. Toxicol.,* 84, 226, 1999.
74. Talman, W.T. et al., Differential involvement of neuronal nitric oxide in mediating cardiovascular responses to activation of NMDA and metabotropic receptors in the nucleus tractus solitarii of rats, *FASEB J.,* 17, A888, 2003.
75. Talman, W.T., Nitschke Dragon, D., and Lin, L.-H., eNOS and nNOS differentially affect cardiovascular regulation through the nucleus tractus solitarii, *Soc. Neurosci. Abst.,* 730.7, 2004.

# 13 Is DOPA a Reliable Neurotransmitter of Baroreflex Neurotransmission in the Brainstem?

*Miao-Kun Sun*

## CONTENTS



## 13.1 INTRODUCTION

L-3,4-Dihydroxyphenylalanine (L-DOPA) is an interesting molecule. It is a precursor of dopamine and other catecholamines and is decarboxylated to form dopamine by action of the enzyme aromatic L-amino acid decarboxylase. L-DOPA can cross the blood–brain barrier, which restricts the access of systemically applied dopamine to the brain. Thus, L-DOPA has been widely used to alleviate the symptoms of Parkinson's disease (a disorder due to insufficient dopaminergic transmission) [1–4], producing a variety of neural responses [5–8]. However, evidence is available that L-DOPA itself may be a signaling molecule in the mammalian brains. Its potential role as a neurotransmitter, especially in the brainstem arterial baroreflex pathway of mammals, was initially suggested by Misu, Goshima, and their colleagues [9,10].

Whether L-DOPA can be accepted as a neurotransmitter is an important issue. In this chapter, we will review the available evidence, either supporting the essential role of L-DOPA as a neurotransmitter of the brainstem baroreflex pathway or suggesting the need for more investigation. The main focus will be on its potential neurotransmitter function in the brainstem in the mediation of the arterial baroreflex.

## 13.2 CRITERIA FOR NOVEL NEUROTRANSMITTERS

To evaluate whether L-DOPA can be accepted as a neurotransmitter in a neural network, we have to first consider the criteria that a candidate substance has to satisfy to qualify. These criteria are: *proof of existence, physiological release*, and *identical responses and antagonism.*

Proof of existence as a neurotransmitter is the evidence that the candidate substance is present in the presynaptic terminals (usually concentrated in the vesicles) and neurons and that the presynaptic neurons synthesize the molecule to be released as a transmitter, and not simply store, accumulate, or use it as an intermediate molecule.

There is no question that L-DOPA exists in neurons but the issue is whether it exists as a separate neurotransmitter; the doubt arises due to the existence of neurons that use catecholamines as their neurotransmitters. The main pathway for biosynthesis of catecholamines is from tyrosine to L-DOPA, to dopamine, to noradrenaline and, finally, to adrenaline. L-DOPA is formed from tyrosine by the action of tyrosine hydroxylase and is used to synthesize dopamine by the action of aromatic L-amino acid decarboxylase. Dopamine can be further converted by dopamine-β-hydroxylase to noradrenaline, from which adrenaline can be synthesized by the action of phenylethanolamine-*N*-methyltransferase. L-DOPA, thus, exists in all catecholaminergic neurons as a precursor molecule but such an existence does not mean that L-DOPA functions as a neurotransmitter in these neurons. Similarly, there is no question that this biosynthetic pathway and these enzymes and mechanisms exist in all catecholaminergic neurons that can synthesize, accumulate, and transport L-DOPA.

However, evidence is available that L-DOPA may exist as an end product in neurons. Jaeger et al. [11] were the first to show that some tyrosine-hydroxylase-containing neurons in the rat dorsal motor nucleus of the vagus lacked immunoreactivity to aromatic L-amino acid decarboxylase, suggesting that DOPA might be the neurotransmitter in these neurons. By using specific DOPA and dopamine antibodies and antibodies against the enzymes involved in the synthesis of catecholamines, tyrosine-hydroxylase-positive, DOPA-positive, aromatic L-amino acid decarboxylase–negative, and dopamine-negative neurons have been identified in the rat hypothalamus and in the dorsal vagal complex/nucleus of the tractus solitarius [12–16], and they have been proposed to be probable DOPAergic neurons [17]. In the house-shrew (*Suncus murinus*) brain, study with electron microscopy and immunocytochemistry shows DOPA-positive reaction products in the axoplasm that is associated with the outer surface of small, clear, or cored vesicles [18]. But, information is currently not available as to whether L-DOPA is accumulated and stored in any particular type of vesicle in the neurons.

Release from the presynaptic terminals to be used to transfer particular types of activity signaling is an essential criterion for any candidate to be accepted as a neurotransmitter. The molecule must be released from the presynaptic terminals upon activation, typically in a voltage- and $Ca^{2+}$-dependent fashion.

Electrical stimulation of the presynaptic inputs in the rat hypothalamic slices and striatal slices has been found to release a molecule, which revealed itself as a peak different from that of the other catecholamine molecules on chromatographic charts, and was subsequently identified as L-DOPA [19]. The release of L-DOPA is decreased by α-methyl-*p*-tyrosine [19–22], a tyrosine hydroxylase inhibitor. The effectiveness of α-methyl-*p*-tyrosine strongly suggests that the evoked release of L-DOPA depends on L-DOPA synthesis from tyrosine. However, it is to be expected that L-DOPA will be accumulated, stored, and released from dopaminergic, noradrenergic, and adrenergic neurons because of the common chemical structures of these molecules. The real issue is whether the release of L-DOPA has functions distinct from that of the release of other neurotransmitters. L-DOPA can be said to function only as a cotransmitter unless separate L-DOPA receptors and antagonism can be demonstrated. The ratio of DOPA/dopamine release in the striatal slices has been shown to range from 1:2 to 1:3. Thus, it is not clear whether the DOPA released in the striatal slices functions as an end product of the neurons. However, aortic depressor nerve stimulation in the brainstem has been reported to evoke an independent L-DOPA release, without an associated release of dopamine, noradrenaline, or adrenaline [23].

Supporting evidence is also available that L-DOPA can be released in a tetrodotoxin-sensitive, $Ca^{2+}$-dependent manner with electrical stimulation, high $K^+$ [19], or nicotine [24]. Similar $Ca^{2+}$-dependent L-DOPA release has also been shown in the dog sympathetic ganglia [25].

Evidence for identical physiological responses and pharmacological antagonism are critical for identifying a novel neurotransmitter system. Neurotransmitters are biologically active. They carry neural signals between neurons or from neurons to other types of cells. Convincing evidence for identical physiological responses is often provided by demonstrating that membrane responses and molecular events are the same whether evoked by presynaptic activation or by the exogenous application of the transmitter candidate. Establishing the pharmacological identity of the receptors, including receptor mRNA, is also in this category of evidence.

Evidence is available that exogenous application of L-DOPA into the brainstem mimics electrical stimulation of the aortic depressor nerve or of local brain regions in producing cardiovascular responses (see the following text). Thus, exogenously applied L-DOPA evokes cardiovascular responses identical to those evoked by stimulating the corresponding neural inputs. In addition, such responses can be blocked by the antagonists that recognize DOPA receptors/recognition sites. DOPA methyl ester and DOPA cyclohexyl ester are reported to be the antagonists for DOPA receptors/recognition sites [21,23,26–33], suggesting the existence of selective DOPA recognition sites or receptors. DOPA methyl ester, for instance, antagonizes the depressor and bradycardic responses induced by microinjections of L-DOPA into the depressor site of the nucleus of tractus solitarius [34] and caudal ventrolateral medulla [28], and the pressor and tachycardic responses to L-DOPA microinjected into the pressor sites of the rostral ventrolateral medulla [21]. The antagonism is specific because DOPA methyl ester does not affect glutamate-induced cardiovascular responses in the nucleus of tractus solitarius [34], the caudal ventrolateral medulla [28], and the rostral ventrolateral medulla [21]. The evidence of an effective blockade within the nucleus of tractus solitarius by these antagonists of aortic depressor nerve-stimulation-evoked cardiovascular responses will be discussed further in the following text.

In addition, there is an obvious $Na^+$-dependent DOPA uptake transporter in the brain [35]. However, it has not been clarified whether the transporter is functionally involved in other catecholaminergic systems or exclusively in the DOPAergic system. Because of the possibility of L-DOPA existing as a precursor molecule in the catecholaminergic neurons, the existence of a DOPA transporter and an uptake system is good evidence, though not sufficient for establishing a neurotransmitter role for DOPA.

## 13.3 BAROREFLEX IN THE BRAINSTEM

Before examining L-DOPA's functional role, i.e., whether or not it serves as a neurotransmitter in the arterial baroreflex pathway in the brainstem, we need to briefly discuss this pathway and its regulatory mechanisms.

A normal cardiovascular function is essential to mammals and is mainly controlled by the central nervous system through several systems that sense, evaluate, control, and execute the pulse-to-pulse cardiovascular functions and regulation [36–41]. Among the cardiovascular reflexes, the arterial baroreflex is the most powerful and important mechanism [42,43], through which the brain regulates cardiovascular function rapidly and within a narrow range.

The arterial baroreflex consists of three major relays, all in the brainstem. The bipolar baroreceptor neurons sense moment-to-moment pressure variations in the circulation with their afferent terminals in the walls of the aortic arch, carotid sinus, the atria, and ventricles. Their activity is transferred through their efferent fibers, including the aortic depressor nerve, in the IXth and Xth cranial nerves, to the brainstem. The first relay for the baroreceptor input is believed to be in the intermediate third of the nucleus of tractus solitarius; the baroreflex information is primarily forwarded from here to the cardiovascular neurons in the caudal ventrolateral medulla, the second relay, as well as to many structures rostral to the brainstem. It is important to note that many neural groups rostral to the brainstem also receive the baroreceptor information but may not be directly involved in the regulation of sympathetic nerve activity or arterial baroreflex. The major neural group through which the cardiovascular neurons of the caudal ventrolateral medulla transduce the baroreflex information for an altered sympathetic nerve activity is the group of sympathetic prevasomotor neurons in the rostral ventrolateral medulla, the third major relay. These neurons are reticulospinal and send their output to the sympathetic preganglionic neurons in the intermediolateral cell column of the spinal cord, thus completing the arterial baroreflex pathway.

The neurotransmitters in the brainstem baroreflex pathway are believed to be either glutamate (in the synapses from the aortic baroreceptor input to the nucleus of tractus solitarius and from the nucleus of tractus solitarius to the neurons in the caudal ventrolateral medulla) or γ-aminobutyric acid (GABA) (in the synapses for baroreflex inputs from the neurons of the caudal ventrolateral medulla to the reticulospinal sympathetic prevasomotor neurons in the rostral ventrolateral medulla). The possibility of the involvement of various neurotransmitters in the baroreflex pathway in the medulla depends on the existence of a number of synaptic transmissions in, for instance, the nucleus of tractus solitarius. If interneurons are locally involved as an essential part of the pathway in processing and transferring the baroreflex information, the door is certainly open for other unidentified neurotransmitters in the baroreflex pathway. In fact, it has been proposed that the second-order nonglutamatergic neurons in the nucleus of tractus solitarius may be involved in the baroreflex control of cardiovascular function, consistent with the observation that neurons in the nucleus of tractus solitarius receive monosynaptic as well as polysynaptic inputs from the aortic depressor nerve terminals. The argument supporting the involvement of additional neurotransmitters in the baroreflex pathway in the nucleus of tractus solitarius is based on the fact that all the studies do not distinguish whether the excitatory amino acid is the transmitter responsible for synaptic transmission between the baroreflex input terminals and the cardiovascular neurons in the nucleus of tractus solitarius or for synaptic transmission beyond the first synapse. The evidence supporting the involvement of DOPAergic synapses in baroreflex regulation is substantial and is reviewed in the text that follows.

## 13.4 EVIDENCE FOR DOPA AS BAROREFLEX NEUROTRANSMITTER IN BRAINSTEM

Potential neurotransmission roles for L-DOPA have been proposed at all the three major relay sites of the arterial baroreflex pathway in the brainstem [44].

Tison et al. [15] demonstrated the existence of DOPA immunoreactivity (DOPA-positive and dopamine-negative) in the rat dorsal vagal complex and the nucleus of tractus solitarius. Such neurons exist in other regions, including the caudal ventrolateral medulla and in the potential DOPAergic pathway from the posterior hypothalamic nucleus to the rostral ventrolateral medulla.

The basal DOPA releases in the nucleus of tractus solitarius, the caudal ventrolateral medulla, and the rostral ventrolateral medulla have all been documented as being partially tetrodotoxin sensitive and $Ca^{2+}$ dependent. This is viewed as evidence of two separate components of DOPA release: a tetrodotoxin-sensitive/$Ca^{2+}$-dependent release as a neurotransmitter and a tetrodotoxin-insensitive/$Ca^{2+}$-independent release as the precursor of dopamine [44]. Stimulation of the aortic depressor nerve releases DOPA ipsilaterally in the nucleus of tractus solitarius [29] and the caudal ventrolateral medulla [23]. $\alpha$-Methyl-*p*-tyrosine, a tyrosine hydroxylase inhibitor, inhibits the release of DOPA in the brainstem [19,23,28,29], indicating that the released DOPA is from a pool that depends on a synthesis from tyrosine. Electrical stimulation of the posterior hypothalamic nucleus elicits a tetrodotoxin-dependent DOPA release in the rostral ventrolateral medulla and hypertension and tachycardia [31]. In addition, no dopamine, noradrenaline, or adrenaline are released by the aortic depressor nerve stimulation in these sites [23], indicating that the observed DOPA release in these sites is not associated with the activity of catecholaminergic neurons.

Cardiovascular responses have been evoked with L-DOPA from all the major relay stations of the baroreflex pathway in the brainstem, including the nucleus of tractus solitarius, the caudal ventrolateral medulla, and the rostral ventrolateral medulla. L-DOPA (10 to 100 ng) microinjected into the depressor sites of the nucleus of tractus solitarius or of the caudal ventrolateral medulla elicits a decrease in arterial blood pressure and bradycardia in the absence and presence of the centrally acting aromatic L-amino acid decarboxylase inhibitor 3-hydroxybenzyl hydrazine (NSD-1015) [23,28,30,32,34,45,46]. L-DOPA (30 to 600 ng) applied into the pressor sites of the rostral ventrolateral medulla has also been reported to elicit hypertension and tachycardia in the absence and presence of NSD-1015 [21,45]. The responses observed in the presence of NSD-1015 suggest that they are unlikely to be the result of the rapid conversion of L-DOPA to dopamine by the aromatic L-amino acid decarboxylase because the enzyme is blocked by NSD-1015. All the cardiovascular responses to L-DOPA microinjections are reported to be dose dependent, stereoselective, and not mimicked by catecholamines at the same dose ranges.

Effects of pharmacological antagonism against DOPA receptors/recognition sites have also been reported. Bilateral microinjection of DOPA methyl ester alone into the depressor region of the nucleus of tractus solitarius elicits hypertension and tachycardia, the responses abolished by α-methyl-*p*-tyrosine [28], suggesting that the effects depend on the activity of tyrosine hydroxylase, i.e., the endogenous formation of DOPA. The application of DOPA esters also blocks or reduces arterial pressure responses evoked by aortic depressor nerve stimulation or DOPA administration in these relay stations [44 for review]. The effects evoked by antagonist administration suggest the presence of a tonic activity of DOPA receptors/recognition sites. Microinjection of DOPA methyl ester alone into the rostral ventrolateral medulla also elicits hypotension and bradycardia in the pressor sites of the rostral ventrolateral medulla [21]. The effectiveness of DOPA and DOPA antagonists in the rostral ventrolateral medulla is viewed as suggesting the presence of a tonic activity from the posterior hypothalamic nucleus to the rostral ventrolateral medulla.

## 13.5 EVIDENCE NEEDED TO ESTABLISH DOPA AS BAROREFLEX NEUROTRANSMITTER IN BRAINSTEM

Evidence that L-DOPA may operate as a neurotransmitter in the arterial baroreflex pathway in the brainstem includes factors such as its existence, release, physiological identity, and pharmacological antagonism [44,47]. Although the available evidence is strong and substantial, several issues remain, requiring further investigation.

One issue is that none of the studies represents a direct evaluation of the effects of the potential neurotransmitter and its receptor antagonism on the singly identified cardiovascular neurons. The barosensitive cardiovascular neurons in the nucleus of tractus solitarius, the caudal ventrolateral medulla, and the rostral ventrolateral medulla can be recorded and studied in greater detail. Likewise, sympathetic activity, the most reliable outflow of the brainstem baroreflex, was not monitored. Other mechanisms may be involved in evoking changes in arterial blood pressure as well as heart rate [43]. For instance, it has been proposed that in the baroreflex pathway, DOPA serves as a neurotransmitter of the primary aortic depressor nerve and glutamate as a neurotransmitter of the second-order neurons or interneurons in the nucleus of tractus solitarius. The DOPAergic terminals innervate these glutamatergic baroreflex neurons, explaining why blocking glutamate receptors in the nucleus of tractus solitarius largely abolishes the arterial baroreflex [48]. Without a direct physiological determination on the cardiovascular neurons, the evidence for L-DOPA remains inconclusive. For instance, the nucleus of tractus solitarius is the site in which the evidence supporting a neurotransmitter role for DOPA is the strongest [44]. The neurons in the nucleus receive either monosynaptic or polysynaptic inputs from the aortic depressor nerve. It has been reported that in the nucleus of tractus solitarius, the neurons that receive monosynaptic glutamatergic (as proved by the complete blockade in the presence of 6-cyano-7-nitroquinoxaline-2,3-dione [CNQX]) excitatory inputs from the afferents directly project to the caudal ventrolateral medulla [49]. The neurons in the nucleus of tractus solitarius that receive a monosynaptic input from the aortic depressor nerve have a peak frequency response that is correlated to the rate of increase in mean arterial pressure [50], consistent with an involvement of the arterial baroreflex pathway and sensitivity to the rate of increase in arterial pressure rather than a net change in the pressure [37]. Those neurons in the nucleus of tractus solitarius that receive polysynaptic inputs from the aortic depressor nerve, on the other hand, exhibit a peak discharge frequency response that is not correlated with the rate of increase in mean arterial pressure [50]. Thus, these neurons that receive polysynaptic inputs from the aortic depressor nerve may not function as part of the essential pathway regulating the baroreflex, although their influence on baroreflex and cardiovascular functions cannot be ruled out entirely. Electrophysiological evidence indicates that in the nucleus of tractus solitarius, the neurons that receive a monosynaptic input from the aortic depressor nerve are most likely the ones that relay baroreflex information to other neurons in the medulla for their correlated peak frequency response to the rate of increase in mean arterial pressure [50]. The monosynaptic excitatory inputs from the aortic nerve are glutamatergic [49]. These neurons that receive the baroreceptor glutamatergic monosynaptic inputs project directly to the caudal ventrolateral medulla [49]. If these results hold, the DOPAergic pathway and DOPA involvement in baroreflex control appear not to be part of the core baroreflex pathway in the nucleus of tractus solitarius. Of course, synaptic inputs can modulate baroreflex and cardiovascular functions without being part of the baroreflex pathway. The cardiovascular neurons in the nucleus of tractus solitarius also receive nonbaroreceptor glutamatergic inputs [51] and GABAergic inputs [49]. These transmitters can also change the operation of the baroreflex control but are not part of the core baroreflex pathway in the brainstem. This may apply to the DOPA transmission in the nucleus of tractus solitarius, functioning as tonic inputs that gate or control activity of the neurons in the baroreflex pathway. It remains to be determined whether this is really the case and also where these DOPAergic pathways originate and what regulatory functions they perform.

Another issue is that more specific DOPA receptor/recognition site agonists and antagonists are needed for a better separation of the effects produced through these sites from those produced through other systems. Detailed data on binding at the receptors/binding sites would be helpful. For instance, at micromolar concentrations and under aromatic L-amino acid decarboxylase inhibition, L-DOPA inhibits dopamine $D_2$ receptors in a sulpiride-sensitive manner. At nanomolar concentrations and under aromatic L-amino acid decarboxylase inhibition, L-DOPA is able to facilitate the release of noradrenaline via presynaptic β-adrenoceptors, as well as dopamine release. Thus, even under complete inhibition of the aromatic L-amino acid decarboxylase, the evoked biological effects may still be mediated by the concerned neurotransmitters. In addition, NSD-1015 is not a specific inhibitor of the aromatic L-amino acid decarboxylase; it is also a potent inhibitor of brain monoamine oxidase and, thus, may not be an appropriate tool for the study of brain aromatic L-amino acid decarboxylase and should not be used to judge the neurotransmitter role of L-DOPA [52,53]. Similarly, the DOPA receptor/recognition site antagonists may not be specific. DOPA esters also act on the ionic channels of the *N*-methyl-D-aspartate (NMDA) receptors (with a millimolar $IC_{50}$) [32,54], especially when the injected solution is about 0.1 *M* (1 μg/50 nl) [23].

One additional important issue is the molecular and pharmacological characterization of specific DOPA receptors/recognition sites. Although the availability of antagonists suggests the existence of specific binding sites, the receptors and their mRNAs need to be defined.

A major role as a neurotransmitter in the brainstem baroreflex pathway would indicate that the application of those agonists that can cross the blood–brain barrier would induce dramatic changes in cardiovascular function and interfere with baroreflex regulation. The receptors would be expected to be affected first because of their high affinity before the applied agent reaches effective concentrations as a precursor. However, levodopa has been used widely and for many years. Cardiovascular side effects are rare and mild to moderate when they do occur. The major cardiovascular side effects of L-DOPA may include orthostatic hypotension probably through its peripheral decarboxylation and activation of vascular dopamine receptors because of the release of dopamine into the circulation. When administered with nonspecific monoamine oxidase inhibitors, which markedly accentuate its actions, L-DOPA may precipitate life-threatening hypertensive crisis. It has been reported that infusion of α-methyl-*p*-tyrosine (for at least 160 min) resulted in no modification of the resting arterial blood pressure and heart rate in laboratory studies [23]. In the same study, an effective blockade of the baroreflex (as judged by the response to aortic depressor nerve stimulation) by DOPA ester microinjection into the caudal ventrolateral medulla did not appear to alter the resting arterial pressure.

## 13.6 CONCLUSION

In light of these findings, are we ready to answer the question, "Is DOPA a neurotransmitter in the baroreflex neurotransmission in the brainstem"? Let us summarize the evidence discussed so far. L-DOPA exists in neurons in several regions and in many species [44], and some of the neurons are either dopamine-negative or do not contain aromatic L-amino acid decarboxylase. Thus, these are not dopaminergic neurons or other catecholaminergic cells, satisfying the criterion for the existence of L-DOPA as a neurotransmitter. Release of L-DOPA from neurons as a neurotransmitter is satisfied because L-DOPA is released when neurons are activated in a $Ca^{2+}$-dependent mechanism. Some studies have shown the release of L-DOPA as occurring independent of the release of dopamine, noradrenaline, and adrenaline. This evidence should be sufficient for making a case for L-DOPA as a neurotransmitter. The criterion of physiological identity and pharmacological antagonism is partially satisfied. The reported evidence points toward a physiologically identical response evoked by stimulation of the different inputs and by application of L-DOPA, and also of a pharmacological antagonism. However, because of the lack of detailed evaluation, lack of specific agonists and antagonists, and lack of receptor identity, other explanations cannot be ruled out.

Obviously, additional studies are needed to define and clone the DOPA receptors. A neurotransmitter role in the lower brainstem is supported by substantial evidence. However, it appears that DOPA is more likely a neurotransmitter of the input pathways that innervate the cardiovascular neurons in the brainstem baroreflex pathway, gating or modifying the operation of the control of baroreflex and sympathetic nerve activity, rather than a neurotransmitter of neurons that are part of the core relays of the baroreflex pathway in the brainstem. Such a role is not less important in terms of cardiovascular regulation and neurotransmission and would be consistent with the electrophysiological data discussed earlier, a relative lack of the cardiovascular side effects due to levodopa medication, and a normal arterial blood pressure control in tyrosine-hydroxylase-deficient patients. Cardiovascular dysfunction is not part of the reported clinical symptoms in tyrosine-hydroxylase-deficient patients [55–59]. Patients with a mutated tyrosine hydroxylase gene cannot produce any functional tyrosine hydroxylase protein due to a lack of the whole catalytic and tetramerization domains in a C-terminal region of the enzyme [57] and can be expected to exhibit dramatic problems in cardiovascular function if the essential relay of arterial baroreflex is impaired. This appears to be most likely, but studies were inconclusive because the functional states of arterial baroreflex in these patients were not evaluated in detail.

## REFERENCES


1. Sweet, R.D. and McDowell, F.H., The "on–off" response to chronic L-DOPA treatment of Parkinsonism, *Adv. Neurol.,* 5, 331, 1974.
2. Marsden, C.D. and Parkes, J.D., Success and problems of long-term levodopa therapy in Parkinson's disease, *Lancet,* 1, 345, 1977.
3. Stocchi, F. et al., The clinical efficacy of a single afternoon dose of levodopa methyl ester: a double-blind cross-over study versus placebo, *Funct. Neurol.,* 9, 259, 1994.
4. Bezard, E., Brotchie, J.M., and Gross, C.E., Pathophysiology of levodopa-induced dyskinesia: potential for new therapies, *Nat. Rev. Neurosci.,* 2, 577, 2001.
5. Doan, V.D. et al., Effects of the selective $D_1$ antagonists SCH 23390 and NNC 01-0112 on the delay, duration, and improvement of behavioral responses to dopaminergic agents in MPTP-treated monkeys, *Clin. Neuropharmacol.,* 22, 281, 1999.
6. Mura, A., Mintz, M., and Feldon, J., Behavioral and anatomical effects of long-term L-dihydroxyphenylalanine (L-DOPA) administration in rats with unilateral lesions of the nigrostriatal system, *Exp. Neurol.,* 177, 252, 2002.
7. Dayan, L. and Finberg, J.P.M., L-DOPA increases noradrenaline turnover in central and peripheral nervous systems, *Neuropharmacology,* 45, 524, 2003.
8. Quik, M. et al., L-DOPA treatment modulates nicotinic receptors in monkey striatum, *Mol. Pharmacol.,* 64, 619, 2003.

9. Goshima, Y., Kubo, T., and Misu, Y., Biphasic actions of L-DOPA on the release of endogenous noradrenaline and dopamine from rat hypothalamic slices, *Br. J. Pharmacol.,* 89, 229, 1986.
10. Misu, Y., Goshima, Y., and Kubo, T., Biphasic actions of L-DOPA on the release of endogenous dopamine via presynaptic receptors in rat striatal slices, *Neurosci. Lett.,* 72, 194, 1986.
11. Jaeger, C.B. et al., Aromatic L-amino acid decarboxylase in the rat brain: immunocyto-chemical localization in neurons of the brain stem, *Neuroscience,* 11, 691, 1984.
12. Meister, B. et al., Do tyrosine hydroxylase-immunoreactive neurons in the ventrolateral arcute nucleus produce dopamine or only L-DOPA?, *J. Chem. Neuroanat.,* 1, 59, 1988.
13. Okamura, H. et al., L-DOPA-immunoreactive neurons in the rat hypothalamic tuberal region, *Neurosci. Lett.,* 95, 42-46, 1988.
14. Okamura, H. et al., Comparative topography of dopamine- and tyrosine hydroxylase-immunoreactive neurons in the rat arcute nucleus, *Neurosci. Lett.,* 95, 347, 1988.
15. Tison, F. et al., Endogenous L-DOPA in the rat dorsal vagal complex: an immunocytochemical study by light and electron microscopy, *Brain Res.,* 497, 260, 1989.
16. Tison, F. et al., Immunohistochemistry of endogenous L-DOPA in the rat posterior hypothalamus, *Histochemistry,* 93, 655, 1990.
17. Misu, Y. et al., Neurobiology of L-DOPAergic systems, *Prog. Neurobiol.,* 49, 415, 1996.
18. Karasawa, N., Isomura, G., and Nagatsu, I., Production of specific antibody against L-DOPA and its ultrastructural localization of immunoreactivity in the house-shrew (*Suncus murinus*) lateral habenular nucleus, *Neurosci. Lett.,* 143, 267, 1992.
19. Goshima, Y., Kubo, T., and Misu, Y., Transmitter-like release of endogenous 3,4-dihydroxyphenylalanine from rat striatal slices, *J. Neurochem.,* 50, 1725, 1988.
20. Nakamura, S. et al., Transmitter-like basal and $K^+$-evoked release of 3,4-dihydroxyphenylalanine from the striatum in conscious rats studied by microdialysis, *J. Neurochem.,* 58, 270, 1992.
21. Yue, J.-L. et al., Evidence for L-DOPA relevant to modulation of sympathetic activity in the rostral ventrolateral medulla of rats, *Brain Res.,* 629, 310, 1993.
22. Yue, J.-L. et al., Endogenously released L-DOPA itself tonically functions to potentiate $D_2$ receptor-mediated locomotor activities of conscious rats, *Neurosci. Lett.,* 170, 107, 1994.
23. Miyamae, T. et al., L-DOPAergic components in the caudal ventrolateral medulla in baroreflex neurotransmission, *Neuroscience,* 92, 137, 1999.
24. Misu, Y. et al., Nicotine releases stereoselectively and $Ca^{2+}$-dependently endogenous 3,4-hydroxyphenylalanine from rat striatal slices, *Brain Res.,* 520, 334, 1990.
25. Chang, W.Y. and Webster, R.A. Effects of 3-*O*-methyl dopa on L-dopa-facilitated synthesis and efflux of dopamine from rat striatal slices, *Br. J. Pharmacol.,* 116, 2637, 1995.
26. Goshima, Y., Nakamura, S., and Misu, Y., L-Dihydroxyphenylalanine methyl ester is a potent competitive antagonist of the L-dihydroxyphenylalanine-induced facilitation of the evoked release of endogenous norepiephrine from rat hypothalamic slices, *J. Pharmacol. Exp. Ther.,* 258, 466, 1991.
27. Goshima, Y. et al., L-DOPA induces $Ca^{2+}$-dependent and tetrodotoxin-sensitive release of endogenous glutamate from rat striatal slices, *Brain Res.,* 617, 167, 1993.
28. Yue, J.-L., Goshima, Y., and Misu, Y., Transmitter-like L-3,4-dihydroxyphenylalanine tonically functions to mediate vasodepressor control in the caudal ventrolateral medulla of rats, *Neurosci. Lett.,* 159, 103, 1993.

29. Yue, J.L. et al., Baroreceptor-aortic nerve-mediated release of endogenous L-3,4-dihydroxyphenylalanine and its tonic function in the nucleus tractus solitarii of rats, *Neuroscience,* 62, 145, 1994.
30. Misu, Y. et al., L-DOPA cyclohexyl ester is a novel stable and potent competitive antagonist against L-DOPA, as compared to L-DOPA methyl ester, *Jpn. J. Pharmacol.,* 75, 307, 1997.
31. Nishihama, M. et al., An L-DOPAergic relay from the posterior hypothalamic nucleus to the rostral ventrolateral medulla and its cardiovascular function in anesthetized rats, *Neuroscience,* 92, 123, 1999.
32. Furukawa, N. et al., L-DOPA cyclohexyl ester is a novel potent and relatively stable competitive antagonist against L-DOPA among L-DOPA ester compounds in rats, *Jpn. J. Pharmacol.,* 82, 40, 2000.
33. Arai, N. et al., DOPA cyclohexyl ester, a competitive DOPA antagonist, protects glutamate release and resultant delayed neuron death by transient ischemia in hippocampus CA1 in conscious rats, *Neurosci. Lett.,* 299, 213, 2001.
34. Kubo, T. et al., Evidence for L-DOPA systems responsible for cardiovascular control in the nucleus tractus solitarii of the rat, *Neurosci. Lett.,* 140, 153, 1992.
35. Sugaya, Y. et al., Autoradiographic studies using L-[$^{14}$C]DOPA and L-[$^{3}$H]DOPA reveal regional $Na^+$-dependent uptake of the neurotransmitter candidate L-DOPA in the CNS, *Neuroscience,* 104, 1, 2001.
36. Reis, D.J. et al., Sympathoexcitatory neurons of the rostral ventrolateral medulla are oxygen sensors and essential elements in the tonic and reflex control of the systemic and cerebral circulations, *J. Hypertens.,* 12, S159, 1994.
37. Sun, M.-K., Central neural organization and control of sympathetic nervous system in mammals, *Prog. Neurobiol.,* 47, 157, 1995.
38. Schreihofer, A.M. and Guyenet, P.G., The baroreflex and beyond: control of sympathetic vasomotor tone by GABAergic neurons in the ventrolateral medulla, *Clin. Exp. Pharmacol. Physiol.,* 29, 514, 2002.
39. Sapru, H.N., Glutamate circuits in selected medullo-spinal areas regulating cardiovascular functions, *Clin. Exp. Pharmacol. Physiol.,* 29, 491, 2002.
40. Blessing, W.W., Lower brainstem pathways regulating sympathetically mediated changes in cutaneous blood flow, *Cell. Mol. Neurobiol.,* 23, 527, 2003.
41. Dampney, R.A. and Hiriuchi, J., Functional organization of central cardiovascular pathways: studies using c-fos gene expression, *Prog. Neurobiol.,* 71, 359, 2003.
42. Spyer, K.M., Annual review prize lecture: central nervous mechanisms contributing to cardiovascular control, *J. Physiol.,* 474, 1, 1994.
43. Sun, M.-K., Pharmacology of reticulospinal vasomotor neurons in cardiovascular regulation, *Pharmacol. Rev.,* 48, 465, 1996.
44. Misu, Y., Kitahama, K., and Goshima, Y., L-3,4-Dihydroxyphenylalanine as a neurotransmitter candidate in the central nervous system, *Pharmacol. Ther.,* 97, 117, 2003.
45. Yue, J.-L. et al., Altered tonic L-3,4-dihydroxypehylalanine system in the nucleus tractus solitarii and the rostral ventrolateral medulla of spontaneously hypertensive rats, *Neuroscience,* 67, 95, 1995.
46. Miyamae, T. et al., Loss of tonic neuronal activity to release L-DOPA in the caudal ventrolateral medulla of spontaneously hypertensive rats, *Neurosci. Lett.,* 198, 37, 1995.
47. Misu, Y., Goshima, Y., and Miyamae, T., Is DOPA a neurotransmitter? *Trends Pharmacol. Sci.,* 23, 262, 2002.
48. Talman, W.T. et al., Evidence for L-glutamate as the neurotransmitter of baroreceptor afferent nerve fibers, *Science,* 209, 813, 1980.

49. Kawai, Y. and Senba, E., Electrophysiological and morphological characteristics of nucleus tractus solitarii neurons projecting to the ventrolateral medulla, *Brain Res.,* 877, 374, 2000.
50. Zhang, J. and Mifflin, S.W., Responses of aortic depressor nerve-evoked neurons in rat nucleus of the solitary tract to change in blood pressure, *J. Physiol. Lond.,* 529, 431, 2000.
51. Zhang, J. and Mifflin, S.W., Subthreshold aortic nerve inputs to neurons in nucleus of the solitary tract, *Am. J. Physiol.,* 278, R1595, 2000.
52. Treseder, S.A., Rose, S., and Jenner, P., The central aromatic amino acid DOPA decarboxylase inhibitor, NSD-1015, does not inhibit L-DOPA-induced circling in unilateral 6-OHDA-lesioned rats, *Eur. J. Neurosci.,* 13, 162, 2001.
53. Treseder, S.A. et al., Commonly used L-amino acid decarboxylase inhibitor block monoamine oxidase activity in rat, *J. Neural Transm.,* 110, 229, 2003.
54. Miyamae, T. et al., Some interactions of L-DOPA and its related compounds with glutamate receptors, *Life Sci.,* 64, 1045, 1999.
55. Bräutigam, C. et al., Biochemical and molecular genetic characteristics of the severe form of tyrosine hydroxylase deficiency, *Clin. Chem.,* 45, 2073, 1999.
56. de Rijk-van, J.F. et al., L-DOPA-responsive infantile hypokinetic rigid Parkinsonism due to tyrosine hydroxylase deficiency, *Neurology,* 55, 1926, 2000.
57. Furukawa, Y. et al., DOPA-responsive dystonia simulating spastic paraplegia due to tyrosine hydroxylase (TH) gene mutations, *Neurology,* 56, 260, 2001.
58. Grattan-Smith, P.J. et al., Tyrosine hydroxylase deficiency: clinical manifestations of catecholamine insufficiency in infancy, *Movement Disord.,* 2, 354, 2002.
59. Hoffman, G.F. et al., Tyrosine hydroxylase deficiency causes progressive encephalopathy and DOPA-nonresponsive dystonia, *Ann. Neurol.,* 54, S56, 2003.

# 14 Levodopa Itself Causes Increased Activity in Rats: The Clinical Implications

*Taizo Nakazato*

## CONTENTS



## 14.1 INTRODUCTION

L-3,4-Dihydroxyphenylalanine (levodopa) therapy remains a primary strategy in the treatment of Parkinson's disease. There are, however, many problems associated with this therapy in long-term use: psychiatric symptoms such as hallucinations [1–6] and levodopa-induced dyskinesia including onset-of-dose, peak-of-dose, and end-of-dose dyskinesia [7–9]. It has been shown [1,2] that high doses of levodopa increase the incidence of dyskinesia, and combination therapy with levodopa and its decarboxylase inhibitor is more hazardous than the use of levodopa alone. Moreover, Lesser et al. [5] have reported that long-term levodopa treatment results in diminished response to the drug and causes the on–off phenomenon and wearing-off phenomenon [10]. There have been many reports that dopamine receptor sensitivity is influenced by levodopa administration. In Parkinson's disease patients, dopamine receptor binding decreases in the striatum with chronic treatment with levodopa [11]. In addition, in rats, with 6-hydroxydopamine or 1-methyl-4-phenyl-1,2,3,6-tetrahydropyridine (MPTP), the sensitivity of dopamine receptors changes with levodopa treatment [12–14], whereas the sensitivity of dopamine receptors in unlesioned rats is not altered with levodopa treatment [15,16]. The nature of the change in dopamine receptors in lesioned rats is controversial, however. Some reports describe a decrease in dopamine receptors (downregulation) [17–19], whereas some describe an increase (upregulation) [20,21].

These changes were thought to be caused by dopamine converted from the administered levodopa. However, Misu et al. [22,23] have reported that levodopa acts as a neurotransmitter; that is, the existence of levodopa neurons has been demonstrated. This finding has also been confirmed immunohistochemically [24]. Moreover, Sharma and Fahn [25] pointed out that levodopa is an agonist at dopamine receptors, and Ogawa and Yamamoto [26] reported that levodopa binds directly to dopamine binding sites. These findings raise the possibility that levodopa itself is involved in the adverse effects of chronic levodopa treatment.

Levodopa is metabolized to dopamine and 3-*O*-methyl-DOPA (3-OM-DOPA) [27–29]. Dopamine is then metabolized to 3,4-dihydroxyphenylacetic acid (DOPAC) and 3-methoxytyramine. DOPAC and 3-methoxytyramine are both further metabolized to homovanillic acid. The effects of the dopamine metabolites have been investigated in Parkinson's disease and rat behavior, but not the effects of levodopa itself. 3-OM-DOPA has been reported to be involved in the on–off phenomenon observed in Parkinson's disease patients [7], and to block levodopa metabolism [19]. The administration of DOPAC causes behavioral changes in rats [30,31]. Furthermore, Yamada et al. (1973) [32] demonstrated that intracerebroventricular administration of DOPAC results in an increase in dopamine content in rat striatal homogenates.

To investigate the action of levodopa itself in the central nervous system, we conducted a series of experiments for recording the concentrations of dopamine and DOPAC in the striatum of rat using *in vivo* voltammetry after intraventricular (i.v.t.) administration of levodopa. At the same time, behavioral changes induced by these drugs were observed. The concentrations of dopamine and DOPAC were also examined after intraperitoneal (i.p.) administration of levodopa following pretreatment with two DOPA decarboxylase inhibitors, peripherally acting benserazide and centrally acting 3-hydroxybenzylhydrazine (NSD-1015) [33,34]. Because the levodopa metabolites 3-OM-DOPA, 3-methoxytyrosine, DOPAC, and homovanillic acid (HVA) are known to cause behavioral effects, these compounds were also administered, and their effects were observed to use as controls for considering the effect of levodopa itself.

## 14.2 METHODS: *IN VIVO* VOLTAMMETRIC MEASUREMENT, INJECTION, AND BEHAVIORAL SCORING

Dopamine and DOPAC are clearly separately measurable [35,36] using the following measurement paradigm. First, the electrode was pretreated using an anodic–cathodic triangular wave (1275 mV, 10 V/sec slope) before every measurement, which has been shown not to cause any behavioral changes, alteration in the sensitivity of the electrode, or change in the concentration of dopamine [37]. Two seconds after the wave, a triple-stepped pulse was applied to determine the concentration of dopamine. The first pulse was 50 mV for 660 msec, the second 150 mV for 660 msec, and the third 250 mV for 22.5 msec to measure dopamine. The current was measured at the end of every pulse. Changes in dopamine concentrations were measured every 3 min. When tested *in vitro*, this measurement protocol yielded a detection limit on the order of several nanomolar. DOPAC was also measured when the potential was raised from 250 to

350 mV [36]. A modified differential pulse voltammogram (DPV) ranging from 100 to 500 mV in 25 mV steps was performed every 30 min for qualitative analysis.

Male Wistar rats were anesthetized with pentobarbital (50 mg/kg) and fixed in a stereotaxic apparatus. A carbon fiber electrode (7 μm in diameter) was inserted unilaterally into the striatum, and a stainless steel cannula was placed contralaterally into the lateral ventricle for the administration of drugs [35,38]. Some rats were pretreated with 200 μg 6-hydroxydopamine (6-OHDA, i.v.t.) dissolved in 24 μl of normal saline for 18 min after desipramine (25 mg/kg, i.p.) [39].

Animals were pretreated with 50 mg/kg benserazide or 100 mg/kg NSD-1015 30 min before injection (i.p.) with 200 mg/kg levodopa. Intraventricular injection was performed using a microcomputer-controlled autoburette. A total of 1 μl of the solution was injected every 45 sec. at a rate of 1/16 μl/2 sec. If a rat's behavior became hyperactive and life threatening because of the drug, injections were discontinued. To construct dose–response curves, various concentrations of the drugs were dissolved in 40 μl of normal saline. Because levodopa is difficult to dissolve at high concentrations, levodopa prepared and donated by Sankyo (Japan) was used when a high concentration was needed.

The rats were placed in a box 30 × 20 × 25 cm in size, behavioral changes were observed, and dopamine currents and DPVs were simultaneously measured. Behavior was scored using the behavioral-rating scale shown in Table 14.1 [40,41].

## 14.3 LEVODOPA-CAUSED BEHAVIORAL CHANGES IN RATS

When 200 mg/kg of levodopa is injected (i.p.) 30 min after pretreatment with benserazide, behavioral activity increases about 10 min after the start of the injection and reaches a maximum level t about 60 min; dopamine current, however, peaked at about 180 min (Figure 14.1A) [8]. When levodopa is administered (i.p.) 30 min after the injection of NSD-1015 (Figure 14.1B), behavioral activity begins to increase

**TABLE 14.1**
**Behavioral-Rating Scale**

| Score | Definition |
|---|---|
| 0 | Asleep, lying down with eyes closed |
| 1 | Lying down with eyes open with little movement |
| 2 | Lying down with head up and slow periodic sniffing |
| 3 | Getting up with periodic sniffing |
| 4 | Some rearing with sniffing, slow head swinging, and some turning |
| 5 | Frequent rearing with prominent sniffing, head swinging, and turning |
| 6 | Continuous rearing with faster head swinging or frequent turning |

*Source:* Reproduced from Nakazato, T. and Akiyama A., Behavioral activity and stereotypy in rats induced by L-DOPA metabolites: a possible role in the adverse effects of chronic L-DOPA treatment of Parkinson's disease, *Brain Res.*, 930, 134, 2002. With permission.

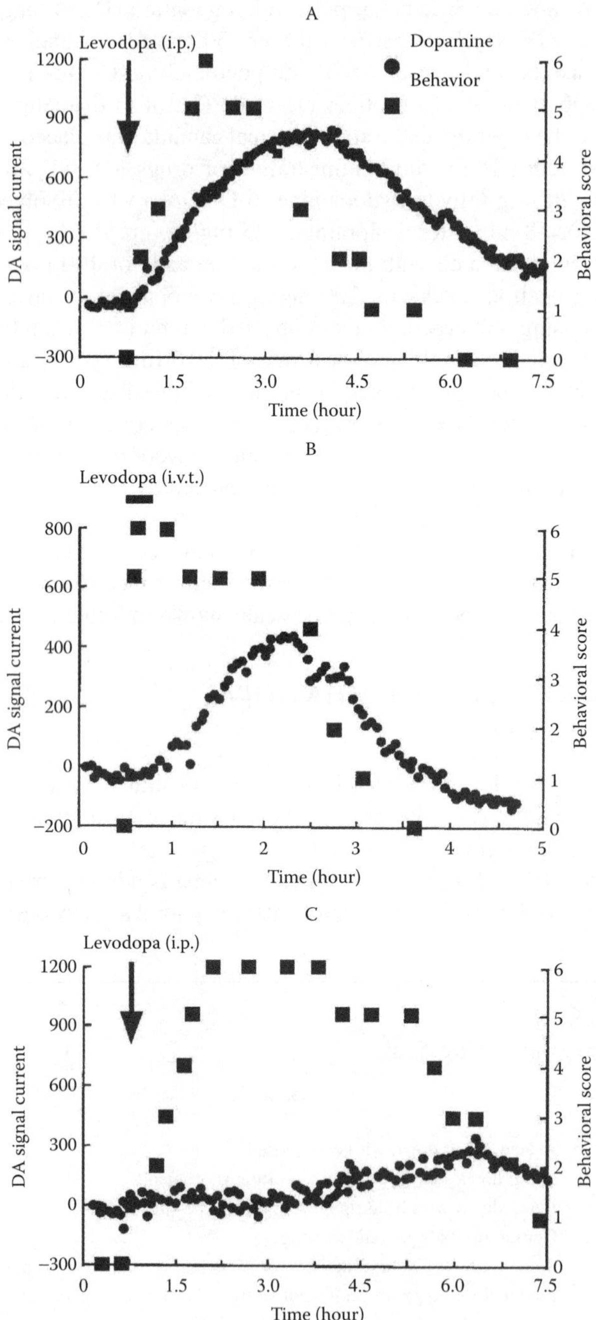


**FIGURE 14.1** Effect of levodopa on extracellular dopamine concentration and behavior. (A) Changes in dopamine currents (●) and behavioral score (■) after intraperitoneal levodopa injection (200 mg/kg, i.p.) after pretreatment with benserazide (50 mg/kg, i.p.). Behavioral change was scored as shown in Table 14.1. (B) Changes in dopamine currents (●)

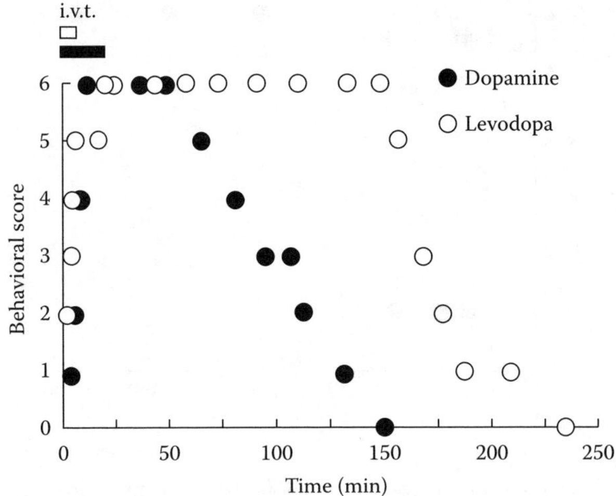


**FIGURE 14.2** Effects of levodopa and dopamine on behavior. Dopamine and levodopa were administered intracerebroventricularly in dopamine-denervated rats. Drugs were injected during the time indicated by the closed and open bars, respectively. (From Nakazato, T. and Akiyama, A., Effects of exogenous L-DOPA on behavior in the rat: an in vivo voltammetric study, *Brain Res.,* 490, 332, 1989. With permission.)

at almost the same time as with benserazide pretreatment, but the effect lasts for a longer period. During this period, dopamine signal current starts to increase slightly at 140 min, reaches a maximum level at 350 min, and then decreases.

When 400 μg (20 μl) of levodopa is injected intracerebroventricularly over 15 min (Figure 14.1C), the rats abruptly become active 3 to 4 min after the injection, but dopamine signal current does not increase in the striatal extracellular fluid at this time. After the levodopa injection is complete, dopamine signal current gradually increases to a maximum level at about 100 min after injection. The rats have already become less active by this time.

Intracerebroventricular injection (80 μg total for each) of dopamine or levodopa in dopamine-denervated rats stimulates behavior in either case (Figure 14.2), but to a much greater degree with levodopa administration than with dopamine administration.

Dose–response curves were made by plotting the maximal behavioral effect induced by different dosages of levodopa and dopamine (Figure 14.3). Comparison of the two curves reveals that a marked behavioral change is produced by administration of more than 100 μg of levodopa. This marked effect on behavior is not observed with dopamine administration.

**FIGURE 14.1 (Continued)** and behavior (■) after intracerebroventricular levodopa (440 μg) injection. Levodopa was injected between the solid vertical bars. (C) Changes in dopamine currents (●) and behavior (■) after levodopa injection (200 mg/kg, i.p.) after pretreatment with NSD-1015 (100 mg/kg, i.p.). Currents of 200 correspond to 1 nA. (From Nakazato, T. and Akiyama, A., Effects of exogenous L-DOPA on behavior in the rat: an in vivo voltammetric study, *Brain Res.,* 490, 332, 1989. With permission.)

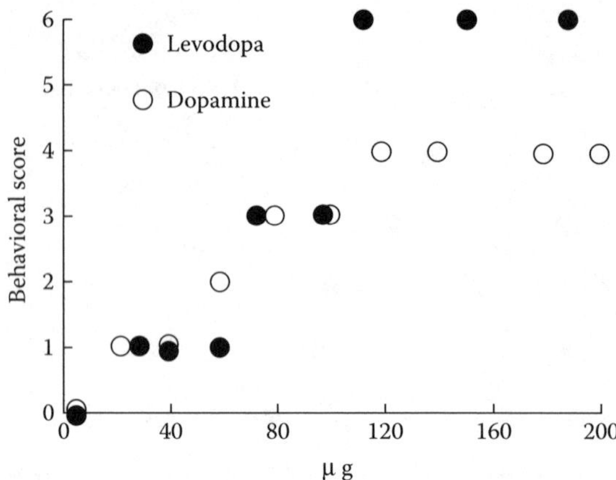


**FIGURE 14.3** Dose–response curves for levodopa (●) and dopamine (○) injections. The behavioral scores resulting from drugs injected at doses of 4 to 200 μg are plotted. Maximal behavioral-rating scores are plotted. A representative case is shown. (From Nakazato, T. and Akiyama, A., Effects of exogenous L-DOPA on behavior in the rat: an in vivo voltammetric study, *Brain Res.,* 490, 332, 1989. With permission.)

When levodopa is administered (i.v.t.), behavioral change is observed within 3 to 4 min. Dopamine signal currents, however, do not increase at this time, but after more than 20 min. Levodopa injections increase behavioral activity, even when any possible increase in dopamine following levodopa injection is suppressed by pretreatment with NSD-1015. This indicates that levodopa-induced behavioral activity is induced with or without an increase in dopamine. In addition, dopamine-denervated rats exhibit a greater degree of behavioral change in response to levodopa administration than to dopamine administration. In lesioned rats, less levodopa should be converted to dopamine than in unlesioned rats. Nevertheless, levodopa administration increases behavioral activity to a greater degree in lesioned vs. unlesioned rats. With dopamine injection (i.v.t.), behavioral response increases gradually with increasing doses, whereas with levodopa, the increase in the response is apparently larger at doses of more than 100 μg. These results suggest that at higher concentrations, both levodopa and the dopamine converted from levodopa are active. Taken together, these results indicate that along with behavioral changes caused by dopamine, levodopa itself also causes increases in behavioral activity, including stereotypy.

There have been many reports that dopamine converted from levodopa causes changes in dopamine receptor sensitivity in dopamine-denervated rats [15,16]. There are conflicting reports, however, about how the receptors change. The receptors are either downregulated [17–19] or upregulated [20,21]. More recently, the molecular changes in the rat striatum have been investigated [42,43]. Treatment with dopamine itself causes a loss of glutamate release in the striatum via glutamatergic presynaptic dopamine receptors, and it causes defective regulation of Fos protein in striatal γ-aminobutyric acid (GABA) neurons via dopamine and adenosine receptors. Meanwhile, there have also been reports concerning the effects of levodopa itself. Levodopa facilitates the release of dopamine via presynaptic adrenoreceptors [22,23,44], and it potentiates postsynaptic-dopamine-receptor-mediated locomotor activity in rats [45]. Levodopa has been reported to act as a neuromodulator [46,47]. There is also immunohistochemical evidence for the existence of levodopa-releasing neurons [24,48]. Taken together with the results from our studies, it is likely that levodopa modulates behavioral activity via dopamine receptors or that levodopa acts as a neurotransmitter.

## 14.4 DOPAC AND 3-METHOXYTYRAMINE ALSO ELICIT BEHAVIORAL CHANGES

Levodopa is metabolized to 3-OM-DOPA via catechol-O-methyltransferase (COMT) [49] and to DOPAC, 3-methoxytyramine, and homovanillic acid [50]. We also examined the behavioral effects of these levodopa metabolites, except dopamine, to determine whether any behavioral effects after levodopa treatment may be attributed to any of these metabolites.

Intraperitoneal administration of 200 mg/kg of 3-OM-DOPA to normal rats has no effect on any of the behavioral measures examined. Similarly, intraventricular injection of 200 μg of 3-OM-DOPA or HVA also fails to produce behavioral change (Table 14.2). After injection of these metabolites, dopamine and DOPAC concentrations in the striatal extracellular fluid remained unchanged.

Intracerebroventricular administration of 3-methoxytyramine (200 μg total over 30 min) causes an increase in behavioral activity (Figure 14.4A), including prominent

**TABLE 14.2**
**Summary of the Behavioral Effects of Levodopa Metabolites**

| Drugs | Behavioral Activity |
|---|---|
| Levodopa | 6 (6) |
| 3-MT | 5 (8)[a] |
| DOPAC | 3 (6)[a] |
| HVA | 0 (5) |
| 3-OM-DOPA | 1 (4) |
| Dopamine | 4 (6)[a] |
| Vehicle | 0 (8) |

*Note:* Each drug was administered intracerebroventricularly over a period of 30 min to unlesioned (control) rats at a dose of 200 μg in 40-μl Krebs–Ringer solution. The behavioral activity was the mean of the maximal behavioral scores during drug injections, rounded to the nearest whole number. Parentheses show the number of rats examined. Krebs–Ringer solution was administered in the vehicle condition.

[a] $P < .01$, compared with vehicle. 3-MT, 3-methoxytyramine; HVA, homovanillic acid.

*Source:* Reproduced from Nakazato, T. and Akiyama A., Behavioral activity and stereotypy in rats induced by L-DOPA metabolites: a possible role in the adverse effects of chronic L-DOPA treatment of Parkinson's disease, *Brain Res.*, 930, 134, 2002. With permission.

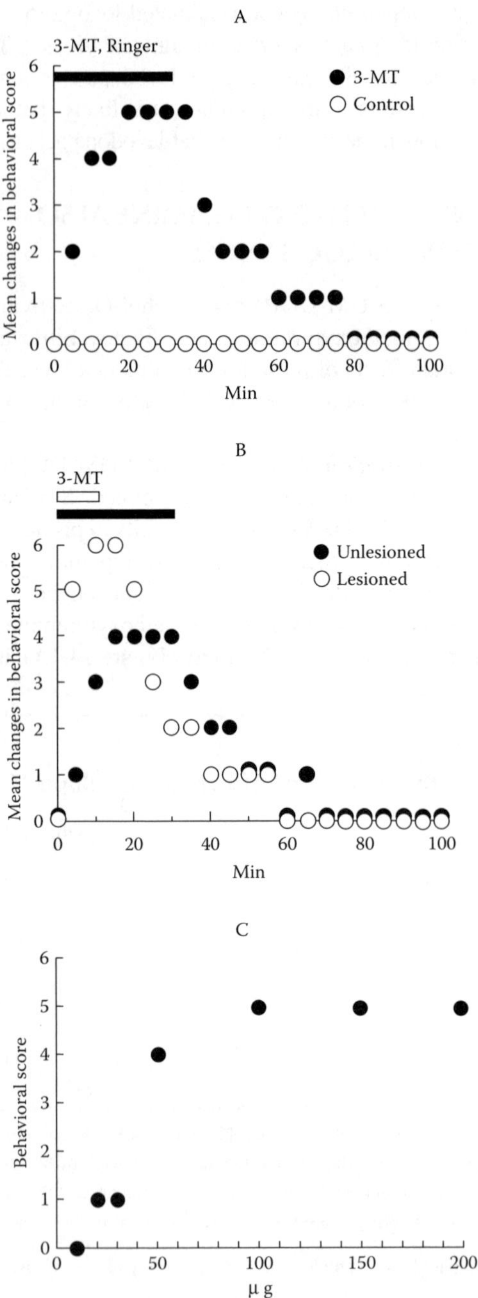


**FIGURE 14.4** Effect of 3-methoxytyramine on behavior. (A) 3-Methoxytyramine-induced behavioral changes. 3-Methoxytyramine (200 μg, ●) was administered intracerebroventricularly over a period of 30 min (n = 8) as indicated by the solid bar. Krebs–Ringer solution (○) was

stereotypy [50,51]. The activity of the rats increases approximately 5 min after the start of the injection, and reaches a peak approximately 15 min after the start of the injection. Activity begins to decrease 5 min after the end of the injection. Extracellular dopamine and DOPAC levels do not increase after 3-methoxytyramine administration. Behavioral activity increases dramatically during the administration of 3-methoxytyramine to 6-hydroxydopamine-lesioned rats, even 9 min after injection time (12 μl total volume and 30 μg total dose administered). In lesioned rats, 3-methoxytyramine injection causes much greater behavioral activity than in unlesioned rats (Figure 14.4B and Figure 14.4C). 3-Methoxytyramine-induced behavioral change is suppressed by the dopamine $D_{1/5}$ receptor antagonist SCH 23390, but not by the dopamine $D_{2/3/4}$ receptor antagonist sulpiride [50]. These data suggest that 3-methoxytyramine induces behavioral changes in rats via dopamine $D_{1/5}$ receptors.

DOPAC administration causes an increase in behavioral activity and stereotypy. However, both effects are mild relative to those following 3-methoxytyramine injection (Table 14.2) [50]. Similar to 3-methoxytyramine, DOPAC does not cause an increase in dopamine levels. Dopamine administration has more of an effect on behavioral activity than DOPAC, but its effects are also to a lesser degree than those observed with 3-methoxytyramine. In dopamine-denervated rats, dopamine administration results in supersensitive behavioral responses. In contrast, DOPAC administration does not result in supersensitive behavioral responses in lesioned rats.

## 14.5 ARE LEVODOPA-INDUCED SIDE EFFECTS IN PARKINSON'S DISEASE PATIENTS RELATED TO LEVODOPA AS WELL AS DOPAMINE?

The dyskinesia due to chronic levodopa treatment in Parkinson's disease patients has been thought to be caused by dopamine acting at supersensitive dopamine receptors [46,47]; however, dopamine receptor sensitivity [17–21] and changes in second messengers [42,43,52,53] have been shown to be affected by levodopa itself.

**FIGURE 14.4 (Continued)** administered as a vehicle control condition (n = 6). Each data point represents mean behavioral activity scores rounded to the nearest whole number. (B) 3-Methoxytyramine-induced behavioral changes in unlesioned (control) and lesioned (dopamine-denervated) rats. In unlesioned rats, 3-methoxytyramine (100 μg) was administered intracerebroventricularly over a period of 30 min (●, n = 6) indicated by the solid bar. In dopamine-denervated rats (○), the same concentration of 3-methoxytyramine was administered at the same rate; however, the injection was discontinued at 9 min (n = 5) indicated by the open bar. A supersensitive response was observed in dopamine-denervated rats. (C) Dose-dependent changes in 3-methoxytyramine-induced behavior. Drug was injected over 30 min at a dose of 10 to 200 μg dissolved in 40-μl solution. Maximal behavioral-rating scores are plotted. A representative case is shown. 3-MT, 3-methoxytyramine. (From Nakazato, T. and Akiyama A., Behavioral activity and stereotypy in rats induced by L-DOPA metabolites: a possible role in the adverse effects of chronic L-DOPA treatment of Parkinson's disease, *Brain Res.*, 930, 134, 2002. With permission.)

Therefore, taken together with the results of the present study, the effects of levodopa and 3-methoxytyramine should be considered in addition to dopamine.

When levodopa is administered orally, it is absorbed from the intestines and partly decarboxylated peripherally outside the brain. It is further decarboxylated to dopamine within the brain. Thus, the concentration of levodopa is very low in the normal human brain. In Parkinson's disease patients, however, the number of dopamine neurons is greatly reduced. Therefore, the rate of levodopa decarboxylation is less than or equal to that in the normal brain. In the advanced stages of Parkinson's disease, there is a severe loss of dopamine terminals in striatal tissue, and it is likely that smaller amounts of levodopa are converted to dopamine. Under these circumstances, the effects of levodopa itself may be more pronounced. Levodopa may have a larger role in dyskinesia in the later stages than in the early stages of Parkinson's disease.

There are three types of chronic levodopa-induced dyskinesia in Parkinson's disease patients: onset-of-dose dyskinesia, peak-of-dose dyskinesia, and end-of-dose dyskinesia [8,9]. Onset-of-dose dyskinesia appears very soon after taking levodopa, when extracellular levels of dopamine are not likely to be high, whereas levodopa levels are relatively high. Therefore, levodopa is more likely to be involved in onset-of-dose dyskinesia, and dopamine is more likely to be involved in peak-of-dose dyskinesia. 3-Methoxytyramine may also be involved in peak-of-dose dyskinesia, because the time course of change in striatal 3-methoxytyramine parallels that of dopamine following levodopa administration. In dopamine-denervated rats, however, levodopa metabolites presumably remain in the brain, because the enzymes necessary for the further metabolism of these metabolites are lost in such rats. Therefore, 3-methoxytyramine may also have a role in end-of-dose dyskinesia.

It has been shown that oral administration of DOPAC results in a reversal of reserpine syndrome in reserpinized rats [30]. DOPAC also causes behavioral changes, although the changes are not of the same degree as those observed with dopamine, levodopa, or 3-methoxytyramine administration. In addition, DOPAC does not cause a supersensitive response in dopamine-denervated rats. DOPAC also does not increase extracellular 3-methoxytyramine levels [50,51], suggesting that DOPAC causes behavioral changes, but it is not related to levodopa-induced dyskinesia.

## 14.6 CONCLUSION

Increasing amounts of evidence indicate that in addition to dopamine, levodopa itself causes an increase in behavioral activity in rats. After pretreatment with NSD-1015, a central decarboxylase inhibitor, intraperitoneal application of levodopa elicits behavioral changes even before there is a measurable increase in striatal dopamine using *in vivo* voltammetry, and its dose–response curve is different from that in the case of dopamine administration. The intracerebroventricular injection of levodopa in rats with striatal dopamine lesions elicits behavioral changes more promptly after the injection, compared with rats without striatal lesions. These results raise the possibility that levodopa itself modulates behavioral activity via dopamine receptors or acts as a neurotransmitter. Injection of the levodopa metabolites 3-methoxytyramine and DOPAC, however, also resulted in behavioral changes. Taken together, these data support the idea that levodopa acts as a neuromodulator.

Levodopa and 3-methoxytyramine injections cause a supersensitive response in dopamine-denervated rats. This suggests that in Parkinson's disease patients with extensive dopamine denervation, in addition to dopamine, levodopa and 3-methoxytyramine are very likely to cause motor effects such as levodopa-induced dyskinesia. These results characterizing the effects of levodopa and the cascade of metabolites following levodopa administration may aid in the refinement of therapeutic strategies in the treatment of Parkinson's disease.

## ACKNOWLEDGMENTS

The author would like to express deep gratitude to Dr. Akitane Akiyama of Rainbow Science Co. in Yokohama, Japan (formerly of the Department of Electronic Chemistry, Graduate School at Nagatsuta, Tokyo Institute of Technology, Yokohama, Japan).

## REFERENCES

1. Barbeau, A., L-DOPA therapy in Parkinson's disease: a critical review of nine years experience, *Can. Med. Assoc. J.*, 101, 791, 1969.
2. Barbeau, A. and Roy, M., Ten-years results of treatment with levodopa plus benserazide in Parkinson's disease, in *Progress in Parkinson's Disease,* Rose, F.C. and Capildeo, R., Eds., Pitman, London, 1981, 241.
3. Diamond, S.G. et al., Multi-center study of Parkinson mortality with early versus later dopa treatment, *Ann. Neurol.,* 22, 8, 1987.
4. Gonzalez, L.P., Alterations in amphetamine stereotypy following acute lesions of substantia nigra, *Life Sci.,* 40, 899, 1987.
5. Lesser, R.P. et al., Analysis of the clinical problems in parkinsonism and the complications of long-term levodopa therapy, *Neurology,* 29, 1253, 1979.
6. Yahr, M.D. et al., Treatment of parkinsonism with levodopa, *Arch. Neurol.,* 21, 343, 1969.
7. Mauradian, M.M. et al., Pathogenesis of dyskinesia in Parkinson's disease, *Ann. Neurol.,* 25, 523, 1989.
8. Muenter, M.D. et al., Patterns of dystonia ('I-D-I' and 'D-I-D') in response to L-DOPA therapy for Parkinson's disease, *Mayo Clin. Proc.,* 52, 163, 1977.
9. Quinn, N., Parkes, J.D., and Marsden, C.D., Control of on/off phenomenon by continuous intravenous infusion of levodopa, *Neurology,* 34, 1131, 1984.
10. Shoulson, I., Glaubiger, G.A., and Chase, T.N., 'On-off' response: clinical and biochemical correlations during oral and intravenous levodopa administration in parkinsonian patients, *Neurology,* 25, 1144, 1975.
11. Lee, T. et al., Receptor basis for dopaminergic supersensitivity in Parkinson's disease, *Nature (Lond.),* 273, 59, 1978.
12. Creese, I., Burt, D.R., and Snyder, S.H., Dopamine receptor binding enhancement accompanies lesion-induced behavioral supersensitivity, *Science,* 197, 596, 1977.
13. Nishi, K., Expression of c-Jun in dopaminergic neurons of the substantia nigra in 1-methyl-4-phenyl-1,2,3,6-tetrahydropyridine (MPTP)-treated mice, *Brain Res.,* 771, 133, 1997.

14. Ungerstedt, U., Postsynaptic supersensitivity after 6-hydroxydopamine induced degeneration of the nigrostriatal dopamine system, *Acta Physiol. Scand.,* 367 (Suppl.), 69, 1971.
15. Hurley, M.J. et al., Dopamine D3 receptors in the basal ganglia of the common marmoset and following MPTP and L-DOPA treatment, *Brain Res.,* 709, 259, 1996.
16. Zeng, B.-Y. et al., Chronic high dose L-DOPA treatment does not alter the levels of dopamine D-1, D-2 or D-3 receptor in the striatum of normal monkeys: an autoradiographic study, *J. Neural. Transm.,* 108, 925, 2001.
17. Gagnon, C., Bédard, P.J., and Di Paolo, T., Effect of chronic treatment of MPTP monkeys with dopamine D-1 and/or D-2 receptor agonists, *Eur. J. Pharmacol.,* 178, 115, 1990.
18. Jenner, P., Boyce, S., and Marsden, C.D., Receptor changes during chronic dopaminergic stimulation, *J. Neural. Transm.*, 27 (Suppl.), 161, 1988.
19. Reches, A. and Fahn, S., 3-*O*-Methyldopa blocks dopa metabolism in rat corpus striatum, *Ann. Neurol.,* 12, 267, 1982.
20. Graham, W.C., Sambrook, M.A., and Crossman, A.R., Differential effect of chronic dopaminergic treatment on dopamine D1 and D2 receptors in the monkey brain in MPTP-induced parkinsonism, *Brain Res.,* 602, 290, 1993.
21. Rioux, L. et al., The effects of chronic levodopa treatment on pre- and postsynaptic marker dopaminergic function in striatum of parkinsonian monkeys, *Movement Disord.,* 12, 148, 1997.
22. Misu, Y., Goshima, Y., and Kubo, T., Biphasic actions of L-DOPA on the release of endogenous dopamine via presynaptic receptors in rat striatal slices, *Neurosci. Lett.,* 72, 194, 1986.
23. Misu, Y., Goshima, Y., and Miyatake, T., Is DOPA a neurotransmitter? *Trends Pharmacol. Sci.,* 23, 262, 2002.
24. Nagastu, I. et al., Quantitative analysis of reduction of aromatic-L-aminoacid decarboxylase- and tyrosine hydroxylase-like immunoreactivities in the nigrostriatal dopaminergic neurons of MPTP-treated mice, *Biogenic Amines,* 6, 263, 1989.
25. Sharma, J.N. and Fahn, S., Microiontophoretic studies with L-DOPA and putative neurotransmitters on the neurons of caudate nucleus of rat, *Soc. Neurosci. Abstr.,* 5, 599, 1979.
26. Ogawa, N. and Yamamoto, M., Neurotransmitters and their receptors in parkinsonism, in *Current Problems in Treatment of Parkinson's Disease,* Kuroiwa, Y. and Toyokura, Y., Eds., DMW Japan, Tokyo, 1985, 23.
27. Bartholini, G. and Pletscher, A., Cerebral accumulation and metabolism of $^{14}$C-DOPA after selective inhibition of peripheral decarboxylase, *J. Pharmacol. Exp. Therap.,* 161, 14, 1968.
28. Bartholini, G., Kuruma, I., and Pletscher, A., 3-*O*-Methyldopa, a new precursor of dopamine, *Nature (Lond.),* 230, 533, 1971.
29. Nutt J.G. and Fellman J.H., Pharmacokinetics of L-DOPA, *Clin Neuropharmacol.,* 7, 35, 1984.
30. Ericsson, A.D. and Wertman, B.E., Sensitivity studies of L-DOPA metabolites in reserpinized rats and their clinical significance, *Neurology (Mineapp.),* 21, 1129, 1971.
31. Tsai T.-H. and Chen, C.-F., Simultaneous measurement of acetylcholine and monoamines by two serial on-line microdialysis system: effects of methamphetamine on neurotransmitters release from the striatum of freely moving rats, *Neurosci. Lett.,* 166, 175, 1994.
32. Yamada, K. et al., Effect of 3,4-dihydroxyphenylacetic acid on catecholamine and serotonin in rat striatum, *J. Neurol. Sci.,* 18, 311, 1973.

33. Bartholini, G. and Pletscher, A., Effect of various decarboxylase inhibitors on the cerebral metabolism of dihydroxyphenylalanine, *J. Pharm. Pharmacol.,* 21, 323, 1969.
34. Melamed, E. et al., Suppression of L-dopa-induced circling in rats with nigral lesions by blockade of central dopa-decarboxylase: implications for mechanism of action of L-dopa in parkinsonism, *Neurology,* 34, 1566, 1984.
35. Nakazato, T., Akiyama, A., and Shimizu, A., Microcomputer-controlled in vivo voltammetry, *Biogenic Amines,* 5, 339, 1988.
36. Nakazato, T. and Akiyama, A., In vivo electrochemical measurement of the long-lasting release of dopamine and serotonin induced by intrastriatal kainic acid, *J. Neurochem.,* 69, 2039, 1997.
37. Nakazato T., Hosoda, S., and Akiyama A., A triangular conditioning voltage wave does not influence spontaneous neuronal activity in the rat striatum, *J. Neurosci. Meth.,* 46, 69, 1993.
38. Nakazato, T. and Akiyama, A., Effects of exogenous L-DOPA on behavior in the rat: an in vivo voltammetric study, *Brain Res.,* 490, 332, 1989.
39. Breese, G.R. and Traylor, T.D., Depletion of brain noradrenaline and dopamine, *Br. J. Pharmacol.,* 42, 88–99, 1971.
40. Creese, I. and Iverson, S.D., Blockade of amphetamine-induced motor stimulation and stereotypy in the adult rat following neonatal treatment with 6-hydroxydopamine, *Brain Res.,* 55, 369, 1973.
41. Goodwin, F.K., Psychiatric side effects of levodopa in man, *JAMA,* 218, 1915, 1971.
42. Calon, F. et al., Dopamine-receptor stimulation: biobehavioral and biochemical consequences, *Trends Neurosci.,* 23 (Suppl.), S92, 2001.
43. Calon, F. et al., Molecular basis of levodopa-induced dyskinesia, *Ann. Neurol.,* 47 (Suppl. 1), S70, 2000.
44. Ponzio, F. et al., Does acute L-DOPA increase active release of dopamine from dopaminergic neurons?, *Brain Res.,* 273, 45, 1983.
45. Nakamura, S. et al., Non-effective dose of exogenously applied L-DOPA itself stereoselectively potentiates postsynaptic $D_2$ receptor-mediated locomotor activities of conscious rats, *Neurosci. Lett.,* 170, 22, 1994.
46. Klawans, H.L. et al., Levodopa-induced dopamine receptor hypersensitivity, *Ann. Neurol.,* 2, 125, 1977.
47. Klawans, H.L. et al., Supersensitivity of d-amphetamine- and apomorphine-induced stereotyped behavior induced by chronic d-amphetamine administration, *J. Neurol. Sci.,* 25, 283, 1975.
48. Kitahama, K. et al., Dopamine- and DOPA-immunoreactive neurons in the cat forebrain with reference to tyrosine hydroxylase-immunohistochemistry, *Brain Res.,* 518, 83, 1990.
49. Rivera-Calimlim, L., Absorption, metabolism and distribution of [$^{14}$C]-*o*-methyldopa and [$^{14}$C]-L-DOPA after oral administration to rats, *Br. J. Pharmacol.,* 50, 259, 1974.
50. Nakazato, T. and Akiyama A., Behavioral activity and stereotypy in rats induced by L-DOPA metabolites: a possible role in the adverse effects of chronic L-DOPA treatment of Parkinson's disease, *Brain Res.,* 930, 134, 2002.
51. Nakazato, T., The medial prefrontal cortex mediates 3-methoxytyramine-induced behavioural changes in rat, *Eur. J. Pharmacol.,* 442, 73, 2002.
52. Dragunow, M. et al., $D_2$ dopamine receptor antagonists induce Fos and related proteins in rat striatal neurons, *Neuroscience,* 37, 287, 1990.
53. Paul, M.L. et al., D1-like and D2-like dopamine receptors synergistically activate rotation and c-fos expression in the dopamine-depleted striatum in a rat model of Parkinson's disease, *J. Neurosci.,* 12, 3729, 1992.

# 15 Behavioral Effects of Exogenously Applied and Endogenously Released DOPA

*Yoshimi Misu, Kaneyoshi Honjo, and Yoshio Goshima*

## CONTENTS



## 15.1 INTRODUCTION

Since the 1950s [1,2], exogenously applied L-3,4-dihydroxyphenylalanine (levodopa) has been generally thought to be an inert amino acid that alleviates the symptoms of Parkinson's disease via its bioconversion to dopamine by the enzyme aromatic L-amino acid decarboxylase (AADC) [3,4]. However, since 1986 [5,6], we have been proposing that endogenous L-3,4-dihydroxyphenylalanine (DOPA) is a neurotransmitter and/or neuromodulator by itself in the central nervous system (CNS), in addition to being a precursor of dopamine [7–9]. Recent evidence has been suggesting that DOPA fulfills the criteria, such as biosynthesis, metabolism, active transport, existence, release, competitive antagonism, and physiological or pharmacological responses, including interactions with the other neurotransmitter systems [10,11], that must be satisfied before a compound is accepted as a neurotransmitter [12]. We have accumulated evidence to support the idea that DOPA is a neurotransmitter of the aortic depressor nerve, one of the primary baroreceptor afferents terminating in the rat nucleus tractus solitarii (NTS) [13–15]. The NTS is the gate of the baroreflex neurotransmission in the lower brainstem.

Neurons showing immunocytochemically tyrosine hydroxylase (TH)-positive, DOPA-positive, AADC-negative, and dopamine-negative reactivity exist in some regions of the CNS including the NTS [14,16,17], which may contain DOPA as an end product [9,11]. During microdialysis of the NTS in anesthetized rats, the basal release of DOPA is partially tetrodotoxin (TTX) sensitive and $Ca^{2+}$ dependent [14,15], suggesting that DOPA is, at least in part, basally released via some spontaneous neuronal activity. Indeed, electrical stimulation of the aortic depressor nerve releases DOPA in a TTX-sensitive manner [14].

In contrast, no immunocytochemical evidence for neurons having DOPA as an end product has been found in the striatum and the nucleus accumbens of rats [9,11]. Notwithstanding, DOPA release evoked by electrical field stimulation from striatal slices is TTX-sensitive and $Ca^{2+}$-dependent in a manner similar to that for the established neurotransmitter dopamine [18]. High $K^+$-induced depolarization and nicotine release DOPA and dopamine in a $Ca^{2+}$-dependent manner from *in vitro* striatal slices [18,19]. Furthermore, both DOPA and dopamine are consistently detectable in dialysates collected during microdialysis of the striatum [20–24] and the shell compartment of the nucleus accumbens [25,26] in conscious rats. DOPA and dopamine are basally released by *in vivo* spontaneous neuronal activity in a TTX-sensitive and $Ca^{2+}$-dependent manner in the striatum [20] and the nucleus accumbens [25]. High $K^+$-evoked $Ca^{2+}$-dependent release of both DOPA and dopamine is also evident in the striatum *in vivo* [20]. Nicotine perfused via microdialysis probes releases DOPA as well as dopamine in a TTX-sensitive and $Ca^{2+}$-dependent manner and this release is antagonized by mecamylamine, an antagonist for central nicotinic receptors [21]. An important finding is that mecamylamine alone perfused via probes inhibits basal DOPA release but not basal dopamine release in the striatum, suggesting that DOPA release, but not dopamine release, is tonically regulated via activation of nicotinic acetylcholine receptors. Endogenously released acetylcholine appears to play an important role in DOPA release, compared with dopamine release. In the nucleus accumbens, nicotine applied systemically or injected into the ventral tegmental area (VTA) releases not only dopamine but also DOPA and this release is antagonized by mecamylamine injected into the VTA [25]. Furthermore, extracellular levels of DOPA as well as dopamine in the nucleus accumbens are elevated in response to discontinuation of electrical foot-shock stress [26]. Thus, the sensitivity of immunocytochemical analysis to identify DOPAergic neurons appears to be lower, compared with biochemical approaches [10,11].

DOPA methyl ester (DOPA ME) was the first competitive antagonist for levodopa to be identified [27,28]. Among the DOPA ester compounds, DOPA cyclohexyl ester (DOPA CHE) is the most potent and relatively stable competitive antagonist [29,30].

These two antagonists antagonize responses to both levodopa [13,27–34] and released DOPA [14,35–37]. Competitive antagonism implies the existence of DOPA recognition sites.

DOPA CHE does not inhibit the $Na^+$-dependent uptake of labeled levodopa into *Xenopus laevis* oocytes [38]. Levodopa and DOPA ester compounds up to 1 m*M* hardly displace the selective binding of tritiated ligands for dopamine $D_1$ and $D_2$ receptors or $\alpha_2$- and β-adrenoceptors in rat brain membrane preparations [27,30]. Furthermore, among the binding sites labeled with tritiated ligands for ionotropic glutamate receptors, DOPA ester compounds act only on *N*-methyl-D-aspartate (NMDA) ion channel domain at millimolar $IC_{50}$, whereas levodopa acts only on DL-α-amino-3-hydroxy-5-methyl-4-isoxazol propionic acid receptors with a low affinity [30,39]. DOPAergic agonists and competitive antagonists should act on the same sites, which appear to differ from DOPA transport sites, catecholamine receptors, and ionotropic glutamate receptors. It is essential that there exist DOPA recognition sites on which DOPA or levodopa acts to elicit physiological or pharmacological responses.

In the striatum, most responses to levodopa are, as is the case for many neurotransmitters, stereoselective [23,28,35,40,41]. Levodopa elicits responses in the absence and presence of a central AADC inhibitor such as 3-hydroxybenzylhydrazine (NSD-1015) [6,28], which are not mimicked by a corresponding concentration of dopamine [28]. Furthermore, responses to released DOPA are potentiated following an increase in DOPA release elicited by NSD-1015 as a result of the inhibition of AADC [24, 35]. In contrast, responses to DOPA are inhibited following a decrease in the biosynthesis of DOPA by α-methyl-*p*-tyrosine (α-MPT), a TH inhibitor, as a result of the inhibition of TH activity [24,37]. Thus, these responses are elicited by levodopa and released DOPA by itself and not by bioconversion to dopamine.

On the other hand, several lines of evidence have suggested that the modification of the mesolimbic dopaminergic activity in the nucleus accumbens plays an essential role in the behavioral effects, including locomotor activity, elicited by nicotine and psychotropic drugs [42–44]. Indeed, systemic nicotine-induced locomotor activity is antagonized by systemically administered dopamine $D_1$ and $D_2$ antagonists in rats [45]. In addition, nicotine locally infused into the VTA induces locomotor activity, which is antagonized by systemic mecamylamine [46]. Systemic nicotine-induced dopamine release [47, 48] and expression of Fos-like immunoreactivity [49] in the nucleus accumbens is antagonized by mecamylamine locally infused into the VTA but not by its infusion into the nucleus accumbens [47].

The nucleus accumbens is also thought to play an important role in adaptive motor responses to psychotropic drugs and environmental stimuli [50]. Actually, extracellular dopamine levels in the nucleus accumbens are elevated after discontinuation of footshock stress in rats [51] and restraint stress in rats [52] and mice [53].

Based on the above findings, we have explored whether or not DOPA released by nicotine [22,25], exogenously applied levodopa [23,34], and basally released DOPA [24] elicit behavioral effects, including locomotor activity, via activation of postsynaptic DOPA recognition sites.

## 15.2 DOPA RELEASED BY NICOTINE APPEARS INVOLVED IN BEHAVIORAL EFFECTS INCLUDING LOCOMOTOR ACTIVITY OF CONSCIOUS RATS

The convincing evidence for a postsynaptic neurophysiological or neuropharmacological response to released DOPA or levodopa is its effect on locomotor activity. Nicotine (0.1 to 1 mg/kg, s.c.) induces locomotor activity in a dose-dependent and stereoselective manner under intact AADC activity [22]. Locomotor activity elicited by nicotine (0.4 mg/kg) was antagonized by mecamylamine (1 mg/kg, s.c.) in accordance with previous findings [44].

At first, we attempted to explore whether or not pretreatment with intracerebroventricular injection of DOPA ME (200 μg) can antagonize nicotine-induced locomotor activity [22]. The result obtained, however, was disappointing: DOPA ME does not antagonize nicotine-induced locomotor activity but even potentiates it, which may be because DOPA ME, a representative prodrug for levodopa, injected in this manner is readily hydrolyzed to DOPA [54].

We found that α-MPT at 0.2 m*M* inhibits DOPA release but not dopamine release, evoked by electrical field stimulation from striatal slices [18]. Thus, in the next step of the experiments, we tried to find a selective low dose of α-MPT, which would inhibit basal DOPA release without inhibition of basal dopamine release. The usual dose of α-MPT (200 mg/kg, i.p.) to inhibit TH activity abolishes the basal release of both DOPA and dopamine during striatal microdialysis [20]. However, among the much lower doses (1 to 10 mg/kg), α-MPT at 3 mg/kg was found to decrease basal DOPA release by one third without showing any tendency to decrease basal dopamine release monitored during striatal microdialysis. This selective low dose of α-MPT inhibits the mecamylamine-sensitive locomotor activity elicited by systemic nicotine (0.4 mg/kg, s.c.) [22], suggesting that the DOPA released is to a certain extent relevant to the elicitation of the locomotor activity induced by nicotine.

In microdialysis experiments [25], we confirmed the previous finding that systemic nicotine (1 mg/kg, s.c.) releases a greater amount of dopamine from the shell compartment of the nucleus accumbens, compared with that from the striatum [42–44]. This is also in the case for DOPA. Nicotine locally infused into the VTA (30 μg) releases not only dopamine but also DOPA in the nucleus accumbens and elicits locomotor activity [25]. All of these responses are antagonized by mecamylamine (100 μg) infused into the VTA. The results of both mecamylamine-sensitive dopamine release and the locomotor activity elicited by nicotine are consistent with previous findings [46–48]. These findings suggest that nicotinic receptors in the somatodendritic region may have greater importance than those located in the terminal area for the stimulatory effects of nicotine on the mesolimbic dopamine system [46–49] and also on the probable mesolimbic DOPA system [25]. As described in Section 15.4, via activation of postsynaptic DOPA recognition sites, exogenously applied levodopa and basally released DOPA, at least, potentiate the ability of a dopamine $D_2$ agonist to activate postsynaptic dopamine $D_2$ receptors involved in locomotor activity of rats [23,24,34]. An interesting finding related to nicotine-induced activation of the probable mesolimbic DOPA system [25] is demonstrated in the human VTA [55,56]. Some neuronal cell bodies show immunocytochemically TH-positive, AADC-negative, DOPA-positive, and dopamine-negative reactivity, and further the reactivity of guanosine triphosphate cyclohydrolase I, the first and rate-limiting enzyme for the biosynthesis of DOPA.

It should be more extensively explored as to whether or not nicotine elicits behavioral effects, including locomotor activity, only through the DOPA-induced direct activation of postsynaptic DOPA recognition sites in the nucleus accumbens. A related interesting finding is shown in destruction experiments of mesolimbic dopamine neurons in rats [57]. Injections of 6-hydroxydopamine (6-OHDA) into the VTA result in complete depletion of dopamine concentrations in the nucleus accumbens. This destruction, however, does not block the acute locomotor response to nicotine (0.4 mg/kg, s.c.). Such depletion also does not prevent the progressive enhancement of nicotine's locomotor effects when the injections were repeated daily for 9 d. It appears likely to us that unknown mesolimbic neurotransmitter(s), other than dopamine, is (are) involved in the elicitation of the locomotor response to systemic nicotine.

Not only dopamine but also DOPA is released in the shell of the nucleus accumbens of conscious rats after discontinuation of electrical foot-shock stress (0.4 mA, 200 msec, 1 Hz, 20 min) [26]. Dopamine release is consistent with previous findings [51–53]. Systemic mecamylamine alone (2 mg/kg, s.c.) tends to decrease basal DOPA release [26]. Pretreatment with mecamylamine completely blocks the DOPA release evoked by foot-shock stress, whereas it only partially blocks dopamine release. These findings suggest that DOPA as well as dopamine is released, at least in part, via activation of central nicotinic receptors in the nucleus accumbens in response to foot-shock stress. The results also suggest the relative importance of some tonic function of these receptors involved in DOPA release, compared with dopamine release. However, such tonic function of central nicotinic receptors does not appear to extend to the blockade of behavioral changes such as rearing, vocalization, jumping, running, crouching, and flinching that occurred during and after foot-shock stress.

Nicotine has been characterized as a weak positive reinforcer, qualitatively similar to the reinforcers having strong addictive properties, such as amphetamine, cocaine, and morphine [58–61]. These reinforcers, including nicotine, increase glucose utilization, and elicit dopamine neurotransmission in the shell of the nucleus accumbens of conscious rats. Thus, we are now exploring whether or not there are similarities or differences between the effects of nicotine and methamphetamine or cocaine on the release of both DOPA and dopamine in the shell of the nucleus accumbens and on behavioral profiles in conscious rats. Preliminary data [Izawa J. et al., to be published] show that the injection of methamphetamine at 0.3 and 1.0 mg/kg (s.c.), doses commonly used in studies of locomotor-stereotypy behaviors, elicits increases in dopamine release from the basal level. In addition, methamphetamine also increases DOPA release, but the drug at 0.3 mg/kg appears to elicit a peak increase. Nicotine at 1 mg/kg (s.c.) increases dopamine and DOPA release [25] to a lesser degree, compared with methamphetamine. Cocaine at 3 and 10 mg/kg (i.p.) also elicits increases in basal dopamine release to a degree similar to that by methamphetamine. In contrast, cocaine at 3 and 10 mg/kg decreases DOPA release in a dose-dependent manner, showing a clear difference from both methamphetamine and nicotine at 1 mg/kg. Aditionally, in rats individually placed in a photo cage and habituated to environment [23,24], we monitored by video camera recording, frequencies/20 min of grooming, chewing, sniffing, rearing, and walking until 80 min after the injection of these psychotropic drugs (n of each measure by each drug = 5 to 6). Methamphetamine at 1 mg/kg (s.c.) increased these behaviors under our experimental conditions. Nicotine 1 mg/kg (s.c.) increased sniffing and walking and tended to increase grooming, chewing, and rearing. Cocaine at 3 mg/kg (i.p.) increased sniffing to a lesser degree, compared with methamphetamine and nicotine at 1 mg/kg.

Importantly, the simultaneous systemic injection of DOPA CHE at 30 mg/kg (i.p.) antagonizes grooming, chewing, and walking and tends to antagonize the sniffing and rearing induced by methamphetamine and also antagonizes the sniffing, and tends to antagonize the grooming and chewing, induced by nicotine. DOPA released by not only nicotine but also methamphetamine in the shell of the nucleus accumbens appears to play some roles in the locomotor-stereotypy behaviors observed. These findings suggest that nicotine has, at least in part, biochemical and pharmacological properties similar to methamphetamine but not to cocaine.

Under the experimental conditions described in this section and in Section 15.3 and Section 15.4, we have to always take into consideration that high concentrations of levodopa releases endogenous glutamate from *in vitro* [28,62] and *in vivo* [35] striata. High concentrations of DOPA CHE also elicit a nonselective antagonistic effect on the NMDA ion channel domain [39]. In the case of nicotine, its stimulatory action on the mesolimbic dopamine system is reported to be, to a considerable extent, mediated via stimulation of NMDA receptors located in the VTA [48,49]. The $IC_{50}$ of DOPA CHE, however, against specific binding of tritiated (+)-5-methyl-10,11-dihydro-5*H*-dibenzo[a,d]cycloheptan-5,10-imine (MK-801), a selective antagonist for NMDA ion channel domain, is 0.68 m*M* in rat brain membrane preparations, which is much higher (by approximately five orders of magnitude) than that of cold MK-801 (10 n*M*) [39]. Furthermore, DOPA CHE up to 1 m*M* fails to displace the specific binding of tritiated ionotropic glutamate ligands to label the NMDA binding site and NMDA glycine site.

## 15.3 LEVODOPA ELICITS LOCOMOTOR ACTIVITY UNDER INHIBITION OF CENTRAL AADC IN CONSCIOUS RATS

Experiments were carried out under AADC inhibition in intact and 6-OHDA-lesioned rats. As shown in Figure 15.1, in intact rats pretreated with NSD-1015 (100 mg/kg, i.p.), levodopa (100 mg/kg, i.p.) elicits marked locomotor activity [23]. This is consistent with the other findings of motor responses to levodopa (200 mg/kg) in

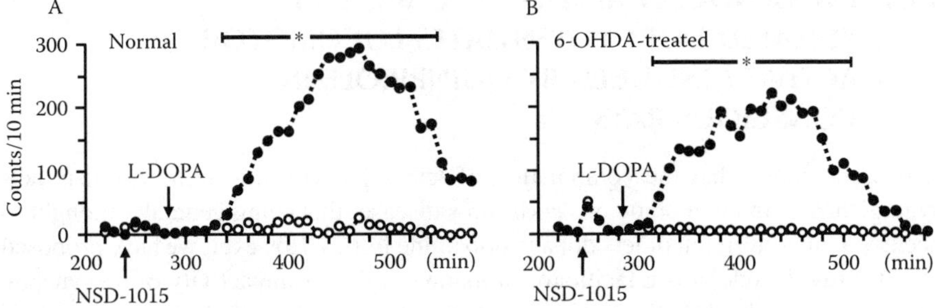


**FIGURE 15.1** Time course of the effect of levodopa (L-DOPA) alone on locomotor activity of normal (A) and 6-OHDA-treated (B) rats under inhibition of AADC by NSD-1015. The counts/10 min of locomotor activity were continuously recorded until 580 min after the start of experiments. NSD-1015 (100 mg/kg) was injected i.p. 240 min after the start of experiments at an upward arrow. Levodopa at 100 (●) and 30 (○) mg/kg in A and 30 (●) and 10 (○) mg/kg in B was i.p. injected at a downward arrow 40 min after NSD-1015 injection. Each value represents mean (n = 5 to 9). SE bars are omitted for clarity. $^*P < .05$ vs. the corresponding value immediately before levodopa injection (Student's *t*-test). (From Nakamura, S. et al., Non-effective dose of exogenously applied L-DOPA itself stereoselectively potentiates postsynaptic $D_2$-receptor-mediated locomotor activities of conscious rats, *Neurosci. Lett.*, 170, 22, 1994. With permission.)

normal rats [63] and in reserpinized rats [64] in the apparent absence of dopamine release under pretreatment with NSD-1015 (100 mg/kg).

Levodopa appears to elicit a "denervation effect" of dopamine neurons, because a noneffective dose of levodopa in intact rats (30 mg/kg) elicits marked locomotor activity following the lesion of dopamine neurons with intracerebroventricular injections of 6-OHDA. This is in part consistent with the other finding [65]: in unilateral 6-OHDA-lesioned rats, NSD-1015 (100 mg/kg, p.o.) increases the total counts of contralateral circling elicited by levodopa (25 mg/kg, i.p.) administered along with a peripheral AADC inhibitor carbidopa (12.5 mg/kg, i.p.).

An important finding is that the systemic administration of high doses of DOPA CHE (40 to 100 mg/kg, i.p.) dose-dependently antagonizes the locomotor activity induced by levodopa (100 mg/kg) in intact rats in the presence of NSD-1015 (100 mg/kg, i.p.) [34]. The highest dose of DOPA CHE antagonizes approximately half the levodopa-induced locomotor activity from 140 min after its simultaneous injection with levodopa. Our findings [23, 34] suggest that levodopa elicits locomotor behavior, at least in part, via activation of postsynaptic DOPA recognition sites, its action being independent of its bioconversion to dopamine. It should be further explored whether or not levodopa elicits any behavioral effects solely through the direct activation of postsynaptic DOPA recognition sites. Interestingly, in rats under inhibition of AADC, levodopa (100 mg/kg) elicits licking behavior (scored by the method of Arnt [66]) which is completely antagonized by the systemic application of DOPA CHE at 100 mg/kg [34].

## 15.4 EXOGENOUSLY APPLIED AND BASALLY RELEASED DOPA POTENTIATES LOCOMOTOR ACTIVITY INDUCED BY QUINPIROLE IN CONSCIOUS RATS

It is well known that the combination of levodopa with dopamine agonists acts synergistically in the therapy of Parkinson's disease, this being generally thought to occur via bioconversion of levodopa to dopamine [4,67]. However, we have proposed the idea that levodopa is a facilitator in its own right for the activity of presynaptic types of catecholaminergic receptors [5,6,31,32]. In rat hypothalamic slices, nanomolar levodopa facilitates noradrenaline release evoked by electrical field stimulation via activation of presynaptic β-adrenoceptors in the absence and presence of AADC inhibition [5]. Noneffective picomolar levodopa potentiates both β-agonist-induced facilitation of noradrenaline release via facilitatory β-adrenoceptors [31] and $\alpha_2$-agonist-induced inhibition of it via presynaptic inhibitory $\alpha_2$-adrenoceptors [32]. Such levodopa-induced potentiation is antagonized by DOPA ME completely — to the level of facilitation or inhibition of noradrenaline release elicited by β- or $\alpha_2$-adrenoceptor agonist alone.

Thus, we have explored whether or not levodopa and basally released DOPA potentiate the locomotor activity induced by quinpirole, an agonist of the dopamine $D_2$ receptor family, via activation of postsynaptic DOPA recognition sites.

As shown in Figure 15.2, the noneffective doses of levodopa (30 mg/kg in intact rats and 10 mg/kg in 6-OHDA-lesioned rats) potentiate the locomotor activity elicited by quinpirole (0.01 to 1 mg/kg, s.c.) in the presence of NSD-1015 [23]. This potentiation is stereoselective and appears to be independent of bioconversion to dopamine.

Figure 15.3 shows the time course of the locomotor activity induced by quinpirole (1 mg/kg) with saline or levodopa (30 mg/kg) in the presence of NSD-1015 (100 mg/kg) (A), recorded simultaneously with the monitoring of the basal release of both DOPA (B) and dopamine (C) during striatal microdialysis in intact rats. Quinpirole alone elicits locomotor activity via postsynaptic dopamine $D_2$ receptors (see Figure 15.2B and Figure 15.4A); ought to decrease basally released DOPA (see Figure 15.4B), probably via activation of presynaptic inhibitory dopamine $D_2$ autoreceptors [68,69]; and decreases basally released dopamine via presynaptic dopamine $D_2$ receptors (see Figure 15.4C). The results shown in Figure 15.3 suggest that levodopa potentiates the locomotor activity elicited by quinpirole following increases in extracellular DOPA but not following increases in extracellular dopamine. Pretreatment with NSD-1015 alone increases basally released DOPA for the initial 40 min, which is followed by a further increase as a result of the addition of levodopa and quinpirole; this increase is not seen following addition of saline and quinpirole (Figure 15.3B). Increases in extracellular levels of DOPA resulting from the application of levodopa can exceed the decreases resulting from the application of quinpirole via presynaptic dopamine $D_2$ autoreceptors. In contrast, NSD-1015 alone decreases basally released dopamine for the initial 40 min, which is followed by a further decrease as a result of the addition of quinpirole and saline, acting via presynaptic inhibitory dopamine $D_2$ receptors (Figure 15.3C). This further decrease in basally released dopamine is not affected by the addition of quinpirole and levodopa.

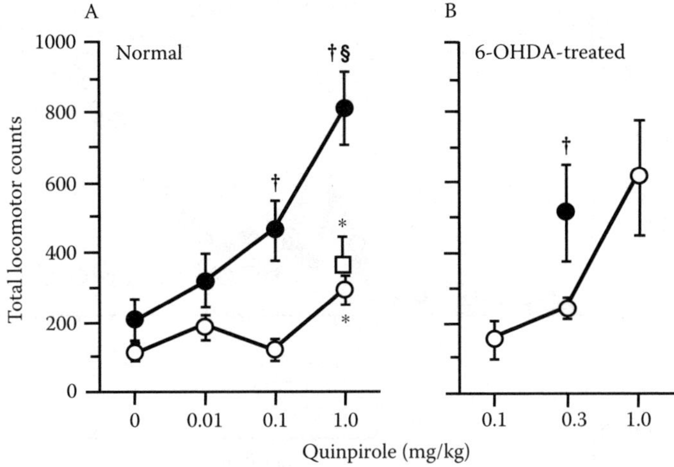


**FIGURE 15.2** Quinpirole-induced increases in locomotor activity of normal (A) and 6-OHDA-treated (B) rats and its potentiation by noneffective doses of levodopa under inhibition of AADC by NSD-1015. Ordinates show the total counts for 140 min after quinpirole injection. Abscissas show the dose of quinpirole, which was injected s.c. 40 min after NSD-1015 injection simultaneously with i.p. levodopa (●, 30 mg/kg, n = 6 to 10), dextrodopa (□, 30 mg/kg, n = 7), or saline (○, n = 4 to 6) in A, and with i.p. levodopa (●, 10 mg/kg, n = 9) or saline (○, n = 4 to 10) in B. Each value represents mean ± SE. *$P < .05$ vs. saline (s.c. and i.p.) at 0 mg/kg, †$P < .05$ vs. corresponding quinpirole plus saline and §$P < .05$ vs. quinpirole plus dextrodopa (Student's *t*-test). (From Nakamura, S. et al., Non-effective dose of exogenously applied L-DOPA itself stereoselectively potentiates postsynaptic $D_2$-receptor-mediated locomotor activities of conscious rats, *Neurosci. Lett.*, 170, 22, 1994. With permission.)

In addition, it appears likely that basally released DOPA functions to potentiate the locomotor activity elicited by quinpirole in a manner similar to exogenously applied levodopa [24]. Quinpirole alone (1 mg/kg) induces locomotor activity (Figure 15.4A) and decreases the basal release of both DOPA (Figure 15.4B) and dopamine (Figure 15.4C) under intact AADC activity in normal rats. Quinpirole-induced locomotor activity is inhibited following a further decrease in basally released DOPA elicited by pretreatment with the selective low dose of α-MPT (3 mg/kg, i.p.) (Figure 15.4B). This dose decreases basally released DOPA by one third without showing any tendency to decrease basally released dopamine [22]. In contrast, quinpirole-induced locomotor activity is potentiated following an increase in basally released DOPA elicited by NSD-1015 (100 mg/kg) as a result of AADC inhibition. NSD-1015 alone markedly increases basally released DOPA [20,35] and this increase can exceed the decrease elicited by quinpirole alone. These modifications of basal DOPA release appear to be independent of its bioconversion from levodopa to dopamine because neither α-MPT (3 mg/kg) nor NSD-1015 (100 mg/kg) alters the decrease in basally released dopamine elicited by quinpirole alone (Figure 15.4C).

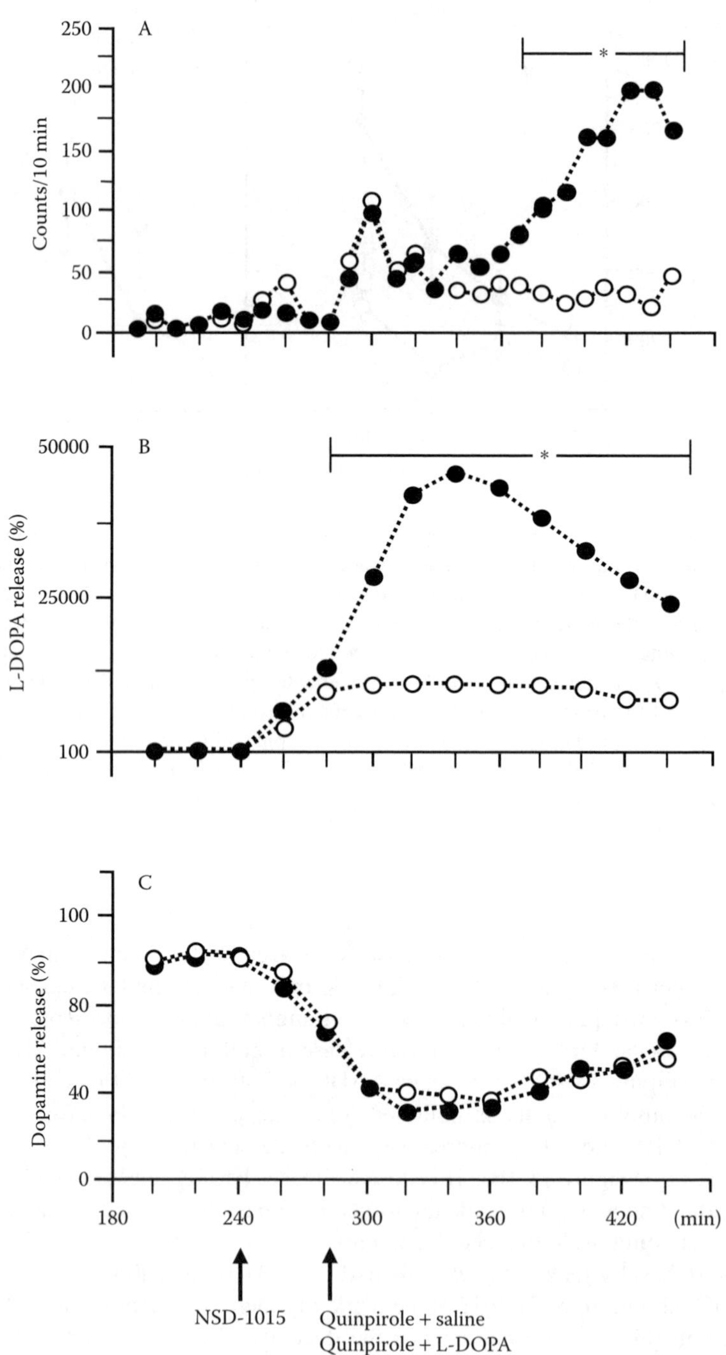


**FIGURE 15.3** Time course of quinpirole-induced increases in locomotor activity and its potentiation by a noneffective dose of levodopa (A), and of the basal release of DOPA (B) and dopamine (C), simultaneously monitored during striatal microdialysis in normal rats under

We further confirmed that a noneffective dose of levodopa (20 mg/kg) potentiates the locomotor activity induced by quinpirole (0.3 mg/kg) from 40 min after the simultaneous injection of both drugs under inhibition of AADC [34]. Importantly, this potentiation is antagonized completely by the systemic application of a low dose of DOPA CHE (10 mg/kg, i.p.) to the level of locomotor activity that is elicited by quinpirole alone. This antagonism is seen from 30 min after the simultaneous injection of DOPA CHE, levodopa, and quinpirole. In addition, levodopa up to 1 m*M* fails to displace the selective binding of tritiated spiperone, a dopamine $D_2$ agonist, in rat brain membrane preparations [27,30] and the $IC_{50}$ value of DOPA CHE against that of spiperone is 1.86 m*M*, which is far higher compared with that of dopamine (2.4 $\mu M$) [30]. These findings provide novel evidence for the synergism between postsynaptic DOPA recognition sites and dopamine $D_2$ receptors in addition to that between postsynaptic dopamine $D_1$ and $D_2$ receptors [70–72]. The results obtained also provide the basis for an argument for the effectiveness of the combination therapy of levodopa and dopamine $D_2$ agonists in Parkinson's disease [4,67,71].

Supporting evidence for levodopa- and DOPA-induced potentiation of the activity of postsynaptic dopamine $D_2$ receptors comes from *in vivo* studies with positron emission tomography and with tritiated spiperone in rats. Under inhibition of AADC, the acute application of levodopa increases the binding of [$^{11}$C]raclopride, an indicator for dopamine $D_2$ receptor density [73,74]. A single application of levodopa increases the number of dopamine $D_2$ binding sites under intact AADC activity [75].

Levodopa appears to be a parent substance not only as a precursor for catecholamines but also as a facilitator for postsynaptic dopamine $D_2$ receptors [23,24,34], in addition to being a presynaptic facilitator for inhibitory dopamine $D_2$ receptors [5,6], inhibitory $\alpha_2$-adrenoceptors [32], and facilitatory $\beta$-adrenoceptors [5,6,31].

## 15.5 EVIDENCE FOR INHIBITORY EFFECTS OF NSD-1015 ON MOTOR RESPONSES TO LEVODOPA

The inhibitory effects of NSD-1015 on levodopa-induced motor responses have traditionally been opposite to the findings reported by Nakazato and Akiyama [63], Nakamura et al. [23], Yue et al. [24], Fisher et al. [64], and Matsushita et al. [34]. Goodale and Moore [76] found that NSD-1015 (100 mg/kg) inhibits contralateral

**FIGURE 15.3 (Continued)** inhibition of AADC by NSD-1015. Quinpirole (1 mg/kg) was injected at an upward arrow with saline (○, n = 6) or levodopa (L-DOPA) (30 mg/kg, ●, n = 5). Dialysates were successively collected every 20 min. The release is expressed as percentage of control, the mean of the initial stable three absolute values. Each value represents mean. SE bars are omitted for clarity. $^*P < .05$ vs. corresponding quinpirole plus saline. Control of DOPA and dopamine (fmol) was 60.9 ± 12.7 and 105.7 ± 25.4 for quinpirole plus saline, and 57.9 ± 3.9 and 102.9 ± 18.4 for quinpirole plus levodopa, respectively. Other details are as in Figure 15.1 and Figure 15.2. (From Nakamura, S. et al., Non-effective dose of exogenously applied L-DOPA itself stereoselectively potentiates postsynaptic $D_2$-receptor-mediated locomotor activities of conscious rats, *Neurosci. Lett.*, 170, 22, 1994. With permission.)

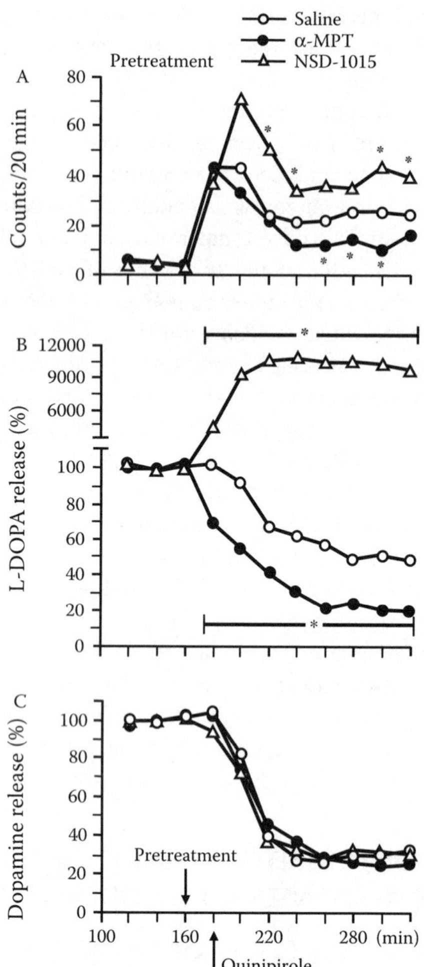


**FIGURE 15.4** Time course of effects of quinpirole alone on locomotor activity of conscious rats, basal DOPA and dopamine release monitored simultaneously during striatal microdialysis, and of their modification by α-MPT and NSD-1015. Pretreatment with α-MPT (3 mg/kg, n = 7), NSD-1015 (100 mg/kg, n = 9), or saline (n = 7) was done i.p. at a downward arrow 160 min after the start of experiments. After 20 min following pretreatment, quinpirole (1 mg/kg) was s.c. injected. Ordinate in A shows the counts/20 min of locomotor activity, which were continuously recorded until 140 min after quinpirole injection. Ordinate in B and C shows release of DOPA and dopamine, respectively. Dialysates were successively collected every 20 min from 120 min after the start of perfusion. $^*P < .05$ vs. corresponding saline plus quinpirole (Student's *t*-test). Control of DOPA and dopamine (fmol) was 47 ± 5 and 118 ± 37 for saline and quinpirole, 43 ± 7 and 95 ± 20 for β-MPT plus quinpirole, and 65 ± 16 and 101 ± 15 for NSD-1015 plus quinpirole, respectively. Other details are as in Figure 15.3. (From Yue, J.-L. et al., Endogenously released L-DOPA itself tonically functions to potentiate $D_2$-receptor-mediated locomotor activities of conscious rats, *Neurosci. Lett.*, 170, 107, 1994. With permission.)

circling responses to levodopa (10 mg/kg) in 6-OHDA-lesioned mice. Melamed et al. [77] showed that NSD-1015 (100 mg/kg) elicits a delay in the onset and decrease in the rate of circling induced by levodopa (50 mg/kg), although this motor response is observed in less than half the rats. Treseder et al. [65] found that NSD-1015 (50 to 150 mg/kg) elicits a dose-dependent delay in the onset of contralateral rotation induced by levodopa (25 mg/kg) with carbidopa (12.5 mg/kg) in 6-OHDA-lesioned rats. Treseder et al. [78] also showed that in common marmosets lesioned with 1-methyl-4-phenyl-1,2,3,6-tetrahydropyridine, NSD-1015 (10 to 50 mg/kg, i.p.) worsened baseline motor deficits. It abolished both locomotor activity and the reversal of disability elicited by levodopa (5 to 18 mg/kg, s.c.). NSD-1015 (25 mg/kg) abolished both the locomotor activity and the improvement in disability elicited by a dopamine $D_1$ agonist or quinpirole. In addition, under inhibition of AADC, levodopa did not affect the locomotor behavior induced by the combined administration of a dopamine $D_1$ agonist and quinpirole.

How can we explain these discrepancies between the two groups? We cannot find suitable explanations. It appears likely that these discrepancies reflect the dual personalities of levodopa and released DOPA. One is the traditionally established bioconversion from levodopa to dopamine [1–4]. The other is the role of DOPA as a neurotransmitter and/or neuromodulator in its own right [7–11]. It has been shown that NSD-1015 inhibits the activity of monoamine oxidase other than AADC [79,80]. From these findings, Treseder et al. concluded that NSD-1015 should not be used as a tool to investigate the neuromodulatory role of levodopa [65,80]. However, we have introduced reasoned approaches in most cases. The responses we obtained to levodopa are primarily seen under intact AADC activity but also in the presence of AADC inhibition [5,6,28,33]. α-MPT was additionally used to study neurotransmitter and/or neuromodulator roles of endogenously released DOPA [14,24,37]. Furthermore, responses to levodopa and released DOPA are antagonized by DOPA ME [13,14,27,28,31,32] and DOPA CHE [29,30,33-37]. However, it should be always taken into consideration that DOPA CHE still has a characteristic as a prodrug for DOPA.

## 15.6 CONCLUSION

DOPA released by the action of nicotine appears to have behavioral, including locomotor, effects on conscious rats. Exogenously applied levodopa and basally released DOPA appear to elicit behavioral effects at least in part via activation of postsynaptic DOPA recognition sites, the effects being independent of its bioconversion to dopamine. There exists a synergistic interaction between postsynaptic DOPA recognition sites and postsynaptic dopamine $D_2$ receptors. These findings provide the basis for an argument for the effectiveness of the combination therapy of levodopa and dopamine $D_2$ agonists in Parkinson's disease. DOPA CHE can be used as a tool to antagonize the responses to exogenously applied levodopa or to endogenously released DOPA in living animals.

## ACKNOWLEDGMENTS

This work was partially supported by Grants-in-Aid for Developmental Scientific Research (No. 06557143) and General Scientific Research (No. 03454146,

No. 03857028, and No. 07407003) from Ministry of Education, Science, Sports and Culture, Japan, and by grants from Mitsui Life Social Welfare Foundation, Uehara Memorial Foundation, and SRF, all in Japan. Many thanks go to Sanae Sato, Department of Pharmacology, Fukushima Medical University School of Medicine, Fukushima, Japan, for remaking the figures.

## REFERENCES

1. Carlsson, A., Lindqvist, M., and Magnusson, T., 3,4-Dihydroxyphenylalanine and 5-hydroxytryptophan as reserpine antagonists, *Nature,* 180, 1200, 1957.
2. Ehringer, H. and Hornykiewicz, O., Verteilung von Noradrenaline und Dopamine (3-Hydroxytyramin) im Gehirn des Menschen und ihr Verhalten bei Erkrankungen des extrapyramidalen Systems, *Klin. Wochenschr.,* 38, 1236, 1960.
3. Bartholini, G. et al., Increase of cerebral catecholamines by 3,4-dihydroxyphenylalanine after inhibition of peripheral decarboxylase, *Nature,* 215, 852, 1967.
4. Hefti, F. and Melamed, E., L-DOPA's mechanism of action in Parkinson's disease, *Trends Neurosci.,* 3, 229, 1980.
5. Goshima, Y., Kubo, T., and Misu, Y., Biphasic actions of L-DOPA on the release of endogenous noradrenaline and dopamine from rat hypothalamic slices, *Br. J. Pharmacol.,* 89, 229, 1986.
6. Misu, Y., Goshima, Y., and Kubo, T., Biphasic actions of L-DOPA on the release of endogenous dopamine via presynaptic receptors in rat striatal slices, *Neurosci. Lett.,* 72, 194, 1986.
7. Misu, Y. and Goshima, Y., Is L-DOPA an endogenous neurotransmitter?, *Trends Pharmacol. Sci.,* 14, 119, 1993.
8. Misu, Y., Ueda, H., and Goshima, Y., Neurotransmitter-like actions of L-DOPA, *Adv. Pharmacol.,* 32, 427, 1995.
9. Misu, Y. et al., Neurobiology of L-DOPAergic systems, *Prog. Neurobiol.,* 49, 415, 1996.
10. Misu, Y., Goshima, Y., and Miyamae, T., Is DOPA a neurotransmitter?, *Trends Pharmacol. Sci.,* 23, 262, 2002.
11. Misu, Y., Kitahama, K., and Goshima, Y., L-3,4-Dihydroxyphenylalanine as a neurotransmitter candidate in the central nervous system, *Pharmacol. Ther.,* 97, 117, 2003.
12. Hoffman, B.B. and Taylor, P., Neurotransmission: the autonomic and somatic motor nervous systems, in *Goodman and Gilman's The Pharmacological Basis of Therapeutics,* Hardman, J.G., Limbird, L.E., and Gilman, A.G., Eds., McGraw-Hill, New York, 2001, p. 115.
13. Kubo, T. et al., Evidence for L-DOPA systems responsible for cardiovascular control in the nucleus tractus solitarii of the rat, *Neurosci. Lett.,* 140, 153, 1992.
14. Yue, J.-L. et al., Baroreceptor-aortic nerve-mediated release of endogenous L-3,4-dihydroxyphenylalanine and its tonic function in the nucleus tractus solitarii of rats, *Neuroscience,* 62, 145, 1994.
15. Yue, J.-L. et al., Altered tonic L-3,4-dihydroxyphenylalanine systems in the nucleus tractus solitarii and the rostral ventrolateral medulla of spontaneously hypertensive rats, *Neuroscience,* 67, 95, 1995.
16. Jaeger, C.B. et al., Aromatic L-amino acid decarboxylase in the rat brain: immunocytochemical localization in neurons of the brain stem, *Neuroscience,* 11, 691, 1984.
17. Tison, F. et al., Endogenous L-DOPA in the rat dorsal vagal complex: an immunocytochemical study by light and electron microscopy, *Brain Res.,* 497, 260, 1989.

18. Goshima, Y., Kubo, T., and Misu, Y., Transmitter-like release of endogenous 3,4-dihydroxyphenylalanine from rat striatal slices, *J. Neurochem.,* 50, 1725, 1988.
19. Misu, Y. et al., Nicotine releases stereoselectively and $Ca^{2+}$-dependently endogenous 3,4-dihydroxyphenylalanine from rat striatal slices, *Brain Res.,* 520, 334, 1990.
20. Nakamura, S. et al., Transmitter-like basal and $K^+$-evoked release of 3,4-dihydroxyphenylalanine from the striatum in conscious rats studied by microdialysis, *J. Neurochem.,* 58, 270, 1992.
21. Nakamura, S. et al., Transmitter-like 3,4-dihydroxyphenylalanine is tonically released by nicotine in striata of conscious rats, *Eur. J. Pharmacol.,* 222, 75, 1992.
22. Nakamura, S. et al., Endogenously released DOPA is probably relevant to nicotine-induced increases in locomotor activities of rats, *Jpn. J. Pharmacol.,* 62, 107, 1993.
23. Nakamura, S. et al., Non-effective dose of exogenously applied L-DOPA itself stereoselectively potentiates postsynaptic $D_2$-receptor-mediated locomotor activities of conscious rats, *Neurosci. Lett.,* 170, 22, 1994.
24. Yue, J.-L. et al., Endogenously released L-DOPA itself tonically functions to potentiate $D_2$ receptor-mediated locomotor activities of conscious rats, *Neurosci. Lett.,* 170, 107, 1994.
25. Goshima, Y. et al., Ventral tegmental injection of nicotine induces locomotor activity and L-DOPA release from nucleus accumbens, *Eur. J. Pharmacol.,* 309, 229, 1996.
26. Yamanashi, K. et al., Tonic function of nicotinic receptors in stress-induced release of L-DOPA from the nucleus accumbens in freely moving rats, *Eur. J. Pharmacol.,* 424, 199, 2001.
27. Goshima, Y., Nakamura, S., and Misu, Y., L-Dihydroxyphenylalanine methyl ester is a potent competitive antagonist of the L-dihydroxyphenylalanine-induced facilitation of the evoked release of endogenous norepinephrine from rat hypothalamic slices, *J. Pharmacol. Exp. Ther.,* 258, 466, 1991.
28. Goshima, Y. et al., L-DOPA induces $Ca^{2+}$-dependent and tetrodotoxin-sensitive release of endogenous glutamate from rat striatal slices, *Brain Res.,* 617, 167, 1993.
29. Misu, Y. et al., L-DOPA cyclohexyl ester is a novel stable and potent competitive antagonist against L-DOPA, as compared to L-DOPA methyl ester, *Jpn. J. Pharmacol.,* 75, 307, 1997.
30. Furukawa, N. et al., L-DOPA cyclohexyl ester is a novel potent and relatively stable competitive antagonist against L-DOPA among several L-DOPA ester compounds, *Jpn. J. Pharmacol.,* 82, 40, 2000.
31. Goshima, Y. et al., Picomolar concentrations of L-DOPA stereoselectively potentiate activities of presynaptic β-adrenoceptors to facilitate the release of endogenous noradrenaline from rat hypothalamic slices, *Neurosci. Lett.,* 129, 214, 1991.
32. Sato, K. et al., L-DOPA potentiates presynaptic inhibitory $\alpha_2$-adrenoceptor- but not facilitatory angiotensin II receptor-mediated modulation of noradrenaline release from rat hypothalamic slices, *Jpn. J. Pharmacol.,* 62, 119, 1993.
33. Akbar, M. et al., Inhibition by L-3,4-dihydroxyphenylalanine of hippocampal CA1 neurons with facilitation of noradrenaline and -amino butyric acid release, *Eur. J. Pharmacol.,* 414, 197, 2001.
34. Matsushita, N., Misu, Y., and Goshima, Y., Antagonism by L-DOPA cyclohexyl ester against behavioral effects of L-DOPA in rats, *J. Pharmacol. Sci.,* 91 (Suppl. I), 202P, 2003.
35. Furukawa, N. et al., Endogenously released DOPA is a causal factor for glutamate release and resultant delayed neuronal cell death by transient ischemia in rat striata, *J. Neurochem.,* 76, 815, 2001.

36. Arai, N. et al., DOPA cyclohexyl ester, a competitive DOPA antagonist, protects glutamate release and resultant delayed neuron death by transient ischemia in hippocampus CA1 of conscious rats, *Neurosci. Lett.,* 299, 213, 2001.
37. Hashimoto, M. et al., DOPA cyclohexyl ester inhibits aglycemia-induced release of glutamate in rat striatal slices, *Neurosci. Res.,* 45, 335, 2003.
38. Ishii, H. et al., Involvement of rBAT in $Na^+$-dependent and -independent transport of the neurotransmitter candidate L-DOPA in *Xenopus laevis* oocytes injected with rabbit small intestinal epithelium poly $A^+$ RNA, *Biochim. Biophys. Acta,* 1466, 61, 2000.
39. Miyamae, T. et al., Some interactions of L-DOPA and its related compounds with glutamate receptors, *Life Sci.,* 64, 1045, 1999.
40. Ueda, H. et al., L-DOPA inhibits spontaneous acetylcholine release from the striatum of experimental Parkinson's model rats, *Brain Res.,* 698, 213, 1995.
41. Cheng, N.-N. et al., Differential neurotoxicity induced by L-DOPA and dopamine in cultured striatal neurons, *Brain Res.,* 743, 278, 1996.
42. Imperato, A., Mulas, A., and Di Chiara, G., Nicotine preferentially stimulates dopamine release in the limbic system of freely moving rats, *Eur. J. Pharmacol.,* 132, 337, 1986.
43. Clarke, P.B. et al., Evidence that mesolimbic dopaminergic activation underlies the locomotor stimulant action of nicotine in rats, *J. Pharmacol. Exp. Ther.,* 246, 701, 1988.
44. Di Chiara, G. and Imperato, A., Drugs abused by humans preferentially increase synaptic dopamine concentrations in the mesolimbic system of freely moving rats, *Proc. Natl. Acad. Sci. USA,* 85, 5274, 1988.
45. O'Neill, M.F., Dourish, C.T., and Iversen, S.D., Evidence for an involvement of D1 and D2 dopamine receptors in mediating nicotine-induced hyperactivity in rats, *Psychopharmacology,* 104, 343, 1991.
46. Reavill, C. and Stolerman, I.P., Locomotor activity in rats after administration of nicotinic agonists intracerebrally, *Br. J. Pharmacol.,* 99, 273, 1990.
47. Nisell, M., Nomikos, G.G., and Svensson, T.H., Systemic nicotine-induced dopamine release in the rat nucleus accumbens is regulated by nicotinic receptors in the ventral tegmental area, *Synapse,* 16, 36, 1994.
48. Schilstrom, B. et al., *N*-methyl-D-aspartate receptor antagonism in the ventral tegmental area diminishes the systemic nicotine-induced dopamine release in the nucleus accumbens, *Neuroscience,* 82, 781, 1998.
49. Schilstrom, B. et al., Nicotine-induced Fos expression in the nucleus accumbens and the medial prefrontal cortex of the rat: role of nicotinic and NMDA receptors in the ventral tegmental area, *Synapse,* 36, 314, 2000.
50. Kalivas, P.W. and Stewart, J., Dopamine transmission in the initiation and expression of drug- and stress-induced sensitization of motor activity, *Brain Res. Rev.,* 16, 223, 1991.
51. Kalivas, P.W. and Duffy, P., Selective activation of dopamine transmission in the shell of the nucleus accumbens by stress, *Brain Res.,* 675, 325, 1995.
52. Doherty, M.D. and Gratton, A., Medial prefrontal cortical D1 receptor modulation of the meso-accumbens dopamine response to stress: an electrochemical study in freely-behaving rats, *Brain Res.,* 715, 86, 1996.
53. Cabib, S. and Puglisi-Allegra, S., Different effects of repeated stressful experiences on mesocortical and mesolimbic dopamine metabolism, *Neuroscience,* 73, 375, 1996.
54. Cooper, D.R. et al., L-DOPA esters as potential prodrugs; behavioural activity in experimental models of Parkinson's disease, *J. Pharm. Pharmacol.,* 39, 627, 1987.

55. Komori, K. et al., Identification of L-DOPA immunoreactivity in some neurons in the human mesencephalic region: a novel DOPA neuron group?, *Neurosci. Lett.*, 157, 13, 1993.
56. Nagatsu, I. et al., Specific localization of the guanosine triphosphate (GTP) cyclohydrolase I-immunoreactivity in the human brain, *J. Neural Transm.*, 106, 607, 1999.
57. Vezina, P. et al., Injections of 6-hydroxydopamine into the ventral tegmental area destroy mesolimbic dopamine neurons but spare the locomotor activating effects of nicotine in the rat, *Neurosci. Lett.*, 168, 111, 1994.
58. Pontieri F.E. et al., Psychostimulant drugs increase glucose utilization in the shell of the rat nucleus accumbens, *Neuroreport*, 5, 2561, 1994.
59. Pontieri, F.E., Tanda, G., and Di Chiara, G., Intravenous cocaine, morphine, and amphetamine preferentially increase extracellular dopamine in the "shell" as compared with the "core" of the rat nucleus accumbens, *Proc. Natl. Acad. Sci. USA*, 92, 12304, 1995.
60. Orzi, F. et al., Intravenous morphine increases glucose utilization in the shell of the rat nucleus accumbens, *Eur. J. Pharmacol.*, 302, 49, 1996.
61. Pontieri, F.E. et al., Effects of nicotine on the nucleus accumbens and similarity to those of addictive drugs, *Nature*, 382, 255, 1996.
62. Maeda, T. et al., L-DOPA neurotoxicity is mediated by glutamate release in cultured rat striatal neurons, *Brain Res.*, 771, 159, 1997.
63. Nakazato, T. and Akiyama, A., Effect of exogenous L-DOPA on behavior in the rat: an *in vivo* voltammetric study, *Brain Res.*, 490, 332, 1989.
64. Fisher, A. et al., Dual effects of L-3,4-dihydroxyphenylalanine on aromatic L-amino acid decarboxylase, dopamine release and motor stimulation in the reserpine-treated rat: evidence that behaviour is dopamine independent, *Neuroscience*, 95, 97, 2000.
65. Treseder, S.A., Rose, S., and Jenner, P., The central aromatic amino acid DOPA decarboxylase inhibitor, NSD-1015, does not inhibit L-DOPA-induced circling in unilateral 6-OHDA-lesioned-rats, *Eur. J. Neurosci.*, 13, 162, 2001.
66. Arnt, J., Antistereotypic effects of dopamine D-1 and D-2 antagonists after intrastriatal injection in rats. Pharmacological and regional specificity, *Naunyn-Schmiedeberg's Arch. Pharmacol.*, 330, 97, 1985.
67. Jackson, D.M., Jenkins, O.F., and Ross, S.B., The motor effects of bromocriptine — a review, *Psychopharmacology*, 95, 433, 1988.
68. Roth, R.H., CNS dopamine autoreceptors: distribution, pharmacology, and function, *Ann. N.Y. Acad. Sci.*, 430, 27, 1984.
69. Koeltzow, T.E. et al., Alterations in dopamine release but not dopamine autoreceptor function in dopamine D3 receptor mutant mice, *J. Neurosci.*, 18, 2231, 1998.
70. Arnt, J., Hyttel, J., and Perregaard, J., Dopamine D-1 receptor agonists combined with the selective D-2 agonist quinpirole facilitate the expression of oral stereotyped behaviour in rats, *Eur. J. Pharmacol.*, 133, 137, 1987.
71. Hagan, J.J. et al., Parkinson's disease: prospects for improved drug therapy, *Trends Pharmacol. Sci.*, 18, 156, 1997.
72. Dziewczapolski, G. et al., Threshold of dopamine content and D1 receptor stimulation necessary for the expression of rotational behavior induced by D2 receptor stimulation under normo and supersensitive conditions, *Naunyn-Schmiedeberg's Arch. Pharmacol.*, 355, 30, 1997.
73. Hume, S.P. et al., Effect of L-DOPA and 6-hydroxydopamine lesioning on [$^{11}$C]raclopride binding in rat striatum, quantified using PET, *Synapse*, 21, 45, 1995.
74. Opacka-Juffry, J. and Brooks, D.J., L-Dihydroxyphenylalanine and its decarboxylase: new ideas on their neuroregulatory roles, *Mov. Disord.*, 10, 241, 1995.

75. Murata, M. and Kanazawa, I., Repeated L-DOPA administration reduces the ability of dopamine storage and abolishes the supersensitivity of dopamine receptors in the striatum of intact rat, *Neurosci. Res.,* 16, 15, 1993.
76. Goodale, D.B. and Moore, K.E., A comparison of the effects of decarboxylase inhibitors on L-DOPA-induced circling behavior and the conversion of DOPA to dopamine in the brain, *Life Sci.,* 19, 701, 1976.
77. Melamed, E. et al., Suppression of L-DOPA-induced circling in rats with nigral lesions by blockade of central dopa-decarboxylase: implications for mechanism of action of L-DOPA in parkinsonism, *Neurology,* 34, 1566, 1984.
78. Treseder, S.A., Jackson, M., and Jenner, P., The effects of central aromatic amino acid DOPA decarboxylase inhibition on the motor actions of L-DOPA and dopamine agonists in MPTP-treated primates, *Br. J. Pharmacol.,* 129, 1355, 2000.
79. Hunter, L.W., Rorie, D.K., and Tyce, G.M., Inhibition of aromatic L-amino-acid decarboxylase under physiological conditions: optimization of 3-hydroxybenzylhydrazine concentration to prevent concurrent inhibition of monoamine oxidase, *Biochem. Pharmacol.,* 45, 1363, 1993.
80. Treseder, S.A. et al., Commonly used L-amino acid decarboxylase inhibitors block monoamine oxidase activity in the rat, *J. Neural Transm.,* 110, 229, 2003.

# 16 Modulatory Effects of Levodopa on D2 Dopamine Receptors in Striatum Assessed Using *In Vivo* Microdialysis and PET

*Jolanta Opacka-Juffry and Susan P. Hume*

**CONTENTS**



## 16.1 INTRODUCTION

In the 1980s and early 1990s, novel evidence was published indicating that (acute or chronic) administration of exogenous L-3,4-dihydroxyphenylalanine (levodopa) might affect dopaminergic neuronal activity through altering dopamine release and/or dopamine receptor status in rat brain tissue. For example, Misu et al. [1] reported that levodopa applied *in vitro* to rat striatal sections caused a biphasic presynaptic regulatory effect, depending on the concentration used. At low doses (30 n*M*), levodopa seemed to facilitate dopamine release via activation of beta-adrenoceptors, whereas at a micromolar level, it caused inhibition of dopamine release via inhibitory D2 presynaptic receptors. These presynaptic effects of levodopa on dopamine release *in vitro* appeared to be independent of dopamine synthesis because they remained unaffected by inhibition of aromatic L-amino acid decarboxylase (AADC). Moreover, it was found that endogenous L-3,4-dihydroxyphenylalanine (DOPA) itself could be released *in vitro* in response to depolarization caused by $K^+$, electrical stimulation, or nicotine [2,3] and that this neurotransmitter-like, impulse-evoked DOPA release was $Ca^{2+}$ dependent and tetrodotoxin sensitive. Interestingly, neurotransmitter-like release of striatal DOPA, similar to that of dopamine, was also observed *in vivo* by means of the microdialysis technique applied in conscious rats [4]. These findings offer good support for the hypothesis that DOPA is a neurotransmitter and/or neuromodulator in the central nervous system and not merely a precursor in dopamine synthesis [5].

Several lines of evidence suggested that exogenously applied levodopa could be neuroactive in multiple ways, depending on the duration of treatment in experimental animals. Long-term administration of levodopa seemed to lead to an inhibition of striatal AADC, which was likely to affect central dopaminergic disposition (for review, see Reference 6). Long-term levodopa treatment also seemed to affect glucose utilization within the basal ganglia, with different effects on D1 and D2 dopamine receptor–mediated striatal outputs [7]. Experimental evidence indicated that exogenous levodopa could also exert effects on the status of striatal dopamine receptors. Thus, acutely administered levodopa was shown to cause an increase in the number of D2 dopamine receptor sites in the rat striatum, as demonstrated *ex vivo* using the radioligand [$^3$H]spiperone in the absence of the central AADC inhibitor, *m*-hydroxybenzylhydrazine (NSD-1015) [8]. D1 dopamine receptors appeared to be affected to an even greater extent: a significant increase in D1 $B_{max}$ lasted for 12 h after a levodopa dose. A significant increase in D1 messenger RNA level in the striatum preceded the changes in D1 receptor binding. The acute levodopa effects on dopamine receptors seemed to habituate in the course of a long-term treatment [8]. Our own early *in vivo* positron emission tomography (PET) studies demonstrated that in rat striatum, acute levodopa administration caused an increase in specific binding of the substituted benzamide antagonist [$^{11}$C]raclopride, a selective D2 receptor ligand [9]. In this instance, the effect of levodopa seemed independent of endogenous dopamine concentration because it remained unaffected by central AADC inhibition achieved using NSD-1015.

## 16.2 ACUTE EFFECTS OF LEVODOPA ON D2 DOPAMINE RECEPTORS

In 1995, our early PET studies with [$^{11}$C]raclopride were reported fully in a paper describing the effects of levodopa in both control (naïve) and 6-hydroxydopamine (6-OHDA) lesioned rats [10], the latter used as a hemiparkinsonian model [11]. At that time, the use of small-animal scanning to effect a "bridge" between human PET and the more conventional *in vitro* or *ex vivo* methodologies used in the laboratory was a relatively novel approach. However, because dedicated small-bore, high-resolution cameras were still at an early stage of development [12], we used a high-sensitivity clinical PET camera. In doing so, we accepted the quantitative constraints imposed by the partial volume effects that resulted from its limited spatial resolution [~5.5 mm, full width at half maximum (FWHM), at the center of the field of view

(FOV)]. Perhaps more importantly, because of the need for rat immobilization, we were also forced to accept the need for anesthesia throughout the scanning procedure and so, in a parallel series of experiments, monitored extracellular dopamine and DOPA in similarly anesthetized animals.

As mentioned earlier, levodopa (L-DOPA methyl ester, given at a dose of 20 mg/kg free base i.p. 60 min after carbidopa, at a dose of 25 mg/kg i.p. and ~100 min before the radioligand) resulted in a statistically significant 15% increase in the binding potential (BP) for "tracer" doses of [$^{11}$C]raclopride in rat striatum. In this case, BP at the D2 receptors was defined as the ratio of rate constants describing the movement of raclopride to and from the specifically bound compartment ($k_3/k_4$) and was estimated from regional radioactivity–time curves using a reference-tissue compartmental model [13]. These early PET data, presented in Figure 16.1, supported

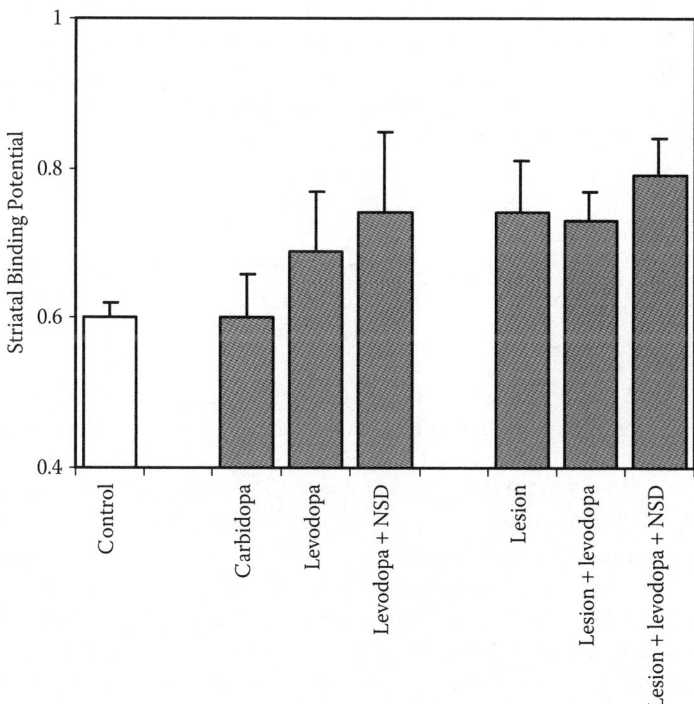


**FIGURE 16.1** Effect of the experimental treatments on striatal BP for [$^{11}$C]raclopride measured *in vivo* in anesthetized rat. Values are means with standard deviation. Groups were as follows: Control (naïve) rats (n = 5); pretreatment with carbidopa alone (n = 4); carbidopa with levodopa (n = 6); carbidopa with levodopa and NSD-1015 (n = 3); 6-OHDA lesioned (n = 7); lesion with carbidopa and levodopa pretreatment (n = 5); lesion with carbidopa, levodopa, and NSD-1015 pretreatment (n = 5). Dosing regimes are given in the text. (Data are taken from Hume, S.P. et al., Effect of L-DOPA and 6-hydroxydopamine lesioning on [$^{11}$C]raclopride binding in rat striatum, quantified using PET, *Synapse,* 21, 45, 1995. With permission.)

an acute neuromodulatory action of levodopa, manifested as supersensitivity of D2 receptors. Also utilizing *in vivo* scanning, but with a different approach using the tracer L-[$^{11}$C]DOPA, Tedroff et al. [14] later described an increased rate of disposition of radioactivity in monkey striatum following intravenous infusion of levodopa (3 or 15 mg/kg/h) and hypothesized that levodopa was acting presynaptically to increase AADC activity. Our own microdialysis data in anesthetized rats showed a marked increase in extracellular DOPA at 40 to 60 min after levodopa injection and a ~60% increase in striatal extracellular dopamine concentration, which peaked at 12 to 140 min [10]. It was, however, unlikely that there was a direct causal relationship between this and the rapid modulation of D2 receptors that we had observed because the latter effect was not prevented by administration of the AADC inhibitor NSD-1015 (100 mg/kg i.p. immediately before levodopa), as shown in Figure 16.1.

In an extension of the original experiment, the dose of stable raclopride coinjected with the radioligand was increased to determine the saturation kinetics in levodopa-treated rat striatum compared with control. Parameters relating to the number of binding sites (apparent $B_{max}$) and the injected dose of raclopride required to occupy half of these sites (apparent $K_d$) were estimated by iterative nonlinear regression, assuming a single-site binding model for ligand–receptor interaction and the Michaelis–Menten-based relationship:

$$BP = [\text{apparent } B_{max}/(C + \text{apparent } K_d)] + NS$$

where C is the mass of injected raclopride (nmol/kg), and NS is a measure of the nonspecific BP, estimated as the maximal blocking obtained with the doses of raclopride used. Both "apparent $B_{max}$" and "apparent $K_d$" are expressed in units of concentration of injected raclopride.

Figure 16.2 shows striatal BP for [$^{11}$C]raclopride binding as a function of stable raclopride in control and levodopa-treated animals and best fits to the data using the relationship described earlier. Fitted parameters (with SE estimated from the fit) were: "apparent $B_{max}$" = 7.7 ± 2.0 nmol/kg in levodopa-treated rats, compared with a control value of 8.7 ± 2.0 nmol/kg, and "apparent $K_d$" = 10.1 ± 3.1 nmol/kg in levodopa-treated rats, compared with a control value of 17.1 ± 4.2 nmol/kg. To confirm that the saturation curves were significantly different from each other, all data were combined and refitted to the model and the F-statistic applied to the sum of squares of the combined fit compared with the sum of squares of the individual fits. In addition, given the degree of scatter seen in Figure 16.2, the data were refitted to the same model, but fixing each of the parameters in turn. The worst fit was obtained for fixed "apparent $K_d$," forcing the solution to an increase in "apparent $B_{max}$." Thus, the data were most consistent with the increase in BP for [$^{11}$C]raclopride following acute levodopa treatment being primarily due to a decrease in "apparent $K_d$" (~40%), that is, an increase in receptor affinity for the radioligand rather than a change in receptor number. In this way, our findings differed from those of Murata and Kanazawa [8] who, as mentioned earlier, had reported a change in D2 receptor number and no change in affinity.

Although the lack of influence of dopamine on the increase of [$^{11}$C]raclopride BP at tracer doses of raclopride was clearly demonstrated by our combining the levodopa

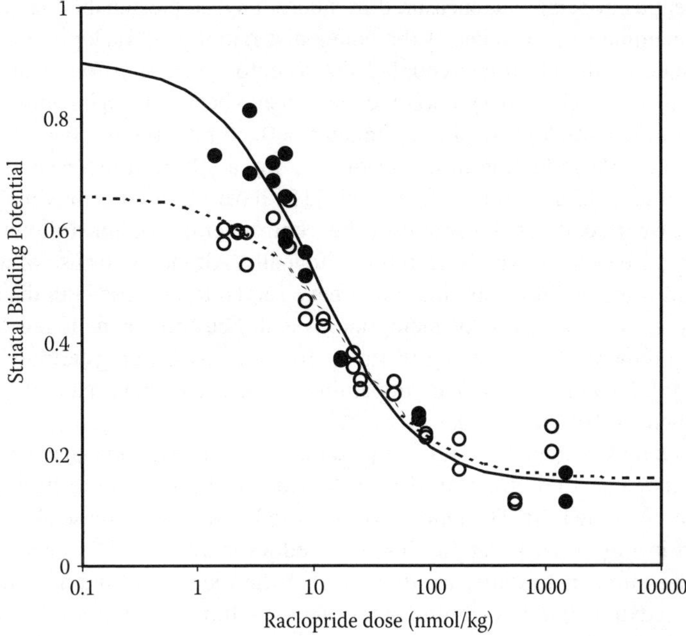


**FIGURE 16.2** [$^{11}$C]Raclopride BP as a function of dose of stable raclopride injected in control (naïve) rats (open circles) or rats given levodopa (closed circles). The dashed line is the best fit through the control data and the solid line is the best fit through the levodopa data, as described in the text. Each pair of data points is from right and left striata from a single rat. (Data are taken from Hume, S.P. et al., Effect of L-DOPA and 6-hydroxydopamine lesioning on [$^{11}$C]raclopride binding in rat striatum, quantified using PET, *Synapse* 21, 45, 1995. With permission.)

treatment with NSD-1015, the possibility exists that as the injected mass of raclopride was increased in the saturation kinetics arm of our study, the D2 antagonist itself caused an increase in release of dopamine via the presynaptic autoreceptors [15]. Thus, in a recent abstract, Schiffer et al. [16], combining *in vivo* microdialysis with PET using one of the new-generation high-resolution cameras designed for small-animal scanning, have reported that [$^{11}$C]raclopride increases extracellular dopamine at only 5% occupancy of rat striatal D2 receptors. Because the mathematical fit to the data in Figure 16.2 cannot distinguish between an increase in receptor affinity and an increase in receptor occupancy by competing endogenous dopamine, our measured "apparent $K_d$" could have reflected a composite *in vivo* effect of both raclopride mass and the pharmacological or neurophysiological response to it. It is then feasible that any direct effect of levodopa at the autoreceptor could have indirectly affected the *in vivo* $K_d$ for raclopride at the D2 receptors, which would, to some extent, explain the difference between our data and the *ex vivo* binding data of Murata and Kanazawa [8].

Whether this is a real rather than a hypothetical issue is less clear because [$^{11}$C]raclopride binding in rat striatum appears to be relatively insensitive to changes in endogenous dopamine as measured by microdialysis, presumably due to the low $K_i$ for dopamine displacement of the antagonist radioligand raclopride at the low-affinity state of the G-protein-coupled D2 receptor [17]. Houston et al. [18], for example, recently reported a monotonic relationship between amphetamine dose and reduction in [$^{11}$C]raclopride binding in anesthetized rat striatum, but with only an approximate 16% reduction in BP accompanying a ~25-fold increase in extracellular dopamine. In awake rats, See et al. [15] have reported a maximal 2.2-fold increase in extracellular dopamine in the ventrolateral striatum following raclopride at a dose of 2 mg/kg, given intraperitoneally. Although our use of anesthesia and the i.v. route of administration might have served to increase both the effectiveness [19] and the local dose of raclopride, it is difficult to reconcile our "apparent $K_d$" measurement of 17 nmol/kg (6 μg/kg free base or 8 μg/kg tartrate salt) with an extracellular concentration of dopamine sufficient to have caused significant displacement of [$^{11}$C]raclopride.

In 6-OHDA lesioned striatum, the same dose of levodopa that was used in control rats caused no effect on [$^{11}$C]raclopride BP in addition to that caused by the lesioning alone [10]. The latter was measured as a 23% increase above control, as shown in Figure 16.1. Again, this reflected an increase in D2 receptor affinity rather than number. Initially, we thought that the increased binding reflected the decrease in extracellular dopamine in lesioned striatum because radiolabeled raclopride binding had been shown to be increased as dopamine levels were decreased [13,20,21]. Interestingly, Tedroff et al. [22] used a similar argument in reverse, suggesting that the reduction in [$^{11}$C]raclopride binding that they observed in human PET studies after levodopa administration to Parkinsonian patients was consequent to the blockade of D2 receptors by dopamine. However, the increase in BP that we measured in the lesioned striatum appeared to be independent of changes in extracellular dopamine. Thus, we detected no change in BP for lesioned rats that were additionally treated with either levodopa (causing a threefold increase in extracellular dopamine) or levodopa with NSD-1015 (causing a 50% decrease in extracellular dopamine).

The most apparent difference that we observed between control and 6-OHDA lesioned striatum in response to the levodopa treatment appeared to be in the extracellular concentration of DOPA measured by microdialysis. The threefold higher concentration in the lesioned tissue is related to a limited ability to biotransform the dopamine precursor and has been discussed in terms of the potentially toxic effects of chronic levodopa treatment [23].

These early data validated the usefulness of PET scanning of rat brain, using [$^{11}$C]raclopride to experimentally monitor changes in D2 receptors *in vivo* and using a methodology that was potentially analogous to that used in human PET studies [24]. Relevant to this review, the results that we obtained supported a direct action of acute levodopa and implicated a postsynaptic response. The PET studies were then extended to include chronic levodopa treatment, as described in the text that follows. In addition, although sensitization of D2-receptor-mediated striatal outflow remained implicated in the induction of levodopa-associated dyskinesias [25–27], microdialysis studies were initiated to monitor neurotransmitter levels downstream from the dopamine system.

## 16.3 LONG-TERM EFFECTS OF LEVODOPA ON D2 DOPAMINE RECEPTORS

It was of interest to know whether the observed *in vivo* sensitization of D2 striatal receptors caused by an acute treatment with levodopa could be affected by long-term levodopa administration. That question was not only curiosity driven but also had clinical connotations because Parkinson's disease management is known to be often complicated by the "wearing off" and "on–off" phenomena reported in chronic levodopa therapy (e.g., Reference 28). In our second study, which addressed the issue of chronic effects of levodopa [29], we employed a dedicated small-animal scanner with an improved spatial resolution of 2.6 mm FWHM at the center of the FOV, a system more appropriate for visualizing and quantifying the bilateral PET signal obtained by the binding of selective dopaminergic ligands such as [$^{11}$C]raclopride to D2 dopamine receptors in the rat striatum [30,31]. With this methodology, we were able to confirm the earlier findings obtained using a clinical PET scanner [10], as discussed earlier. Acute levodopa treatment did indeed cause an increase in striatal BP for [$^{11}$C]raclopride in naïve rats at two doses of levodopa: the lower dose of 20 mg/kg i.p. used in our earlier study and at a higher dose of 100 mg/kg i.p. (In all acute experiments, L-DOPA methyl ester was given at the selected dose 60 min after carbidopa, 25 mg/kg i.p.) Predictably, these two doses of levodopa had different effects on the levels of striatal extracellular dopamine, as measured using *in vivo* microdialysis in anesthetized rats, under conditions similar to those of the PET studies. The higher levodopa dose caused an approximately eightfold increase in extracellular striatal dopamine, whereas the lower dose caused only a mild elevation up to 60% in striatal dopamine [29]. Pretreatment with NSD-1015 was introduced to half of the animals of both "low" (20 mg/kg i.p.) and "high" (100 mg/kg i.p.) levodopa groups. PET studies showed that all four acute treatment groups (low or high levodopa, with or without NSD-1015) were characterized by significantly elevated striatal BP values with the maximal values consistent with an approximately 40% increase in BP as compared with controls (Figure 16.3a). It should be emphasized that the effects of levodopa on the values of the BP of [$^{11}$C]raclopride at each dose tested were not affected by pretreatment with NSD-1015. Admittedly, although the analyses indicated a significant increase in striatal binding, it was not possible to tease out which of the kinetic parameters, "apparent $B_{max}$," "apparent $K_d$," or both, contributed to the observed D2 receptor sensitization following an acute dose of levodopa.

In the long-term experiments, rats were given either carbidopa as a vehicle (control) or levodopa and carbidopa ("chronic DOPA" group) in drinking water for up to 5 weeks. The intake of levodopa was 170 ± 18 mg/kg/d and that of carbidopa was 21 ± 2 and 24 ± 3 mg/kg/d in levodopa and carbidopa control groups, respectively. Each rat was scanned twice: once for chronic levodopa effects and then for chronic + acute effects. At the end of week 4, a PET scan was performed, followed by another a week later. The second scan was coupled with an acute levodopa dose (20 mg/kg i.p.) administered after acute carbidopa (as described earlier). Two days before each scan, the chronic levodopa/carbidopa administration was discontinued to ensure a drug-free period. The experimental design gave the following "chronic"

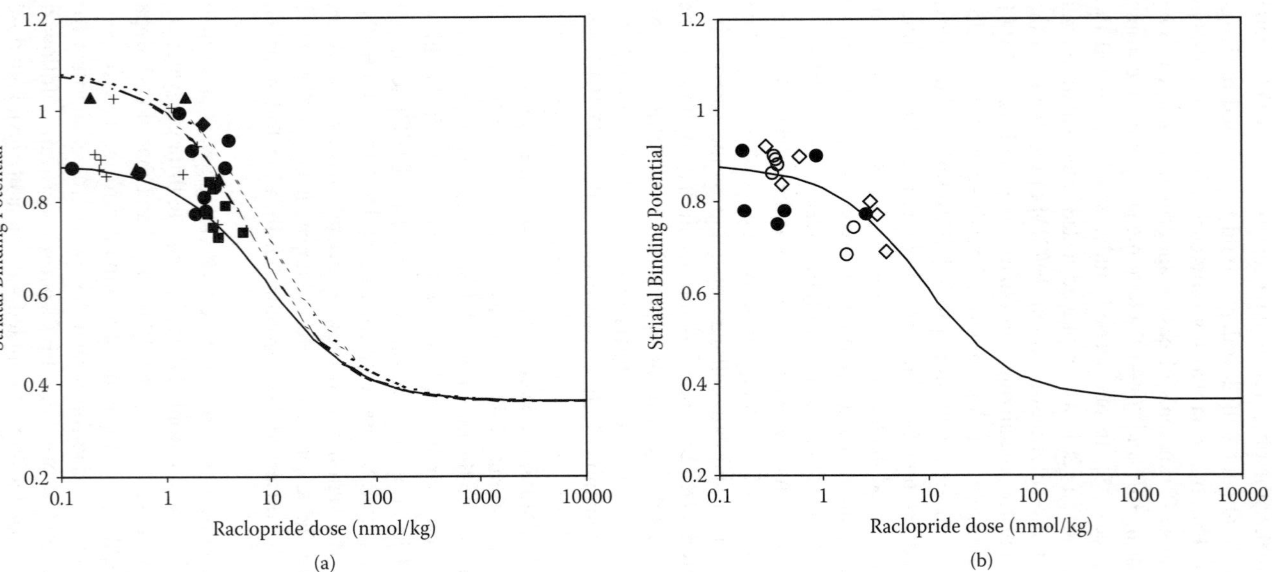


**FIGURE 16.3** Striatal binding potential values for [$^{11}$C]raclopride expressed as a function of stable raclopride. Shown in (a) are data from those levodopa treatment groups that had BP values on or above the control saturation curve (solid line). The dashed and dashed–dot lines represent, respectively, the effect of either a 40% simulated increase in $B_{max}$ or an equivalent decrease in $K_d$ to cause the same increase in BP. Groups are: (●) acute levodopa 20 mg/kg; (♦) acute levodopa 20 mg/kg + NSD; (■) acute levodopa 100 mg/kg; (▶) acute levodopa 100 mg/kg + NSD; (+) chronic carbidopa + acute levodopa 100 mg/kg. Shown in (b) are data from those treatment groups that had values within or below the control range. Groups are: (◊) chronic carbidopa; (○) chronic levodopa; (●) chronic levodopa + acute levodopa 20 mg/kg. Each datum point is an average of BP calculated separately for left and right striatal ROI from a single rat. (From Opacka-Juffry, J. et al., Modulatory effects of L-DOPA on D2 dopamine receptors in rat striatum, measured using in vivo microdialysis and PET, *J. Neural Transm.*, 105, 349, 1998. With permission.)

groups: chronic carbidopa (vehicle control); chronic levodopa; chronic carbidopa + acute levodopa; chronic levodopa + acute levodopa. It may need explaining that the lower levodopa dose was selected for the acute treatment because of its limited effect on striatal dopamine release. Because the direct dopamine change was very mild at the lower levodopa dose, and to restrict the severity of pharmacological manipulations, no NSD-1015 was added in this experimental design. The concern about the complexity of the pharmacological manipulations was justified because the PET experimental protocol necessarily included general anesthesia.

Neither chronic levodopa (plus carbidopa) nor carbidopa treatment seemed to affect striatal D2 receptor binding properties because striatal BP for [$^{11}$C]raclopride was unaffected by those treatments (Figure 16.3b). According to the reports on *in vitro* effects on rat striatal D2 receptors caused by long-term levodopa administration (duration varying from 14 d to 12 months), there may be no change, a mild increase, or a reduction in D2 receptor density in intact rat striatum [32–36]. In contrast, chronic levodopa treatment in rats with a unilateral nigrostriatal lesion as a model of Parkinson's disease most often leads to a reversal of the D2 supersensitivity caused by the lesion [37–39]. In the rats that had been given levodopa chronically for 5 weeks, an acute dose of levodopa did not cause any measurable increase in striatal BP for [$^{11}$C]raclopride (Figure 16.3a). It is worth emphasizing that long-term treatment with carbidopa alone (vehicle) did not alter the sensitizing effect of acute levodopa on D2 striatal receptors (Figure 16.3b). We concluded that levodopa-dependent sensitization of striatal D2 receptors could be prevented by a long-term exposure to levodopa. These *in vivo* PET findings were similar to that obtained *ex vivo* by Murata and Kanazawa [8]. Considering the significance of D2 striatal dopamine receptors in the movement control executed by the basal ganglia circuitry, our *in vivo* findings that the D2 receptor sensitization caused by an acute levodopa dose could be prevented by a longer term levodopa treatment seemed relevant to the clinical situation of fluctuating responses to levodopa seen in advanced Parkinson's disease. Although the study findings describe an intact (nonlesioned) rat striatum, they may bring us closer to the explanation of why continuous administration of levodopa achieves therapeutic effects with relatively reduced side effects in the form of dyskinesias [40,41].

## 16.4 FUNCTIONAL CONSEQUENCES OF LEVODOPA EFFECTS ON D2 DOPAMINE RECEPTORS

It was appropriate to ask the question whether the levodopa-dependent increase in the ligand binding ability of striatal D2 receptors could translate into functional implications within the basal ganglia. We hypothesized that the levodopa sensitizing influence on D2 receptors in the striatum should affect neuronal activity within the D2-controlled pathways of the basal ganglia. D2 receptor-like protein has been found in the medium spiny projection neurons and the medium- and large-sized aspiny interneurons, as well as in presynaptic terminal boutons and in small thinly myelinated axons [42–44]. This localization indicates D2 postsynaptic involvement in striatal γ-aminobutyric acid (GABA)-projection neurons and striatal cholinergic interneurons, as well as in presynaptic nigrostriatal axon endings terminating in the striatum [43]. In addition, there is evidence of D2-like immunoreactivity, representing heteroreceptors associated with corticostriatal and thalamostriatal asymmetric synapses [42,43], and D2-dependent presynaptic inhibition of glutamate release from corticostriatal projections appears to be well documented in both physiological and pharmacological studies [45–47]. In our study, we chose to consider putative functional effects of levodopa-dependent D2 supersensitivity at the level of D2-regulated corticostriatal glutamatergic terminals and at the level of GABAergic striatopallidal projection neurons, which undergo postsynaptic inhibition exerted by D2 receptors and form the so-called "indirect" striatal pathway [48,49].

We performed *in vivo* microdialysis experiments to monitor extracellular amino acid neurotransmitters in the striatum (for the cortical glutamatergic input) and external globus pallidus (for the GABAergic output) during acute levodopa administration in rats. Microdialysis probes were stereotaxically implanted in the brain regions of interest and extracellular fluid was sampled on the following day, after postoperative recovery in the animals, which were lightly reanesthetized to mimic the conditions of PET scanning. High-performance liquid chromatography with fluorescent detection was employed to analyze extracellular levels of glutamate and GABA among the amino acids in the striatum and external globus pallidus [29]. A bolus of levodopa at a dose of 20 mg/kg i.p. was given 1 h after carbidopa (25 mg/kg i.p.) and following a couple of hours of stable baseline sampling. This treatment did not significantly affect extracellular striatal glutamate (or GABA) when compared with the carbidopa-injected control. However, extracellular levels of GABA measured in the globus pallidus appeared to be significantly reduced in response to the acute levodopa treatment. The reduction, which remained within the range of 20% of the baseline, was observable for approximately 2 h, starting 1 h after the levodopa bolus (Figure 16.4). This finding was consistent with an activation of the D2 dopamine receptors expressed on striatopallidal neurons [50,51]. Because PET experiments showed no effect of central AADC inhibition on acute levodopa effects on D2 receptor binding, our microdialysis studies were performed on rats that were not treated with NSD-1015, to restrict the complexity of the pharmacological manipulations. Hence, the issue of selectivity of the levodopa effects observed may be seen as disputable, even though the low dose of levodopa used (20 mg/kg) had only limited effect on striatal dopamine release (leading to a maximum 60% increase in striatal extracellular dopamine reached at 2 h after the dose [29]). However, evidence suggesting that the observed effects on GABA levels in the globus pallidus were caused by DOPA rather than dopamine action comes from a study performed in free-moving rats, in which a single dose of levodopa (following carbidopa, without NSD-1015) did not alter striatal extracellular dopamine (unpublished data). Nevertheless, GABA reduction in the globus pallidus of those animals (unpublished data) was very similar to the changes analyzed in the preceding text. A recent study by Ochi et al. [52] showing no influence of levodopa (50 mg/kg p.o., without NSD-1015) on extracellular pallidal GABA in 6-OHDA lesioned rats does not necessarily contradict our findings [29], considering the fact that nigrostriatal dopamine depletion is likely to lead to D2 receptor upregulation in the "indirect" pathway. Hence, further D2 receptor sensitization may not be readily effected by levodopa.

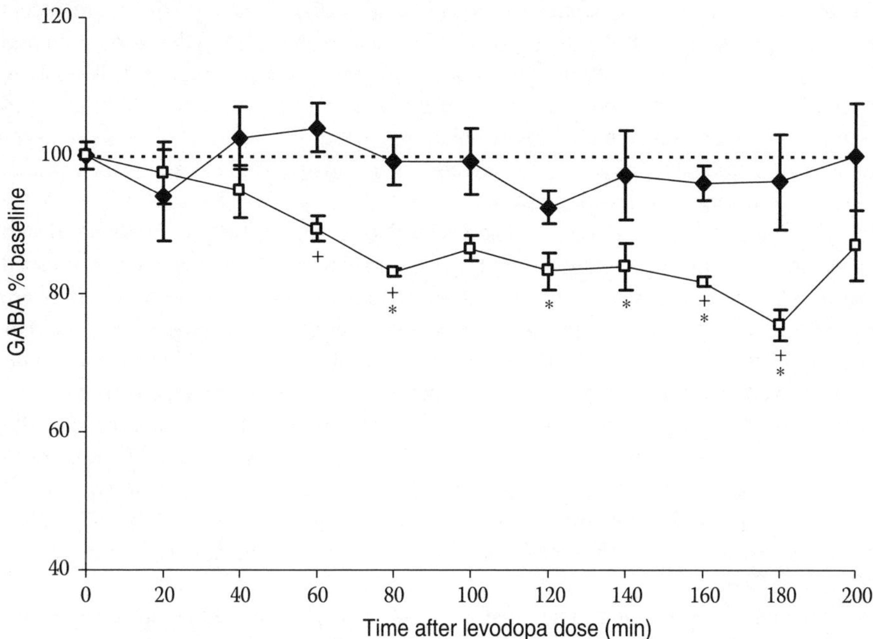


**FIGURE 16.4** Extracellular levels of GABA measured in globus pallidus of isoflurane anesthetized rat after injections of levodopa 20 mg/kg i.p. (open squares), given 1 h after carbidopa (25 mg/kg, i.p.). Closed diamonds represent carbidopa-only control. The data (mean ± SD, n =4 per group) are expressed as a percentage of the baseline, collected for 2 h prior to the levodopa injection. The basal levels of GABA in the globus pallidus were 0.623 ± 0.120 pmol/20 min sample (mean ± SD, n = 12). A two-way ANOVA with repeated measures indicated significant effects; for treatment differences, F = 35,459, $P = .001$; for time, F = 3,133, $P = .003$. *Significantly different from the last baseline sample ($P < .01$, post hoc *t*-test). +Significantly different from carbidopa-only control at the time point shown ($P < .05$, post hoc *t*-test). (From Opacka-Juffry, J. et al., Modulatory effects of L-DOPA on D2 dopamine receptors in rat striatum, measured using in vivo microdialysis and PET, *J. Neural Transm.*, 105, 349, 1998. With permission.)

The fact that the acute levodopa administration significantly affected striatopallidal GABA output in intact rats [29], without apparently changing corticostriatal glutamatergic disposition, might indicate a degree of selectivity of acute levodopa at the level of D2 receptors regulating striatopallidal projection neurons. However, this attractive conclusion should be considered with some caution because the apparent lack of effect of levodopa on corticostriatal glutamate may be a net effect of several factors. One of them is a possibility that only stimulated release of corticostriatal glutamate (as opposed to basal, studied in our experiments) is likely to be regulated by D2 receptors [53]. Additionally, striatal extracellular glutamate represents a mixed pool, with some of the amino acid originating from calcium-dependent synaptic release and some being of metabolic origin; the latter possibly masking neurotransmitter fluctuations caused by synaptic activity [47]. It is of interest that the more recent *in vitro* findings have shown that the corticostriatal pathway may develop an altered form of plasticity, which seems to be associated with levodopa-induced dyskinesia in 6-OHDA lesioned rats [54]. Admittedly, caution should be exercised when extrapolating from intact brains into a Parkinsonian-type condition because different postsynaptic transcriptional regulatory mechanisms may be involved in DA -denervated striatal neurons in response to levodopa [55].

In our studies of levodopa effects on D2 striatal receptors [29], we were interested in finding out whether levodopa might cause sensitization of D2 receptors expressed in GABA neurons projecting to the external globus pallidus. These projections are directly involved in the motor control executed by the basal ganglia, and GABA deficits in the striatopallidal pathway may be implicated in dyskinesias [51]. Our PET-derived observation that levodopa-dependent D2 receptor supersensitivity could be abolished by a chronic treatment with levodopa might give grounds for speculations that the D2-regulated striatopallidal GABAergic activity of the indirect pathway could be a vulnerable target in levodopa treatment in advanced Parkinson's disease, although there are lines of evidence that the D1 receptor-controlled "direct" striatal pathway seems to respond predominantly to levodopa therapy (for review, see Reference 56).

Molecular and cellular mechanisms of levodopa-induced D2 receptor supersensitivity remain to be elucidated. Our PET findings [10] tentatively suggested that an acute dose of levodopa was likely to affect D2 receptor affinity rather than the receptor number ($B_{max}$). Hence, we would be inclined to speculate that D2 receptor affinity should also be the factor affected by longer term levodopa treatment. The more recent autoradiographic study on the effects of chronic levodopa administration on dopamine receptor (D1, D2, and D3) binding in the caudate and putamen of normal monkeys has shown no changes in the level of receptor binding in animals with or without dyskinesias when compared with untreated controls [57]. Zeng et al. [57] suggest that a functional coupling of D2 receptor with enkephalin and adenosine A2a receptor activity rather than dopamine receptor expression may be implicated in the development of levodopa-induced dyskinesias. The complexity of levodopa and DOPA interactions with dopamine receptors seem to be reflected in a recent study on parkinsonian monkeys reported by Bezard et al. [58]. These authors have found that levodopa-induced dyskinesia was associated with overexpression of the dopamine D3 receptor, which appeared to have a dual involvement in both dyskinesia induced by levodopa and its therapeutic action.

## 16.5 CONCLUSION

*In vivo* derived evidence indicates that acute treatment with levodopa may cause sensitization of dopamine D2 receptors in rat striatum. The sensitization, independent of endogenous dopamine, manifests itself through an increase in specific binding of the selective D2 antagonist, [$^{11}$C]raclopride, as assessed using PET. It seems to be coupled with a postsynaptic functional effect within the D2 receptor-regulated striatopallidal GABA pathway. It also appears that levodopa-dependent sensitization of striatal D2 receptors can be prevented by a long-term exposure to levodopa. These findings, while of relevance to the clinical aspects of levodopa treatment in Parkinson's disease, support the theory that DOPA may act as a modulator in the central nervous system.

## ACKNOWLEDGMENT

This study was supported by the U.K. Medical Research Council.

## REFERENCES

1. Misu, Y., Goshima, Y., and Kubo, T., Biphasic actions of L-DOPA on the release of endogenous dopamine via presynaptic receptors in rat striatal slices, *Neurosci. Lett.,* 72, 194, 1986.
2. Goshima, Y., Kubo, T., and Misu, Y., Transmitter-like release of endogenous 3,4-dihydroxyphenylalanine from rat striatal slices, *J. Neurochem.,* 50, 1725, 1988.
3. Misu, Y. et al., Nicotine releases stereoselectively and $Ca^{2+}$-dependently endogenous 3,4-dihydroxyphenylalanine from rat striatal slices, *Brain Res.,* 520, 334, 1990.
4. Nakamura, S. et al., Transmitter-like basal and $K^+$-evoked release of 3,4-dihydroxyphenylalanine from the rat striatum in conscious rats studied by microdialysis, *J. Neurochem.,* 58, 270, 1992.
5. Misu, Y. and Goshima, Y., Is L-dopa an endogenous neurotransmitter?, *Trends Pharmacol. Sci.,* 14, 119, 1993.
6. Opacka-Juffry, J. and Brooks, D.J., L-Dihydroxyphenylalanine and its decarboxylase: new ideas on their neuroregulatory roles, *Movement Disord.,* 10, 241, 1995.
7. Engber, T.M. et al., Chronic levodopa treatment alters basal and dopamine agonist-stimulated cerebral glucose utilization, *J. Neurosci.,* 10, 3889, 1990.
8. Murata, M. and Kanazawa, I., Repeated L-DOPA administration reduces the ability of dopamine storage, *Neurosci. Res.,* 16, 15, 1993.
9. Hume, S.P. et al., Effect of L-DOPA and 6-OHDA lesioning on [$^{11}$C]raclopride binding in rat striatum, quantified using PET, *J. Cerebr. Blood F. Met.,* 13, S295, 1993.
10. Hume, S.P. et al., Effect of L-DOPA and 6-hydroxydopamine lesioning on [$^{11}$C]raclopride binding in rat striatum, quantified using PET, *Synapse,* 21, 45, 1995.
11. Perese, D.A. et al., A 6-hydroxydopamine-induced selective parkinsonian rat model, *Brain Res.,* 495, 285, 1989.
12. Myers, R. and Hume, S., Small animal PET, *Eur. Neuropsychopharmacol.,* 12, 545, 2002.
13. Hume, S.P. et al., Quantitation of carbon-11 labelled raclopride in rat striatum using positron emission tomography, *Synapse,* 12, 47, 1992.
14. Tedroff, J. et al., L-DOPA modulates striatal dopaminergic function in vivo: evidence from PET investigations in nonhuman primates, *Synapse,* 25, 56, 1997.
15. See, R.E. et al., In vivo assessment of release and metabolism of dopamine in the ventrolateral striatum of awake rats following administration of dopamine D1 and D2 receptor agonists and antagonists, *Neuropharmacology,* 30, 1269, 1991.
16. Schiffer, W.K. et al., Understanding the fundamentals of radiotracer binding in the rodent dopaminergic system using simultaneous microPET and microdialysis techniques, *J. Nucl. Med.,* 44 (Suppl.), 70P, 2003.
17. Lidow, M.S. et al., Dopamine D2 receptors in the cerebral cortex: distribution and pharmacological characterization with [$^{3}$H]raclopride, *Proc. Natl. Acad. Sci. USA,* 86, 6412, 1989.

18. Houston, G. et al., Temporal characterisation of amphetamine-induced dopamine release assessed with [$^{11}$C]raclopride in anaesthetised rodents, *Synapse,* 51, 206 2004.
19. Opacka-Juffry, J. et al., Nomifensine-induced increase in extracellular striatal dopamine is enhanced by isoflurane anaesthesia, *Synapse,* 7, 169, 1991.
20. Seeman, P., Guan, H.-C., and Niznik, H.B., Endogenous dopamine lowers the dopamine D2 receptor density as measured by [$^{3}$H]raclopride: implications for positron emission tomography of the human brain, *Synapse,* 3, 96, 1989.
21. Young, L.T. et al., Effects of endogenous dopamine on kinetics of [$^{3}$H]N-methylpiperone and [$^{3}$H]raclopride binding in the rat brain, *Synapse,* 9, 188, 1991.
22. Tedroff, J. et al., Levodopa-induced changes in synaptic dopamine in patients with Parkinson's disease as measured by [$^{11}$C]raclopride displacement and PET, *Neurology,* 46, 1430, 1996.
23. Abercrombie, E.D., Bonatz, A.E., and Zigmond, M.J., Effects of L-DOPA on extracellular dopamine in striatum of normal and 6-hydroxydopamine-traeted rats, *Brain Res.,* 525, 36, 1990.
24. Playford, E.D. and Brooks, D.J., In vivo and in vitro studies of the dopaminergic system in movement disorders, *Cerebrovasc. Brain Metab. Rev.,* 4, 144, 1992.
25. Blanchet, P.J., Gomez-Mancilla, B., and Bedard, B.J., DOPA-induced 'peak dose' dyskinesia: clues implicating D2 receptor-mediated mechanisms using dopaminergic agonists in MPTP monkeys, *J. Neural Transm.,* 45 (Suppl.), 103, 1995.
26. Blanchet, P.J. et al., Is striatal dopaminergic receptor imbalance responsible for levodopa-induced dyskinesia?, *Fundam. Clin. Pharmacol.,* 9, 434, 1995.
27. Verhagen Metman, L. et al., Apomorphine responses in Parkinson's disease and the pathogenesis of motor complications, *Neurology,* 48, 369, 1997.
28. Marsden, C.D. and Parkes, J.D., Success and problems of long-term levodopa therapy in Parkinson's disease, *Lancet,* 12, 345, 1977.
29. Opacka-Juffry, J. et al., Modulatory effects of L-DOPA on D2 dopamine receptors in rat striatum, measured using in vivo microdialysis and PET, *J. Neural Transm.,* 105, 349, 1998.
30. Bloomfield, P.M. et al., The design and physical characteristics of a small animal positron emission tomograph, *Phys. Med. Biol.,* 40, 1105, 1995.
31. Hume, S.P. et al., The potential of high-resolution positron emission tomography to monitor striatal dopaminergic function in rat models of disease, *J. Neurosci. Methods,* 67, 103, 1996.
32. Jackson, D.M. et al., Chronic L-DOPA treatment of rats and mice does not change the sensitivity of post-synaptic dopamine receptors, *Arch. Pharmacol.,* 324, 271, 1983.
33. Jenner, P. and Marsden, C.D., Chronic pharmacological manipulation of dopamine receptors in brain, *Neuropharmacology,* 26, 931, 1987.
34. Reches, A. et al., The effect of chronic L-DOPA administration on supersensitive pre- and postsynaptic dopaminergic receptors in rat brain, *Life Sci.,* 31, 37, 1982.
35. Rouillard, C. et al., Behavioural and biochemical evidence for a different effect of repeated administration of L-DOPA and bromocriptine on denervated versus non-denervated striatal dopamine receptors, *Neuropharmacology,* 26, 1601, 1987.
36. Wilner, K.D. et al., Biochemical alterations of dopamine responses following chronic L-DOPA therapy, *Biochem. Pharmacol.,* 29, 701, 1980.
37. Parenti, M. et al., Differential effects of repeated treatment with L-DOPA on dopamine-D1 or -D2 receptors, *Neuropharmacology,* 25, 331, 1986.
38. Reches, A. et al., Chronic levodopa or pergolide administration induces down regulation of dopamine receptors in denervated striatum, *Neurology,* 34, 1208, 1984.

39. Schneider, M.B. et al., Dopamine receptors: effects of chronic L-DOPA and bromocriptine treatment in an animal model of disease, *Clin. Neuropharmacol.,* 7, 247, 1984.
40. Nutt, J.G., Obeso, J.A., and Stocchi, F., Continuous dopamine-receptor stimulation in advanced Parkinson's disease, *Trends Neurosci.,* 23, S109, 2000.
41. LeWitt, P.A. and Nyholm, D., New developments in levodopa therapy, *Neurology,* 62, S9, 2004.
42. Levey, A.I. et al., Localization of D1 and D2 dopamine receptors in rat, monkey, and human brain with subtype-specific antibodies, *Proc. Natl. Acad. Sci. USA,* 90, 8861, 1993.
43. Sesack, S.R., Aoki, C., and Pickel, V.M., Ultrastructural localization of D2 receptor-like immunoreactivity in midbrain dopamine neurons and their striatal target, *J. Neurosci.,* 14, 88, 1994.
44. Fisher, R.S. et al., D-2 dopamine receptor protein location: Golgi impregnation-gold toned and ultrastructural analysis of the rat neostriatum, *J. Neurosci. Res.,* 38, 551, 1994.
45. Mitchell, P.R. and Doggett, N.S., Modulation of striatal [$^3$H]-glutamic acid release by dopaminergic drugs, *Life Sci.,* 26, 2073, 1980.
46. Calabresi, P. et al., Chronic neuroleptic treatment: D2 dopamine receptor supersensitivity and striatal glutamatergic transmission, *Ann. Neurol.,* 31, 366, 1992.
47. Yamamoto, B.K. and Davy, S., Dopaminergic modulation of glutamate release in striatum as measured by microdialysis, *J. Neurochem.,* 58, 1736, 1992.
48. Gerfen, C.R., The neostriatal mosaic: multiple levels of compartmental organization in the basal ganglia, *Ann. Rev. Neurosci.,* 15, 285, 1992.
49. Gerfen, C.R., Molecular effects of dopamine on striatal-projection pathways, *Trends Neurosci.,* 23, S64, 2000.
50. Linderfors, N., Dopaminergic regulation of glutamic acid decarboxylase, mRNA expression and GABA release in the striatum: a review, *Prog. Neuropsychopharmacol. Biol. Psychiatry,* 17, 887, 1993.
51. Scheel-Krüger, J., Dopamine–GABA interactions: evidence that GABA modulates and mediates dopaminergic functions in the basal ganglia and the limbic system, *Acta Neurol. Scand.,* 73 (Suppl.), 1, 1986.
52. Ochi, M., Shiozaki, S., and Kase, H., L-DOPA-induced modulation of GABA and glutamate release in substantia nigra pars reticulata in a rodent model of Parkinson's disease, *Synapse,* 52, 163, 2004.
53. Daly, D.R. and Moghaddam, B., Actions of clozapine and haloperidol on the extracellular levels of excitatory amino acids in the prefrontal cortex and striatum of conscious rats, *Neurosci. Lett.,* 152, 61, 1993.
54. Picconi, B. et al., Loss of bidirectional striatal synaptic plasticity in L-DOPA-induced dyskinesia, *Nature Neurosci.,* 6, 501, 2003.
55. Andersson, M., Konradi, C., and Cenci, M.A., cAMP response element-binding protein is required for dopamine-dependent gene expression in the intact but not the dopamine-denervated striatum, *J. Neurosci.,* 21, 9930, 2001.
56. Graybiel, A.M., Canales, J.J., and Capper-Loup, C., Levodopa-induced dyskinesia and dopamine-dependent stereotypies: a new hypothesis, *Trends Neurosci.,* 23, S71, 2000.
57. Zeng, B.Y. et al., Chronic high dose L-dopa does not alter the levels of dopamine D-1, D-2 or D-3 receptor in the striatum of normal monkeys: an autoradiographic study, *J. Neural Transm.,* 108, 925, 2001.
58. Bezard, E. et al., Attenuation of levodopa-induced dyskinesia by normalizing dopamine D3 receptor function, *Nature Med.,* 9, 762, 2003.

# 17 Dopamine-Independent Inhibition of Hippocampal CA1 Neurons Produced by Nanomolar Levodopa with Facilitation of Noradrenaline and GABA Release in Rats

*Masashi Sasa, Kumatoshi Ishihara, Muhammad Akbar, and Yoshimi Misu*

## CONTENTS



## 17.1 INTRODUCTION

L-3,4-Dihydroxyphenylalanine (levodopa) has been believed to be an inert amino acid. It has been widely used for the treatment of Parkinson's disease since the 1960s, and the efficacy is most likely attributed to the conversion of levodopa to dopamine by the enzyme aromatic L-amino acid decarboxylase (AADC). On the other hand, accumulated evidence has supported levodopa as a distinct neurotransmitter in the brain, in which a series of relevant enzymes for the synthesis and metabolism of levodopa have been found to exist. Furthermore, the $Ca^{2+}$-dependent release and $Na^{+}$-dependent transport of levodopa as well as stereoselective and antagonist-reversible responses to levodopa have further reinforced the neurotransmitter hypothesis, although the receptor sequence for levodopa has not yet been identified [1]. Recent intensive studies in particular have suggested that levodopa is a candidate of neurotransmitters involved in the neurotransmission from the first-order neuron of the primary baroreceptor afferents to the second-order neurons in the nucleus tractus solitarii and in the central blood pressure regulation in the lower brainstem [2–6]. Interestingly, very low concentrations in the nanomolar order of exogenously applied levodopa facilitate the release of neurotransmitters such as noradrenaline and dopamine from the nerve terminals in rat hypothalamic slices [7,8]. These findings also suggest that levodopa acts as a neurotransmitter and/or neuromodulator on synaptic nerve terminals. In doing so, levodopa may also affect the release of other neurotransmitters in other brain regions.

The neuronal circuit in the hippocampus, including monoaminergic control, is relatively well understood. In the hippocampus, CA1 pyramidal neurons are activated by glutamate released from Schaffer fibers originating from the CA3 region and are inhibited by $\gamma$-aminobutyric acid (GABA) released from the interneurons via activation of $GABA_A$ receptors located within the hippocampus. The same CA1 pyramidal neurons are also excited by acetylcholine released from fibers originating from the septum via muscarinic receptors [9]. In addition, serotonin (5-HT) released from serotonergic fibers derived from the dorsal raphe nucleus induces inhibition of the CA1 neurons via $5\text{-}HT_{1A}$ receptors and excitation of the neurons via $5\text{-}HT_4$ receptors [10–12]. Release of neurotransmitters such as glutamate and GABA is mediated via $5\text{-}HT_3$ receptors. Furthermore, the hippocampus is known to receive dense noradrenergic innervation from the locus coeruleus. It is also well documented that both $\alpha$- an $\beta$-adrenoceptors are located in the hippocampus, mediating inhibition and excitation, respectively [13–15]. Noradrenaline involved in neurotransmission from the locus coeruleus to the hippocampus plays an important role in producing long-term potentiation (LTP). If noradrenaline is depleted from the hippocampus, LTP could dramatically decrease, as observed in the CA3 region of reserpine-treated rats [16].

Based on our findings to date [17], this chapter focuses on the presynaptic action of levodopa, viz., very low concentrations of the drug induce the release of neurotransmitters involved in neurotransmission to the CA1 neurons, possibly acting on certain recognition sites for levodopa.

## 17.2 METHODS

Extracellular recording from the CA1 region of hippocampal slice preparations obtained from Wistar rats was performed using a glass micropipette filled with 3-*M* NaCl. A 450-μm-thick hippocampal slice was placed in a bath continuously perfused with artificial cerebrospinal fluid oxygenated with a gas mixture of 95% $O_2$ and 5% $CO_2$. Drugs tested were applied in the bath through the perfusion system.

Population spikes were induced by stimulation of the Schaffer collateral/commissural fibers using a bipolar electrode at 5-min intervals. The population spikes of 6 in 11 consecutive responses were averaged. The amplitude of the population spikes was taken as the differences between the spike-peak negativity and the average of the preceding and following positive spike-peaks determined by the height and the vertical line shown in inset of Figure 17.1. The drug effects were expressed as the percentage of the control just before the respective drug application. The results were expressed as mean ± SE. The statistical significance was determined using the one-way analysis of variance (ANOVA) test followed by the Tukey honestly significant difference (HSD) test as a post hoc analysis.

## 17.3 RESULTS AND DISCUSSION

### 17.3.1 Inhibition by Levodopa of Population Spikes via Levodopa Recognition Sites

When levodopa at concentrations of 100 p*M* to 10 μ*M* were applied in the bath (Table 17.1), picomolar to nanomolar levodopa produced a concentration-dependent inhibition of population spikes induced by stimulation of the Schaffer collateral/commissural fibers. Levodopa at 100 n*M* elicited the maximum inhibition, and its time course is shown in Figure 17.1. The inhibition was observed within 5 min after levodopa application, the peak inhibition was seen 10 min after the application, and complete recovery was observed 5 min after washing. Such a strong potency of levodopa at nanomolar concentrations in the hippocampus is equivalent to that in the hypothalamus [7,8].

Interestingly, the concentration–response relationship of levodopa-induced inhibitory effects on the population spikes assumed a bell-shaped pattern (Table 17.1). Levodopa at micromolar concentrations produced a less pronounced inhibition of population spikes, compared with the maximum inhibition at 100 n*M*, suggesting that higher micromolar concentrations of levodopa may elicit different modes of responses from those seen at nanomolar concentrations.

The enzyme AADC converts levodopa to dopamine. 3-Hydroxybenzylhydrazine (NSD-1015), a central AADC inhibitor, is reported to increase the levodopa level in the striatum [18]. In the presence of NSD-1015 at 20 μ*M*, a concentration sufficient to block the enzyme activity [19], the inhibition of population spikes by levodopa at 1 μ*M* was still observed in a similar manner as in the absence of the inhibitor [17]. Therefore, this inhibition is considered to be due to levodopa itself and not due to dopamine converted from levodopa. Such failure of effects of NSD-1015 on levodopa-induced responses has been observed in the hypothalamus and the striatum [7,20].

Furthermore, the inhibition of population spikes by 100-n*M* levodopa was completely antagonized by 30-n*M* DOPA cyclohexyl ester (DOPA CHE), a potent and relatively stable competitive levodopa antagonist (Figure 17.2) [21,22]. However, DOPA CHE alone up to 30 n*M* did not affect the population spikes. Such antagonism against levodopa-induced facilitation of noradrenaline release by another levodopa antagonist, DOPA methyl ester (DOPA ME), has been found in the hypothalamus [8,23]. DOPA CHE does not inhibit the uptake of labeled levodopa into *Xenopus laevis*

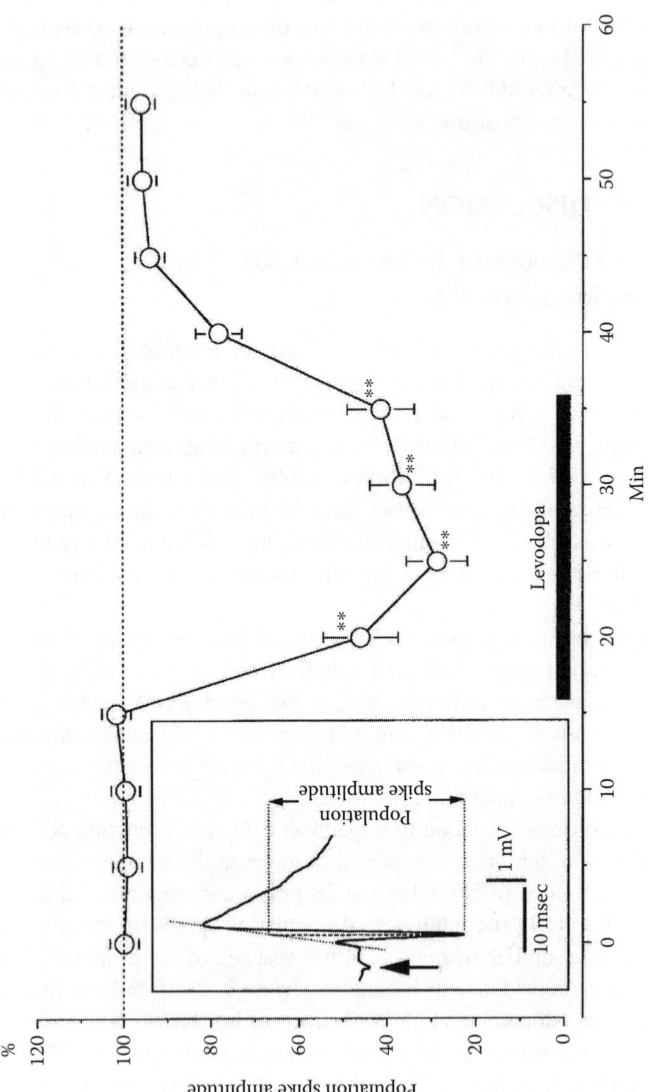


**FIGURE 17.1** Effects of 100-n*M* levodopa on population spikes elicited by stimulation of Schaffer collateral/commissural fibers in hippocampal CA1 region of rats. Inset indicates the method of measurement of actual population spike amplitude. Stimulus was given at the arrow. Open circles and vertical bars show means and SE, respectively (n = 15). ***P* < .01, significantly different from the value before drug application. (From Akbar, M. et al., Inhibition by L-3,4-hydroxyphenylalanine of hippocampal CA1 neurons with facilitation of noradrenaline and γ-aminobutyric acid release, *Eur. J. Pharmacol.*, 414, 197, 2001. With permission.)

**TABLE 17.1**
**Concentration–Response Relationship of the Inhibitory Effects of Levodopa on Population Spikes Induced by Stimulation of the Schaffer Collateral/Commissural Fibers in Hippocampal CA1 Region of Rats**

| Concentration of Levodopa | n | Amplitude of Population Spikes (%) |
|---|---|---|
| 100 p*M* | 7 | 96.6 ± 1.7 |
| 1 n*M* | 7 | 71.4 ± 16.1 |
| 10 n*M* | 12 | 68.2 ± 12.3 |
| 100 n*M* | 15 | 28.7 ± 6.8[a] |
| 1 μ*M* | 14 | 44.9 ± 9.8[a] |
| 10 μ*M* | 10 | 73.6 ± 6.1[b,c] |

*Note:* Levodopa-induced inhibition 10 min after application is expressed as percentage of the control amplitude of the population spikes before application of respective concentration of levodopa.

[a] $P < .01$ vs. control.
[b] $P < .05$ vs. control.
[c] $P < .05$ vs. 100 n*M* (one-way ANOVA and Tukey HSD tests).

oocytes [24], suggesting the existence of levodopa recognition sites different from the transport sites. Furthermore, displacement studies using levodopa and levodopa antagonists including DOPA CHE and DOPA ME against selective binding of labeled ligands in rat membranes suggest that levodopa recognition sites may differ from dopamine $D_1$ and $D_2$ receptors, $\alpha_2$- and β-adrenoceptors, GABA receptors, and ionotropic glutamate receptors [1,22,23,25,26]. Therefore, it is strongly suggested that the levodopa-induced inhibition of the hippocampal population spikes is mediated by activation of levodopa-specific recognition sites.

### 17.3.2 Levodopa-Induced Inhibition via Noradrenaline and GABA Systems

Interestingly, 1-μ*M*-levodopa-induced inhibition of the population spikes was antagonized by 10-μ*M* phentolamine, an α-adrenoceptor antagonist (Figure 17.3), although phentolamine alone did not affect the population spikes. This finding suggests that the levodopa-induced inhibition is probably mediated, at least in part, via activation of α-adrenoceptors induced by released noradrenaline. This idea is further supported by the finding that 10-μ*M* 6-fluoronorepinephrine, an α-adrenoceptor agonist, conversely inhibited the population spikes (Figure 17.3). It is most unlikely, however, that β-adrenoceptors are involved in this inhibition, because the α-adrenoceptor antagonist phentolamine completely blocked the levodopa-induced inhibition without residual inhibitory effects. The levodopa-induced facilitation of noradrenaline release has been observed in the hypothalamus, although presynaptic

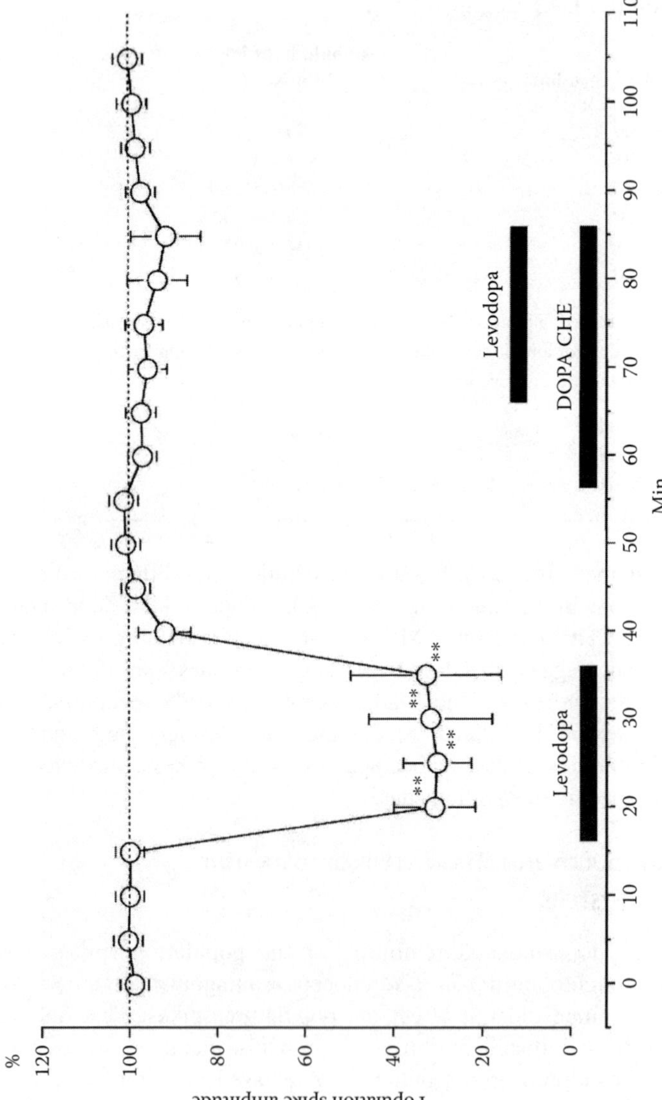


**FIGURE 17.2** Antagonizing effects of 30-n*M* DOPA CHE, a competitive levodopa antagonist, against the inhibitory effects of 100-n*M* levodopa on population spikes. Open circles and vertical bars show means and SE, respectively (n = 5). $^{**}P < .01$, significantly different from the value before drug application. (From Akbar, M. et al., Inhibition by L-3,4-hydroxyphenylalanine of hippocampal CA1 neurons with facilitation of noradrenaline and γ-aminobutyric acid release, *Eur. J. Pharmacol.*, 414, 197, 2001. With permission.)

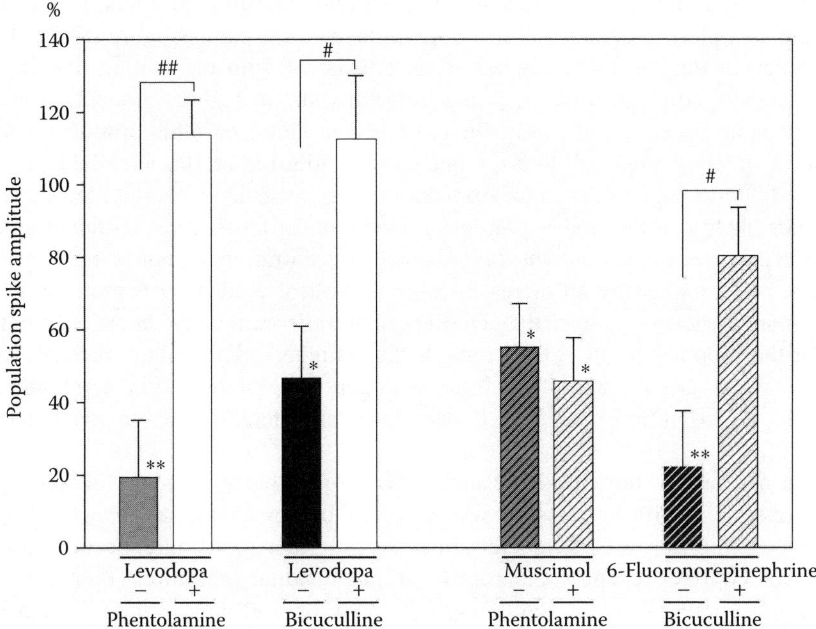


**FIGURE 17.3** Antagonizing effects of 10-μ*M* phentolamine (n = 3) and 1-μ*M* bicuculline (n = 6) on 1-μ*M*-levodopa-induced inhibition, and 10-μ*M* phentolamine (n = 5) on 1-μ*M*-muscimol-induced inhibition, as well as those of 1-μ*M* bicuculline (n = 4) on 10-μ*M*-6-fluoronorepinephrine-induced inhibition. Each column and vertical bar indicates mean and SE, respectively. $^*P < .05$, $^{**}P < .01$, compared with control, the value before the drug application. #$P < .05$, ##$P < .01$, compared with the values in the absence of each antagonist.

β-adrenoceptors located on noradrenergic nerve terminals appear to contribute to the facilitation [8,19,23,27].

More interestingly, 1-μ*M*-levodopa-induced inhibition was completely antagonized by 1-μ*M* bicuculline, a $GABA_A$ receptor antagonist (Figure 17.3). Conversely, 1-μ*M* muscimol, a $GABA_A$ receptor agonist, inhibited population spikes induced by stimulating the Schaffer collateral/commissural fibers (Figure 17.3). Furthermore, 6-fluoronorepinephrine-induced inhibition of the population spikes was also significantly reduced by bicuculline (Figure 17.3). Therefore, levodopa is also considered to release GABA to inhibit the population spikes. However, muscimol-induced inhibition was not affected by phentolamine, suggesting that $GABA_A$ receptors involved in the levodopa-induced inhibition appear to be a downstream factor of noradrenergic neurotransmission in the hippocampus.

### 17.3.3 Effects of Levodopa on Glutamatergic and Serotonergic Neurotransmission

When levodopa at 1 μ*M* was applied in the bath concomitantly with phentolamine or bicuculline, there was a tendency of enhancement of population spikes in response to stimulation of the Schaffer collateral/commissural fibers, in addition to the complete blockade of levodopa-induced inhibition (Figure 17.3). This tendency of enhancement appears to be consistent with the finding that in the bell-shaped concentration–response relationship, a higher concentration of levodopa at 10 μ*M* significantly elicited a less pronounced inhibition of the population spikes, compared with the maximum inhibition at 100 n*M* (Table 17.1). These findings suggest that some excitatory events occur in response to levodopa. It appears likely that the tendency of enhancement under blockade of α-adrenoceptors or $GABA_A$ receptors and the bell-shaped concentration–response relationship might be explained by an increase in the release of excitatory neurotransmitter glutamate by levodopa from the Schaffer collateral/commissural fibers originating from the hippocampal CA3 neurons. In fact, higher micromolar concentrations of levodopa are reported to release endogenous glutamate via activation of levodopa recognition sites in a dopamine-independent manner in the striatum [28,29].

In contrast to noradrenergic and GABAergic fibers, it seems unlikely that serotonergic neurotransmission was affected by levodopa, because levodopa-induced inhibition was completely blocked by either the α-adrenoceptor antagonist or $GABA_A$ receptor antagonist without residual inhibition (Figure 17.3). Although there are dense serotonergic innervation and many subtypes of 5-HT receptors including $5\text{-HT}_{1A}$, $5\text{-HT}_2$, $5\text{-HT}_3$, and $5\text{-HT}_4$ in the hippocampus, levodopa does not appear to act on the serotonergic nerve terminals originating from the raphe nucleus. However, the possibilities that levodopa releases 5-HT and that the net effects of 5-HT on the population spikes are nearly canceled due to opposite actions via activation of 5-HT receptor subtypes could not be completely excluded [10–12].

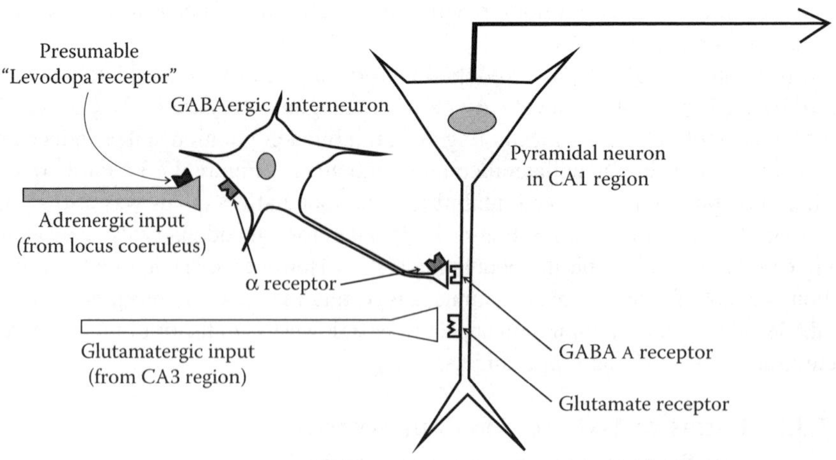


**FIGURE 17.4** Proposed sites of actions of levodopa, noradrenaline, GABA, and glutamate in hippocampal interneurons and CA1 pyramidal neurons.

## 17.4 CONCLUSION

Levodopa appears to act on levodopa-specific recognition sites located on noradrenergic nerve terminals from the locus coeruleus to release noradrenaline, which in turn acts on α-adrenoceptors located on the GABA-containing interneurons to release GABA, leading to inhibition of CA1 pyramidal neurons via $GABA_A$ receptors in the hippocampus (Figure 17.4).

## ACKNOWLEDGMENTS

This work was supported in part by a grant-in-aid for scientific research from the Ministry of Education, Culture, Sports, Science, and Technology, Japan, and grants from the Japanese Smoking Research Association and National Center of Neurology and Psychiatry (NCNP) of the Ministry of Health and Welfare, Japan.

## REFERENCES

1. Misu, Y., Goshima, Y., and Miyamae, T., Is DOPA a neurotransmitter?, *Trends Pharmacol. Sci.,* 23, 262, 2002.
2. Kubo, T. et al., Evidence for L-DOPA systems responsible for cardiovascular control in the nucleus tractus solitarii of the rat, *Neurosci. Lett.,* 140, 153, 1992.
3. Yue, J.-L. et al., Baroreceptor-aortic nerve-mediated release of endogenous L-3,4-dihydroxyphenylalanine and its tonic function in the nucleus tractus solitarii of rats, *Neuroscience,* 92, 145, 1994.
4. Yue, J.-L. et al., Altered tonic L-3,4-dihydroxyphenylalanine systems in the nucleus tractus solitarii and rostral ventrolateral medulla of spontaneously hypertensive rats, *Neuroscience,* 67, 95, 1995.
5. Misu, Y. et al., Neurobiology of L-DOPAergic systems, *Prog. Neurobiol.,* 49, 415, 1996.
6. Miyamae, T. et al., L-DOPAergic components in the caudal ventrolateral medulla in baroreflex neurotransmission, *Neuroscience,* 92, 137, 1999.
7. Goshima, Y., Kubo, T., and Misu, Y., Biphasic actions of L-DOPA on the release of endogenous noradrenaline and dopamine from rat hypothalamic slices, *Br. J. Pharmacol.,* 89, 229, 1986.
8. Goshima, Y., Nakamura, S., and Misu, Y., Picomolar concentrations of L-DOPA stereoselectively potentiate activities of presynaptic β-adrenoceptors to facilitate the release of endogenous noradrenaline from rat hypothalamic slices, *Neurosci. Lett.,* 129, 214, 1991.
9. Hirose, A. et al., Inhibition of hippocampal CA1 neurons by 5-hydroxytryptamine derived from the dorsal raphe nucleus and the 5-hydroxytryptamine 1A agonist, SM-3997, *Neuropharmacology,* 29, 93,1990.
10. Andrade, R. and Nicoll, R.A., Pharmacologically distinct actions of serotonin on single pyramidal neurons of the rat hippocampus recorded in vitro, *J. Physiol. (London),* 394, 99, 1987.
11. Ishihara, K. and Sasa, M., Potentiation of $5\text{-HT}_3$ receptor functions in the hippocampal CA1 region of rats following repeated electroconvulsive shock treatments, *Neurosci. Lett.,* 307, 37, 2001.
12. Ishihara, K. and Sasa, M., Failure of repeated electroconvulsive shock treatment on $5\text{-HT}_4$-receptor-mediated depolarization due to protein kinase A system in young rat hippocampal CA1 neurons, *J. Pharmacol. Sci.,* 95, 329, 2004.

13. Curet, O. and De Montigny, C., Electrophysiological characterization of adrenoceptors in the rat dorsal hippocampus: II. Receptors mediating the effect of synaptically released norepinephrine, *Brain Res.,* 475, 47, 1988.
14. Frankhuyzen, A.L. and Mulder, A.H., Pharmacological characterization of presynaptic $\alpha$-adrenoceptors modulating [$^3$H]noradrenaline and [$^3$H]5-hydroxytryptamine release from slices of the hippocampus of the rat, *Eur. J. Pharmacol.,* 81, 97, 1982.
15. Mueller, A.L. et al., Hippocampal noradrenergic responses in vivo and in vitro: characterization of $\alpha$ and $\beta$ components, *Naunyn-Schmiedberg's Arch. Pharmacol.,* 318, 259, 1982.
16. Sakai, N. et al., Effects of L-*threo*-DOPS, a noradrenaline precursor, on the long-term potentiation in the rat hippocampal mossy fiber-CA3 region, *Brain Res.,* 567, 267, 1991.
17. Akbar, M. et al., Inhibition by L-3,4-hydroxyphenylalanine of hippocampal CA1 neurons with facilitation of noradrenaline and $\gamma$-aminobutyric acid release, *Eur. J. Pharmacol.,* 414, 197, 2001.
18. Nakamura, S. et al., Transmitter-like basal and $K^+$-evoked release of 3,4-dihydroxyphenylalanine from the striatum in conscious rats studied by microdialysis, *J. Neurochem.,* 58, 270, 1992.
19. Goshima, Y., Nakamura, S., and Misu, Y., L-DOPA facilitates the release of endogenous norepinephrine and dopamine via presynaptic $\beta_1$- and $\beta_2$-adrenoceptors under essentially complete inhibition of L-aromatic amino acid decarboxylase in rat hypothalamic slices, *Jpn. J. Pharmacol.,* 53, 47, 1990.
20. Misu, Y. et al., Biphasic actions of L-DOPA on the release of endogenous dopamine via presynaptic receptors in rat striatal slices, *Neurosci. Lett.,* 72, 194, 1986.
21. Misu, Y. et al., L-DOPA cyclohexyl ester is a novel stable and potent competitive antagonist against L-DOPA, as compared to L-DOPA methyl ester, *Jpn. J. Pharmacol.,* 75, 307, 1997.
22. Furukawa, N. et al., L-DOPA cyclohexyl ester is a novel potent and relatively stable competitive antagonist against L-DOPA among L-DOPA ester compounds in rats, *Jpn J. Pharmacol.,* 82, 40, 2000.
23. Goshima, Y. et al., L-Dihydroxyphenylalanine methyl ester is a potent competitive antagonist of the L-dihydroxyphenylalanine-induced facilitation of the evoked release of endogenous norepinephrine from rat hypothalamic slices, *J. Pharmacol. Exp. Ther.,* 258, 466, 1991.
24. Ishii, H. et al., Involvement of rBAT in $Na^+$-dependent and -independent transport of the neurotransmitter candidate L-DOPA in *Xenopus laevis* oocytes injected with rabbit small intestinal epithelium poly $A^+$RNA, *Biochem. Biophys. Acta,* 1466, 61, 2000.
25. Honjo, K. et al., GABA may function tonically via $GABA_A$ receptors to inhibit hypotension and bradycardia by L-DOPA microinjected into depressor sites of the nucleus tractus solitarii in anesthetized rats, *Neurosci. Lett.,* 261, 93, 1999.
26. Miyamae, T. et al., Some interactions of L-DOPA and its related compounds with glutamate receptors, *Life Sci.,* 64, 1045, 1999.
27. Sato, K. et al., L-DOPA potentiates presynaptic inhibitory $\alpha_2$-adrenoceptor- but not facilitatory angiotensin II receptor-mediated modulation of noradrenaline release from rat hypothalamic slices, *Jpn. J. Pharmacol,* 62, 119, 1993.
28. Goshima, Y. et al., L-DOPA induces $Ca^{2+}$-dependent and tetrodotoxin-sensitive release of endogenous glutamate from rat striatal slices, *Brain. Res.,* 617, 167, 1993.
29. Maeda, I. et al., L-DOPA neurotoxicity is mediated by glutamate release in cultured rat striatal neurons, *Brain Res.,* 771, 159, 1997.

# Part V

## Neurotoxicology

# 18 Levodopa-Induced Vesicular Release of Glutamate and Mechanisms of Levodopa-Induced Neurotoxicity in Primary Neuron Cultures in Rat Striata

*Akinori Akaike and Takehiko Maeda*

## CONTENTS



## 18.1 INTRODUCTION

The nigrostriatal dopamine system plays an important role in the regulation of striatal function [1,2], and a number of studies have been performed to elucidate the physiological role of dopamine in the striatum. L-3,4-Dihydroxyphenylalanine (levodopa), a precursor substance of dopamine in catecholaminergic neurons, is the most effective therapeutic agent for Parkinson's disease. Contrary to such a generally accepted idea, it has been proposed that endogenous levodopa is a neurotransmitter or neuromodulator [3–6]. For example, exogenously applied levodopa at nano- or picomolar concentrations stereoselectively potentiates activities of presynaptic β-adrenoceptors to facilitate the impulse-evoked release of noradrenaline and dopamine in the absence and even in the presence of an L-aromatic amino acid decarboxylase (AADC) inhibitor in rat brain slices [7–9].

Though catecholamines function as neurotransmitters or neuromodulators, it has been proposed that dopamine and related substances have glutamatergic properties, which may cause neural excitation and excitotoxicity. It has been suggested that the dopaminergic system plays a role in the neuronal injury and death accompanying cerebral ischemia [10,11], a process mediated at least in part by glutamate receptor activation. In another line of investigation, it has been shown that a methamphetamine-induced loss of dopaminergic fiber in the striatum was blocked by an *N*-methyl-D-aspartate (NMDA) receptor antagonist, (+)-5-methyl-10,11-dihydro-5*H*-dibenzol[a,d]cyclohepten-5,10-imine maleate (MK-801) [12]. Moreover, 1-methyl-4-phenyl-1,2,3,6-tetrahydropyridine (MPTP)-induced injury of dopaminergic neurons in the substantia nigra cytotoxicity was blocked by several NMDA antagonists [13]. These results implied a link between dopaminergic systems and glutamate toxicity, and led us to consider the possibility that levodopa, dopamine, their metabolites, or substances derived from them by normal or abnormal degradation pathways might have glutamatergic properties.

Therefore, we investigated the effects of levodopa and related compounds on glutamate release and neuron survival by using slices and cultured neurons derived from rat striatum.

## 18.2 LEVODOPA-INDUCED VESICULAR RELEASE OF GLUTAMATE

The fact that levodopa facilitates the release of dopamine and noradrenaline led us to examine whether or not levodopa affects the release of glutamate in the striatum. *In vivo* perfusion studies have revealed that the stimulation of dopamine receptors in the rat striatum with push–pull cannula [14] or in the rat substantia nigra with microdialysis [15] increases the release of glutamate. We further investigated whether or not levodopa itself induces release of endogenous glutamate from rat striatal slices [16]. We found that micromolar concentrations of levodopa released endogenous glutamate from superfused rat striatal slices. The levodopa-induced glutamate release was neurotransmitter-like because this was in part $Ca^{2+}$ dependent and tetrodotoxin sensitive. Because the effect was neither mimicked by dopamine nor affected by AADC inhibition, we concluded that the action of levodopa is due to levodopa itself. Furthermore, the effect of levodopa was antagonized in a competitive fashion by L-DOPA methyl ester, the compound that we have evaluated as a potent competitive antagonist for the facilitatory effect of levodopa on the impulse-evoked release of noradrenaline from superfused rat hypothalamic slices. It has been reported that the site of action of L-DOPA methyl ester appears to be different from the carrier proteins for its transport system and from adrenergic and dopaminergic receptors [17]. The specificity of L-DOPA methyl ester was also confirmed on a cardiovascular response in anesthetized rat. L-DOPA methyl ester showed no antagonism against the cardiovascular response induced by glutamate microinjected into the nucleus tractus solitarii, whereas the same dose of this ester compound completely antagonized that by levodopa [18]. Therefore, it is likely that levodopa releases endogenous glutamate from *in vitro* striatal slices via a recognition site for levodopa itself. This site was stereoselective in nature in common with many receptors, as dextrodopa produced no effect.

$Ca^{2+}$ deprivation from the medium decreased the levodopa-induced release by 70%, which is analogous to kainite-induced release of glutamate from rat striatal slices. This suggests that glutamate is released by levodopa via a $Ca^{2+}$-dependent excitation–secretion coupling process similar to that involved in neurotransmitter release. Because levodopa-induced glutamate release showed tetrodotoxin sensitivity, it is plausible that levodopa in part depolarizes neuronal cell soma and/or dendrites that in turn lead to conduction of nerve impulses, terminal depolarization, and release of glutamate. However, tetrodotoxin at the same concentration (0.3 μ*M*), which almost completely blocks the impulse-evoked release of glutamate from slices of rat medulla oblongata, decreased the release by only ~40%. This suggests that levodopa could also induce the release via a tetrodotoxin-insensitive process.

The effective concentration range of levodopa to evoke the release of glutamate was comparable to that of kainite to evoke the release of glutamate from rat striatal slices [19]. The absolute amount of glutamate released by levodopa (0.3 m*M*), in terms of percentage of increase over the spontaneous release, was approximately 30%, which was almost comparable to that obtained with kainite (0.5 m*M*) in rat striatal slices. It could be possible that the release of glutamate caused by exogenously applied levodopa affects the progress of Parkinson's disease during the chronic therapy because endogenous glutamate has been implicated in the pathogenesis of neuronal cell death [20]. The therapeutic plasma concentrations of levodopa are estimated to be 1 to 10 μ*M* in parkinsonian patients [21]. This concentration is about one tenth of the $ED_{50}$ value (140 μ*M*) of levodopa obtained in this study. It is, therefore, unlikely that levodopa at the therapeutic doses could produce neuroexcitatory and/or neurotoxic actions on the striatal neurons. However, this might occur if the tissue levels of AADC activity are extremely low. In fact, the mean activities of AADC in the caudate putamen of patients with Parkinson's disease are only 5 to 15% of those of control patients [22]. This warrants further study to investigate whether or not the neuroexcitatory action of levodopa could have some adverse influence on the neurodegenerative process in the long-term therapy in Parkinson's disease.

## 18.3 GLUTAMATE NEUROTOXICITY IN CULTURED STRIATAL NEURONS

There are a number of studies indicating the crucial role of glutamate and related excitatory amino acids in the neurodegeneration caused by brain ischemia [23]. Neurodegeneration is reportedly induced following injection of excitatory amino acids in the striatum *in vivo* [24]. In some forebrain regions including the cerebral cortex and hippocampus, NMDA receptor is considered to be the predominant route of glutamate neurotoxicity [25,26]. On the other hand, studies using primary cultures have demonstrated that both the non-NMDA and NMDA receptors mediate cytotoxic effects of glutamate in the striatum [27,28]. However, there was still insufficient information concerning the receptor subtype that mediates glutamate neurotoxicity in the subcortical regions including the striatum when we began the research on glutamate neurotoxicity in the striatum.

Thus, we performed an *in vitro* study to elucidate the receptor subtypes of glutamate receptors mediating excitotoxicity in the striatum [29]. Primary cultures were prepared from fetal rats on the 16- to 18-d gestation and were maintained for 10 to 14 d *in vitro*. Viability of cultured cells was assessed by Trypan blue exclusion. A 24-h exposure to either glutamate (1 m*M*) or kainate (1 m*M*) significantly reduced cell viability. Glutamate induced neurotoxicity in a time-dependent manner. Statistically significant reduction of cell viability was observed with glutamate treatment for more than 4 h. Thus, the striatal cultures were relatively resistant to a brief glutamate exposure in contrast to the cortical cultures, which were vulnerable to the brief (10 min) glutamate exposure when cultures were incubated in glutamate-free media for more than 1 h following glutamate treatment. These findings are in line with those that striatal cultures are less susceptible to glutamate neurotoxicity than cortical neurons, although Gallarrge et al. [28] have demonstrated that coculturing with neocortical neurons potentiated the glutamate toxicity in striatal cultures.

To determine the glutamate receptor subtypes that mediate glutamate cytotoxicity in the striatum, the cytotoxic effects of selective agonists of ionotropic glutamate receptors were examined (Figure 18.1). The cultures were exposed to 1 m*M* of excitatory amino acid (EAA) for 24 h, and the viability was compared with cultures incubated with EAA-free medium. Glutamate and kainate significantly reduced cell viability. On the other hand, α-amino-3-hydroxy-5-methyl-4-isoxazole-propionate (AMPA) did not affect cell viability. NMDA significantly reduced cell viability when added to a normal medium that contains $Mg^{2+}$ ($+Mg^{2+}$ in Figure 18.1).

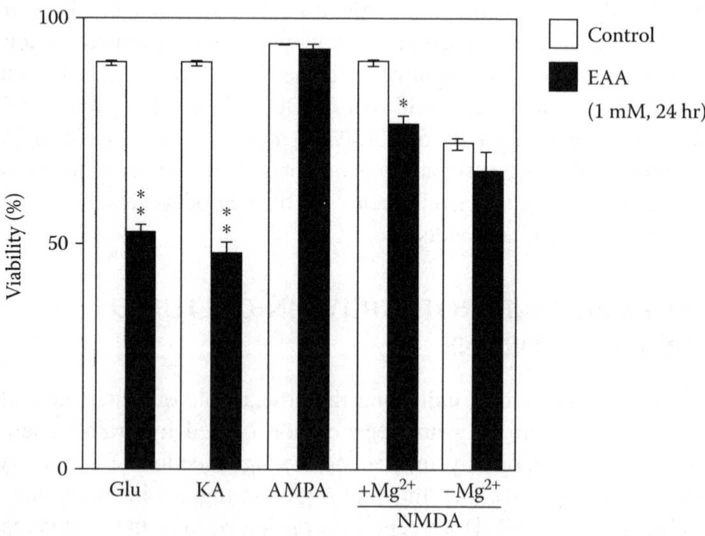


**FIGURE 18.1** Effects of excitatory amino acids (EAAs) on viability of cultured striatal cells. Glu, glutamate; KA, kainate; and AMPA, α-amino-3-hydroxy-5-methyl-4-isoxazole-propionate. Control shows viability of untreated cultures. $^{*}P < .05$ and $^{**}P < .01$, compared with the respective control. (From Amano, T. et al., Dopamine-induced protection of striatal neurons against kainate-receptor-mediated glutamate cytotoxicity in vitro, *Brain Res.*, 655, 61, 1994. With permission.)

The NMDA-induced reduction of viability was much smaller than the reduction induced by glutamate or kainate. The viability was significantly lower in the cultures maintained in $Mg^{2+}$-free medium for 24 h than in those maintained in $Mg^{2+}$-containing solution. However, there were no significant alterations of viability of NMDA-treated cultures in $Mg^{2+}$-free medium, compared with that of untreated cultures in the $Mg^{2+}$-free medium. Moreover, glutamate neurotoxicity was antagonized by kynurenate but not by MK-801. Kynurenate is an antagonist of non-NMDA receptors such as kainate and AMPA receptors although this drug possesses an antagonistic effect on strychnine-insensitive glycine modulatory sites of NMDA receptors. These findings indicate that glutamate neurotoxicity in the striatal cultures is mediated by kainate receptors among the glutamate receptors.

## 18.4 NEUROTOXICITY INDUCED BY LEVODOPA AND DOPAMINE

Dopamine and its related substances have been implicated as the cause of neuronal injury. The mechanism underlying neuronal toxicity has been thought to be the oxidative damage by degradation products of catecholamines, including quinone derivatives [30] and oxygen-free radicals [31]. In addition, it has been proposed that dopamine and related substances have glutamatergic properties [11,14,15]. Therefore, we examined the effects of levodopa and dopamine on the survival of cultured striatal neurons [32]. The exposure of the striatal cultures derived from fetal rats in 10 d in culture (10 DIC) to either levodopa or dopamine elicited marked reduction of cell viability as revealed by Trypan blue exclusion. As shown in Figure 18.2, exposing cultures to levodopa or dopamine (30 to 300 $\mu M$) for 6 to 24 h reduced the viability of the striatal neurons in a concentration- and time-dependent manner.

One of the proposed mechanisms of the neurotoxic actions of levodopa and dopamine is their autoxidation into reactive free radicals and quinones [33,34]. The generated radicals have potent toxicity to the respiratory chain in the mitochondria; the radicals cause oxidation of mitochondrial pyridine nucleotides and thereby stimulate $Ca^{2+}$ release from intact mitochondria [35]. Thus, an energy crisis evoked by mitochondrial dysfunction may result in cell death. However, in the present study, we observed that the susceptibility of striatal neurons to levodopa and dopamine toxicity varied with an increase in the number of days in culture. The younger cells (3 DIC) were more vulnerable to their toxicity than their elder counterparts (10 DIC). Cultures were exposed to 100 $\mu M$ of either levodopa or dopamine for 24 h. The younger (3 DIC) cells were more vulnerable to the cytotoxic actions of levodopa and dopamine than were the elder (10 DIC) cells (Figure 18.3). It was interesting that the neurotoxic profile of kainate was markedly distinct from that of levodopa and dopamine: a 24-h exposure to 1-m$M$ kainate evoked a marked decrease in the viability of the elder cultures, whereas no cytotoxicity was observed in the younger cultures. These findings indicate that levodopa and dopamine are cytotoxic to cultured striatal neurons via different mechanisms.

Therefore, we examined the effects of ascorbic acid on levodopa and dopamine toxicity, because preceding papers employed ascorbic acid to prevent the autoxidation

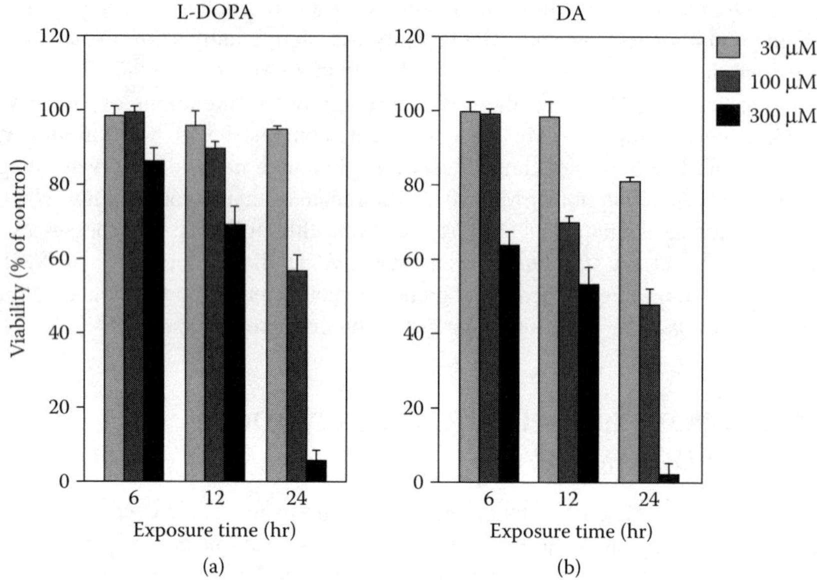


**FIGURE 18.2** Time- and concentration-dependent cytotoxicity elicited by levodopa and dopamine (DA) on cultured striatal neurons (10 d in culture, 10 DIC). Ordinate shows the ratio (%) of the viability to that of sham treatment cultures. (From Cheng, N.-N. et al., Differential neurotoxicity induced by L-DOPA and dopamine in cultured striatal neurons, *Brain Res.*, 743, 278, 1996. With permission.)

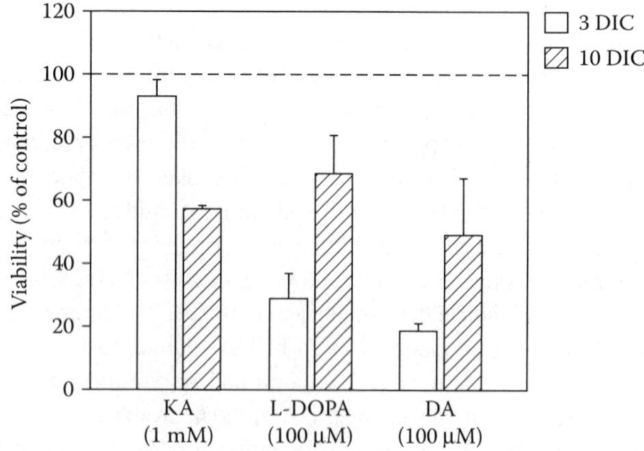


**FIGURE 18.3** Maturation-dependent cytotoxicity of kainate (KA), levodopa, and dopamine (DA) on cultured striatal neurons. Both cultures, 3 d in culture (3 DIC) and 10 d in culture (10 DIC), were exposed to neurotoxins for 24 h. Ordinate shows the ratio (%) of the viability to that of sham treatment cultures. (From Cheng, N.-N. et al., Differential neurotoxicity induced by L-DOPA and dopamine in cultured striatal neurons, *Brain Res.*, 743, 278, 1996. With permission.)

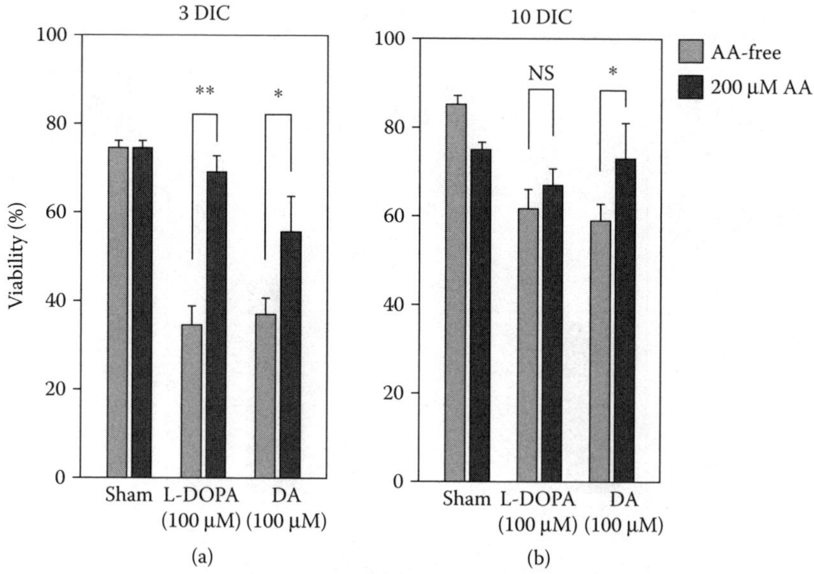


**FIGURE 18.4** Effects of ascorbic acid (AA) on levodopa and dopamine (DA) cytotoxicity in cultured striatal neurons. AA was applied simultaneously with levodopa or DA. $^{*}P < .05$ and $^{**}P < .01$. NS, no significance. (From Cheng, N.-N. et al., Differential neurotoxicity induced by L-DOPA and dopamine in cultured striatal neurons, *Brain Res.*, 743, 278, 1996. With permission.)

of levodopa (Figure 18.4). The younger cultures were also more protected by ascorbic acid against the toxicity, implying that the autoxidation products of levodopa and dopamine contribute more to their toxicity on younger cultures than to that on elder ones. This is further supported by the fact that dextrodopa, which is an inactive enantiomer of levodopa, lacking antiparkinsonism activity but undergoing autoxidation during incubation, also produced the cytotoxicity in the younger cells but not in the elder cultures (Figure 18.5). These observations may reflect a deficiency in the younger neurons of the defensive mechanism against the neurotoxicity of free radicals and quinones generated from levodopa and dopamine autoxidation. Maturation dependency was particularly prominent in kainate neurotoxicity. Kainate was neurotoxic to the elder cultures but inert to the younger ones. This is in agreement with the result of a Northern analysis of kainate/AMPA receptor subunit mRNAs of the rat striatum; the quantity of mRNAs was augmented with the development of the rat during postnatal days 1 to 14 [36]. The maturation-dependent toxicity of kainate is possibly based on the difference between the number of kainate receptors expressed in younger cultures and in elder ones.

To elucidate whether levodopa and dopamine neurotoxicity involves glutamatergic properties, either 6-cyano-7-nitroquinoxaline-2,3-dione (CNQX), a non-NMDA receptor antagonist, or MK-801, an NMDA receptor antagonist, was administered simultaneously with levodopa or dopamine in 10-DIC cultures (Figure 18.6). Levodopa neurotoxicity was exclusively inhibited by a simultaneous application of CNQX (10 μ*M*)

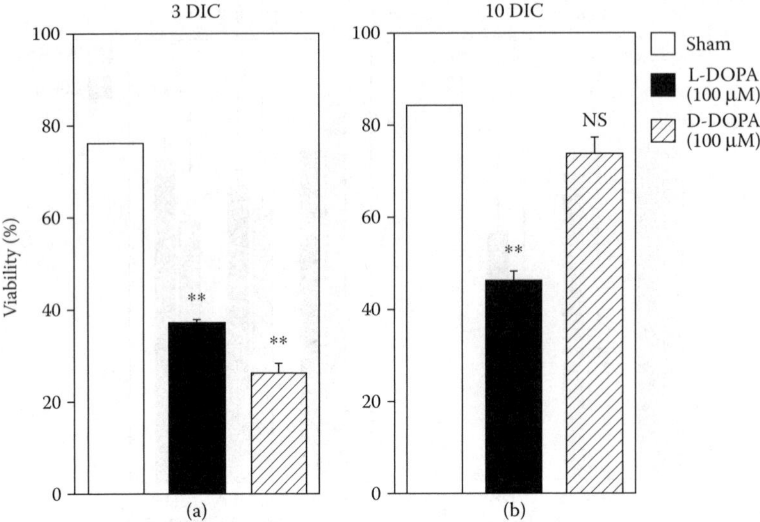


**FIGURE 18.5** Cytotoxicity induced by levodopa and dextrodopa on cultured striatal neurons. Both cultures, 3 d in culture (3 DIC) and 10 d in culture (10 DIC), were exposed to neurotoxins for 24 h. Ordinate shows the viability of cultures. $^{**}P < .01$ vs. sham. NS, no significance. (From Cheng, N.-N. et al., Differential neurotoxicity induced by L-DOPA and dopamine in cultured striatal neurons, *Brain Res.*, 743, 278, 1996. With permission.)

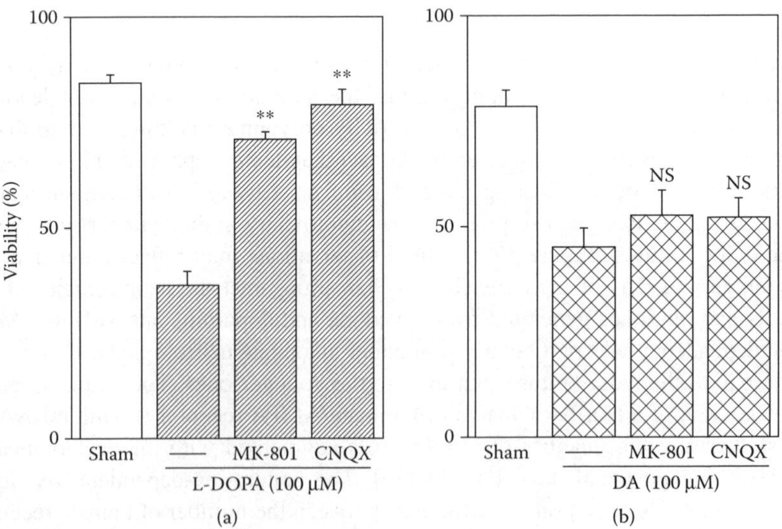


**FIGURE 18.6** Effects of glutamate receptor antagonists on levodopa and dopamine (DA) cytotoxicity in cultured striatal neurons. CNQX (10 μ*M*) and MK-801 (10 μ*M*) were applied simultaneously with levodopa or DA. $^{**}P < .01$ vs. sham. NS, no significance when compared with DA alone. (From Cheng, N.-N. et al., Differential neurotoxicity induced by L-DOPA and dopamine in cultured striatal neurons, *Brain Res.*, 743, 278, 1996. With permission.)

or MK-801 (10 μ*M*). In contrast, these drugs did not affect dopamine neurotoxicity. This suggests that levodopa-induced neuronal death is mediated by endogenous glutamate released from synaptic vesicles upon stimulation by levodopa. The proposed mechanism of levodopa cytotoxicity is as follows: exposure of the neurons to levodopa stimulates glutamate release from the neurons. The released glutamate acts on non-NMDA receptors to depolarize the neurons. The depolarized neurons facilitate the relief of NMDA receptor channel function from $Mg^{2+}$ block. The glutamate acts on NMDA receptors to induce $Ca^{2+}$ influx to the neurons. The $Ca^{2+}$, forming the complex with calmodulin, activates the neuronal type of nitric oxide synthase (nNOS). The nNOS, with increased activity, produces NO with a detrimental property to the cell and mitochondrial membrane, leading to neuronal death.

## 18.5 ROLE OF GLUTAMATE RELEASE IN LEVODOPA-INDUCED NEUROTOXICITY

As mentioned in the preceding text, levodopa-induced cytotoxicity was blocked by the simultaneous application of antagonists to both NMDA and non-NMDA receptors, indicating that endogenous glutamate is involved in levodopa neurotoxicity. In the course of studies to analyze the mechanism underlying levodopa neurotoxicity, we could demonstrate that facilitated glutamate release in the presence of levodopa plays a crucial role in levodopa neurotoxicity [37]. Glutamate concentration in the culture medium was determined by high-performance liquid chromatography (HPLC) with fluorometric detection. When the striatal cultures (10 DIC) were incubated with levodopa-containing media for 6 h, the glutamate content of the incubation media increased with increasing levodopa concentrations (Figure 18.7). The exposure for 6 h to 100-μ*M* levodopa elicited a significant increase in the glutamate content compared with that of media without levodopa. Then, we examined whether levodopa neurotoxicity was affected by a blockade of the neurotransmitter release from striatal cultures. When the cultures were incubated with 100-μ*M* levodopa in $Ca^{2+}$-free culture medium, the viability was significantly greater than in cultures treated with levodopa alone (Figure 18.8). The incubation with levodopa in culture medium containing 10-m*M* $Mg^{2+}$ also prevented levodopa cytotoxicity. The concomitant application of 0.3-μ*M* tetrodotoxin and levodopa prevented levodopa cytotoxicity. The presence of levodopa in the cultures induced an increase in the glutamate content of their incubation media.

These results suggest that levodopa facilitates glutamate release from the cultures in a neurotransmitter-releasing manner and that the glutamate elicits cell death via non-NMDA and NMDA receptors. The mean concentration of glutamate in the cultured media after 6-h levodopa exposure was approximately 1 μ*M*. Although this value was less than the concentration of glutamate exogenously applied to produce cytotoxicity [32], the difference can be explained by the reuptake of glutamate by cultured cells and the dilution of released glutamate by the medium. If the glutamate concentration at the synaptic region is much higher than that in the medium, it is possible that the glutamate released by levodopa application causes the overstimulation of NMDA and/or non-NMDA receptors located at the postsynaptic membrane.

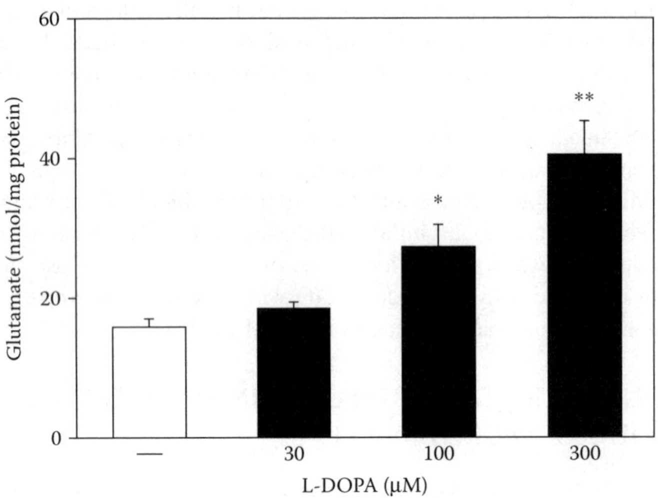


**FIGURE 18.7** Effect of levodopa on glutamate release from striatal cultures. Cultures were incubated with levodopa-containing medium for 6 h, and then the incubation medium and cultures were subjected to HPLC analysis and measurement for protein content, respectively. $^{*}P < .05$ and $^{**}P < .01$ vs. sham (open column). (From Maeda, T. et al., L-DOPA neurotoxicity is mediated by glutamate release in cultured rat striatal neurons, *Brain Res.*, 771, 159, 1997. With permission.)

## 18.6 CONCLUSION

Levodopa may have a role as a neurodegenerator under pathologic conditions in addition to its physiological role. Levodopa appears to be an upstream factor for glutamate release and the resultant delayed neuron death in the striatum. A common series of cascades might be the activation of levodopa recognition sites, glutamate release, activation of ionotropic glutamate receptors, nNOS activation, NO production, and, then, delayed neuron death. Moreover, the interactions between the released glutamate and compounds produced as a result of levodopa autoxidation might start amplifying cycles of neurotoxic cascades.

Though levodopa is well established as a medicine for symptomatic treatment of patients with Parkinson's disease, its influence on the progression of the disease is debated [38,39]. The toxic concentration of levodopa used in these studies was about ten times as high as the therapeutic plasma concentrations in Parkinson's disease patients [21]. However, neurotoxicity cannot be ruled out in patients who receive long-term levodopa therapy, because they are chronically exposed to levodopa and may have particular susceptibility to neurotoxic agents. Thus, there is concern that long-term levodopa therapy has the potential to accelerate the progression of Parkinson's disease [40], and it favors the notion that increasing the dose in an attempt to compensate the inevitable decline in the efficacy of levodopa therapy should be limited. Olney et al. [41] have proposed that the cytotoxicity of levodopa and its orthohydroxylated derivative might be implicated in Parkinson's disease and Huntington's disease. If these assumptions are correct,

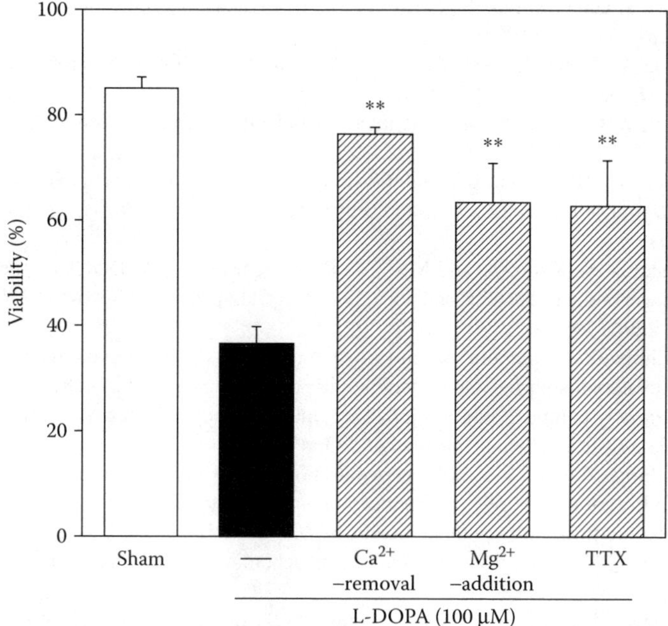


**FIGURE 18.8** Involvement of neurotransmitter release in levodopa-induced cell death. The cultures were incubated for 24 h in the absence (sham) or presence of 100-$\mu M$ levodopa. $Ca^{2+}$ removal: $Ca^{2+}$ was removed from levodopa-containing medium. $Mg^{2+}$ addition: 10-m$M$ $MgSO_4$ was added to levodopa-containing medium. TTX: 0.3-$\mu M$ tetrodotoxin was added to levodopa-containing medium. Ordinate shows the viability of cultures. $^{**}P < .01$ vs. L-DOPA alone (filled column). (From Maeda, T. et al., L-DOPA neurotoxicity is mediated by glutamate release in cultured rat striatal neurons, *Brain Res.*, 771, 159, 1997. With permission.)

the duration and effectiveness of levodopa therapy might be reinforced by the combination with agents that protect neurons against neurotoxicity induced by glutamate and/or radicals [42–45].

## ACKNOWLEDGMENT

This study was supported in part by grants-in-aid from the Ministry of Education, Science, Sports, and Culture, Japan.

## REFERENCES

1. Moore, R.E. and Bloom, F.E., Central catecholamine neuron system: anatomy and physiology of the dopamine system, *Annu. Rev. Neurosci.*, 1, 129, 1978.
2. Roberston, G.S. and Robertson, H.A., $D_1$ and $D_2$ dopamine agonist synergism: separate site of action?, *Trends Pharmacol. Sci.*, 8, 295, 1987.

3. Misu, Y. and Goshima, Y., Is L-dopa an endogenous neurotransmitter?, *Trends Pharmacol. Sci.,* 14, 119, 1993.
4. Misu, Y., Ueda, H., and Goshima, Y., Neurotransmitter-like actions of L-DOPA, *Adv. Pharmacol.,* 32, 427, 1995.
5. Misu, Y., Goshima, Y., and Miyamae, T., Is DOPA a neurotransmitter?, *Trends Pharmacol. Sci.,* 23, 262, 2002.
6. Misu, Y., Kitahama, K., and Goshima, Y., L-3,4-Dihydroxyphenylalanine as a neurotransmitter candidate in the central nervous system, *Pharmacol. Ther.,* 97, 117, 2003.
7. Goshima, Y., Kubo, T., and Misu, Y., Biphasic actions of L-DOPA on the release of endogenous noradrenaline and dopamine from rat hypothalamic slices, *Br. J. Pharmacol.,* 89, 229, 1986.
8. Goshima, Y., Nakamura, S., and Misu, Y., L-DOPA facilitates the release of endogenous norepinephrine and dopamine via presynaptic β1- and β2-adrenoceptors under essentially complete inhibition of L-aromatic amino acid decarboxylase in rat hypothalamic slices, *Jpn. J. Pharmacol.,* 53, 47, 1990.
9. Goshima, Y. et al., Picomolar concentrations of L-DOPA stereoselectively potentiate activities of presynaptic β-adrenoceptors to facilitate the release of endogenous noradrenaline from rat hypothalamic slices, *Neurosci. Lett.,* 129, 214, 1991.
10. Globus, M.Y. et al., Role of dopamine in ischemic striatal injury: metabolic evidence, *Neurology,* 37, 1712, 1987.
11. Globus, M.Y. et al., Effect of ischemia on the in vivo release of striatal dopamine, glutamate, and γ-aminobutyric acid studied by intracerebral microdialysis, *J. Neurochem.,* 51, 1455, 1988.
12. Sonslla, P.K., Nicklas, W.J., and Heikkila, R.E., Role for excitatory amino acids in methamphetamine-induced nigrostriatal dompaminergic toxicity, *Science,* 243, 398, 1989.
13. Turski, L. et al., Protection of substantia nigra from $MPP^+$ neurotoxicity by *N*-methyl-D-aspartate antagonist, *Nature,* 349, 414, 1991.
14. Porras, A. and Mora, F., Dopamine-glutamate-GABA interaction and aging: studies in the striatum of the conscious rat, *Eur. J. Neurosci.,* 7, 2183, 1995.
15. Abarca, J. et al., Changes in extracellular level of glutamate and aspartate in rat substantia nigara induced by dopamine receptor ligands: in vivo microdialysis studies, *Neurochem. Res.,* 20, 159, 1995.
16. Goshima, Y. et al., L-DOPA induces $Ca^{2+}$-dependent and tetrodotoxin-sensitive release of endogenous glutamate from rat striatal slices, *Brain Res.,* 617, 167, 1993.
17. Goshima, Y., Nakamura, S., and Misu, Y., L-Dihydroxyphenylalanine methyl ester is a potent competitive antagonist of the L-dihydroxyphenylalamine-induced facilitation of the evoked release of endogenous norepinephrine from rat hypothalamic slices, *J. Pharmacol. Exp. Ther.,* 258, 466, 1991.
18. Kubo, T. et al., Evidence of L-DOPA systems responsible for cardiovascular control in the nucleus tractus solitarii of the rat, *Neurosci. Lett.,* 140, 153, 1992.
19. Ferkany, J.W. and Coyle, J.T., Kainic acid selectively stimulates release of endogenous excitatory amino acids, *J. Pharmacol. Exp. Ther.,* 225, 399, 1983.
20. Olney, J.W., Excitotoxic amino acids and neuropsychiatric disorders, *Annu. Rev. Pharmacol. Toxicol.,* 30, 47, 1990.
21. Rossor, M.N. et al., Plasma levodopa, dopamine and therapeutic response following levodopa therapy of parkinsonian patients, *Brain Res.,* 46, 385, 1980.
22. Lloyd, K.G., Davidson, L., and Hornykiewicz, O., The neurochemistry of Parkinson's disease: effect of L-DOPA therapy, *J. Pharmacol. Exp. Ther.,* 195, 453, 1975.

23. Choi, D.W., Glutamate neurotoxicity and diseases of the nervous system, *Neuron,* 1, 623, 1988.
24. Ferrer, I. et al., Both apoptosis and necrosis occur following intrastriatal administration of excitotoxins, *Acta Neuropathol. (Berl.),* 90, 504, 1995.
25. Meldrum, B. and Garthwaite, J., Excitatory amino acid neurotoxicity and neurodegenerative disease, *Trends Pharmacol. Sci.,* 11, 379, 1990.
26. Akaike, A. et al., Regulation by neuroprotective factors of NMDA receptor mediated nitric oxide synthesis in the brain and retina, *Prog. Brain Res.,* 103, 391, 1994.
27. Freese, A. et al., Characterization and mechanism of glutamate neurotoxicity in primary cultures, *Brain Res.,* 521 254, 1990.
28. Galarrage, E., Surmeier, D.J., and Kitai, S.T., Quinolinate and kainate neurotoxicity in neostriatal cultures is potentiated by co-culturing with neocortical neurons, *Brain Res.,* 512, 269, 1990.
29. Amano, T. et al., Dopamine-induced protection of striatal neurons against kainate receptor-mediated glutamate cytotoxicity in vitro, *Brain Res.,* 655, 61, 1994.
30. Graham, D.G., Oxidative pathways for catecholamines in the genesis of neuromelanin and cytotoxic quinones, *Mol. Pharmacol.,* 14, 633, 1978.
31. Smith, T.S., Parker, W.D., and Bennett, J.P., L-DOPA increases nigral production of hydroxyl radicals in vivo: potential role of L-DOPA toxicity?, *NeruroReport,* 5, 1009. 1994.
32. Cheng, N.-N. et al., Differential neurotoxicity induced by L-DOPA and dopamine in cultured striatal neurons, *Brain Res.,* 743, 278, 1996.
33. Basma, A.N. et al., DOPA cytotoxicity in PC12 cells in culture is via its autoxidation, *J. Neurochem.,* 64, 718, 1995.
34. Ben-Sachar, D., Zuk, R., and Glinka, Y., Dopamine neurotoxicity: inhibition of mitochondrial respiration, *J. Neurochem.,* 64, 718, 1995.
35. Richter, C. et al., Oxidants in mitochondria: from physiology to disease, *Biochem. Biophys. Acta,* 1271, 67, 1995.
36. Guylaine, M.D. and Zulkin, R.S., Developmental regulation of mRNAs encoding rat brain kainate/AMPA receptors: a northern analysis study, *J. Neurochem.,* 61, 2239, 1993.
37. Maeda, T. et al., L-DOPA neurotoxicity is mediated by glutamate release in cultured rat striatal neurons, *Brain Res.,* 771, 159, 1997.
38. Bilin, J., Bonnet, A.M., and Agid, Y., Does levodopa aggravate Parkinson's disease?, *Neurology,* 38, 1410, 1988.
39. Fahn, S. and Bressman, S.B., Should levodopa therapy for parkinsonism be started early or late? Evidence against early treatment, *Can. J. Neurol. Sci.,* 11, 200, 1984.
40. Forenstedt, B., Role of catechol autoxidation in the degeneration of dopamine neurons, *Acta Neurol. Scand. Suppl.,* 129, 12, 1990.
41. Olney, J.W. et al., Excitotoxicity of DOPA and 6-OH-DOPA: implications for Parkinson's and Huntington's diseases, *Exp. Neurol.,* 108, 269, 1990.
42. Sawada, H. et al., Dopamine $D_2$ type agonists protect mesencephalic neurons from glutamate neurotoxicity: mechanism of neuroprotective treatment against oxidative stress, *Ann. Neurol.,* 44, 110, 1998.
43. Kume, T. et al., Isolation of a diterpenoid substance with potent neuroprotective activity from fetal calf serum, *Proc. Natl. Acad. Sci. USA,* 99, 3288, 2002.
44. Sawada, H. et al., Estradiol protects dopaminergic neurons in a $MPP^+$ Parkinson's disease model, *Neuropharmacology,* 42, 1056, 2002.
45. Osakada, F. et al., Neuroprotective effects of α-tocopherol on oxidative stress in rat striatal cultures, *Eur. J. Pharmacol.,* 465, 15, 2003.

# 19 Neuroprotective and Neurotoxic Effects of L-DOPA in Rat Midbrain Dopaminergic Neurons in Culture

*María Angeles Mena, María José Casarejos, Eulalia Rodríguez-Martín, Rosa María Solano, Jaime Menéndez, and Justo García de Yébenes*

**CONTENTS**

## 19.1 INTRODUCTION

L-3,4-Dihydroxyphenylalanine (L-DOPA) is toxic for dopamine (DA) neurons in culture [1–6] but its toxicity has not been proven in animals [7] or in patients with Parkinson's disease (PD) [8]. In rats with moderate nigrostriatal lesions, chronic L-DOPA is not toxic for the surviving DA neurons but it does, instead, promote their recovery [9]. Because most experiments *in vitro* showing L-DOPA toxicity were performed in neurons cultured in the absence of glia, we hypothesized that the discrepancy between the effects of L-DOPA *in vivo* and *in vitro* may be related to the presence or absence, respectively, of glia [10]. Fetal midbrain neuronal cultures were treated with L-DOPA, 200 $\mu M$, in the presence or absence of mesencephalic glial conditioned medium (GCM). In the absence of GCM, L-DOPA greatly reduced the number of tyrosine hydroxylase (TH)-immunoreactive neurons and increased the levels of quinones in the medium; GCM prevented these effects of L-DOPA and increased the length and arborization of neurites of the TH-immunoreactive cells [10–12]. Furthermore, L-DOPA has neurotrophic effects in postnatal midbrain DA neuron–cortical astrocyte coculture by upregulation of glutathione (GSH) [13,14]; L-DOPA increases nerve growth factor (NGF) effects and promotes NGF-dependent outgrowth and synaptic vesicle quantal release in PC12 cells [15]. L-DOPA and GCM have synergistic effects on TH protein expression in human catecholamine-rich neuroblastoma cells [16] and GCM activates extracellular signal-regulated protein kinase (ERK/MAPK) pathway [17,18].

## 19.2 MECHANISMS OF L-DOPA NEUROTOXICITY ON CATECHOLAMINE-RICH NEURONS AND FETAL DOPAMINE NEURONS *IN VITRO*

L-DOPA, at concentrations of 50 $\mu M$ or higher, is toxic for catecholamine-rich human neuroblastoma cells in cultures grown with serum. The toxicity is dose and time dependent and is associated with a decrease in the number of total and viable cells, reduction of protein and DNA levels and $^3$H-thymidine uptake, and elevation of quinones [1].

L-DOPA toxicity is shared by other catechols but not by other large amino acids, such as leucine and tryptophan, in the mentioned range of concentrations. L-DOPA toxicity to DA neurons may depend on several factors, including dose, length of exposure, penetration across the blood–brain barrier, pharmacological and metabolic activity of DA neurons, age of the culture, culture medium, and presence of glial cells.

Several mechanisms play a role in L-DOPA-induced neurotoxicity on human neuroblastoma cells, NB69 [1], and fetal DA neurons [2]. Enhanced production of quinone derivatives and of free radicals related to the metabolism of DA via monoamine oxidase (MAO) type B [3–5] are well documented; inhibition of mitochondrial activity, production of complex tetrahydroquinolines and papaverolines, DNA damage by nucleophilic derivatives, and other mechanisms are postulated.

## 19.3 EFFECTS OF L-DOPA ON ELECTRON CHAIN TRANSPORT IN CATECHOLAMINE-RICH NEURONS

Several studies have investigated the effects of L-DOPA on the activities of enzyme complexes in the electron transport chain (ETC). In homogenate preparations from the human neuroblastoma cell line NB69 [6], complex I activity was dose-dependently inhibited by 1-methyl-4-phenylpyridinium ion with an $EC_{50}$ around 150 μ*M*. L-DOPA, 250 μ*M*, reduced complex IV activity to 74% of control values but did not change either complex I or citrate synthase. Przedborski et al. [19] found inhibition of complex I in substantia nigra of rats after chronic treatment with L-DOPA methyl ester; the control group was not treated with equimolar doses of methyl alcohol and, therefore, these results are difficult to interpret because the changes observed in complex I could be related to the methanol released after the hydrolysis of the methyl ester, rather than with the effects of L-DOPA.

Ascorbic acid (AA), 1 m*M*, which protects NB69 cells from L-DOPA neurotoxicity, increases complex IV activity to 133% of the control but it does not change other ETC complexes. AA also reverses L-DOPA-induced reduction of complex IV activity in NB69 cells and prevents L-DOPA autooxidation. These observations might indicate that the protection observed with AA is related to complex IV activation, in addition to its antioxidating properties.

Free radicals could be responsible for the inhibition of ETC by catechols. Inhibition of complex IV by L-DOPA may be critical for the global ETC function because this complex produces the most important part, the energy-rich phosphates of the ETC. Therefore, the possibility of using AA as a cotreatment with L-DOPA in PD should be investigated.

## 19.4 PREVENTIVE EFFECTS OF ANTIOXIDANTS, MONOAMINE OXIDASE INHIBITORS, AND GROWTH FACTORS

L-DOPA at concentrations of 250 μ*M* or greater is toxic for NB69 cells grown in the presence of serum. Toxicity is associated with high levels of quinones. Deprenyl, a selective inhibitor of type-B MAO, which does not alter the production of quinones, has a partial protective effect. Tocopherol, 23 and 115 μ*M*, lacks significant protective effect on L-DOPA toxicity, but AA, 1 m*M*, prevents L-DOPA toxicity and quinone formation. Deprenyl, 0.1 m*M*, provides additional protection in cultures treated with L-DOPA and AA [3]. These results indicate that AA and deprenyl prevent L-DOPA neurotoxicity by unrelated mechanisms. Both compounds should be considered as complementary drugs that need to be tested for their potential for slowing the progression of Parkinson's disease.

Deprenyl is a type-B MAO inhibitor that blocks the conversion of DA to dihydroxyphenylacetic acid (DOPAC) and $H_2O_2$. Its prevention of L-DOPA toxicity was thought to be related to the blockade of this free-radical-generating pathway [20], although this does not exclude other alternative explanations [21].

Pargyline, from 10 to 100 μ*M*, is not a selective MAO inhibitor and, therefore, it blocks both the conversion of DA to DOPAC and serotonin (5HT) to 5-hydroxy-indole-acetic acid (5HIAA). Pargyline, however, does not protect from L-DOPA toxicity (Mena et al., unpublished results). On the other hand, deprenyl has, in addition to its MAO inhibitory effects, other pharmacological properties. Deprenyl induces the expression of neurotrophic factors, such as ciliary neurotrophic factor (CNTF), and prevents degeneration of motoneurons after axotomy [21]. Furthermore, deprenyl has been found to close the high-energy mitochondrial pore, one of the events that triggers irreversible cell death in apoptotic neurons [22,23].

Such a limited protective role of deprenyl in this cellular system suggests that the toxic effects of L-DOPA could be mediated, in part, through other mechanisms. In other experimental models, L-DOPA toxicity is correlated with the rate of quinone formation [24,25]. Deprenyl did not decrease quinone production in L-DOPA-treated cells, but AA did.

Antioxidants and free-radical scavengers are promising chemicals that may help to prevent the putative toxic effects of L-DOPA or other chemicals on DA neurons in patients with PD. It is known that AA exerts an important role in neuronal function, working as a neuromodulator, antioxidant, and enzyme cofactor in catecholamine biosynthesis [26]. Because of its antioxidant properties, AA scavenges oxygen-containing radicals and it plays a role in the enzymatic reduction of lipid peroxides. Kalir and Mytilineou [27] showed that AA, 0.2 m*M*, increases neurite growth and TH activity and enhances DA uptake and catecholamine levels in mesencephalic cultures. Also, these authors observed a marked increase in glial proliferation. It is unknown whether these neurotrophic effects of AA on DA neurons are direct neuronal effects, mediated by increased glial proliferation, or both.

The neurotrophic factors glial cell-derived neurotrophic factor (GDNF) and brain-derived neurotrophic factor (BDNF) totally prevented L-DOPA toxicity in fetal midbrain DA neurons in culture [11,12], whereas NGF and basic fibroblast neurotrophic factor (bFGF) were less effective. NGF is, however, the only neurotrophic factor known to reduce the elevation of quinones induced by L-DOPA [11,12]. Several neurotrophic factors increase the activity of important enzymes of the free-radical-scavenging system. NGF induces superoxide dismutase (SOD) [28]. SOD activity may represent the first line of defense of the cells against oxidative stress mediated by oxygen free radicals. Consistent with a protective role for SOD, both *in vivo* and *in vitro* studies showed that increased Cu/Zn–SOD activity is associated with greater resistance of different types of cells against free-radical-generating insults [14,28].

## 19.5 NEUROTROPHIC EFFECTS OF L-DOPA IN POSTNATAL DA NEURON–CORTICAL ASTROCYTE COCULTURES OR IN PRESENCE OF GCM

Whereas glial cells surround and protect neurons *in vivo*, neurons are usually cultured *in vitro* in the absence of glia. We treated fetal midbrain rat neurons with L-DOPA, mesencephalic GCM, and L-DOPA + GCM. L-DOPA reduced the number of TH+

cells and $^{3}$H-DA uptake, and increased quinone levels. L-DOPA + GCM restored $^{3}$H-DA uptake and quinone levels to normal, and increased the number of TH+ cells and terminals to 170% of control. GCM greatly increased the number of TH+ cells and $^{3}$H-DA uptake (Figure 19.1). Mesencephalic glia, therefore, produced soluble factors, namely, small antioxidants, such as AA and GSH, as well as peptidic growth factors, which are neurotrophic for DA neurons and which protect these neurons from the toxic effects of L-DOPA (Figure 19.2A and Figure 19.2B).

In order to test the effect of glia on L-DOPA toxicity in DA neurons, we used postnatal ventral midbrain neuron–cortical astrocyte cocultures in serum-free GCM. L-DOPA (50 $\mu M$) protected against the cell death of DA neurons and increased the number and branching of DA processes. In contrast to embryonically derived glia-free cultures in which L-DOPA is toxic, the presence of glia protected postnatal midbrain cultures from L-DOPA in concentrations up to 400 $\mu M$ (Figure 19.3A and Figure 19.3B). The stereoisomer D-DOPA (50 to 400 $\mu M$) was not neurotrophic. The aromatic amino acid decarboxylase inhibitor carbidopa (25 $\mu M$) did not block the neurotrophic effect (Figure 19.3C). These data suggest that the neurotrophic effect of L-DOPA is stereospecific but independent of the production of DA. In a different set of experiments it was shown that L-DOPA increased the levels of glutathione. Inhibition of glutathione synthesis by L-buthionine sulfoximine (BSO), 3 $\mu M$ for 24 h, blocked the neurotrophic action of L-DOPA (Figure 19.3D). L-*N*-acetylcysteine (250 $\mu M$ for 48 h), which promotes GSH synthesis, had a neurotrophic effect similar to L-DOPA. These data suggest that the neurotrophic effect of L-DOPA may be mediated, at least in part, by the elevation of GSH.

## 19.6 MECHANISMS OF L-DOPA NEUROTROPHIC EFFECTS

L-DOPA has neurotrophic effects on DA neurons in cocultures with glia, stimulates elongation of neurites, and protects DA neurons from cell death [13]. These results contrast with previous studies using embryonic neurons cultured without an astrocyte layer, in which L-DOPA was toxic at relatively low concentrations [2,4,5]. In a coculture system using postnatal neurons and glia, L-DOPA is neurotrophic. That may provide an explanation for the discrepancies between the effects of L-DOPA in different models.

Previous studies indicate that DA added to culture medium increases the expression of TH and aromatic amino acid decarboxylase in primary cultures of fetal neurons [29], and the neurotransmitter itself has been postulated as a neurotrophic factor [30]. If conversion of L-DOPA to DA is required for its neurotrophic effects, the response should be blocked by the aromatic amino acid decarboxylase inhibitor carbidopa. Our study reveals, however, that the mechanism responsible for the neurotrophic effect of L-DOPA is independent of its conversion to DA [13].

Astrocytes produce several neurotrophic and neurite-promoting agents that influence development, survival, neurite extension, and neurotoxin resistance. Furthermore, GCM increases cyclic adenosine monophosphate (cAMP) intracellular levels and activates p-ERK/MAPK pathway (Figure 19.2C and Figure 19.2D) [2,17,18].

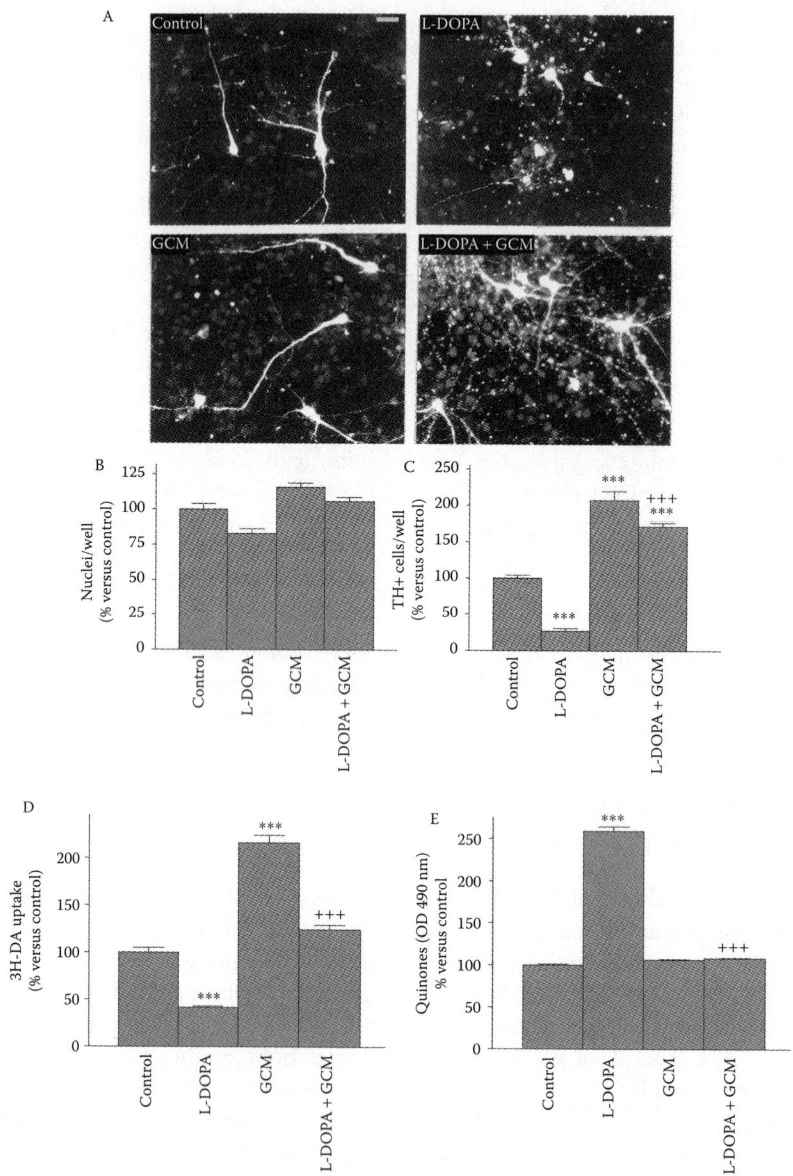


**FIGURE 19.1** GCM protects from L-DOPA-induced toxicity in fetal rat midbrain neuronal cultures. Cultures were treated, at 5 d *in vitro*, with serum-free-defined medium (DM) or GCM and/or L-DOPA 200 μ*M* for 24 h. (A) TH immunofluorescence of rat midbrain fetal neurons. Scale bar = 25 μm. (B) Nuclei per well, (C) TH+ cells per well, (D) $^{3}$H-DA uptake, and (E) quinone levels. Control values are: nuclei/well, 28137 ± 2739; TH+ cells/well, 1005 ± 43; $^{3}$H-DA uptake, 29913 ± 654 cpm/well; quinones, 0.147 ± 0.001 (OD at 490 nm). Statistical analysis was performed by one-way analysis of variance followed by the Student's *t*-test. ***$P < .001$, L-DOPA or GCM-treated groups vs. controls; +++$P < .001$, L-DOPA plus GCM vs. L-DOPA-treated group.

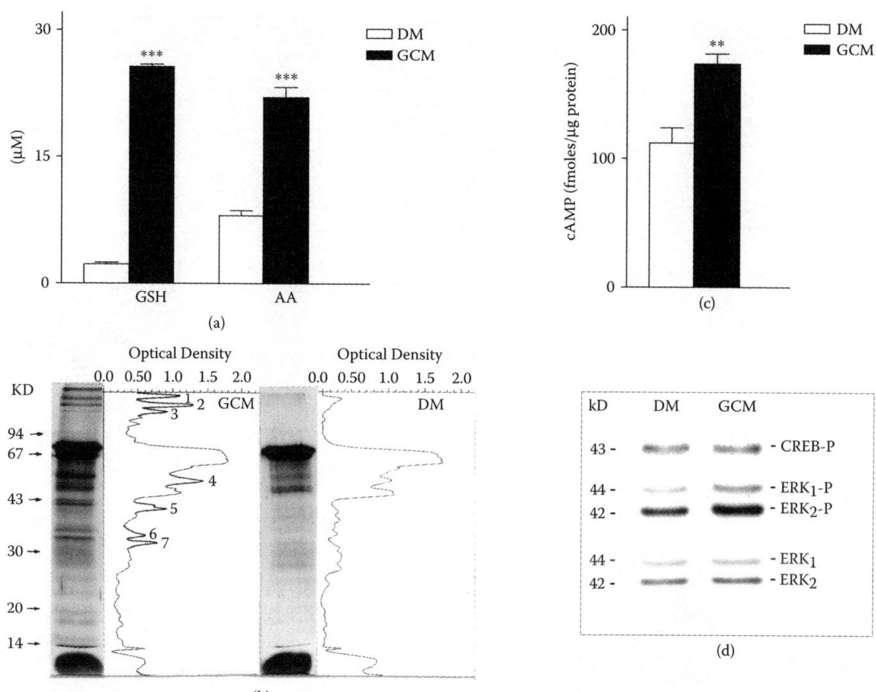


**FIGURE 19.2** Mechanisms of fetal midbrain GCM on DA phenotype expression. (A) Effects of GCM on GSH and AA levels in the medium of fetal midbrain cultures. Cells were treated, at 5 d *in vitro*, with DM or GCM for 24 h. (B) Gel electrophoresis of GCM and defined medium (DM) and their corresponding densitometric scanning. Numbers 1 to 7 on left panel, indicate the 7 bands, that are exclusive, or present at greater concentrations in GCM than in DM, with relative molecular weight of 181, 158, 143, 55, 42, 35, and 33 kDa, respectively. (C) Effects of GCM treatment for 24 h on cAMP levels, determined by the method of Gilman [45]. Control values: 112 ± 11.8 fmoles cAMP/μg of protein. Results are expressed as the mean ± SE (n = 6). Statistical analysis was performed by ANOVA followed by the Student's *t*-test. $^{**}P < .01$, $^{***}P < .001$, GCM vs. DM. (D) Effects of 30 min treatment with GCM on ERK-1/2 MAP kinases and CREB-P expression. Western blot analysis of MAPK-P and CREB-P proteins from control (DM) and GCM-treated groups. Control of charge with total MAPK. (From Mena, M.A. et al., The role of astroglia on the survival of DA neurons, *Mol. Neurobiol.*, 25, 245, 2002. With permission.)

The protective role of the astroglia is, in part, played by a glutamate uptake system that protects against excitotoxicity [31] and stimulation of the synthesis of GSH and the activity of the enzymes involved in the cycle of GSH metabolism [28]. GSH plays a major role in protection against oxidative stress and removes the free radicals generated by MAO [32]. GSH peroxidase is mainly a mitochondrial enzyme exclusively located in the glia in the human midbrain [33]. GSH is produced by GSH peroxidase in culture [34]. GSH levels decrease with age in neurons but remain

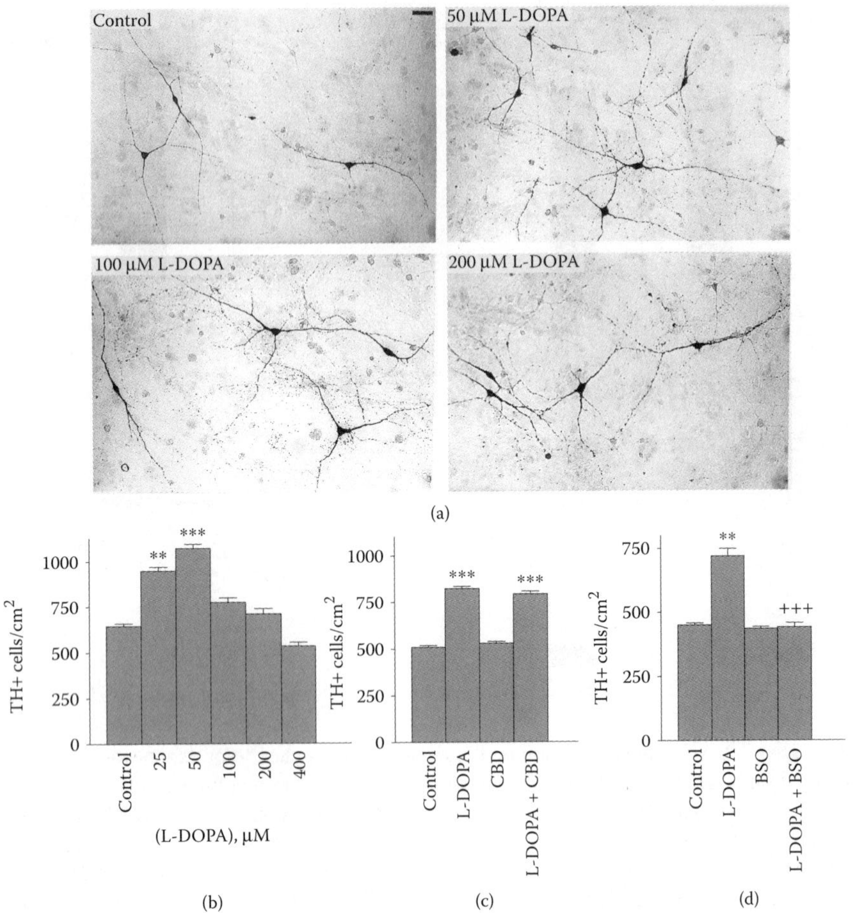


**FIGURE 19.3** Neurotrophic effects of L-DOPA in postnatal cortical astrocytes–ventral midbrain DA neurons cocultures. (A) TH+ DAB-immunostaining of DA neurons. Cultures were treated, at 5 d *in vitro*, with L-DOPA 50, 100, and 200 μ*M* for 48 h. Scale bar = 30 μm. (B) Effects of L-DOPA (50, 100, 200, and 400 μ*M*), (C) L-DOPA 50 μ*M* and/or L-aromatic amino acid decarboxylase inhibitor carbidopa (CBD) 25 μ*M*, and (D) L-DOPA 50 μ*M* and/or GSH synthesis inhibitor BSO, 3 μ*M*, on TH+ cell number. Statistical analysis was performed by one-way analysis of variance followed by the Student's *t*-test. $^{**}P < .01$, $^{***}P < .001$, L-DOPA-treated groups vs. controls; $^{+++}P < .001$, 50 μ*M* L-DOPA plus 3 μ*M* BSO vs. 50 μ*M* L-DOPA-treated group. (From Mena, M.A. et al., The role of astroglia on the survival of DA neurons, *Mol. Neurobiol.*, 25, 245, 2002. With permission.)

stable in astrocytes in culture [35]. However, neurons can maintain their intracellular GSH pool by taking up cysteine provided by glial cells [34].

Protection from oxidation enhances the survival of mesencephalic cultures [2,3,10], and the number of TH+ neurons are increased if SOD, GSH peroxidase, or L-*N*-acetylcysteine (L-NAC) are added to the media [36]. We have shown that

TH+ neurons that overexpress SOD display increased neurite outgrowth and survival [14]. Han et al. [37] reported that L-DOPA raises GSH levels in cultures of fetal rat mesencephalon, a mouse neuroblastoma line (Neuro-2A), a human neuroblastoma (SKNMC), and glia from newborn rat brain, but not in midbrain neuronal cultures grown without glia.

L-DOPA effects in postnatal neuron–cortical astrocytes coculture were not associated with an elevation of quinones [13], in contrast with the observations in embryonically derived glial-free cultures [9], suggesting that astrocytes may inhibit L-DOPA quinone formation by providing an antioxidant system. Moreover, we found that the GSH synthesis promoter L-NAC, and GSH itself, reproduced L-DOPA's effects, whereas the GSH synthase inhibitor BSO, blocked L-DOPA's effect. This suggests that L-DOPA promotes its neurotrophic effect by stimulating GSH peroxidase systems. In particular, it is unlikely that L-DOPA acts as an antioxidant itself because high doses of L-DOPA do not provide neuroprotection and, even more to the point, increase quinone formation; a cautionary note is provided by the finding that L-NAC, previously assumed to act as an antioxidant and as a substrate for GSH synthesis, may protect PC12 cells by acting directly as an oxygen radical scavenger [38]. L-NAC is a pluripotent protector that increases intracellular GSH, protects against neuronal apoptosis, and enhances trophic factor-mediated cell survival [39]. Indeed, depletion of brain GSH by BSO in rats is accompanied by impaired mitochondrial function and reduced complex IV activity, which leads to energy crisis as a mechanism of nigral cell death [40].

Postmortem studies in humans have shown that GSH is decreased in the substantia nigra of patients with idiopathic PD [41]. *In vitro*, L-DOPA treatment reverses GSH depletion [42]. In normal subjects, the nigrostriatal DA neurons, the most important disease target associated with PD, are normally surrounded by a low density of glia [33,35], suggesting that these neurons are less protected from oxidative stress. Strikingly, in PD the density of GSH peroxidase-positive glia surrounding the surviving DA neurons is increased, suggesting a compensatory proliferation of glia that protects surviving neurons against pathological death [33].

Although L-DOPA is toxic for catecholamine neurons in other *in vitro* systems [1–6], catecholamines are required for the development of the nervous system and survival. Inactivation of the TH gene produced the death of 90% of mutant mice fetuses, whereas administration of L-DOPA to pregnant females resulted in a complete rescue of the mutant mice [43]. Transgenic animals with mutations of the TH gene have a poorer prognosis than those with mutations of the dopamine-beta-hydroxylase (DBH) gene [44], suggesting that DA is essential for development of the nervous system. It is likely that L-DOPA and catecholamines are critical factors for promoting the survival and differentiation of DA neurons, and that the effects of L-DOPA on DA cells *in vivo* might be neurotrophic or neurotoxic, depending on the local environment produced by glial cells.

Conclusions from our studies indicate that a long-term potentiation of DA release by L-DOPA may be due to promotion of neuritic arborization and protection against cell death. The neurotrophic effects are independent of conversion to DA, dependent on the presence of astrocytes, and may result from antioxidant mechanisms such as upregulation of GSH. Significant questions that need to be addressed include whether the effect is due to L-DOPA itself, acting as an antioxidant; whether it is due to upregulation of a cellular system, such as the GSH peroxidase pathway; whether sequestration of DA by synaptic vesicles plays a role in protection from L-DOPA toxicity; and also what the precise pathway is for exchange of cysteine, or other protective interactions, between glia and DA neurons.

## 19.7 SYNERGISTIC NEUROTROPHIC RESPONSE BETWEEN L-DOPA AND NGF

To characterize the long-term effects of L-DOPA, we used a pheochromocytoma (PC12) line that extends neurites on exposure to NGF. L-DOPA potentiated the outgrowth of processes that was elicited by NGF. This response did not require conversion of L-DOPA to DA, was not caused by an agonist acting at DA receptors, and was not blocked by the tyrosine kinase inhibitor genistein. However, similar results were found after exposure to L-NAC or apomorphine, a DA receptor agonist that produces a quinone metabolite and increases GSH synthesis.

Long-term process elaboration was blocked by BSO, consistent with mediation by an antioxidant mechanism. L-DOPA potentiation of NGF response was important functionally, as seen by increased quantal neurotransmitter release from L-DOPA/ NGF-treated neurite varicosities, which displayed both twofold greater quantal size and frequency of quantal release. These results demonstrate potentiation by L-DOPA of morphological and physiological responses to neurotrophic factors as well as synergistic induction of antioxidant pathways. Together with effects on transmitter synthesis, these properties seem to provide a basis for the compound's long-term presynaptic potentiation of DA release and therapeutic action [15].

## 19.8 L-DOPA AND GCM HAVE ADDITIVE EFFECTS ON TH PROTEIN EXPRESSION

The aim of this study was to investigate the effect of L-DOPA and GCM on cell viability, TH expression, DA metabolism, and GSH levels in NB69 cells. L-DOPA (200 $\mu M$) induced differentiation of NB69 cells with more than 4 weeks *in vitro* growth in serum-free medium, as shown by phase-contrast microscopy (Figure 19.4A) and TH immunocytochemistry, and decreased their replication, as shown by 5-bromodeoxyuridine immunostaining. L-DOPA did not increase the number of necrotic or apoptotic cells, as shown by morphological features, trypan blue, lactate dehydrogenase (LDH) activity, bisbenzimide staining, and TUNEL assay [16].

Furthermore, L-DOPA, 200 $\mu M$, increased Bcl-xL protein expression (Figure 19.4B). Incubation of cells with L-DOPA, 50, 100, and 200 $\mu M$, for 24 h resulted in an increase of TH protein levels to 174, 196, and 212% of the control, respectively. Neither carbidopa, an inhibitor of L-aromatic amino acid decarboxylase enzyme,

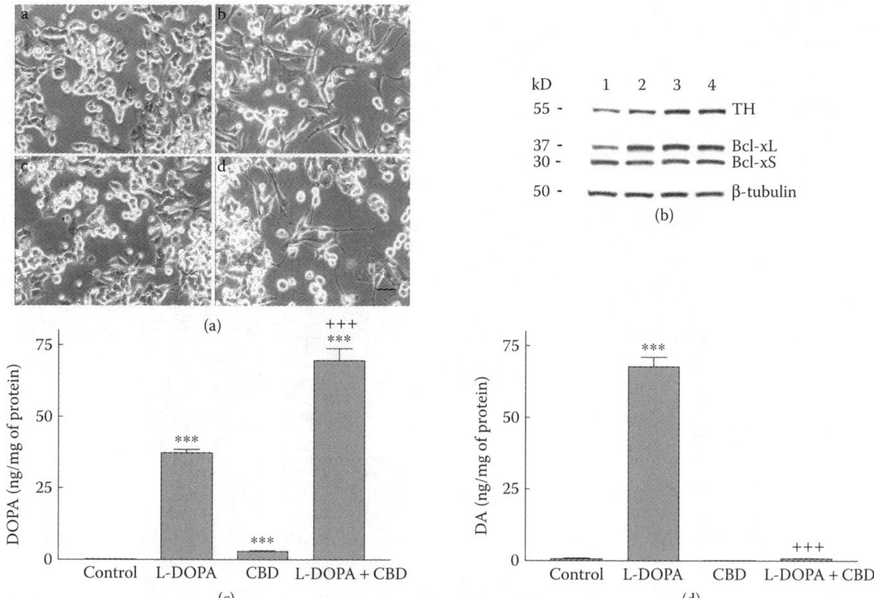


**FIGURE 19.4** Effect of carbidopa on L-DOPA response in NB69 cells. (A) Phase-contrast microscopy of NB69 human neuroblastoma cells after 6 d in culture. On the fourth day, the cultures were treated with L-DOPA (200 μ*M*) or vehicle (serum-free-defined medium) and/or carbidopa (25 μ*M*) for 48 h. a: Control cells treated with vehicle; b: Cells treated with L-DOPA, 200 μ*M*; c: Cells treated with carbidopa; d: Cells treated with 200 μ*M* L-DOPA plus carbidopa. Scale bar = 25 μm. (B) Western blot analysis of total TH and Bcl-xL/S proteins from control, 200 μ*M* L-DOPA-treated cells, 25 μ*M* carbidopa, and 200 μ*M* L-DOPA plus 25 μ*M* carbidopa (lanes 1 to 4, respectively). TH protein increased to 168 ± 3, 189 ± 14, and 205 ± 2%, respectively. Control of charge with B-tubulin. (n = 3). (C) DOPA and (D) DA endogenous levels determined by HPLC: values are expressed in ng/mg protein as the mean ± SE (n = 4 to 6). DA control levels were 0.71 ± 0.25 and DOPA control levels were at the limit of detection (≤0.15 ± 0.03 ng/well). Statistical analysis was performed by ANOVA followed by Student's *t*-test. ****P* < .001 vs. controls; +++*P* < .001, L-DOPA plus CBD- vs. L-DOPA-treated group. (From Rodriguez-Martin, E. et al., L-DOPA and glia-conditioned medium have additive effects on TH expression in human catecholamine-rich neuroblastoma NB69 cells, *J. Neurochem.*, 78, 535, 2001. With permission.)

nor L-BSO, that inhibits GSH synthesis, nor AA, an antioxidant, blocked the L-DOPA-induced effect on TH protein expression. GCM, without or with L-DOPA, 50, 100, and 200 μ*M*, increased the amount of TH protein by 346, 446, 472, and 424%, respectively (Figure 19.5). L-DOPA, 200 μ*M*, increased TH protein levels to 132, 191, and 245% of controls after incubation for 24, 48, and 72 h, respectively. DA metabolism in NB69 cells was increased in cultures treated with either L-DOPA, 200 to 300 μ*M*, or GCM, and these two agents had a synergistic effect on DA metabolism. In addition, L-DOPA, 200 μ*M*, and GCM-treated cells increased their extracellular GSH after 48 h of treatment (Figure 19.6).

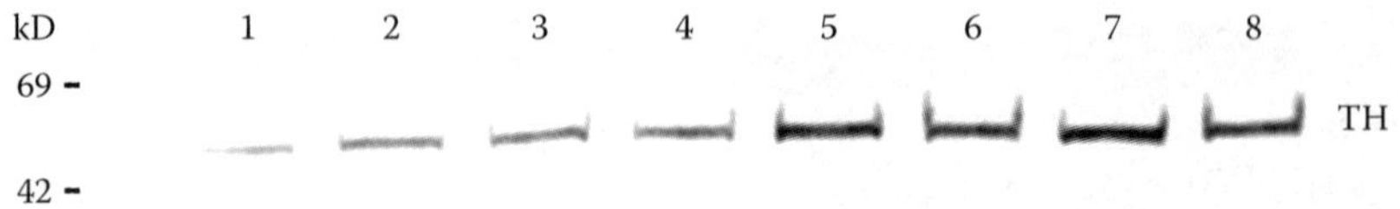


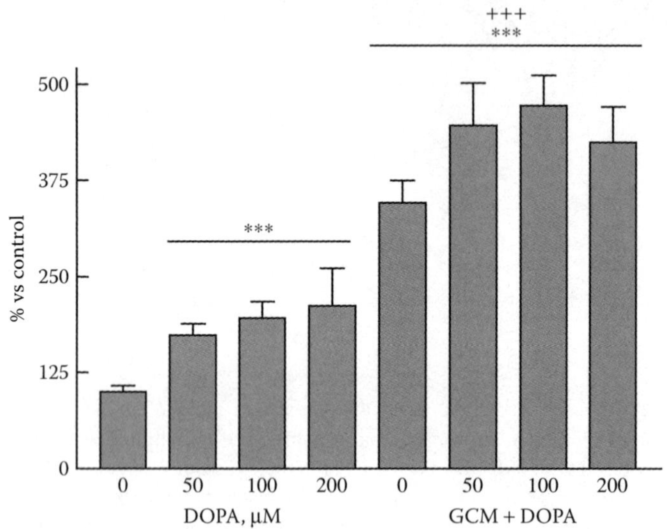


**FIGURE 19.5** Dose-dependent effect of L-DOPA and GCM on TH protein expression from control (lane 1), L-DOPA-treated cells (50, 100, and 200 μ*M* L-DOPA, lanes 2 to 4), GCM-treated cells (lane 5) and GCM plus L-DOPA (50, 100, and 200 μ*M* L-DOPA, lanes 6 to 8)-treated cells (n = 4). Values are expressed as percentages of the control blot (174, 196, 212, 346, 446, 472, and 424%). Statistical analysis was performed by one-way analysis of variance followed by the Student's *t*-test. ***$P < .001$ vs. controls, +++$P < .001$, L-DOPA plus GCM- vs. L-DOPA-treated group. The levels of TH protein were determined by computer-assisted videodensitometry from Western blot films. Each lane contains 20 μg of total protein. Control of charge with B-tubulin. (From Rodriguez-Martin, E. et al., L-DOPA and glia-conditioned medium have additive effects on TH expression in human catecholamine-rich neuroblastoma NB69 cells, *J. Neurochem.*, 78, 535, 2001. With permission.)

The L-DOPA-induced increase of TH protein expression in NB69 cells was independent of DA production, free radicals, and GSH upregulation [16]. Furthermore, L-DOPA decreased cAMP levels and carbidopa did not block those effects (Figure 19.7B). GCM increased cAMP levels and this increase persisted after cotreatment of L-DOPA with GCM (Figure 19.7A). GCM induces *de novo* synthesis of TH and increases DA cell survival by ERK/MAPK pathway activation [17,18].

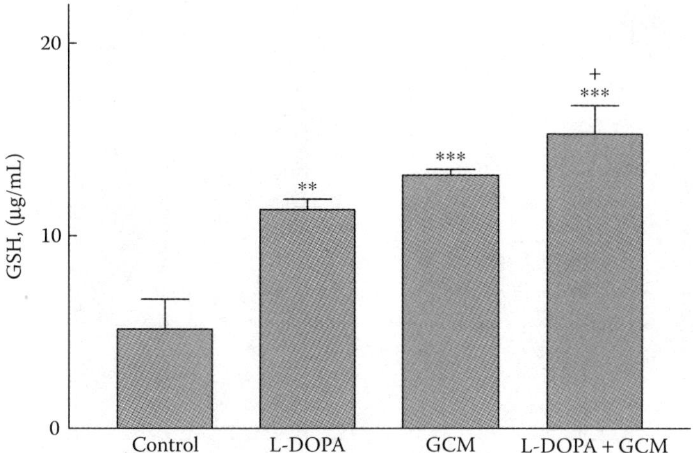


**FIGURE 19.6** GSH levels in the culture medium induced by L-DOPA 200 μ*M* and/or GCM after 48 h of treatment (n = 5). Results are expressed as percentage vs. control ± SE. Control values: 5.14 ± 1.58 μg/mL. Statistical analysis was performed by one-way analysis of variance followed by the Student's *t*-test. $^{**}P < .01$, $^{***}P < .001$ vs. controls, $+P < .05$, L-DOPA plus GCM- vs. L-DOPA-treated group. (From Rodriguez-Martin, E. et al., L-DOPA and glia-conditioned medium have additive effects on TH expression in human catecholamine-rich neuroblastoma NB69 cells, *J. Neurochem.*, 78, 535, 2001. With permission.)

## 19.9 CONCLUSION

L-DOPA may be toxic for human neuroblastoma cells, NB69, and for fetal DA neurons, and its toxicity is related to several mechanisms including quinone formation and enhanced production of free radicals related to the metabolism of DA via MAO B. L-DOPA is toxic for DA neurons in culture but its toxicity has not been proven in animals or in patients with Parkinson's disease. Chronic L-DOPA is not toxic for the remaining DA neurons but instead, promotes their recovery in rats with partial nigrostriatal lesions. Because most experiments *in vitro* showing L-DOPA toxicity were performed in neurons cultured in the absence of glia, we hypothesized that the discrepancy between the effects of L-DOPA *in vivo* and *in vitro* may be due to the presence or absence of glia, respectively. Fetal midbrain neuronal cultures were treated with L-DOPA, 200 μ*M*, in the presence or absence of mesencephalic GCM. In the absence of GCM, L-DOPA greatly reduced the number of TH-immunoreactive neurons and increased the levels of quinones in the medium; GCM prevented these effects of L-DOPA and increased the length and arborization of the neurites of the TH-immunoreactive cells.

Therefore, L-DOPA, 50 μ*M*, has neurotrophic effects in postnatal midbrain DA neuron–cortical astrocyte coculture by upregulation of GSH, and potentiates NGF-dependent outgrowth and synaptic vesicle quantal release in PC12 cells. Furthermore, L-DOPA increases TH protein expression and intracellular cAMP levels in the presence of GCM in human neuroblastoma cell line. In conclusion, the critical factors for L-DOPA neurotrophism or toxicity are glial function and GSH homeostasis.

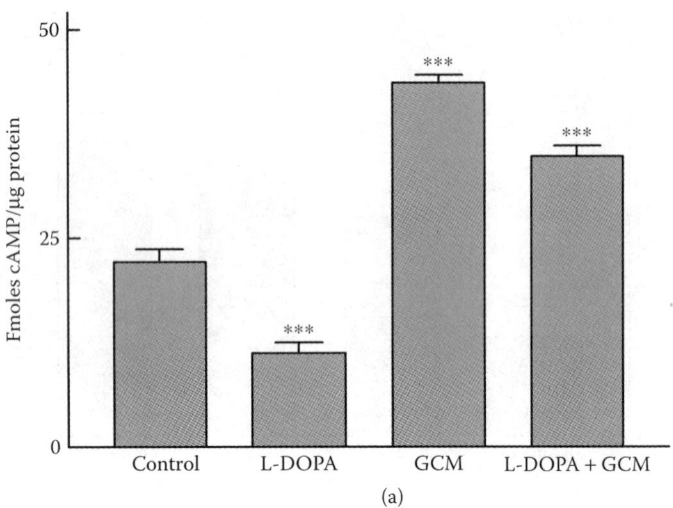


(a)

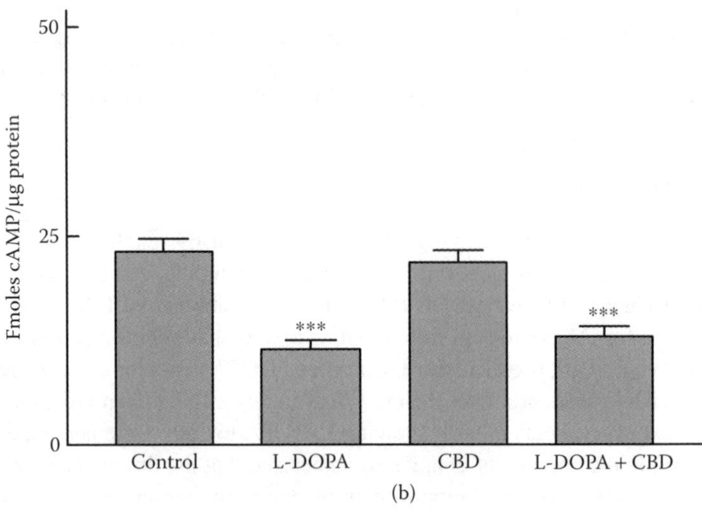


(b)

**FIGURE 19.7** Effect of L-DOPA 200 µ*M* and GCM on cAMP levels in NB69 cells after 6 d in culture. On the fifth day, the cells were treated with L-DOPA 200 µ*M* or vehicle (serum-free-defined medium) and/or glia-conditioned medium (GCM) for 24 h (A), or with L-DOPA 200 µ*M* or vehicle (serum-free-defined medium) and/or carbidopa 25 µ*M* for 24 h (B). Results are representative of two experiments and are expressed as fmoles cAMP/µg of protein ± SE. (n = 6 to 9). Statistical analysis was performed by one-way analysis of variance followed by the Student's *t*-test. ***$P < .001$ vs. controls.

## ACKNOWLEDGMENTS

This study was supported in part by grants FIS 2002/PI020265 and FIS 2004/PI40360 of the Spanish Ministry of Health and CAM 8.5/49/2001 of the Madrid Community. The authors thank R. Villaverde for excellent technical assistance.

## REFERENCES

1. Mena, M.A. et al., Neurotoxicity of L-DOPA on catecholamine-rich neurons, *Movement Disord.,* 7, 441, 1992.
2. Mena, M.A. et al., L-DOPA toxicity in foetal rat midbrain neurons in culture: modulation by AA, *NeuroReport,* 4, 438, 1993.
3. Pardo, B. et al., AA protects against L-DOPA-induced neurotoxicity on a catecholamine-rich human neuroblastoma cell line, *Movement Disord.,* 8, 268, 1993.
4. Mytilineou, C., Han, S., and Cohen, G., Toxic and protective effects of L-DOPA on mesencephalic cell cultures, *J. Neurochem.,* 61, 1470, 1993.
5. Pardo, B. et al., Toxic effects of L-DOPA on mesencephalic cell cultures: protection with antioxidants, *Brain Res.,* 682, 133, 1995.
6. Pardo, B., Mena, M.A., and G. de Yebenes, J., L-DOPA inhibits complex IV of the electron transport chain in catecholamine-rich human neuroblastoma NB69 cells, *J. Neurochem.,* 64, 576, 1995.
7. Hefti, F. et al., Long-term administration of L-DOPA does not damage DArgic neurons in the mouse, *Neurology,* 31, 1194, 1981.
8. Fahn, S., Welcome news about L-DOPA, but uncertainty remains, *Ann. Neurol.,* 43, 551, 1998.
9. Murer, M.G. et al., Chronic L-DOPA is not toxic for remaining DA neurons, but instead promotes their recovery, in rats with moderate nigrostriatal lesions, *Ann. Neurol.,* 43, 561, 1998.
10. Mena, M.A. et al., Glial conditioned medium protects fetal rat midbrain neurons in culture from L-DOPA toxicity, *NeuroReport,* 7, 441, 1996.
11. Mena, M.A. et al., Glia protects midbrain DA neurons in culture from L-DOPA toxicity through multiple mechanisms, *J. Neural Transm.,* 104, 317, 1997.
12. Mena, M.A. et al., The critical factor for L-DOPA toxicity on DA neurons is glia, in *Understanding Glial Cells,* Castellano, B., Gonzalez, B., and Nieto-Sampedro, M., Eds., Kluwer Academic Publishers, Boston, 1998, 213.
13. Mena, M.A., Davila, V., and Sulzer, D., Neurotrophic effects of L-DOPA in postnatal midbrain DA neuron/cortical astrocyte cocultures, *J. Neurochem.,* 69, 1398, 1997.
14. Mena, M.A. et al., Effects of wild-type and mutated copper/zinc superoxide dismutase on neuronal survival and L-DOPA-induced toxicity in postnatal midbrain culture, *J. Neurochem.,* 69, 21, 1997.
15. Mena, M.A. et al., A synergistic neurotrophic response to L-dihydroxyphenylalanine and nerve growth factor, *Mol. Pharmacol.,* 54, 678, 1998.
16. Rodriguez-Martin, E. et al., L-DOPA and glia-conditioned medium have additive effects on TH expression in human catecholamine-rich neuroblastoma NB69 cells, *J. Neurochem.,* 78, 535, 2001.
17. Mena, M.A. et al., The role of astroglia on the survival of DA neurons, *Mol. Neurobiol.,* 25, 245, 2002.
18. de Bernardo, S. et al., Glia-conditioned medium induces de novo synthesis of TH and increases DA cell survival by differential signaling pathways, *J. Neurosci. Res.,* 73, 818, 2003.

19. Przedborski, S. et al., Chronic levodopa administration alters cerebral mitochondrial respiratory chain activity, *Ann. Neurol.,* 34, 715, 1993.
20. Cohen, G. and Spina, M.B., Deprenyl suppresses the oxidant stress associated with increased DA turnover, *Ann. Neurol.,* 26, 689, 1989.
21. Tatton, W.G. et al., Selegiline induces "trophic-like" rescue of dying neurons without MAO inhibition, *Adv. Exp. Med. Biol.,* 363, 15, 1995.
22. Marruyama, W. et al., Neuroprotection by propargylamines in Parkinson's disease: supression of apoptosis and induction of prosurvival genes, *Neurotoxicol. Teratol.* 24, 675, 2002.
23. De Marchi, U. et al., L-Deprenyl as an inhibitor of menadione-induced permeability transition in liver mitochondria, *Biochem. Pharmacol.,* 66, 1749, 2003.
24. Mytilineou, C. and Danias P., 6-HydroxyDA toxicity to DA neurons in culture: potentiation by the addition of superoxide dismutase and N-acetylcysteine, *Biochem. Pharmacol.,* 38, 1872, 1989.
25. Graham, D.G. et al., Autoxidation versus covalent binding of quinones as the mechanism of toxicity of DA, 6-hydroxyDA, and related compounds toward C1300 neuroblastoma cells in vivo, *Mol. Pharmacol.,* 14, 644, 1978.
26. Rice, M.E., Ascorbate regulation and its neuroprotective role in brain, *Trends Neurosci.,* 23, 209, 2000.
27. Kalir, H.H. and Mytilineou, C., AA in mesencephalic cultures: effects on DArgic neuron development, *J. Neurochem.,* 57, 458, 1990.
28. Nistico, G. et al., NGF restores decrease in catalase activity and increases superoxide dismutase and GSH peroxidase activity in the brain of the aged rats, *Free Radic. Biol. Med.,* 12, 177, 1992.
29. De Vitry, F. et al., DA increases the expression of TH and aromatic amino acid decarboxylase in primary cultures of fetal neurons, *Dev. Brain Res.,* 59, 123, 1991.
30. Leslie, F.M., Neurotransmitters as neurotrophic factors, in *Neurotrophic Factors,* Hefti, F. et al., Eds., Academic Press, New York, 1993, 565.
31. Rosenberg, P.A. et al., 2,4,5-Trihydroxyphenylalanine in solution forms a non N-methyl-D-aspartate glutaminergic agonist and neurotoxin, *Proc. Natl. Acad. Sci. USA.,* 88, 4865, 1991.
32. Werner, P. and Cohen G., Glutathione disulfide (GSSG) as a marker of oxidative injury to brain mitochondria, *Ann. N.Y. Acad. Sci.,* 679, 364, 1993.
33. Damier, P. et al., Protective role of GSH peroxidase against neuronal death in Parkinson's disease, *Neuroscience,* 52, 1, 1993.
34. Raps, S.P. et al., GSH is present in high concentrations in cultured astrocytes but not in cultured neurons, *Brain Res.,* 493, 398, 1989.
35. Sagara, J., Miura, K., and Bannai, S., Maintenance of neuronal GSH by glial cells, *J. Neurochem.,* 61, 1672, 1993.
36. Colton, C.A. et al., Protection from oxidation enhances the survival of cultured mesencephalic neurons, *Exp. Neurol.,* 132, 54, 1995.
37. Han, S.-K., Mytilineou, C., and Cohen, G., L-DOPA up-regulates GSH and protects mesencephalic cultures against oxidative stress, *J. Neurochem.,* 66, 501, 1996.
38. Yan, C.Y.I., Ferrari, G., and Greene, L.A., N-Acetylcysteine-promoted survival of PC12 cells is glutathione-independent but transcription-dependent, *J. Biol. Chem.,* 270, 26827, 1995.
39. Ferrari, G., Irene, C.Y., and Greene, L.A., *N*-Acetylcysteine (D- and L-stereoisomers) prevents apoptotic death of neuronal cells, *J. Neurosci.,* 15, 2857, 1995.
40. Jenner, P., Altered mitochondrial function, iron metabolism and GSH levels in Parkinson's disease, *Acta Neurol. Scand.,* 145, 6, 1993.

41. Sian, J. et al., Alterations in GSH levels in Parkinson's disease and other neurodegenerative disorders affecting basal ganglia, *Ann. Neurol.,* 36, 348, 1994.
42. Tohgi, H. et al., Reduced and oxidized forms of glutathione and alpha-tocopherol in the cerebrospinal fluid of parkinsonian patients: comparison between before and after L-DOPA treatment, *Neurosci. Lett.,* 184, 21, 1995.
43. Zhou, Q.Y., Quaife, C.J., and Palmiter, R.D., Target disruption of the TH gene reveals that catecholamines are required for mouse fetal development, *Nature,* 374, 640, 1995.
44. Thomas, S.A., Matsumoto, A.M., and Palmiter, R.D., Noradrenaline is essential for mouse fetal development, *Nature,* 374, 643, 1995.
45. Gilman, A.G., A protein binding assay for adenosine 3′5′-cyclic monophosphate, *Proc. Natl. Acad. Sci. USA,* 67, 305, 1970.

# 20 DOPA: An Upstream Factor for Glutamate Release and Delayed Neuron Death by Transient Ischemia in Striatum and Hippocampal CA1 Region of Conscious Rats

*Yoshimi Misu and Yoshio Goshima*

**CONTENTS**

## 20.1 INTRODUCTION

Neurons in the dorsolateral striata and hippocampal CA1 region are vulnerable to ischemia, which can precipitate cell death [1–3]. Excitotoxicity due to increases in extracellular glutamate is proposed to be one of causal factors for neuron death [4,5] and is implicated in a final common pathway for neurologic disorders [6,7]. The mechanisms of increase in external glutamate are postulated to be due to $Ca^{2+}$-dependent vesicular release [8] and neuronal and/or nonneuronal $Ca^{2+}$-independent increase through reversed operation of glutamate transporters [9,10]. No endogenous substance has been proved to induce, by itself, glutamate release and resultant delayed neuron death due to brain ischemia. Endogenous L-3,4-dihydroxyphenylalanine (DOPA) may be such a neuroactive substance involved in an upstream process for these events.

Since the 1950s, [11,12] exogenously applied L-3,4-dihydroxyphenylalanine (levodopa) has been traditionally thought to be an inert amino acid that alleviates the symptoms of Parkinson's disease via its conversion to dopamine by aromatic L-amino acid decarboxylase (AADC) [13,14]. Since 1986 [15,16], however, we have been proposing that DOPA is a neurotransmitter and/or neuromodulator in the central nervous system (CNS), in addition to being a precursor of dopamine [17–19]. Recent evidence suggests that DOPA fulfills several criteria such as biosynthesis, metabolism, active transport, existence, physiological release, competitive antagonism, and physiological or pharmacological responses, including interactions with the other neurotransmitter systems [20–22]; these criteria must be satisfied before a compound is accepted as a neurotransmitter. We have accumulated evidence to support the idea that DOPA is a neurotransmitter of the aortic depressor nerve (ADN), one of the primary baroreceptor afferents terminating in the rat nucleus tractus solitarii (NTS) [23–25]. The NTS is the gate of the baroreflex neurotransmission in the lower brainstem.

Neurons that may contain DOPA as an end product exist mainly in the CNS including the NTS [19,22,24,26,27]. These neurons show tyrosine hydroxylase (TH)-(+), DOPA-(+), AADC-(−), and dopamine-(−) immunocytochemical reactivity. During microdialysis of the NTS in anesthetized rats, basal DOPA release is partially tetrodotoxin (TTX) sensitive and $Ca^{2+}$-dependent, suggesting that DOPA release is evoked by spontaneous neuronal activity [24,25]. $Ca^{2+}$ seems to be primarily involved in the process of DOPA release rather than that of TH activation induced by nerve depolarization [28]. Indeed, electrical stimulation of the ADN consistently releases DOPA in a neurotransmitter-like manner [24].

In contrast, no immunocytochemical evidence is available for neurons having DOPA as an end product in rat striata and hippocampal CA1 pyramidal cell layers [19,22]. Notwithstanding, DOPA is released by electrical field stimulation in a TTX-sensitive and $Ca^{2+}$-dependent manner from striatal slices [28]. During striatal microdialysis of conscious rats, the basal release of DOPA is in part TTX sensitive and $Ca^{2+}$ dependent, and DOPA is released by high $K^+$ in a $Ca^{2+}$-dependent manner [29]. During microdialysis of the hippocampal CA1 region, the basal release of DOPA is not evident in the usual samples collected for 10 min but is detected in samples collected for 20 min [30]. These findings suggest that the sensitivity of immunocytochemical analysis to find DOPAergic neurons is lower compared with that of biochemical approaches [21,22].

DOPA methyl ester (DOPA ME) was the first competitive antagonist for levodopa to be identified [31,32]. Among the DOPA ester compounds, DOPA cyclohexyl ester (DOPA CHE) is the most potent and relatively stable competitive antagonist for levodopa [33,34]. These two antagonists antagonize responses to both levodopa [32,35] and released DOPA [30,36] in the striatum and the hippocampal CA1 region of rats. Competitive antagonism suggests the existence of DOPA recognition sites.

DOPA CHE does not inhibit the $Na^+$-dependent uptake of labeled levodopa into *Xenopus laevis* oocytes [37]. In addition, levodopa and DOPA ester compounds fail to displace, or hardly displace, the specific binding of tritiated ligands for dopamine $D_1$ and $D_2$ receptors or $\alpha_2$- and β-adrenoceptors in rat brain membrane preparations [31,34]. Furthermore, among the binding sites labeled with tritiated ionotropic glutamate receptor ligands, DOPA ME and DOPA CHE act only on the ion channel of *N*-methyl-D-aspartate (NMDA) receptors with an $IC_{50}$ in the millimolar range [34,38]. In contrast, levodopa acts only on DL-α-amino-3-hydroxy-5-methyl-4-isoxazol propionic acid (AMPA) receptors with low affinity. Levodopa may not interact with the competitive antagonists on these ionotropic glutamate receptors. DOPAergic agonists and competitive antagonists should act at the same sites, which appear to differ from DOPA transport sites, catecholamine receptors, and ionotropic glutamate receptors. It is essential that DOPA recognition sites exist on which DOPA or levodopa acts to elicit physiological or pharmacological responses.

In rat striata and hippocampal CA1 region, most responses to levodopa, similar to that for many neurotransmitters, are stereoselective [32,36,39–41]. Levodopa elicits responses in the absence and presence of 3-hydroxybenzylhydrazine (NSD-1015), a central AADC inhibitor [16,32,35]. These pharmacological responses are elicited by levodopa itself and not by bioconversion to dopamine. Furthermore, responses to released DOPA are potentiated following an increase in the basal release of DOPA elicited by NSD-1015 as a result of the inhibition of AADC [36,42]. In contrast, responses to DOPA are inhibited following a decrease in the biosynthesis of DOPA induced by α-methyl-*p*-tyrosine, an inhibitor of TH [42]. In addition, pressor responses to DOPA released by electrical stimulation of the posterior hypothalamic nucleus are inhibited following the blockade of $Na^+$ channels elicited by TTX microinjected into the rostral ventrolateral medulla (RVLM) [43]. The RVLM is the exit of baroreflex neurotransmission from the lower brainstem to the thoracic spinal cord. These findings suggest that physiological functions are induced following basally released and impulse-evoked DOPA.

In the hippocampal CA1 region, nanomolar levodopa inhibits population spikes elicited by electrical stimuli applied to the Schaffer collateral/commissural fibers in the absence and presence of NSD-1015 and in a DOPA CHE-sensitive manner [35].

In the striatum, nanomolar levodopa potentiates the activity of presynaptic β-adrenoceptors [44] to facilitate dopamine release [16] and inhibits the basal release of acetylcholine in a model of Parkinson's disease [40]. Both levodopa [39] and released DOPA [42] potentiate the activity of the postsynaptic $D_2$ receptors that are involved in locomotion. All of these responses to levodopa or DOPA itself may supplement the effectiveness of levodopa in alleviating the symptoms, especially at the early stages of Parkinson's disease, in addition to the effects following its bioconversion to dopamine.

On the other hand, in relation to possible side effects or neurotoxicological cell death, clear relationships are seen between levodopa and glutamate in striata *in vitro*. Levodopa releases, by itself, neuronal glutamate from striatal slices with the $ED_{50}$ of 140 $\mu M$ [32], which appears to be at least related to the neuroexcitatory side effects such as dyskinesia, encountered during chronic therapy of Parkinson's disease. This levodopa-induced glutamate release is stereoselective, partially TTX sensitive and $Ca^{2+}$ dependent, antagonized by DOPA ME in a competitive manner, and not inhibited by NSD-1015. In contrast, a corresponding concentration of dopamine (300 $\mu M$) releases no glutamate.

In primary striatal neuron cultures, levodopa and dopamine elicit differential neurotoxicity [41]. Micromolar dopamine elicits neuron death in 3 and 10 d in culture. This type of neuron death is prevented by ascorbic acid, an antioxidant, but not by NMDA and non-NMDA glutamate antagonists. Many studies show that reactive free radicals or dopamine quinones derived from enzymatic oxidation or autoxidation of dopamine play an important role in its neurotoxicity [41,45] (see Chapter 2, Chapter18, and Chapter 19).

However, levodopa elicits neurotoxicity via two different pathways. One is derived from autoxidation. Micromolar levodopa elicits antioxidant-sensitive neuron death in 3 d in culture [41]. Levodopa itself, and/or converted dopamine, can produce this type of neurotoxicity. Dextrodopa also elicits neuron death in a similar manner. In addition, levodopa autoxidation yields 3,4,6-trihydroxyphenylalanine and its quinone derivative, a non-NMDA agonist and excitotoxin [46–48]. Furthermore, another type of neuron death is induced stereoselectively by micromolar levodopa in 10 d in culture [41], which results from glutamate release [49]. It is protected by TTX application, $Ca^{2+}$ deprivation, and $Mg^{2+}$ addition [49], and is antagonized by NMDA and non-NMDA glutamate antagonists, but not by the antioxidant [41]. This type of neuron death appears to be mediated by vesicular release of glutamate [32]. Increased glutamate elicits excessive $Ca^{2+}$ entry [6,50,51] mainly via activation of postsynaptic NMDA [52] and non-NMDA [53] receptors. The cascades that follow appear to be: activation of neuronal nitric oxide (NO) synthase, production of NO [54], and formation of peroxynitrite by the reaction of NO with superoxide anion, leading to neuronal death [51,55].

Herein we have tried to clarify whether or not transient brain ischemia releases DOPA during microdialysis and whether DOPA, when released, functions by itself to cause glutamate release and resultant delayed neuron death in the striata and hippocampal CA1 region of conscious rats. If this is the case, the inhibition of AADC should exaggerate these events because it markedly increases extracellular levels of DOPA [29]. In contrast, DOPA CHE, a potent and relatively stable competitive antagonist [33,34], ought to protect against these events.

## 20.2 EXTRACELLULAR LEVELS OF DOPA, DOPAMINE, AND GLUTAMATE INCREASED BY TRANSIENT ISCHEMIA DURING STRIATAL MICRODIALYSIS OF CONSCIOUS RATS

At first we attempted to clarify whether or not endogenous DOPA is released by a 10-min ischemia during striatal microdialysis and, further, whether or not DOPA, when released, functions to elicit glutamate release and resultant delayed neuron death by ischemia [36,56].

Using conscious male Wistar rats (300 to 350 g), DOPA and dopamine or glutamate in striatal dialysates are consistently detectable by high-performance liquid chromatography with an electrochemical detector or with a spectrofluorometer. In Figure 20.1, the extracellular levels of these substances are stabilized 2 to 3 h after perfusion with Ringer solution. Ischemia of 10-min duration due to four-vessel occlusion increases extracellular levels of DOPA, dopamine, and glutamate. Increase in DOPA in the first 10-min sample during ischemia is apparently slower than that of dopamine and glutamate. Peak increases in these substances are seen in the second sample, immediately after ischemia. Some parallelism is seen between DOPA and glutamate because the peak release ratio of DOPA:dopamine:glutamate is 6:220:8. This ischemia releases a 16-fold higher amount of glutamate, compared with the basal level.

## 20.3 DELAYED NEURON DEATH IN THE STRIATUM AND THE HIPPOCAMPAL CA1 REGION OF CONSCIOUS RATS CAUSED BY TRANSIENT ISCHEMIA

Figure 20.2a and Figure 20.2e shows sham ischemia in the dorsolateral striatum and the hippocampal CA1 pyramidal cell layers, respectively. Four days after reperfusion, the striatum (Figure 20.2b) and the hippocampal CA1 region (Figure 20.2f) show tissue injuries such as reactive astrocytosis, atrophic acidophilic cytoplasm, and pyknotic nuclei in preparations stained by hematoxylin–eosin and Klüver–Barrera methods. Ischemic damage is mild to moderate in the striatum and severe in the hippocampal CA1 region. By the quantitative analysis shown in Table 20.1, approximately 20 and 50% of the neurons show cellular injuries in the striatum and the hippocampal CA1 region, respectively.

## 20.4 INCREASES IN EXTRACELLULAR GLUTAMATE AND DELAYED NEURON DEATH BY TRANSIENT ISCHEMIA IN THE STRIATUM OF CONSCIOUS RATS EXAGGERATED BY INHIBITION OF STRIATAL AADC AND PROTECTED BY THE PERFUSION OF DOPA ANTAGONIST

Released DOPA may function to elicit glutamate release and resultant delayed neuron death by ischemia. Inhibition of intrastriatal AADC by perfusion of NSD-1015 (30 μ*M*) markedly increases the basal release of DOPA, triples the peak glutamate release by ischemia, exaggerates delayed neuron death (Figure 20.2c), and increases the density of ischemic neurons (Table 20.1) [36]. In contrast, intrastriatal perfusion of 30 to 100 n*M* DOPA CHE decreases by 80% the peak glutamate release by ischemia. This DOPA antagonist, at 100 n*M*, protects neurons from delayed cell death (Figure 20.2d) and decreases the density of ischemic neurons

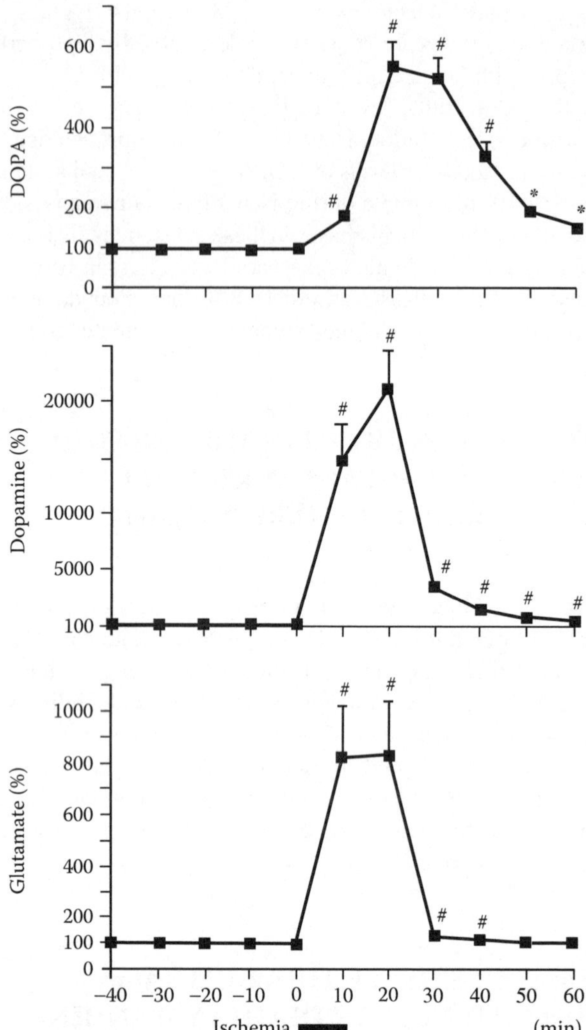


**FIGURE 20.1** Time course of DOPA, dopamine, and glutamate released by 10-min transient ischemia due to four-vessel occlusion (n = 38 to 40) during microdialysis of the right striata in conscious rats. The bilateral vertebral arteries were electrocauterized and a dummy cannula was implanted in the striatum. Two to three days later, the dummy cannula was replaced by a dialysis probe. Ringer solution was perfused at a rate of 2 μl/min. Dialysates were collected every 10 min. After the extracellular basal levels became stable, the bilateral carotid arteries were occluded at a horizontal bar. Carotid clamps were removed to achieve postischemic reperfusion. Data are means ± SE. $^*P < .05$, $\#P < .01$, vs. the value immediately before ischemia (paired *t*-test). (From Misu, Y. et al., DOPA causes glutamate release and delayed neuron death by brain ischemia in rats, *Neurotoxicol. Teratol.*, 24, 629, 2002. With permission.)

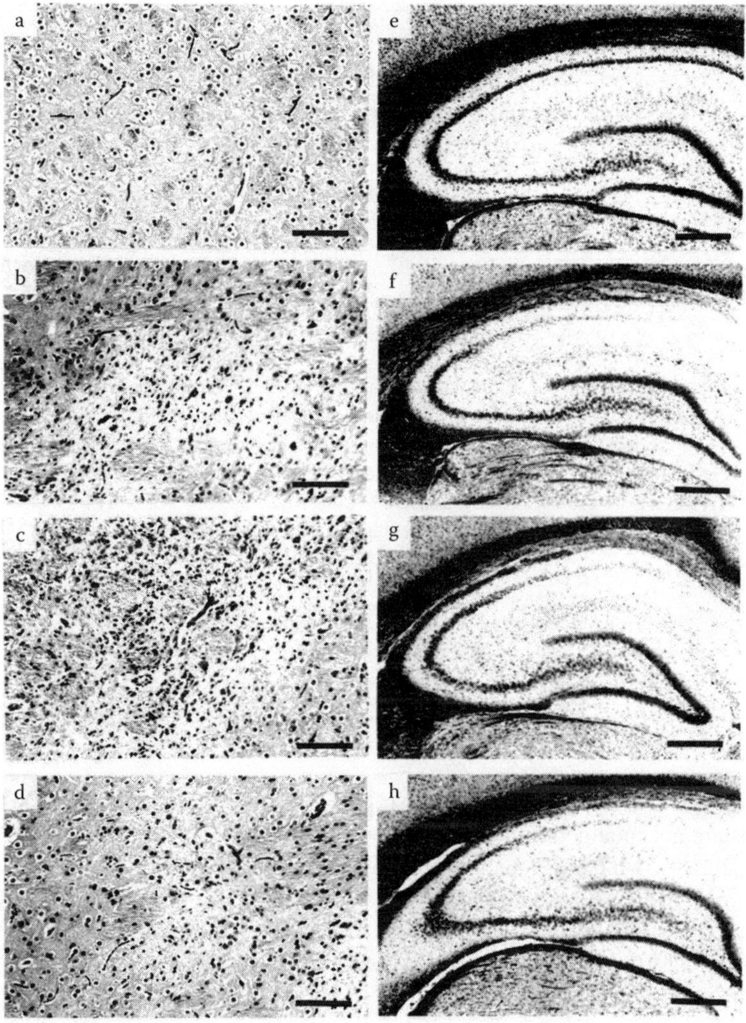


**FIGURE 20.2** Photomicrographs of the dorsolateral striatum (a to d) and the hippocampal CA1 region (e to h), showing sham ischemia (a and e) and typical neuropathological changes of delayed cell death by 10-min transient ischemia (b and f). Exaggeration of the changes by intrastriatal perfusion of 30 ν*M* NSD-1015 10 min before ischemia until 1 h after postischemic reperfusion (c) and protection by 100 n*M* DOPA CHE (d) are seen in the striatum, but there are no modifications in the hippocampal CA1 region (g and h). Four days after ischemia, rats were decapitated and the brains were removed, fixed with 10% formalin, cut, embedded in paraffin, and stained by hematoxylin–eosin and Klüver–Barrera methods. Scale bar, 100 νm in a to d and 500 νm in e to h. (From Misu, Y. et al., DOPA causes glutamate release and delayed neuron death by brain ischemia in rats, *Neurotoxicol. Teratol.*, 24, 629, 2002. With permission.)

**TABLE 20.1**
**Delayed Neuron Death by 10-Min Ischemia in Striatum and Hippocampal CA1 Region and Its Modifications by NSD-1015 and DOPA CHE**

| | | Density of Ischemic Neurons/mm$^2$ | |
|---|---|---|---|
| **Groups** | **n** | **Striatum** | **Hippocampus** |
| Ischemia alone | 17 | 34.9 ± 1.2 | 194.0 ± 2.0 |
| NSD-1015 30 μ*M* + Ischemia | 10 | 49.4 ± 0.8[a] | 195.2 ± 1.8 |
| DOPA CHE 100 n*M* + Ischemia | 12 | 22.0 ± 1.2[a] | 192.0 ± 1.8 |

*Note:* Intrastriatal perfusion of NSD-1015 or DOPA CHE was done 10 min before ischemia until 1 h after postischemic reperfusion. The brain was dissected 4 d after the acute experiments. Values are means ± SE of ischemic neurons/mm$^2$ in the dorsolateral striatum and the hippocampal CA1 region from n estimations. Intact neurons/mm$^2$ in sham ischemia (n = 5) are 179.5 ± 8.7 in the striatum and 366.7 ± 14.0 in the hippocampal CA1 region.

[a] $P < .01$, vs. ischemia alone (Student's *t*-test).

*Source:* From Misu, Y. et al., DOPA causes glutamate release and delayed neuron death by brain ischemia in rats, *Neurotoxicol. Teratol.*, 24, 629, 2002. With permission.

(Table 20.1). In the hippocampal CA1 region, the degree and density of ischemic neurons are not modified by intrastriatal perfusion of either drug (Figure 20.2g and Figure 20.2h). Although dopamine has been implicated in glutamate release and resultant delayed neuron death by ischemia [57,58], the dopamine released by ischemia tends to be decreased by NSD-1015 and is not modified by DOPA CHE. Dopamine does not appear to be involved in these events induced by ischemia under our experimental conditions. This idea is supported by the findings that aglycemia-induced increases in glutamate release from rat striatal slices are antagonized by DOPA CHE but not by dopamine $D_1$ and $D_2$ antagonists [59].

Judging by the time course of increases in DOPA, dopamine, and glutamate elicited by ischemia, it is not clear that DOPA is an upstream factor for glutamate release. One issue is that the increase in DOPA in the first sample is approximately one third of the peak, whereas the increase in glutamate has already reached its peak (Figure 20.1). It appears likely that there is a threshold for extracellular DOPA to trigger glutamate release by ischemia and an apparently low level of DOPA clears the threshold. Extremely low concentrations of levodopa elicit pharmacological responses [60]. Noneffective 3 to 10 p*M* levodopa potentiates the facilitation of noradrenaline release by 1 to 3 n*M* isoproterenol via activation of presynaptic β-adrenoceptors in rat hypothalamic slices. We confirmed that exogenously perfused levodopa (0.3 to 1 m*M*) stereoselectively releases glutamate in striata *in vivo* [36]. Another issue is a rapid decline from the peak increase in glutamate release seen in the process of the recovery, accompanied by the still high levels of DOPA (Figure 20.1). There are also clear discrepancies between the continuously raised DOPA levels resulting from exposure to NSD-1015 and the absence of further increases in glutamate release by ischemia. These findings may be explained by a downregulation of the DOPA recognition sites that trigger glutamate release following exposure to increased DOPA levels. This idea is supported by the finding that the highest dose of exogenously perfused levodopa (1 m*M*) tends to elicit desensitization of glutamate release in a time course similar to that after ischemic insult [36]. It appears likely that NSD-1015 triples the peak glutamate release at a critical time before the occurrence of such desensitization.

## 20.5 PERFUSION OF DOPA ANTAGONIST PROTECTING AGAINST INCREASES IN EXTRACELLULAR GLUTAMATE DURING HIPPOCAMPAL MICRODIALYSIS AND DELAYED NEURON DEATH BY MILD TRANSIENT ISCHEMIA IN THE HIPPOCAMPAL CA1 REGION OF CONSCIOUS RATS

We tried to clarify whether or not DOPA CHE antagonizes glutamate release and delayed neuron death by ischemia in the hippocampal CA1 region most vulnerable to brain ischemia [1,2,5,6]. The antagonist is neuroprotective under mild ischemic conditions [30]. DOPA CHE is a mother compound for developing neuroprotectants. Ischemia of 5- and 10-min duration causes mild glutamate release in the 10-min samples during microdialysis (Figure 20.3A and Figure 20.3B) and approximately 20% (Table 20.2) [30] to 50% (Table 20.1) [36] neuron injuries 4 d after reperfusion. Compared with striata [36], the glutamate released by 10-min ischemia is lower (one fourth), but the density of ischemic neurons is higher (2.5-fold), showing the high vulnerablity of the hippocampal CA1 region [1,2,5,6]. DOPA and dopamine are below assay sensitivity in this design but are basally detected in 20-min samples and released by 20-min ischemia. The result seen with dopamine is consistent with previous findings [5]. We cannot measure an extracellular level of DOPA lower than 0.4 n*M* because its assay sensitivity is 0.05 fmol/μl and the recovery of probes used is 12% [36].

Intrahippocampal perfusion of 100 n*M* DOPA CHE abolishes glutamate release (Figure 20.3B) and protects neurons from cell death elicited by 5-min ischemia (Table 20.2). The antagonist ought to exert neuroprotection via antagonism of DOPA. Released DOPA that is lower than the measurable extracellular level (0.4 n*M*) appears to clear over a threshold to elicit glutamate release and to cause resultant delayed neuron death by ischemia in the hippocampal CA1 region. Extremely low concentrations of levodopa (3 to 10 p*M*) elicit pharmacological responses that potentiate the activity of presynaptic β-adrenoceptors [60]. These findings suggest that the sensitivity of a biochemical analysis used, the release of DOPA, to find DOPAergic components to be identified as a neurotransmitter and/or neuromodulator, is lower,

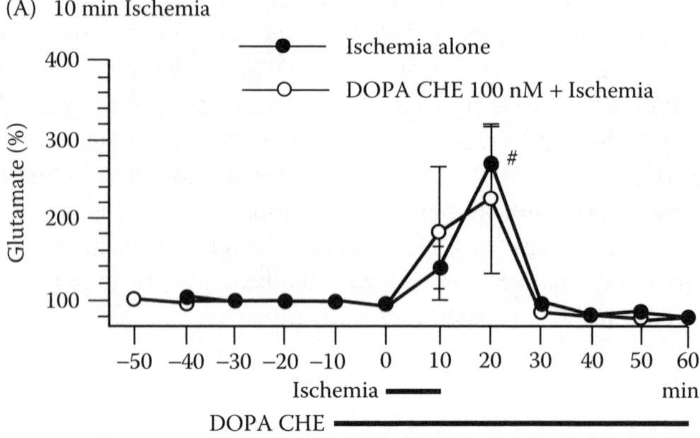


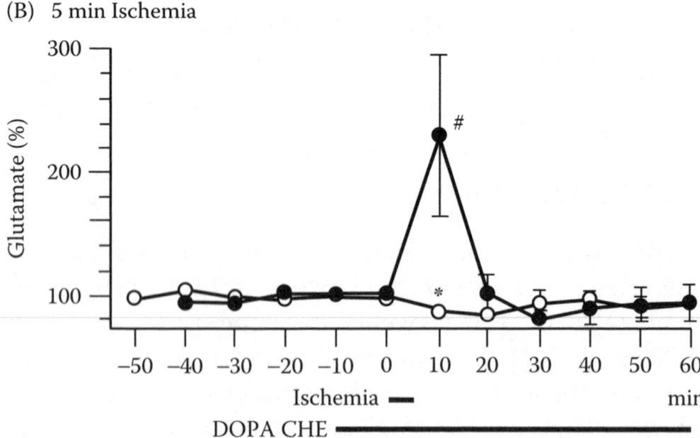


**FIGURE 20.3** Effect of intrahippocampal perfusion with 100 n*M* DOPA CHE on glutamate release by 10-min (A) and 5-min (B) ischemia at short horizontal bars during microdialysis of the right hippocampal CA1 region in conscious rats. DOPA CHE was perfused 10 min before ischemia until 1 h after postischemic reperfusion at long horizontal bars. #$P < .05$ vs. the value immediately before ischemia (paired $t$-test), $^{*}P < .01$ vs. ischemia alone (Mann–Whitney $U$ test). Other details are as in Figure 20.1. (From Misu, Y. et al., DOPA causes glutamate release and delayed neuron death by brain ischemia in rats, *Neurotoxicol. Teratol.*, 24, 629, 2002. With permission.)

compared with pharmacological approaches by means of DOPAergic agonist and antagonist.

Although DOPA CHE acts on the NMDA ion channel domain with millimolar $IC_{50}$ [38], DOPA CHE is unlikely to elicit neuroprotection via blockade of this site because the nanomolar dose itself is effective. Neuroprotection by systemic (+)-5-methyl-10,11-dihydro-5*H*-dibenzol[*a*,*d*]cyclohepten-5,10-imine maleate (MK-801), an NMDA receptor ion channel antagonist, is largely attributed to hypothermia [61]. DOPA CHE, however, elicits no hypothermia during and after ischemia, at least, following intrastriatal perfusion [36].

**TABLE 20.2**
**Delayed Neuron Death by 5-Min Ischemia in the Hippocampal CA1 Region and Its Protection by DOPA CHE**

| Groups | n | Density of Ischemic Neurons/mm$^2$ |
|---|---|---|
| Ischemia alone | 8 | 78.5 ± 7.0 |
| DOPA CHE 100 n*M* + Ischemia | 6 | 11.6 ± 2.8[a] |

*Note:* Intrahippocampal perfusion of DOPA CHE was done 10 min before ischemia until 1 h after postischemic reperfusion. Other details are as in Table 20.1.

[a] $P < .01$ vs. ischemia alone (Student's *t*-test).

*Source:* From Misu, Y. et al., DOPA causes glutamate release and delayed neuron death by brain ischemia in rats, *Neurotoxicol. Teratol.*, 24, 629, 2002. With permission.

In contrast, DOPA CHE at 100 n*M* does not inhibit the increase in glutamate release elicited by 10-min ischemia, compared with 5-min ischemia (Figure 20.3A). Different mechanisms might underlie the effects of mild and severe ischemia. We suggest that glutamate released by mild ischemia is largely vesicular [32], but the glutamate increase elicited by severe ischemia is mainly mediated by the reversed operation of neuronal glutamate transporters [10]. DOPA might not be involved in this process. This idea is further supported by the findings that DOPA CHE antagonizes glutamate release by mild insult such as glucose deprivation alone in rat striatal slices but does not antagonize it by the severe insult of aglycemia plus oxygen deprivation [59]. However, it should be always taken into consideration that DOPA CHE still has a characteristic as a prodrug of DOPA [34].

## 20.6 IS LEVODOPA NEUROTOXIC IN LONG-TERM THERAPY OF PARKINSON'S DISEASE?

Although levodopa still remains the most effective drug for treatment of Parkinson's disease, disease progression does not appear to be altered. Adverse effects such as decreased control of symptoms, increased dyskinesia, increased diurnal fluctuations, episodes of akinetic freezing and crisis, increased fatigue and neurasthenia, and alteration in mentation are common during long-term therapy of more than 5 yr. Furthermore, there has been concern that chronic monotherapy with levodopa in patients with Parkinson's disease might accelerate the degeneration process of nigrostriatal dopaminergic neurons [18,19].

Levodopa by itself, or accumulated cytoplasmic dopamine following the long-term administration of levodopa, is likely to be related to the abnormal motor responses and might be related to acceleration of dopaminergic degeneration. A lot of studies have shown that levodopa and dopamine produce neurotoxic disorders in animal studies *in vitro* and *in vivo* [41,45,62]. Exogenously applied levodopa [32,49] and endogenously derived DOPA [36,59] release endogenous glutamate in their own right and produce delayed neuron death in rat striata *in vitro* [41,49] and *in vivo* [36]. Furthermore, the degradation products of levodopa and dopamine converted from levodopa, such as reactive free radicals, DOPA quinones, and dopamine quinones, may produce nigrostriatal excitotoxic damages [63–69].

Therefore, several agents have been developed as a part of a levodopa-sparing strategy, such as agonists of the dopamine $D_2$ receptor subfamily including $D_2$, $D_3$, and $D_4$ receptors. These are ergot derivatives such as bromocriptine, pergolide, and cabergoline, and nonergot compounds such as ropinirole, pramipexole, and talipexole. Growing evidence has suggested that these dopamine agonists have probable neuroprotective properties: to decrease biosynthesis and turnover of dopamine following activation of presynaptic dopamine $D_2$ autoreceptors, to scavenge reactive free radicals, and to increase endogenous antioxidant and neurotrophic substances, in addition to stimulation of postsynaptic dopamine receptors [69–73].

Bromocriptine inhibits glutamate-induced death in rat-cultured mesencephalic neurons in a dopamine $D_2$ antagonist-sensitive manner [74]. Bromocriptine and pergolide are suggested to directly scavenge hydroxyl radicals [75] and NO radicals [76]. Pramipexole and talipexole reduce 1-methyl-4-phenylpyridinium ion-induced production of reactive oxygen species in human neuroblastoma SH-SY5Y cells [77,78] and in rat striatum *in vivo* [77], and inhibit the opening of the mitochondrial transition pore [77] and apoptotic neuron death [78] in these cells. In mice Parkinson's models, bromocriptine [79] and ropinirole [80] scavenge reactive free radicals and also increase the level of the endogenous antioxidant glutathione. Furthermore, in cultured mouse astrocytes, pergolide and cabergoline increase the levels of nerve growth factor (NGF) and glial cell line-derived neurotrophic factor, and bromocriptine increases the level of NGF [81]. These findings suggest that dopamine $D_2$ agonists are able to be neuroprotectants in addition to their levodopa-sparing characteristic in patients with Parkinson's disease.

In recent years, new clinical approaches have been attempted, such as monotherapy with $D_2$ receptor subfamily agonists and early combination therapy with levodopa [72,82]. Furthermore, comparative studies of $D_2$ agonist vs. levodopa on the possible modification of disease progression in early Parkinson's disease, by dopamine transporter imaging using single-photon emission computed tomography (SPECT) with $[^{123}I]$2β-carboxymethoxy-3β(4-iodophenyl)tropane (β-CIT) [83] and by DOPA uptake with $[^{18}F]$DOPA positron emission tomography (PET) [84], have shown that the rate of loss of striatal β-CIT uptake until 46 months of follow-up is reduced for those subjects initially treated with pramipexole compared with those initially treated with levodopa ($n = 82$) [83] and that the reductions in $[^{18}F]$DOPA uptake in the putamen and substantia nigra for 24 months of follow-up are lesser with ropinirole compared with levodopa ($n = 186$) [84]. These findings likely reflect a relative reduction by $D_2$ agonists, or acceleration by levodopa, of the progressive loss of striatal dopamine neuronal function, although these radiotracers do not measure the exact number or density of dopaminergic neurons. Another clinical trial with levodopa and a matching placebo for 40 weeks in 361 patients with early Parkinson's disease suggested that levodopa either slows the progression of the disease or has a prolonged effect on the symptoms of the disease. In contrast, the neuroimaging data with [$^{123}$I]β-CIT suggest either that levodopa accelerates the loss of nigrostriatal dopamine nerve terminals or that its pharmacological effects modify the dopamine transporter [85].

A criticism is stated that in Parkinson's disease these biomarkers must have not only biological relevance but also a strong linkage to the clinical outcome [86]. No radiotracers fulfill these criteria, and current evidence does not support the use of imaging as a diagnostic tool in clinical practice or as a surrogate endpoint in clinical trials. Thus, the findings with radiotracers [83–85] suggest the need to compare these imaging markers of dopaminergic neuronal loss with multiple meaningful clinical endpoints of disease progression in larger and potentially long-term studies to fully assess the clinical relevance.

## 20.7 CONCLUSION

Endogenously released DOPA appears to be an upstream factor for glutamate release and resultant delayed neuron death due to transient brain ischemia in the striatum and the hippocampal CA1 region of conscious rats. The common sequence of the cascades might be: the activation of DOPA recognition sites, glutamate release, activation of ionotropic glutamate receptors, nNOS activation, NO production, and, finally, delayed neuron death. However, interactions between the released glutamate and compounds produced as a result of autoxidation of levodopa, and autoxidation or enzymatic oxidation of dopamine converted from levodopa, might start amplifying cycles of neurotoxic cascades. Thus, at present, there is some concern that levodopa therapy might accelerate the neuronal degeneration process, especially at the later stages of Parkinson's disease [19,21,22,56]. Levodopa-induced glutamate release [32,36,49,59] appears to be at least involved in the adverse effects such as neuroexcitatory motor responses [19,21,22,56]. In addition, it is possible that some glutamate antagonists can slow the progression of Parkinson's disease. Clinical trials of glutamate antagonists (e.g., [87]) should preferably be done to get safe and tolerable adjuncts to the combination therapy of levodopa and dopamine agonists for patients with Parkinson's disease (see Chapter 22).

## ACKNOWLEDGMENTS

This study was in part supported by Grants-in-Aid for Developmental Scientific Research (No. 06557143); Scientific Research (No. 07407003, 09877022, 09280280, 10176229, and 10470026) from the Ministry of Education, Science, Sports, and Culture, Japan; and by grants from the Mitsubishi Foundation, Japan; the Uehara Memorial Foundation, Japan; and SRF, Japan. Many thanks go to Sanae Sato, Department of Pharmacology, Fukushima Medical University School of Medicine, Fukushima, Japan, for remaking the figures.

## REFERENCES


1. Pulsinelli, W.A., Brierley, J.B., and Plum, F., Temporal profile of neuronal damage in a model of transient forebrain ischemia, *Ann. Neurol.,* 11, 491, 1982.
2. Schmidt-Kastner, R. and Freund, T.F., Selective vulnerability of the hippocampus in brain ischemia, *Neuroscience,* 40, 599, 1991.
3. Pulsinelli, W.A., Pathophysiology of acute ischaemic stroke, *Lancet,* 339, 533, 1992.
4. Olney, J.W. and Sharpe, L.G., Brain lesions in an infant rhesus monkey treated with monosodium glutamate, *Science,* 166, 386, 1969.
5. Obrenovitch, T.P. and Richards, D.A., Extracellular neurotransmitter changes in cerebral ischemia, *Cerebrovasc. Brain Metab. Rev.,* 7, 1, 1995.
6. Choi, D.W., Glutamate neurotoxicity and diseases of the nervous system, *Neuron,* 1, 623, 1988.
7. Lipton, S.A. and Rosenberg, P.A., Excitatory amino acids as a final common pathway for neurologic disorders, *N. Engl. J. Med.,* 330, 613, 1994.
8. Drejer, J. et al., Cellular origin of ischemia-induced glutamate release from brain tissue in vivo and in vitro, *J. Neurochem.,* 45, 145, 1985.
9. Szatkowski, M., Barbour, B., and Attwell, D., Non-vesicular release of glutamate from glial cells by reversed electrogenic glutamate uptake, *Nature,* 348, 443, 1990.
10. Rossi, D.J., Oshima, T., and Attwell, D., Glutamate release in severe brain ischaemia is mainly by reversed uptake, *Nature,* 403, 316, 2000.
11. Carlsson, A., Lindqvist, M., and Magnusson, T., 3,4-Dihydroxyphenylalanine and 5-hydroxytryptophan as reserpine antagonists, *Nature,* 180, 1200, 1957.
12. Ehringer, H. and Hornykiewicz, O., Verteilung von Noradrenaline und Dopamine (3-Hydroxytyramin) im Gehirn des Menschen und ihr Verhalten bei Erkrankungen des extrapyramidalen Systems, *Klin. Wochenschr.,* 38, 1236, 1960.
13. Bartholini, G. et al., Increase of cerebral catecholamines by 3,4-dihydroxyphenylalanine after inhibition of peripheral decarboxylase, *Nature,* 215, 852, 1967.
14. Hefti, F. and Melamed, E., L-DOPA's mechanism of action in Parkinson's disease, *Trends Neurosci.,* 3, 229, 1980.
15. Goshima, Y., Kubo, T., and Misu, Y., Biphasic actions of L-DOPA on the release of endogenous noradrenaline and dopamine from rat hypothalamic slices, *Br. J. Pharmacol.,* 89, 229, 1986.
16. Misu, Y., Goshima, Y., and Kubo, T., Biphasic actions of L-DOPA on the release of endogenous dopamine via presynaptic receptors in rat striatal slices, *Neurosci. Lett.,* 72, 194, 1986.
17. Misu, Y. and Goshima, Y., Is L-dopa an endogenous neurotransmitter?, *Trends Pharmacol. Sci.,* 14, 119, 1993.
18. Misu, Y., Ueda, H., and Goshima, Y., Neurotransmitter-like actions of L-DOPA, *Adv. Pharmacol.,* 32, 427, 1995.
19. Misu, Y. et al., Neurobiology of L-DOPAergic systems, *Prog. Neurobiol.,* 49, 415, 1996.
20. Hoffman, B.B. and Taylor, P., Neurotransmission: the autonomic and somatic motor nervous systems, in *Goodman & Gilman's The Pharmacological Basis of Therapeutics,* Hardman, J.G., Limbird, L.E., and Gilman, A.G., Eds., McGraw-Hill, New York, 2001, 115.
21. Misu, Y., Goshima, Y., and Miyamae, T., Is DOPA a neurotransmitter?, *Trends Pharmacol. Sci.,* 23, 262, 2002.
22. Misu, Y., Kitahama, K., and Goshima, Y., L-3,4-Dihydroxyphenylalanine as a neurotransmitter candidate in the central nervous system, *Pharmacol. Ther.,* 97, 117, 2003.

23. Kubo, T. et al., Evidence for L-DOPA systems responsible for cardiovascular control in the nucleus tractus solitarii of the rat, *Neurosci. Lett.,* 140, 153, 1992.
24. Yue, J.-L. et al., Baroreceptor-aortic nerve-mediated release of endogenous L-3,4-dihydroxyphenylalanine and its tonic function in the nucleus tractus solitarii of rats, *Neuroscience,* 62, 145, 1994.
25. Yue, J.-L. et al., Altered tonic L-3,4-dihydroxyphenylalanine systems in the nucleus tractus solitarii and the rostral ventrolateral medulla of spontaneously hypertensive rats, *Neuroscience,* 67, 95, 1995.
26. Jaeger, C.B. et al., Aromatic L-amino acid decarboxylase in the rat brain: immunocytochemical localization in neurons of the brain stem, *Neuroscience,* 11, 691, 1984.
27. Tison, F. et al., Endogenous L-DOPA in the rat dorsal vagal complex: an immunocytochemical study by light and electron microscopy, *Brain Res.,* 497, 260, 1989.
28. Goshima, Y., Kubo, T., and Misu, Y., Transmitter-like release of endogenous 3,4-dihydroxyphenylalanine from rat striatal slices, *J. Neurochem.,* 50, 1725, 1988.
29. Nakamura, S. et al., Transmitter-like basal and $K^+$-evoked release of 3,4-dihydroxyphenylalanine from the striatum in conscious rats studied by microdialysis, *J. Neurochem.,* 58, 270, 1992.
30. Arai, N. et al., DOPA cyclohexyl ester, a competitive DOPA antagonist, protects glutamate release and resultant delayed neuron death by transient ischemia in hippocampus CA1 of conscious rats, *Neurosci. Lett.,* 299, 213, 2001.
31. Goshima, Y., Nakamura, S., and Misu, Y., L-Dihydroxyphenylalanine methyl ester is a potent competitive antagonist of the L-dihydroxyphenylalanine-induced facilitation of the evoked release of endogenous norepinephrine from rat hypothalamic slices, *J. Pharmacol. Exp. Ther.,* 258, 466, 1991.
32. Goshima, Y. et al., L-DOPA induces $Ca^{2+}$-dependent and tetrodotoxin-sensitive release of endogenous glutamate from rat striatal slices, *Brain Res.,* 617, 167, 1993.
33. Misu, Y. et al., L-DOPA cyclohexyl ester is a novel stable and potent competitive antagonist against L-DOPA, as compared to L-DOPA methyl ester, *Jpn. J. Pharmacol.,* 75, 307, 1997.
34. Furukawa, N. et al., L-DOPA cyclohexyl ester is a novel potent and relatively stable competitive antagonist against L-DOPA among L-DOPA ester compounds in rats, *Jpn. J. Pharmacol.,* 82, 40, 2000.
35. Akbar, M. et al., Inhibition by L-3,4-dihydroxyphenylalanine of hippocampal CA1 neurons with facilitation of noradrenaline and $\gamma$-amino butyric acid release, *Eur. J. Pharmacol.,* 414, 197, 2001.
36. Furukawa, N. et al., Endogenously released DOPA is a causal factor for glutamate release and resultant delayed neuronal cell death by transient ischemia in rat striata, *J. Neurochem.,* 76, 815, 2001.
37. Ishii, H. et al., Involvement of rBAT in $Na^+$-dependent and -independent transport of the neurotransmitter candidate L-DOPA in *Xenopus laevis* oocytes injected with rabbit small intestinal epithelium poly $A^+$ RNA, *Biochim. Biophys. Acta,* 1466, 61, 2000.
38. Miyamae, T. et al., Some interactions of L-DOPA and its related compounds with glutamate receptors, *Life Sci.,* 64, 1045, 1999.
39. Nakamura, S. et al., Non-effective dose of exogenously applied L-DOPA itself stereoselectively potentiates postsynaptic $D_2$-receptor-mediated locomotor activities of conscious rats, *Neurosci. Lett.,* 170, 22, 1994.
40. Ueda, H. et al., L-DOPA inhibits spontaneous acetylcholine release from the striatum of experimental Parkinson's model rats, *Brain Res.,* 698, 213, 1995.
41. Cheng, N.-N. et al., Differential neurotoxicity induced by L-DOPA and dopamine in cultured striatal neurons, *Brain Res.,* 743, 278, 1996.

42. Yue, J.-L. et al., Endogenously released L-DOPA itself tonically functions to potentiate $D_2$ receptor-mediated locomotor activities of conscious rats, *Neurosci. Lett.*, 170, 107, 1994.
43. Nishihama, M. et al., An L-DOPAergic relay from the posterior hypothalamic nucleus to the rostral ventrolateral medulla and its cardiovascular function in anesthetized rats, *Neuroscience,* 92, 123, 1999.
44. Misu, Y. and Kubo, T., Presynaptic β-adrenoceptors, *Med. Res. Rev.,* 6, 197, 1986.
45. Fahn, S., Is levodopa toxic?, *Neurology,* 47 (Suppl. 3), S184, 1996.
46. Olney, J.W. et al., Excitotoxicity of L-DOPA and 6-OH-DOPA: implications for Parkinson's and Huntington's diseases, *Exp. Neurol.,* 108, 269, 1990.
47. Rosenberg, P.A. et al., 2,4,5-Trihydroxyphenylalanine in solution forms a non-*N*-methyl-D-aspartate glutamatergic agonist and neurotoxin, *Proc. Natl. Acad. Sci. USA,* 88, 4865, 1991.
48. Newcomer, T.A., Rosenberg, P.A., and Aizenman, E., TOPA quinone, a kainate-like agonist and excitotoxin, is generated by a catecholaminergic cell line, *J. Neurosci.,* 15, 3172, 1995.
49. Maeda, T. et al., L-DOPA neurotoxicity is mediated by glutamate release in cultured rat striatal neurons, *Brain Res.,* 771, 159, 1997.
50. Siesjö, B.K. and Bengtsson, F., Calcium fluxes, calcium antagonists, and calcium-related pathology in brain ischemia, hypoglycemia, and spreading depression: a unifying hypothesis, *J. Cereb. Blood Flow Metabol.,* 9, 127, 1989.
51. Lipton, P., Ischemic cell death in brain neurons, *Physiol. Rev.,* 79, 1431, 1999.
52. Gill, R., Foster, A.C., and Woodruff, G.N., Systemic administration of MK-801 protects against ischemia-induced hippocampal neurodegeneration in the gerbil, *J. Neurosci.,* 7, 3343, 1987.
53. Sheardown, M.J. et al., 2,3-Dihydroxy-6-nitro-7-sulfamoyl-benzo (F) quinoxaline: a neuroprotectant for cerebral ischemia, *Science,* 247, 571, 1990.
54. Dawson, V.L. et al., Mechanisms of nitric oxide-mediated neurotoxicity in primary brain cultures, *J. Neurosci.,* 13, 2651, 1993.
55. Lipton, S.A. et al., A redox-based mechanism for the neuroprotective and neurodestructive effects of nitric oxide and related nitroso-compounds, *Nature,* 364, 626, 1993.
56. Misu, Y. et al., DOPA causes glutamate release and delayed neuron death by brain ischemia in rats, *Neurotoxicol. Teratol.,* 24, 629, 2002.
57. Globus, M.Y.-T. et al., Effect of ischemia on the in vivo release of striatal dopamine, glutamate, and γ-aminobutyric acid studied by intracerebral microdialysis, *J. Neurochem.,* 51, 1455, 1988.
58. Prado, R., Busto, R., and Globus, M.Y.-T., Ischemia-induced changes in extracellular levels of striatal cyclic AMP: role of dopamine neurotransmission, *J. Neurochem.,* 59, 1581, 1992.
59. Hashimoto, M. et al., DOPA cyclohexyl ester potently inhibits aglycemia-induced release of glutamate in rat striatal slices, *Neurosci. Res.,* 45, 335, 2003.
60. Goshima, Y. et al., Picomolar concentrations of L-DOPA stereoselectively potentiate activities of presynaptic β-adrenoceptors to facilitate the release of endogenous noradrenaline from rat hypothalamic slices, *Neurosci. Lett.,* 129, 214, 1991.
61. Buchan, A. and Pulsinelli, W.A., Hypothermia but not the *N*-methyl-D-aspartate antagonist, MK-801, attenuates neuronal damage in gerbil subjected to transient global ischemia, *J. Neurosci.,* 10, 311, 1990.
62. Jenner, P.G. and Brin, M.F., Levodopa neurotoxicity: experimental studies versus clinical relevance, *Neurology,* 50 (Suppl. 6), S39, 1998.

63. Graham, D.G., Oxidative pathways for catecholamines in the genesis of neuromelanin and cytotoxic quinones, *Mol. Pharmacol.,* 14, 633, 1978.
64. Halliwell, B., Reactive oxygen species and the central nervous system, *J. Neurochem.,* 59, 1609, 1992.
65. Fahn, S. and Cohen, G., The oxidant stress hypothesis in Parkinson's disease. Evidence supporting it, *Ann. Neurol.,* 32, 804, 1992.
66. Smith, T.S., Parker, W.D., and Bennett, J.P., L-DOPA increases nigral production of hydroxyl radicals in vivo: potential role of L-DOPA toxicity?, *NeruroReport,* 5, 1009, 1994.
67. Basma, A.N. et al., DOPA cytotoxicity in PC12 cells in culture is via its autoxidation, *J. Neurochem.,* 64, 718, 1995.
68. Metodiewa, D. and Koska, C., Reactive oxygen species and reactive nitrogen species: relevance to cyto(neuro)toxic events and neurologic disorders. An overview, *Neurotoxicity Res.,* 1, 197, 2000.
69. Asanuma, M., Miyazaki, I., and Ogawa, N., Dopamine or L-DOPA-induced neurotoxicity: the role of dopamine quinone formation and tyrosinase in a model of Parkinson's disease, *Neurotoxicity Res.,* 5, 165, 2003.
70. Hagan, J.J. et al., Parkinson's disease: prospects for improved drug therapy, *Trends Pharmacol. Sci.,* 18, 307, 1997.
71. Tolosa, E. et al., History of levodopa and dopamine agonists in Parkinson's disease treatment, *Neurology,* 50 (Suppl. 6), S44, 1998.
72. Hubble, J.P., Pre-clinical studies of pramipexole: clinical relevance, *Eur. J. Neurol.,* 7 (Suppl. 1), 15, 2000.
73. Kitamura, Y. et al., Neuroprotective mechanisms of antiparkinsonian dopamine $D_2$-receptor subfamily agonists, *Neurochem. Res.,* 28, 1035, 2003.
74. Sawada, H. et al., Dopamine D2-type agonists protect mesencephalic neurons from glutamate neurotoxicity: mechanisms of neuroprotective treatment against oxidative stress, *Ann. Neurol.,* 44, 110, 1998.
75. Yoshikawa, T. et al., Antioxidant properties of bromocriptine, a dopamine agonist, *J. Neurochem.,* 62, 1034, 1994.
76. Nishibayashi, S. et al., Scavenging effects of dopamine agonists on nitric oxide radicals, *J. Neurochem.,* 67, 2208, 1996.
77. Cassarino, D.S. et al., Pramipexole reduces reactive oxygen species production in vivo and in vitro and inhibits the mitochondrial permeability transition produced by the parkinsonian neurotoxin methylpyridinium ion, *J. Neurochem.,* 71, 295, 1998.
78. Kitamura, Y. et al., Protective effects of the antiparkinsonian drugs talipexole and pramipexole against 1-methyl-4-phenylpyridinium-induced apoptotic death in human neuroblastoma SH-SY5Y cells, *Mol. Pharmacol.,* 54, 1046, 1998.
79. Muralikrishnan, D. and Mohanakumar, K.P., Neuroprotection by bromocriptine against 1-methyl-4-phenyl-1,2,3,6-tetrahydropyridine-induced neurotoxicity in mice, *FASEB J.,* 12, 905, 1998.
80. Tanaka, K. et al., Molecular mechanism in activation of glutathione system by ropinirole, a selective dopamine D2 agonist, *Neurochem. Res.,* 26, 31, 2001.
81. Ohta, K. et al., Effects of dopamine agonists bromocriptine, pergolide, cabergoline, and SKF-38393 on GDNF, NGF, and BDNF synthesis in cultured mouse astrocytes, *Life Sci.,* 73, 617, 2003.
82. Olanow, C.W. and Koller, W.C., An algorithm (decision tree) for the managemant of Parkinson's disease: treatment guideline, *Neurology,* 50 (Suppl. 3), S1, 1998.

83. Marek, K. et al., The Parkinson Study Group, Dopamine transporter brain imaging to assess the effects of pramipexole vs levodopa on Parkinson's disease progression, *JAMA,* 287, 1653, 2002.
84. Whone, A.L. et al., Slower progression of Parkinson's disease with ropinirole versus levodopa: the REAL-PET study, *Ann. Neurol.,* 54, 93, 2003.
85. Fahn, S. et al., The Parkinson study Group, Levodopa and the progression of Parkinson's disease, *N. Engl. J. Med.,* 351, 2498, 2004.
86. Ravina, B. et al., The role of radiotracer imaging in Parkinson's disease, *Neurology,* 64, 208, 2005.
87. Shoulson, I. et al., The Parkinson Study Group, A randomized, controlled trial of remacemide for motor fluctuations in Parkinson's disease, *Neurology,* 56, 455, 2001.

# 21 Effect of Neonatal Treatment with Monosodium Glutamate on Dopamine and DOPA Neurons in the Medial Basal Hypothalamus of Rats

*Ibolya Bodnár, Daniel Szekács, Márk Oláh, Hitoshi Okamura, Miklós Vecsernyés, Marton I.K. Fekete, and György M. Nagy*

## CONTENTS

## 21.1 INTRODUCTION

It is well known that the hypothalamic neuroendocrine dopamine (NEDA) neurons consist of three morphologically distinct and functionally independent systems. Neurons arising from the rostral periventricular regions of the hypothalamus and terminating in the intermediate lobe (IL) of the pituitary gland form the periventriculo–hypophyseal dopamine (PHDA) system [1–4]. These neurons are responsible for control of the α-melanocyte–stimulating hormone (α-MSH) secretion of the IL [3–7]. The second subpopulation of neurons originates from the rostral portion of the hypothalamic arcuate nucleus (ARC) dorsally, adjacent to the PHDA system, and terminates in both the neural lobe (NL) and the IL of the pituitary gland [8,9]. It is called the tuberohypophyseal dopamine (THDA) system and forms the osmosensitive hypothalamic dopaminergic (DAergic) system [10–12] and also participates in the control of anterior lobe prolactin (PRL) secretion [8,13–16]. The third group of cells, the tuberoinfundibular dopamine (TIDA) system, is located in the mid- and caudal portions of the ARC and is well accepted as a major tonic regulator of the adenohypophyseal PRL secretion [1,2,17,18]. Recently, it has been demonstrated that only a part of the TIDA neurons contains the two enzymes, tyrosine hydroxylase (TH) and aromatic L-amino acid decarboxylase (AADC), that are indispensable for L-3,4-dihydroxyphenylalanine (DOPA) and dopamine (DA) synthesis, respectively. It has been shown in fetal [17,18], perinatal [19], and adult [20–24] rats that neurons located in the dorsomedial (DM) part of TIDA system contain both enzymes. Thus, they may be termed as "the genuine" DAergic neurons, whereas neurons originating from the ventrolateral (VL) part of the ARC contain TH but not AADC and, therefore, they can be considered as a DOPAergic neuronal system [22].

In our present studies, we have investigated the sensitivity of the rostrocaudal subdivisions of hypothalamic NEDA neurons, i.e., PHDA, THDA, and TIDA systems, and their terminal fields in the median eminence (ME), the IL, and the NL of the pituitary gland, respectively, to monosodium L-glutamate (MSG) with special emphasis on the DM DAergic and the VL DOPAergic neurons of the TIDA system. To obtain information on the effect of MSG, the number of TH and AADC-immunoreactive neurons were counted. We have also determined the levels of DA and DOPA in the ME and the ARC, as well as the changes in the concentration of these catecholamines following MSG treatment. Basal levels of plasma PRL and α-MSH in both groups of animals were also tested.

## 21.2 EFFECT OF MONOSODIUM GLUTAMATE (MSG) ON THE DISTRIBUTION OF TH AND AADC-IMMUNOREACTIVE NEURONS IN MEDIOBASAL HYPOTHALAMUS

At the middle and posterior levels of the ARC (TIDA system) of control animals, intensely stained TH-immunoreactive cells could be detected in both the DM and VL parts of the nucleus, whereas AADC-immunoreactive perikarya could be found only in the DM portion of the nuclei (Figure 21.1A and Figure 21.1C). About half of the TH-immunoreactive neurons was also AADC immunoreactive and the other half was AADC-negative, indicating that half of the TIDA neurons is DAergic and the other half is DOPAergic (Figure 21.1A, Figure 21.1C, and Figure 21.2B). In the rostral periventricular region (PHDA/THDA systems) of intact rats, the great majority of the TH-immunopositive neurons also exhibited AADC immunoreactivity, indicating that most of these neurons are DAergic (Figure 21.2A).

MSG treatment resulted in about a 35 to 40% reduction of the TIDA neurons immunoreactive for TH (Figure 21.2B). The TH-immunopositive neurons almost completely disappeared from the VL part of the cell group without significant change in the presence of such nerve cells in the DM part of the nucleus (Figure 21.1B and Figure 21.2B). At the same time, at these levels of the ARC, the number of AADC-immunoreactive cells in the MSG-treated rats was very similar to the controls (Figure 21.1C, Figure 21.1D, and Figure 21.2B). This indicates that the marked reduction in TH-immunoreactive neurons was primarily due to the reduction of neurons expressing TH only, i.e., the DOPAergic neurons. Treatment with MSG did not cause considerable cell damage of the neurons in the rostral periventricular region (PHDA/THDA systems, Figure 21.2A). The number of immunoreactive cells for both TH and AADC of this region was almost identical to that of untreated controls.

It should be emphasized that there were no sexual differences in the morphological appearance of DAergic and DOPAergic neurons in the TIDA system, as well as in the PHDA/THDA systems, in intact as well as MSG-treated animals [25].

## 21.3 EFFECT OF MSG ON DISTRIBUTION OF TH- AND AADC-IMMUNOREACTIVE NEURONAL TERMINALS IN PITUITARY GLAND

In agreement with the findings in the hypothalamic PHDA/THDA neurons, there were no appreciable differences in the amount and distribution of TH- or AADC-immunopositive fibers at the terminal fields of these neuronal systems in the IL and the NL of the pituitary gland of MSG-treated rats (Figure 21.3B and Figure 21.3D), compared to untreated controls (Figure 21.3A and Figure 21.3C).

## 21.4 PLASMA PROLACTIN AND A-MSH IN MSG-TREATED ANIMALS

There were no changes [25] in the basal level of plasma PRL of MSG-treated animals (Plasma PRL was measured by RIA using materials kindly provided by the National Hormone and Pituitary Program). It has been also shown previously that inhibition of TH with α-methyl-*p*-tyrosine (α-MPT, 8 mg/kg body weight, i.v.) increased plasma PRL concentrations in both sexes, and MSG treatment did not interfere with PRL responses to inhibition of TH [25]. At the same time, α-MPT treatment did not cause any alterations of plasma α-MSH levels (α-MSH was determined by specific RIA developed in our laboratory [26]) in controls or in MSG-treated rats [25].

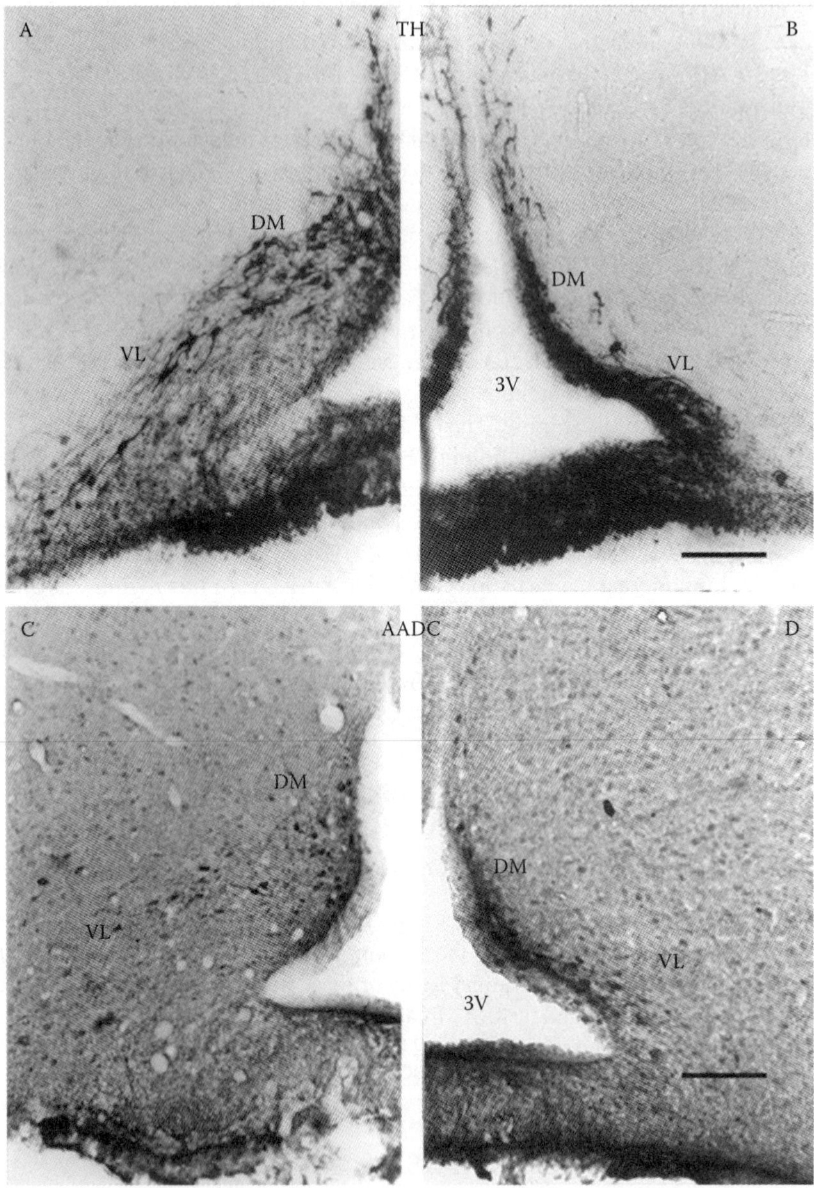


**FIGURE 21.1** Photomicrographs showing tyrosine hydroxylase (TH)-immunoreactive cells (A, B) and L-amino acid decarboxylase (AADC)-immunoreactive cells (C, D) in the mid-arcuate nucleus region (representing TIDA neurons) obtained from intact (A, C) and monosodium L-glutamate (MSG)-treated (B, D) rats. Newborn Sprague-Dawley rats obtained from both sexes were injected with MSG (4 mg/g body weight, s.c.) on days 2, 4, 6, 8, and 10 of age [24,42]. The control group of rats received NaCl in solution on the same days. Following perfusion, 50 μm coronal sections of brain (between the levels of optic chiasma and premammillary region) were cut using a cryostat or freezing microtome. Sections were chosen

## 21.5 EFFECT OF MSG TREATMENT ON CATECHOLAMINE LEVELS IN ME AND ARC OF HYPOTHALAMUS

DA concentration (Figure 21.4A), 3,4-dihydroxyphenylacetic acid (DOPAC)/DA ratio (Figure 21.4B), and DOPA concentration (Figure 21.4C) in the ME and the ARC have been also measured in both intact and MSG-treated animals using high performance liquid chromatography with electrochemical detector (HPLC-ECD). There was a significant decrease in the DOPAC/DA ratio (Figure 21.4B) in the ARC but not in the ME obtained from MSG-treated rats. More importantly, the concentration of DOPA was significantly lower in the ARC of MSG-treated animals (Figure 21.4C). These results are completely in agreement with the changes in the morphological appearance of TH- and AADC-immunoreactive neurons of the ARC following MSG treatment.

## 21.6 SUMMARY

First of all, our observations confirm previous findings showing that only the DM part of TIDA neurons produces DA in adult rats, whereas the VL part most likely synthesizes only DOPA [20–24]. In addition, we have provided evidence that the majority of PHDA and THDA neurons express both enzymes, i.e., TH and AADC; consequently they can be considered to be "genuine" DAergic systems like the DM part of TIDA neurons.

Second, it is clear that TH-positive neurons located in the VL part of the ARC are much more sensitive to the neurotoxic effect of MSG. However, this is accompanied by shrinkage and dislocation of the DM part toward the ME (Figure 21.1A to Figure 21.1D). There seems to be no significant change in the amount of TH-positive fibers of the ME; however, it should be pointed out that quantitative analysis of densely innervated areas, like the ME, is very difficult.

It is well known that neonatal treatment with MSG results in the disappearance of a large number of nerve cells from the mid- and posterior levels of the hypothalamic ARC [16,27–29]. There is a loss of several chemically identified hypophysiotrophic

**FIGURE 21.1 (Continued)** from two regions of this area, which were identified by using the atlas of Paxinos and Watson [41]. TH and AADC immunreactivity were visualized by using specific TH and AADC antibodies. The polyclonal antibodies for TH (Eugen Tech International, Inc., Ridgefield Park, NJ) and AADC were raised in rabbits [24] and used at dilution of 1:2000 or 1:5000, respectively. Following 48 h incubation in a solution containing the TH or AADC antiserum, the free-floating sections were incubated for 1 h in biotinylated antirabbit IgG (Vector Laboratories, Burlingame, CA). The sections then were incubated for 1 h with avidin–biotin–horseradish peroxidase complex according to the Vector kit instructions (Vector Laboratories). At the final step, the sections were exposed to solution containing 3,3-diaminobenzidine and 0.03% $H_2O_2$ for 15 min at room temperature. Between each transfer of solutions, the sections were always rinsed with 0.1 *M* phosphate buffer solution several times. The stained sections were mounted on gel-dipped slides. DM, the dorsomedial part; VL, the ventrolateral part; 3V, the third ventricle. Bars represent 100 μm.

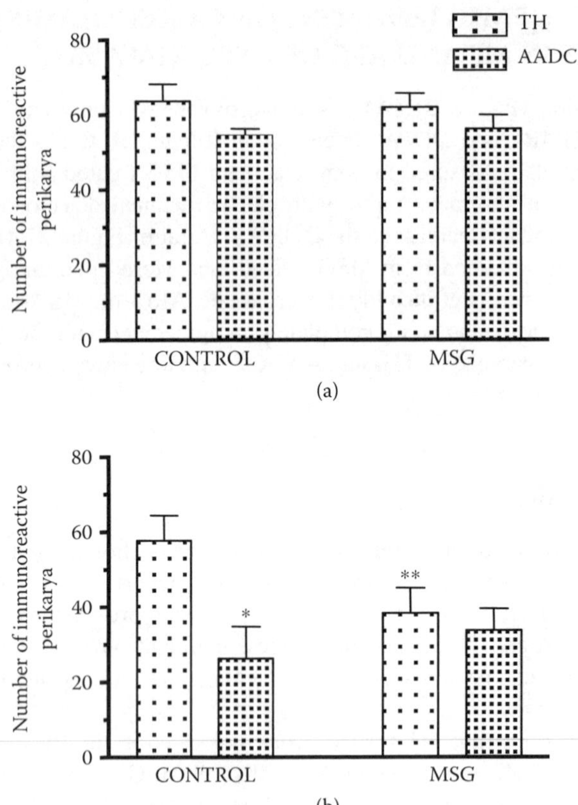


**FIGURE 21.2** Effect of MSG treatment on the number of TH- and AADC-immunoreactive neurons in the rostral periventricular–arcuate region (PHDA/THDA neurons) (A) and in the midportion of the arcuate nucleus (TIDA neurons) (B) of the hypothalamic neuroendocrine DA system (NEDA). Number of TH- and AADC-immunoreactive perikarya was counted on 3 to 5 frontal sections, 50 μm thick, in the PHDA/THDA neurons and in the TIDA neurons of each rat. Digital photo system was used for counting. Each value is the mean ± SE obtained from three rats. Statistical analysis was performed with ANOVA followed by Dunnett's multiple comparison tests. *Values of AADC-immunoreactive perikarya are significantly different ($P < .05$) from those of TH-immunoreactive neurons counted in the same region. **Values are significantly different from control.

neurons [6], but the majority of them have been found to be catecholaminergic cells [16]. The relative resistance of TH- and AADC-positive neurons located in the DM part of the ARC to MSG treatment may be due to the maturation process of the blood–brain barrier (BBB) at the ME–ARC region [30]. The existence of a barrier between the DM and VL parts of the ARC is strongly supported by the finding that horseradish peroxidase injected into the ventromedial nucleus diffuses into the adjacent DM area of the ARC but the VL part of it remains free of label [31]. Permeability

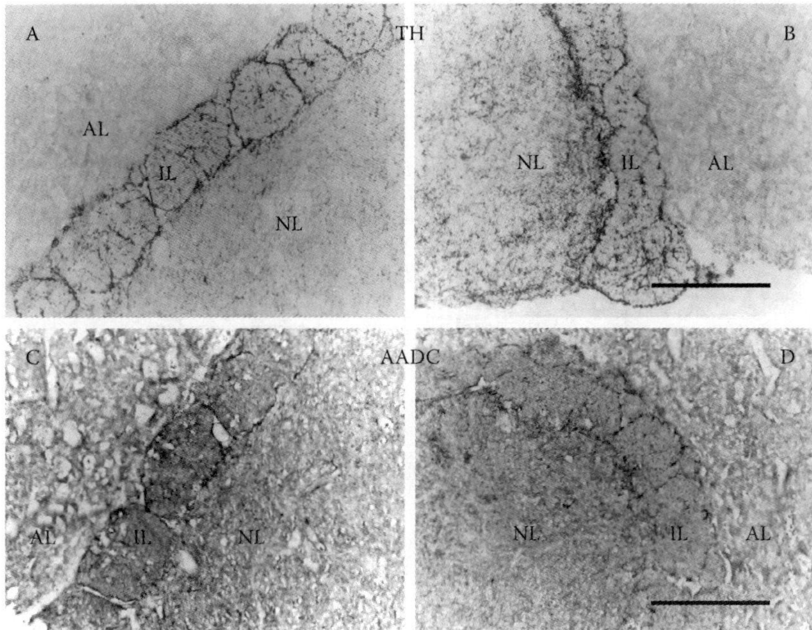


**FIGURE 21.3** Photomicrographs demonstrating TH (A, B) and AADC (C, D) immunoreactivity in the intermediate lobe (IL) and the neural lobe (NL) of the pituitary gland (representing terminals of PHDA and THDA hypothalamic neurons) obtained from intact (A, C) and MSG-treated (B, D) male rats. Following perfusion, 10 μm horizontal sections of the pituitary gland were cut using a cryostat or freezing microtome. Note the absence of immunostaining in the anterior lobe (AL). Bars represent 200 μm.

studies of capillaries to horseradish peroxidase and immunocytochemical analyses using BBB markers also indicate that in the rat, the BBB develops during the first postnatal week [32,33]. The nature and the functional significance of this barrier during early postnatal life, however, need to be further investigated.

It has been recently demonstrated by Ugrumov and his coworkers [34] that the majority (almost 99%) of ARC neurons in fetal rats (at the end of the intrauterine development) expressing enzymes of DA synthesis are monoenzymatic, i.e., TH neurons or AADC neurons [19,35]. However, neurons of the ARC are capable of synthesizing and releasing DA *in vivo* in a "cooperative" manner [34], which is sufficient to provide the inhibitory control for PRL secretion at least in fetuses. These observations further support our suggestion that monoenzymatic TH neurons might have higher functional significance, because without synthesis of DOPA by TH neurons, monoenzymatic AADC neurons do not have substrate produced locally for synthesize of DA. The reduced but still high proportion of TH-only neurons during the early postnatal period compared to adult rats [19,35] might also be responsible for their preferential sensitivity to MSG treatment.

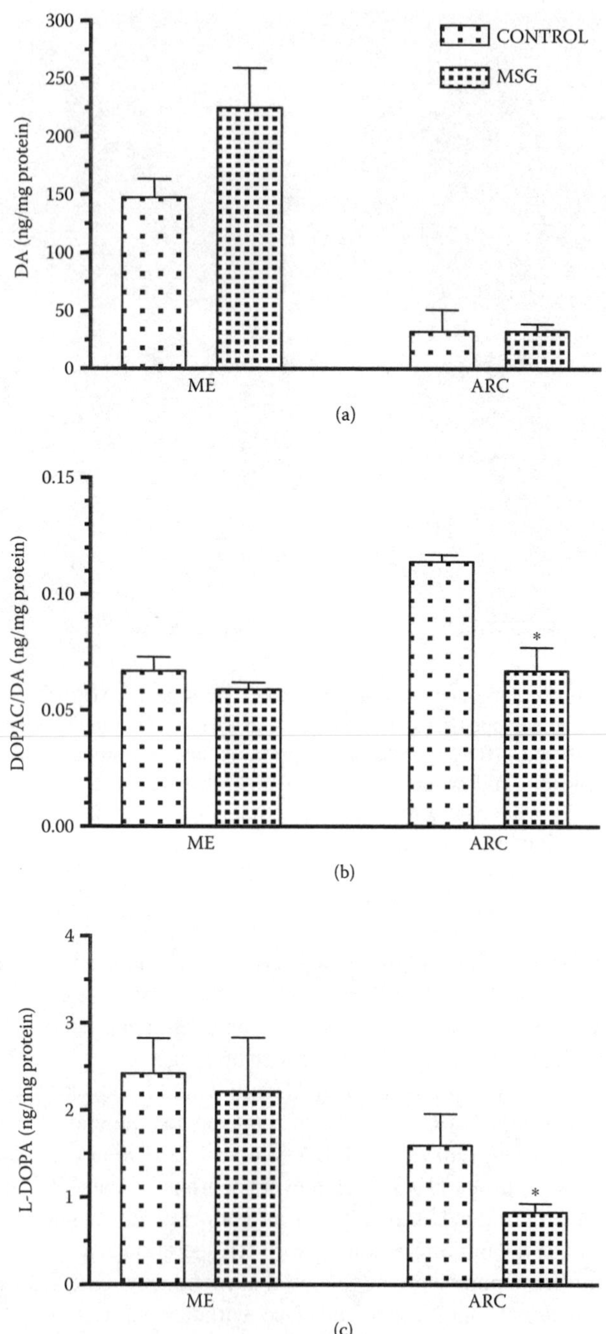


**FIGURE 21.4** DA concentration (A), DOPAC/DA ratio (B), and DOPA (L-DOPA) concentration (C) in the median eminence (ME) and in the arcuate nucleus (ARC). Tissue

At the same time, our observations clearly indicate that MSG does not affect other DAergic perikarya located rostrally from the TIDA neurons, namely, in the PHDA and THDA systems [16,29,36] that contain only bienzymatic neurons i.e., expressing both TH and AADC. In agreement with this, our previous data also show that MSG does not alter α-MSH secretion. As far as the observed resistance of THDA neurons to the toxic effects of MSG is concerned, it is also in agreement with previous findings that DA levels in the posterior pituitary (including IL and NL) are not significantly altered by MSG treatment [36,37].

There were no changes in plasma PRL levels of MSG-treated rats. This finding is in accordance with published data [16,38,39]. There is only one report [40] on elevated plasma PRL levels of MSG-treated rats. Taking into account that MSG primarily affected DOPAergic neurons at the mid- and posterior levels of the ARC and that, at the same time, this treatment did not cause alterations in plasma PRL levels, it may be concluded that these DOPAergic neurons are less involved in the control of basal PRL secretion than DAergic TIDA neurons.

As far as the possible function of DOPA neurons is concerned, they probably participate in hypothalamic neuroendocrine regulatory mechanisms other than the inhibitory regulation of PRL secretion. This population of neurons may have significance during changes in the secretory pattern of pituitary PRL, as in the estrous cycle or lactation. DOPA itself may be a substrate of further, and presently unknown, enzymatic processes resulting in a different hypothalamic hypophysiotrophic factor. Besides our previous and present data, this hypothesis is supported by the fact that there are no cycling changes in anterior lobe hormone secretion, including that of PRL, following MSG treatment of the animals [27,39]. Moreover, it has been also demonstrated that the subpopulation of cells in the ARC that specifically bind estrogen is destroyed by neonatal treatment of MSG [27]. These hypotheses, as well as other possible regulatory functions of DOPAergic neurons, need to be further investigated.

**FIGURE 21.4 (Continued)** catecholamines were measured by HPLC-ECD. Tissue samples were placed into cryogenic vials (Nalgen) containing 250 ml of homogenization buffer (0.2 N perchloric acid (Merck), 26 m*M* ethyleneglycotetraacetic acid (EGTA, Sigma), 700 p*M* dihydroxybenzylamine (DHBA, internal standard, Sigma) and sonicated (3 × 10 sec). Then they were centrifuged (10 min at 10000 g) and 20 μl of the supernatant was placed into autosampler vials. Twenty microliters of each sample were injected. Catecholamines were separated on a reverse phase C18 column (Wacosil II, SGE), oxidized on a conditioning cell, and then reduced on a dual channel analytical cell. The change in current on the second analytical electrode was measured by a coulometric detector (ESA coulochem II, ESA). The sensitivity of the assay was 30 pg of DA, 30 pg of DOPAC, and 30 pg of DOPA [43]. Catecholamine concentrations are reported as ng/mg protein. Twenty microliters of homogenates were removed for protein assay. Protein content of each sample was measured using Lowry's method. Each value is the mean ± SE (n = 5). Statistical analysis was performed with analysis of variance (ANOVA) followed by Dunnett's multiple comparison tests. *$P < .05$.

## 21.7 CONCLUSION

The effect of neonatal treatment with MSG on the DAergic systems of the medial basal hypothalamus was investigated using TH and AADC immunocytochemistry. Changes in plasma levels of PRL and α-MSH were also determined in intact and in MSG-treated rats following inhibition of TH by α-MPT or without inhibition of enzyme activity. MSG resulted in a 40% reduction in the number of TH-immunopositive TIDA neurons but no change in the number of AADC-positive TIDA nerve cells, indicating that this reduction occurred mainly in TH-positive but AADC-negative elements, i.e., in DOPAergic neurons. In contrast, MSG did not cause changes in the number of TH- and AADC-immunoreactive neurons of the PHDA and THDA systems and it did not influence basal or α-MPT-induced plasma PRL concentrations. None of the treatments had any effect on plasma α-MSH levels. More importantly the concentration of DOPA, but not DA, was significantly lower in the ARC following MSG treatment. These findings suggest that (1) MSG affects primarily DOPAergic neurons located in the VL part of the ARC but not the DAergic neurons situated in the DM part of the ARC; (2) perikarya located in the rostral arcuate–periventricular regions (PHDA/THDA systems) of the hypothalamus and their terminal fields at the posterior pituitary are not influenced by this treatment; and (3) neither PRL nor α-MSH secretion is altered.

## ACKNOWLEDGMENTS

This study was supported by a grant from the Hungarian National Research Fund (OTKA 043370 to GMN and ETT18/2001 to MV) and by a grant from the Ph.D. program of the Semmelweis University to Daniel Szekács and Márk Oláh. We also thank the expert technical assistance of Mária Mészáros.

## REFERENCES

1. Fuxe, K. and Hökfelt, T., Further evidence for the existence of tubero-infundibular dopamine neurons, *Acta Physiol. Scand.,* 66, 245, 1966.
2. Fuxe, K., Cellular localization of monoamines in the median eminence and in the infundibular stem of some mammals, *Acta Physiol. Scand.,* 58, 383, 1964.
3. Goudreau, J.L. et al., Evidence that hypothalamic periventricular dopamine neurons innervate the intermediate lobe of the rat pituitary, *Neuroendocrinology,* 56, 100, 1992.
4. Goudreau, J.L. et al., Periventricular-hypophysial dopaminergic neurons innervate the intermediate but not the neural lobe of the rat pituitary gland, *Neuroendocrinology,* 62, 147, 1995.
5. Bower, A., Hadley, M.E., and Hruby, V.J., Biogenic amines and control of melanophore stimulating hormone release, *Science,* 184, 70, 1974.
6. Taleisnik, S., Control of melanocyte-stimulating hormone (MSH) secretion, in *The Endocrine Hypothalamus,* Jeffcoate, S.L. and Hutchinson, J.S.M., Eds, Academic Press, London, 1978, 421.

7. Tilders, F.J.H., Berkenbosch, F., and Smelik, P.G., Control of secretion of peptides related to adrenocorticotropin, melanocyte-stimulating hormone and endorphin, in *Frontiers in Hormone Research,* van Wimersma Greidanus, T.J.B., Ed., Karger, Basel, 1985, 161.
8. Björklund, A. et al., The organization of tubero-hypophysial and reticulo-infundibular catecholamine neuron systems in the rat brain, *Brain Res.,* 51, 171, 1973.
9. Lindley, S.E. et al., Effects of alterations in the activity of tuberohypophysial dopaminergic neurons on the secretion of α-melanocyte stimulating hormone, *Proc. Soc. Exp. Biol. Med.,* 188, 282, 1988.
10. Alper, R.H., Demarest, K.T., and Moore, K.E., Dehydration selectively increases dopamine synthesis in tuberohypophyseal dopaminergic neurons, *Neuroendocrinology,* 31, 112, 1980.
11. Holzbauer, M. and Rack C.K., The dopaminergic innervation of the intermediate lobe and of the neural lobe of the pituitary gland, *Med. Biol.,* 63, 97, 1985.
12. Racke, K. et al., Dehydration increases the electrically evoked dopamine release from the neural and intermediate lobes of the rat hypophysis, *Neuroendocrinology,* 43, 6, 1986.
13. DeMaria, J.E., Léránt, A., and Freeman, M.E., Prolactin activates all three populations of hypothalamic neuroendocrine dopaminergic neurons in ovariectomized rats, *Brain Res.,* 837, 236, 1999.
14. DeMaria, J.E., Livingstone, J.D., and Freeman, M.E., Characterization of the dopaminergic input to the pituitary gland throughout the estrous cycle of the rat, *Neuroendocrinology,* 67, 377, 1998.
15. DeMaria, J.E. et al., The effect of neurointermediate lobe denervation on hypothalamic neuroendocrine dopaminergic neurons, *Brain Res.,* 806, 89, 1998.
16. Nemeroff, C.B. et al., Analysis of the disruption in hypothalamic-pituitary regulation in rats treated neonatally with monosodium L-glutamate: evidence for the involvement of tuberoinfundibular cholinergic and dopaminergic systems in neuroendocrine regulation, *Endocrinology,* 101, 613, 1977.
17. Ben-Jonathan, N., Dopamine: a prolactin-inhibiting hormone, *Endocr. Rev.,* 6, 564, 1985.
18. Ben-Jonathan, N. et al., Dopamine in hypophysial portal plasma of the rat during the estrous cycle and throughout pregnancy, *Endocrinology,* 100, 452, 1977.
19. Balan, I.S. et al., Tyrosine hydroxylase-expressing and/or aromatic L-amino acid decarboxylase-expressing neurons in the mediobasal hypothalamus of perinatal rats: differentiation and sexual dimorphism, *J. Comp. Neurol.,* 425, 167, 2000.
20. Jaeger, C.B. et al., Aromatic L-amino acid decarboxylase in the rat brain: immunocytochemical localization in neurons of the brain stem, *Neuroscience,* 11, 691, 1984.
21. Meister, B. et al., Do tyrosine hydroxylase-immunoreactive neurons in the ventrolateral arcuate nucleus produce dopamine or only L-dopa?, *J. Chem. Neuroanat.,* 1, 59, 1988.
22. Misu, Y. et al., Neurobiology of L-DOPAergic systems, *Prog. Neurobiol.,* 49, 415, 1996.
23. Okamura, H. et al., Comparative topography of dopamine- and tyrosine hydroxylase-immunoreactive neurons in the rat arcuate nucleus, *Neurosci. Lett.,* 95, 347, 1988.
24. Okamura, H. et al., Aromatic L-amino acid decarboxylase (AADC)-immunoreactive cells in the tuberal region of the rat hypothalamus, *Biomed. Res.,* 9, 261, 1988.
25. Bodnar, I. et al., Effect of neonatal treatment with monosodium glutamate on dopaminergic and L-DOPAergic neurons of the medial basal hypothalamus and on prolactin and MSH secretion of rats, *Brain Res. Bull.,* 55, 767, 2001.

26. Vecsernyés, M. and Julesz, J., Specific radioimmunoassay of α-melanocyte-stimulating hormone in rat plasma, *Exp. Clin. Endocrinol.,* 93, 45, 1989.
27. Grant, L.D. et al., Monosodium glutamate destruction of estrogen-feedback neurons in the hypothalamic arcuate nucleus-median eminence, *Fed. Proc.,* 37, 297, 1978.
28. Manzanares, J. et al., Sexual differences in the activity of periventricular-hypophysial dopamineric neurons in rats, *Life Sci.,* 51, 995, 1992.
29. Seress, L., Divergent effects of acute and chronic monosodium L-glutamate treatment on the anterior and posterior parts of the arcuate nucleus, *Neuroscience,* 7, 2207, 1982.
30. Peruzzo, B. et al., A second look at the barriers of the medial basal hypothalamus, *Exp. Brain Res.,* 132, 10, 2000.
31. Réthelyi, M., Diffusional barrier around the hypothalamic arcuate nucleus in the rat, *Brain Res.,* 307, 355, 1984.
32. Cassella, J.P. et al., Ontogeny of four blood-brain barrier markers: an immunocytochemical comparison of pial and cerebral cortical microvessels, *J. Anat.,* 189, 407, 1996.
33. Risau, W. and Wolburg, H., Development of the blood-brain barrier, *Trends Neurosci.,* 13, 174, 1990.
34. Ugrumov, M.V. et al., Dopamine synthesis by non-dopaminergic neurons expressing individual complementary enzymes of the dopamine synthetic pathway in the arcuate nucleus of fetal rats, *Neuroscience,* 124, 629, 2004.
35. Ershov, P.V. et al., Neurons possessing enzymes of dopamine synthesis in the mediobasal hypothalamus of rats. Topographic relations and axonal projections to the median eminence in ontogenesis, *J. Chem. Neuroanat.,* 24, 95, 2002.
36. Meister, B. et al., Neurotransmitters, neuropeptides and binding sites in the rat mediobasal hypothalamus: effects of monosodium glutamate lesions, *Exp. Brain Res.,* 76, 343, 1989.
37. Dawson, R., Valdes, J.J., and Annau, Z., Tuberohypophyseal and tuberoinfundibular dopamine systems exhibit differential sensitivity to neonatal monosodium glutamate treatment, *Pharmacology,* 31,17, 1985.
38. Antoni, F.A. et al., Neonatal treatment with monosodium-L-glutamate: differential effect on growth hormone and prolactin release induced by morphine, *Neuroendocrinology,* 35, 231, 1982.
39. Clemens, J.A. et al., Changes in luteinizing hormone and prolactin control mechanisms produced by glutamate lesions of the arcuate nucleus, *Endocrinology,* 103, 1304, 1978.
40. Heiman, M.L. and Ben-Jonathan, N., Increase in pituitary dopaminergic receptors after monosodium glutamate treatment, *Am. J. Physiol.,* 245, 261, 1983.
41. Paxinos, G. and Watson, C., *The Rat Brain in Stereotaxic Coordinates,* Academic Press, Sydney, 1982.
42. Saiardi, A. et al., Antiproliferative role of dopamine: loss of D2 receptors causes hormonal dysfunction and pituitary hyperplasia, *Neuron,* 19, 115, 1997.
43. Nagy, G.M., DeMaria, J.E., and Freeman, M.E., Changes in the local metabolism of dopamine in the anterior and neural lobes but not in the intermediate lobe of the pituitary gland during nursing, *Brain Res.,* 790, 315, 1998.

# 22 Molecular Mechanism of Memantine in Treatment of Alzheimer's Disease and Other Neurologic Insults

*Stuart A. Lipton*

## CONTENTS

## 22.1 OVERVIEW

Memantine is the first clinically tolerated neuroprotective drug to be approved by both the European Union and the U.S. Food and Drug Administration (FDA). Recent phase-III clinical trials have shown that memantine is effective in the treatment of moderate-to-severe Alzheimer's disease. Here, we review the molecular mechanism of memantine's action, as first described by my laboratory, and also the basis for the drug's use in neurologic diseases mediated at least in part by excitotoxicity, defined as excessive exposure to the neurotransmitter glutamate or overstimulation of its membrane receptors, leading to neuronal injury or death. Excitotoxic neuronal cell death is due, at least in part, to excessive activation of *N*-methyl-D-aspartate (NMDA)-type glutamate receptors and, hence, excessive $Ca^{2+}$ influx through the receptor's associated ion channel. Physiological NMDA receptor activity, however, is also essential for normal neuronal function. This means that potential neuroprotective agents that block virtually all NMDA receptor activity will very likely have unacceptable clinical side effects. For this reason, many NMDA receptor antagonists have disappointingly failed advanced clinical trials for a number of diseases, including stroke and neurodegenerative disorders such as Huntington's disease. In contrast, studies in our laboratory have shown that the adamantane derivative, memantine, preferentially blocks excessive NMDA receptor activity without disrupting normal activity. Memantine does this through its uncompetitive mode of action as a low-affinity, open-channel blocker; it enters the receptor-associated ion channel preferentially when it is excessively open and, most importantly, its off-rate is relatively fast so that it does not substantially accumulate in the channel to interfere with normal synaptic transmission. Clinical use has demonstrated that memantine is well tolerated. Besides Alzheimer's disease, memantine is currently in trials for additional neurologic disorders, including other forms of dementia, glaucoma, severe neuropathic pain, and depression. A series of second-generation memantine derivatives are currently in development and may prove to have even greater neuroprotective properties than memantine. These second-generation drugs take advantage of the fact that the NMDA receptor has other modulatory sites, in addition to its ion channel, that potentially could also be used for safe and effective clinical intervention.

## 22.2 INTRODUCTION

Acute and chronic neurologic diseases are among the leading causes of death, disability, and economic expense in the world. For example, cerebrovascular ischemia (stroke) and Alzheimer's disease rank third and fourth as causes for mortality in the United States. In fact, it has been estimated that as the population continues to age, treatment of patients with dementia from Alzheimer's disease and vascular causes will consume our entire gross national product by the latter decades of this century. Excitotoxic cell death, as described in the following text, is thought to contribute to neuronal cell injury and death in these and other neurodegenerative disorders. Excitotoxicity is due, at least in part, to excessive activation of NMDA-type glutamate receptors and, hence, excessive $Ca^{2+}$ influx through the receptor's associated ion channel. Relevant to the theme of this book, DOPA and dopamine can be an upstream factor for glutamate release and neuronal cell death, in addition to the formation of hydroxyl radicals and TOPA quinones. The cascade to neuronal cell death can be triggered by activation of dopamine receptors, vesicular glutamate release, $Ca^{2+}$ influx via NMDA and non-NMDA receptors, neuronal nitric oxide synthase (nNOS) activation, and excessive nitric oxide (NO) production. Additionally, NMDA receptor activity can be upregulated by dopamine D1 receptors [1]. Moreover, the interaction can be reciprocal, with upregulation of dopamine D1 receptors by NMDA receptor activation.

In addition to the toxic affects of excessive stimulation of glutamate receptors, glutamate-mediated synaptic transmission is also critical for normal functioning of the nervous system. Glutamate is the major excitatory neurotransmitter in the brain. There are three classes of glutamate-gated ion (or ionotropic) channels, known as α-amino-3-hydroxy-5-methyl-4-isoxazole propionic acid (AMPA), kainate, and NMDA receptors. Among these, the ion channels coupled to classical NMDA receptors are generally the most permeable to $Ca^{2+}$. Excessive activation of the NMDA receptor, in particular, leads to the production of damaging free radicals and other enzymatic processes contributing to cell death [2,3]. With the disruption of energy metabolism during acute and chronic neurodegenerative disorders, glutamate is not cleared properly and may even be inappropriately released. Moreover, energetically compromised neurons become depolarized (more positively charged) because in the absence of energy they cannot maintain their ionic homeostasis; this depolarization relieves the normal $Mg^{2+}$ block of NMDA receptor-coupled channels because the relatively positive charge in the cell repels positively charged $Mg^{2+}$ from the channel pore. Hence, during periods of ischemia and in many neurodegenerative diseases, excessive stimulation of glutamate receptors is thought to occur. These neurodegenerative diseases, including Alzheimer's disease, Parkinson's disease, Huntington's disease, HIV-associated dementia, multiple sclerosis, amyotrophic lateral sclerosis (ALS or Lou Gehrig's disease), and glaucoma, are caused by different mechanisms but may share a final common pathway to neuronal injury due to the overstimulation of glutamate receptors, especially of the NMDA subtype [2]. Hence, NMDA receptor antagonists could potentially be of therapeutic benefit in a number of neurologic disorders manifesting excessive NMDA receptor activity, including stroke, dementia, and neuropathic pain. NMDA receptors are made up of different subunits: NR1 (whose presence is mandatory), NR2A–D, and in some cases, NR3A or B subunits. The receptor is probably composed of a tetramer of these subunits. The subunit composition determines the pharmacology and other parameters of the receptor–ion channel complex. Alternative splicing of some subunits, such as NR1, further contributes to the pharmacological properties of the receptor. The subunits are differentially expressed both regionally in the brain and temporally during development. For this reason, some authorities have suggested developing antagonists selective for particular subunits, such as NR2B, which is present predominantly in the forebrain (reviewed in reference [4]).

### 22.2.1 Searching for Clinically Tolerated NMDA Receptor Antagonists

Excitotoxicity is a particularly attractive target for neuroprotective efforts because it is implicated in the pathophysiology of a wide variety of acute and chronic neurodegenerative disorders [2]. The challenge facing those trying to devise strategies for combating excitotoxicity is that the same processes which, in excess, lead to excitotoxic cell death are, at lower levels, absolutely critical for normal neuronal function. Until recently, all drugs that showed promise as inhibitors of excitotoxicity also blocked normal neuronal function and consequently had severe and unacceptable side effects and so clinical trials for stroke, traumatic brain injury, and Huntington's disease failed [5–7].

Recently, however, the well-tolerated but underappreciated drug memantine has been rediscovered to be capable not only of blocking excitotoxic cell death [8] but, most importantly, also of doing it in a clinically tolerated, nontoxic manner [9–12]. Memantine was recently approved by the European Union and the U.S. FDA for the treatment of moderate-to-severe Alzheimer's disease, and it may also show efficacy for vascular dementia [13,14]. Previous anticholinergic drugs only offered symptomatic relief from Alzheimer's disease, whereas memantine is believed to be the first neuroprotective drug to achieve clinical approval. The drug is also under investigation as a potential treatment for other neurodegenerative disorders, including HIV-associated dementia, neuropathic pain, and glaucoma, as well as depression and movement disorders.

The purpose of this chapter is to provide a brief and perhaps somewhat surprising primer on excitotoxicity as a promising target of neuroprotective strategies and to present a scientific and clinical overview of the excitotoxicity blocker memantine. Some preliminary information on second-generation memantine derivatives, termed *NitroMemantines*, is also provided.

## 22.3 EXCITOTOXICITY

### 22.3.1 Definition and Clinical Implications

The ability of the nervous system to rapidly convey sensory information and complex motor commands from one part of the body to another and to form thoughts and memories is largely dependent on a single powerful excitatory neurotransmitter, glutamate. There are other excitatory neurotransmitters in the brain but glutamate is the most common and widely distributed. Most neurons (and also glia) contain high concentrations of glutamate (~10 m*M*) [2]; after sequestration inside synaptic vesicles, glutamate is released for very brief amounts of time (milliseconds) to communicate with other neurons via synaptic endings. Because glutamate is so powerful, however, its presence in excessive amounts or for excessive periods of time can literally excite cells to death. This phenomenon was first documented when Lucas and Newhouse [15] observed that subcutaneously injected glutamate selectively damaged the inner layer of the retina (representing primarily the retinal ganglion cells) [15]. John Olney later coined the term "excitotoxicity" to describe this phenomenon [16,17].

A large variety of insults can lead to the excessive release of glutamate within the nervous system and, thus, excitotoxic cell death. When the nervous system suffers a severe mechanical insult, as in head or spinal cord injury, large amounts of glutamate are released from injured cells. These high levels of glutamate reach thousands of nearby cells that had survived the original trauma, causing them to depolarize, swell, lyse, and die by necrosis. The lysed cells release more glutamate, leading to a cascade of autodestructive events and progressive cell death that can continue for hours or even days after the original injury. A similar phenomenon occurs in stroke; the ischemic event deprives many neurons of the energy they need to maintain ionic homeostasis, causing them to depolarize, lyse, die, and propagate the same type of autodestructive events that are seen in traumatic injury (reviewed in [2,18]). This acute form of cell death occurs by a necrotic-like mechanism, although a slower component leading to an apoptotic-like death can also be present, as well as a continuum of events somewhere between the two (see following text).

A slower, subtler form of excitotoxicity is implicated in a variety of slowly progressing neurodegenerative disorders as well as in the penumbra of stroke damage. In disorders such as Huntington's disease, Parkinson's disease, Alzheimer's disease, multiple sclerosis, HIV-associated dementia, ALS, and glaucoma, it is hypothesized that chronic exposure to moderately elevated glutamate concentrations, or glutamate receptor hyperactivity for longer periods of time than occur during normal neurotransmission, trigger cellular processes in neurons that eventually lead to apoptotic-like cell death, a form of cell death related to the programmed cell death that occurs during normal development [3,19–25]. More subtle or incipient insults can lead to synaptic and dendritic damage in the early stages of disease. This process may be reversible and so is of considerable therapeutic interest.

Importantly, elevations in extracellular glutamate are not necessary to invoke an excitotoxic mechanism. Excitotoxicity can come into play even with normal levels of glutamate if NMDA receptor activity is increased, e.g., when neurons are injured and thus become depolarized (more positively charged); this condition relieves the normal block of the ion channel by $Mg^{2+}$ and, thus, abnormally increases NMDA receptor activity [26].

Increased activity of the enzyme NOS is associated with excitotoxic cell death. The neuronal isoform of the enzyme is physically tethered to the NMDA receptor and activated by $Ca^{2+}$ influx via the receptor-associated ion channel, and increased levels of NO have been detected in animal models of stroke and several neurodegenerative diseases.

### 22.3.2 Possible Links between Excitotoxicity and Alzheimer's Disease

There are several potential links between excitotoxic damage and the primary insults of Alzheimer's disease, which, based on rare familial forms of the disease, are believed to involve toxicity from misfolded mutant proteins (recently reviewed by [27]). These proteins include soluble oligomers of β-amyloid peptide (Aβ) and hyperphosphorylated tau proteins [28]. For example, oxidative stress and increased intracellular $Ca^{2+}$

generated by Aβ have been reported to enhance glutamate-mediated neurotoxicity *in vitro*. Additional experiments suggest that Aβ can increase NMDA responses and, thus, excitotoxicity [29–31]. Another potential link comes from recent evidence that glutamate transporters are downregulated in Alzheimer's disease and that Aβ can inhibit glutamate reuptake or even enhance its release [32,33]. Finally, excessive NMDA receptor activity has been reported to increase the hyperphosphorylation of tau, which contributes to neurofibrillary tangles [34]. The NMDA receptor antagonist memantine has been found to offer protection from intrahippocampal injection of Aβ [35]. Moreover, memantine improved performance on behavioral tests (T-maze and Morris water maze) in a transgenic mouse model of Alzheimer's disease consisting of a mutant form of amyloid precursor protein and presenilin 1 [36]. Additionally, memantine was recently found to reduce tau hyperphosphorylation, at least in culture [37].

### 22.3.3 Pathophysiology of Excitotoxicity: Role of NMDA Receptor

Apoptotic-like excitotoxicity is caused in part by excessive stimulation of the NMDA subtype of glutamate receptor (Figure 22.1). When activated, the NMDA receptor opens a channel that allows $Ca^{2+}$ (and other cations) to move into the cell. In some areas of the brain, this activity is important for long-term potentiation (LTP), which is thought to be a cellular/electrophysiological correlate of learning and memory formation. Under normal conditions of synaptic transmission, the NMDA receptor channel is blocked by $Mg^{2+}$ sitting in the channel and only activated for brief periods of time. Under pathological conditions, however, overactivation of the receptor causes an excessive amount of $Ca^{2+}$ influx into the nerve cell, which then triggers a variety of processes that can lead to necrosis, apoptosis, or dendritic/synaptic damage. These processes include $Ca^{2+}$ overload of mitochondria, resulting in oxygen free-radical formation and activation of caspases; $Ca^{2+}$-dependent activation of nNOS, leading to increased NO production and the formation of toxic peroxynitrite ($ONOO^-$); and stimulation of mitogen-activated protein kinase p38 (MAPK p38), which activates transcription factors that can enter the nucleus to influence neuronal injury and apoptosis [19,38–44].

As mentioned earlier, conventional NMDA receptors consist of at least two types of subunits (NR1 and NR2A–D), and more rarely NR3A or B subunits. There are binding sites for glutamate, the endogenous agonist, and glycine, which is required as a coagonist for receptor activation (Figure 22.2) [45]. NMDA is generally not thought to be an endogenous substance in the body; it is an experimental tool that is highly selective for this subtype of glutamate receptor and therefore became the source of its name. When glutamate and glycine bind and the cell is depolarized to remove the $Mg^{2+}$ block, the NMDA receptor channel opens with consequent influx of $Ca^{2+}$ and $Na^+$ into the cell, the amount of which can be altered by higher levels of agonists and by substances binding to one of the modulatory sites on the receptor. The two modulatory sites that are most relevant to this discussion are the magnesium ($Mg^{2+}$) site within the ion channel and an *S*-nitrosylation site located toward the N-terminus (and hence extracellular region) of the receptor. Note that *S*-nitrosylation reactions represent transfer of NO to a thiol or sulfhydryl group (-SH) of a critical cysteine residue. This reaction modulates protein function, in this case decreasing the channel

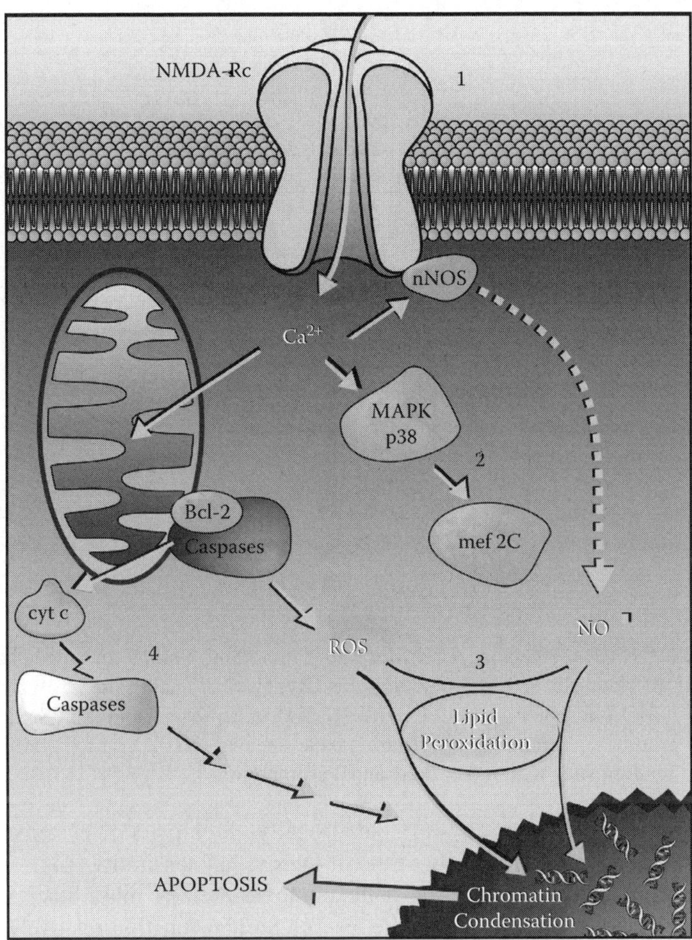


**FIGURE 22.1** Schematic illustration of the apoptotic pathways triggered by excessive NMDA receptor activity. The cascade of steps leading to neuronal cell death include: (1) NMDA receptor (NMDA-Rc) hyperactivation; (2) activation of the p38-mitogen-activated kinase (MAPK)–MEF2C (transcription factor) pathway (MEF2 is subsequently cleaved by caspases to form an endogenous dominant-interfering form that contributes to neuronal cell death [33]); (3) toxic effects of free radicals such as nitric oxide (NO) and reactive oxygen species (ROS); and (4) activation of apoptosis-inducing enzymes including caspases and apoptosis-inducing factor (AIF). nNOS: nitric oxide synthase; Cyt c: cytochrome c (adapted from the Lipton Website at www.burnham.org).

activity associated with stimulation of the NMDA receptor. Each of these sites can be considered as targets for therapeutic intervention to block excitotoxicity, as explained in the following text. Moreover, other modulatory sites also exist on the NMDA receptor and may in the future prove to be of therapeutic value. These include binding sites for $Zn^{2+}$, polyamines, the drug ifenprodil, and a pH (i.e., proton)-sensitive

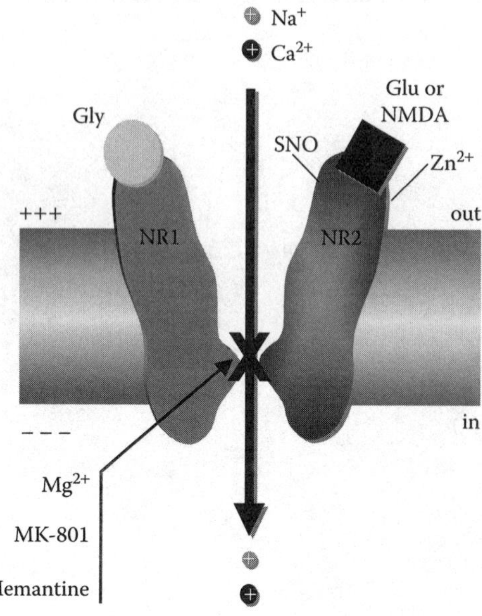


**FIGURE 22.2** NMDA receptor model illustrating important binding and modulatory sites. Glu or NMDA: glutamate or NMDA binding site. Gly: glycine binding site. $Zn^{2+}$: zinc binding site. NR1: NMDA receptor subunit 1. NR2: NMDA receptor subunit 2A. SNO: cysteine sulfhydryl group (-SH) reacting with nitric oxide species (NO). X: $Mg^{2+}$, MK-801, and memantine binding sites within the ion channel pore region.

site (for a review see reference [46]). Additionally, three pairs of cysteine residues can modulate channel function by virtue of their redox sensitivity [47].

To be clinically acceptable, an antiexcitotoxic therapy must block excessive activation of the NMDA receptor while leaving normal function relatively intact in order to avoid side effects. Drugs that simply compete with glutamate or glycine at the agonist-binding sites block normal function and, therefore, do not meet this requirement and have thus failed in clinical trials to date because of their side effects (drowsiness, hallucinations, and even coma) [2,9,13,48–52]. Competitive antagonists compete one-for-one with the agonist (glutamate or glycine) and, therefore, will block healthy areas of the brain (where lower, more physiological, levels of these agonists exist) before they can affect pathological areas (where higher levels of agonist accumulate). Thus, in fact, such drugs would preferentially block normal activity and would most likely be displaced from the receptor by the prolonged high concentrations of glutamate that can exist under excitotoxic conditions.

### 22.3.4 Off-Rate from Channel Block Important for Clinical Tolerability

As a useful analogy, the NMDA receptor can be compared to a television set. The agonist sites are similar to the on/off switch of the television. Drugs that act here cut off all normal NMDA receptor function. What we need to find is the equivalent of the volume control (or in biophysical terms, the gain) of the NMDA receptor. Then, when excessive $Ca^{2+}$ fluxes through the NMDA receptor-associated ion channel, we could simply turn down the "volume" of the $Ca^{2+}$ flux more toward normal. A blocker that acts at the $Mg^{2+}$ site within the channel could act in such a manner. However, in the case of $Mg^{2+}$ itself, the block is too ephemeral, a so-called "flickery block," and the cell continues to depolarize until the $Mg^{2+}$ block is totally relieved. Hence, in most cases $Mg^{2+}$ does not effectively block excessive $Ca^{2+}$ influx to the degree needed to prevent neurotoxicity. If, on the other hand, a channel blocker binds too tightly or works too well at low levels of receptor activation, it will block normal as well as excessive activation and be clinically unacceptable. Following the television set analogy, turning the volume all the way down is as bad as turning off the on/off switch and blocking normal functioning of the television. This is the case with (+)-5-methyl-10,11-dihydro-5*H*-dibenzol[a,d]cyclohepten-5,10-imine maleate (MK-801); it is a very good excitotoxicity blocker, but because its "dwell time" in the ion channel is so long (reflecting its slow off-rate) due to its high affinity for the $Mg^{2+}$ site, it also blocks critical normal functions. A human taking a neuroprotective dose of MK-801 would not only become drowsy but would lapse into a coma. Drugs with slightly shorter but still excessive dwell times (off-rates) would make patients hallucinate (e.g., phencyclidine, also known as "angel dust") or so drowsy that they could serve as anesthetic agents (e.g., ketamine).

A clinically tolerated NMDA receptor antagonist would not make a patient drowsy, hallucinate, or comatose, and would spare normal neurotransmission while blocking the ravages of excessive NMDA receptor activation. In fact, a drug that would do this and preferentially block higher (pathological) levels of glutamate over normal (physiological) levels would be an "uncompetitive" antagonist. An uncompetitive antagonist is distinct from a noncompetitive antagonist (which simply acts allosterically at a noncompetitive site, i.e., a site other than the agonist-binding site). An uncompetitive antagonist is defined as an inhibitor whose action is contingent upon prior activation of the receptor by the agonist. Hence, the same amount of antagonist blocks higher concentrations of agonist better than lower concentrations of agonist. This uncompetitive mechanism of action coupled with a longer dwell time in the channel (and consequently a slower off-rate from the channel) than $Mg^{2+}$, but a substantially shorter dwell time (faster off-rate) than MK-801, would yield a drug that blocks NMDA receptor-operated channels only when they are excessively open while relatively sparing normal neurotransmission. Our experiments suggested that memantine is such a drug [2,9–12].

Note that our principal discovery that led to the testing of a clinically successful NMDA receptor antagonist was the kinetics of the drug in the NMDA receptor-associated ion channel [2,9–12]: we found that the dwell time in (or off-rate from) the channel is the major determinant of clinical tolerability of an open-channel blocker because excessive dwell time (associated with a slow off-rate) causes the drug to accumulate in the channels, interfere with normal neurotransmission, and produce unacceptable adverse effects (as is the case with MK-801). In contrast, too short a dwell time (reflecting too fast an off-rate) yields a relatively ineffectual blockade, especially with membrane depolarization which relieves the block of positively charged molecules (as seen with $Mg^{2+}$). The apparent affinity of a positively charged channel blocker such as memantine at a particular membrane voltage is related to its off-rate divided by its on-rate (derived from Scheme 1, in the Appendix). The on-rate is a property of the channel's open probability and the drug's diffusion rate and concentration. On the other hand, the off-rate is an intrinsic property of the drug–receptor complex, unrelated to drug concentration (see reference [10] for a detailed quantitative discussion of these points). A relatively fast off-rate is a major contributor to a drug's (such as memantine's) low affinity for the channel pore.

Thus, memantine represents a class of relatively low-affinity, open-channel blockers, i.e., drugs that only enter the channel when it is opened by an agonist. In the case of memantine, at the concentrations administered to patients, the drug appears to enter the channel preferentially when it is (pathologically) activated for long periods of time, e.g., under conditions of excessive glutamate exposure. The relatively fast off-rate prevents the drug from accumulating in the ion channels and interfering with normal synaptic transmission. Hence, as we showed previously [10], memantine has favorable kinetics in the channel to provide neuroprotection, while displaying minimal adverse effects (e.g., occasional restlessness or, in rare cases, slight dizziness at higher dosages) [2,9].

## 22.4 MEMANTINE

### 22.4.1 Pharmacology and Mechanism: Uncompetitive Open-Channel Block

Memantine was first synthesized by Eli Lilly and Company and patented in 1968, as documented in the Merck Index, as a derivative of amantadine, an anti-influenza agent. Memantine has a three-ring (adamantane) structure with a bridgehead amine (-$NH_2$) that, under physiological conditions, carries a positive charge (-$NH_3^+$) and binds at or near the $Mg^{2+}$ site in the NMDA receptor-associated channel (Figure 22.3) [9–12,53].

Memantine (MEM)

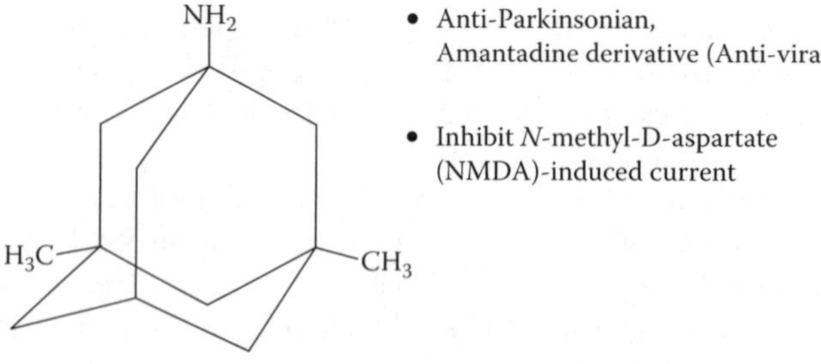


**FIGURE 22.3** Chemical structure of memantine. Several important features are: (1) the three-ring structure; (2) the bridgehead amine (-$NH_2$ group), which is charged at the physiological pH of the body (-$NH_3^+$) and represents the region of memantine that binds at or near the $Mg^{2+}$ binding site in the NMDA receptor-associated ion channel; and (3) the methyl group (-$CH_3$) side chains (unlike amantadine), which serve to stabilize memantine's interaction in the channel region of the NMDA receptor.

Unlike amantadine, memantine has two methyl (-$CH_3$) side groups that prolong its dwell time in the channel (hence, slowing the off-rate and increasing the affinity for the channel somewhat, compared to amantadine). The reported efficacy of amantadine and memantine in Parkinson's disease, which was discovered by serendipity in a patient taking amantadine for influenza, led scientists to believe that these compounds were dopaminergic or possibly anticholinergic drugs. It was not until the late 1980s and early 1990s that memantine was found to be neither dopaminergic nor anticholinergic at its clinically used dosage but, rather, an NMDA receptor antagonist. Work at a small German company named Merz first suggested that the drug was a quite potent NMDA receptor inhibitor [54], which in fact it is not. Instead, my laboratory found that under physiological conditions the drug is of such weak potency at the NDMA receptor (affinity in the micromolar range rather than nanomolar or higher) that big pharma initially thought that it was a poor candidate for use as a neuroprotective drug. However, one should not confuse affinity with selectivity; as long as a drug acts selectively and specifically on the target of interest and the effective concentration can be achieved, a high affinity *per se* is not the key issue. In fact, work in my laboratory first showed why memantine could be clinically tolerated as an NMDA receptor antagonist; namely, it was an uncompetitive open-channel blocker with a dwell time/off-rate from the channel that limited pathological activity of the NMDA receptor while sparing normal synaptic activity [10–12]. These findings led to a number of U.S. and worldwide patents on the use of memantine for NMDA receptor-mediated disorders and spurred several successful clinical trials with the drug, as discussed later in this chapter. (By way of disclosure, my laboratory was then at Harvard Medical School, and so patents on which I am the named inventor are assigned to Harvard-affiliated institutions, including Children's Hospital of Boston).

To illustrate the blockade of NMDA-induced ionic currents by memantine, a sample experiment is shown in Figure 22.4, in which the membrane voltage of a neuron was held at the resting potential. The concentration of memantine used in this experiment is similar to the level that can be achieved in the human brain when the drug is used clinically. At such concentrations, memantine greatly reduces pathologically high levels of NMDA-induced current to near zero within approximately 1 sec. Once the memantine application stops, the NMDA response returns to previous levels over a period of about 5 sec. This indicates that memantine is an effective, but temporary, NMDA receptor blocker.

Perhaps the most astonishing property of memantine is illustrated in Figure 22.5 [9,11]. In this experiment, the concentration of memantine was held constant (at a clinically achievable level of 1 $\mu M$) while the concentration of NMDA was increased over a wide range. It was found that the degree to which this fixed concentration of memantine blocked NMDA receptor activity actually increased as the NMDA concentration was increased to pathological levels. In fact, memantine was relatively ineffective at blocking the low levels of receptor activity associated with normal neurologic function but became exceptionally effective at higher concentrations. This is classical "uncompetitive" antagonist behavior.

Further studies indicate that memantine exerts its effect on NMDA receptor activity by binding at or near the $Mg^{2+}$ site within the ion channel [9–12,53]. This information and the pharmacological/kinetic data presented earlier suggest that

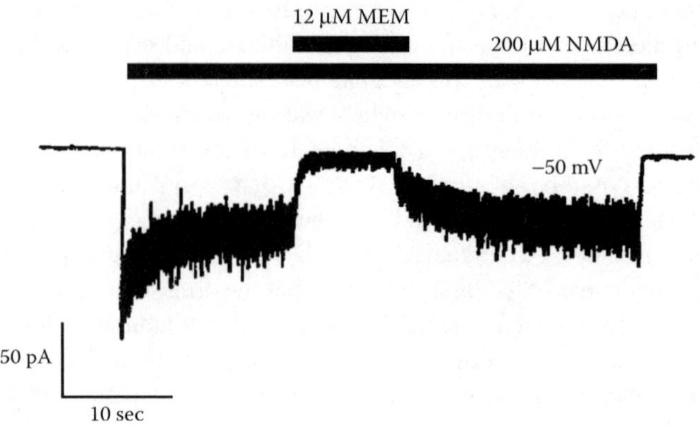


**FIGURE 22.4** Blockade of NMDA current by memantine. At a holding potential of approximately –50 mV, whole-cell recording of NMDA-evoked current from a solitary neuron revealed that the on-time (time until peak blockade) of micromolar memantine was approximately 1 sec, while the off-time (recovery time) from the effect was ~5.5 sec. The application of memantine produced an effective blockade only during NMDA receptor activation, consistent with the notion that its mechanism of action is open-channel block. (From Chen, H.S.-V. et al., Open-channel block of N-methyl-D-aspartate (NMDA) responses by memantine: therapeutic advantage against NMDA receptor-mediated neurotoxicity, *J. Neurosci.,* 12, 4427, 1992. With permission.)

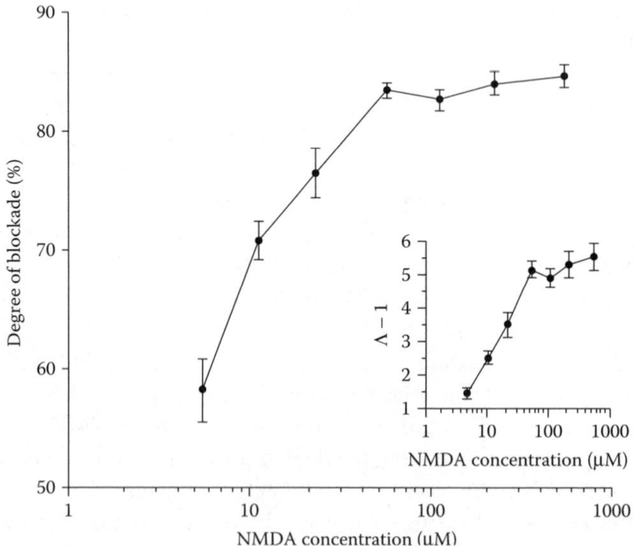


**FIGURE 22.5** Paradoxically, a fixed dose of memantine (i.e., 1 μ*M*) blocks the effect of increasing concentrations of NMDA to a greater degree than lower concentrations of NMDA. This finding is characteristic of an uncompetitive antagonist. (From Chen, H.S.-V. et al., Open-channel block of N-methyl-D-aspartate (NMDA) responses by memantine: therapeutic advantage against NMDA receptor-mediated neurotoxicity, *J. Neurosci.,* 12, 4427, 1992. With permission.)

memantine preferentially blocks NMDA receptor activity if the ion channel is excessively open. During normal synaptic activity, the channels are open on average for only several milliseconds and memantine is unable to act or accumulate in the channels; hence, synaptic activity continues essentially unabated. In technical terms, the component of the excitatory postsynaptic current due to activation of NMDA receptors is inhibited by only 10% or less [12]. During prolonged activation of the receptor, however, as occurs under excitotoxic conditions, memantine becomes a very effective blocker. In essence, memantine only acts under pathological conditions without much affecting normal function, thus, relatively sparing synaptic transmission, preserving LTP, and maintaining physiological function on behavioral tests such as the Morris water maze [12]. The kinetics of memantine action in the NMDA receptor-associated ion channel explain the favorable clinical safety profile that has been seen to date.

### 22.4.2 Voltage Dependence, Partial Trapping in the Ion Channel, and Other Possible Effects of Memantine

We were the first to find that memantine block is voltage dependent like $Mg^{2+}$ [10,11] (recall that memantine is positively charged at physiological pH). Hence, during the depolarization caused by excitatory neurotransmission (excitatory postsynaptic potentials), memantine block might be partially relieved. Nonetheless, we have shown that even when the neuron is voltage-clamped to avoid such depolarization, memantine does not block synaptic activity to any substantial degree [12]. Hence, the voltage dependence of memantine is not the major factor for its lack of effect on synaptic activity. Rather, the relatively fast off-rate from the ion channel is the critical feature, as explained earlier. Additionally, our group [10] and Blanpied et al. [55] demonstrated that memantine can be trapped in the NMDA receptor-associated ion channel (i.e., the channel could shut behind memantine and trap it in the closed channel). However, since memantine does not block a significant proportion of the synaptic activity (and thus is not in the channels for a prolonged period during normal neurotransmission), we do not believe that trapping in the channel is a major factor under physiological conditions. Additionally, closed-channel block (entering the channel before it has opened) has been proposed as a mechanism of action of memantine [56], but this occurs only at very high concentrations of the drug and so we think that this action is not pathophysiologically relevant under the conditions in which memantine is clinically used [10]. Finally, memantine has provisionally been found to block 5-$HT_3$ receptor channels at concentrations similar to those that block the NMDA receptor channels; at somewhat higher concentrations, memantine may also block nicotinic receptor channels [10–12,57,58, reviewed in 27]. Memantine's effect on 5-$HT_3$ receptors may possibly further enhance cognitive performance.

### 22.4.3 Neuroprotective Effects of Memantine

The neuroprotective properties of memantine have been studied in a large number of *in vitro* and *in vivo* animal models by several laboratories (for a recent review, see reference [59]). Among the neurons protected in this manner, both in culture and *in vivo*, are cerebrocortical neurons, cerebellar neurons, and retinal neurons [9–12,22,60–63]. Additionally, in a rat model of stroke, memantine, given as long as 2 h after the ischemic event, reduced the amount of brain damage by approximately 50% [11,12].

A series of human clinical trials have been launched to investigate the efficacy of memantine for the treatment of Alzheimer's disease, vascular dementia, HIV-associated dementia, diabetic neuropathic pain, as well as glaucoma. Some of these studies have only recently been completed and remain unpublished at this time except in abstract form. One recent, high-profile publication reported the results of a U.S. phase III (final) clinical study showing that memantine (20 mg/d) is efficacious for moderate-to-severe Alzheimer's disease [13]. Another recent study reported that when combined with donepezil (Aricept), memantine treatment actually improves memory and function in moderate-to-severe Alzheimer's patients [64]. The results of clinical studies for Alzheimer's disease were sufficiently positive to convince the European Union to approve memantine for the treatment of this form of dementia last year, and the U.S. FDA also recently granted approval to the drug. One full-length European publication of a multicenter, randomized controlled trial reported that memantine was beneficial in severely demented patients, probably representing both Alzheimer's disease and vascular dementia [65]. Still another recent publication of a randomized, placebo-controlled clinical trial described significant benefit from memantine therapy (20 mg/d) in mild to moderate vascular dementia [14]. These clinical trials are summarized in Table 22.1. Most trials have reported minimal adverse effects from use of memantine. In those trials reporting adverse effects, the only memantine-induced side effects encountered were a rare dizziness and occasional restlessness/agitation at higher doses (40 mg/d), but these effects were mild and dose related.

Importantly, as outlined above, the antiintuitive aspects of memantine action that we discovered show that more glutamate receptor activity is blocked better than less activity at a fixed dose of memantine. Hence, memantine would be expected to work better for severe conditions, e.g., excessive glutamate receptor activity to the point of causing cell death, than milder conditions manifest only by slightly elevated synaptic transmission. Bearing this out, recent studies suggest that memantine may have a larger or more consistent effect in moderate-to-severe dementia than in mild dementia. Another case in point is neuropathic pain, which is currently thought to be mediated at least in part by excessive NMDA receptor activity.

**TABLE 22.1**
**Clinical Trials with Memantine**

| | |
|---|---|
| German/Merz phase III trial for vascular dementia and Alzheimer's disease | + |
| Karolinska/Italian phase III trial for vascular dementia | + |
| Three U.S. multicenter phase III trials for Alzheimer's disease | + |
| U.K. phase III trial for vascular dementia | + |
| French phase III trial for vascular dementia | + |
| U.S. phase II trials for neuropathic pain and HIV-associated dementia | ± |

Given the mode of action that we described for memantine, more severe pain, e.g., the nocturnal pain of diabetic neuropathy, might be expected to benefit from memantine to a greater extent than milder forms of neuropathic pain. In fact, a phase IIB clinical trial suggested that this is indeed the case: milder pain conditions were not statistically benefited by memantine in phase II/III clinical trials. Along these same lines, one could predict that a higher concentration of memantine would be needed to combat pain than to prevent neuronal cell death because greater NMDA receptor activity is associated with cell death (recall that the on-rate of channel block by memantine can be increased by increasing the drug's concentration. Thus, a greater proportion of channels will be blocked even with less total channel activity). Again, clinical trials have suggested that this is indeed the case, because 40 mg/d of memantine was needed in successful pain studies but only 20 mg/d in dementia studies. However, further clinical trials will be necessary to prove the efficacy of memantine for severe neuropathic pain. The data thus far suggest that mild cases of neuropathic pain may not be helped by the drug because of its uncompetitive mechanism of action, whereby more intense NMDA channel activity is blocked to a relatively greater degree than mild increases in activity.

As promising as the results with memantine are, we are continuing to pursue ways to use additional modulatory sites (the "volume controls") on the NMDA receptor to safely block excitotoxicity even more effectively than memantine alone. New approaches in this regard are explored in the following section.

## 22.5 NITROMEMANTINES

NitroMemantines are second-generation memantine derivatives that were designed to have enhanced neuroprotective efficacy without sacrificing safety. As mentioned earlier, a nitrosylation site is located on the N-terminus or extracellular domain of the NMDA receptor, and *S*-nitrosylation of this site (NO reaction with the sulfhydryl group of the cysteine residue) downregulates receptor activity (Figure 22.2). The drug nitroglycerin, which generates NO-related species, can act at this site to limit excessive NMDA receptor activity. In fact, in rodent models, nitroglycerin can limit ischemic damage [66], and there is some evidence that patients taking nitroglycerin for other medical reasons may be resistant to glaucomatous visual-field loss [67].

From crystal structure models and electrophysiological experiments, we have found that NO binding to the NMDA receptor at the major *S*-nitrosylation site apparently induces a conformational change in the receptor protein that makes glutamate and $Zn^{2+}$ bind more tightly to the receptor. The enhanced binding of glutamate and $Zn^{2+}$, in turn, causes the receptor to desensitize and, consequently, the ion channel to close [47]. Electrophysiological studies have demonstrated this effect of NO on the NMDA channel [40,68,69].

Unfortunately, nitroglycerin is not very attractive as a neuroprotective agent. The same cardiovascular vasodilator effect that makes it useful in the treatment of angina could cause dangerously large drops in blood pressure in stroke, traumatic injury, or glaucoma patients. Consequently, we carefully characterized *S*-nitrosylation sites on the NMDA receptor in order to determine whether we could design a nitroglycerin-like drug that could be more specifically targeted to the NMDA receptor.

In brief, we found that five different cysteine residues on the NMDA receptor could interact with NO. One of these, located at cysteine residue 399 on the NR2A subunit of the NMDA receptor, mediates approximately 90% of the effect of NO under our experimental conditions. Using this kind of information, we created modified memantine molecules called NitroMemantines that will interact with both the memantine site within the NMDA receptor-associated ion channel and the nitrosylation site. Two sites of modulation are analogous to having two volume controls on your television set for fine-tuning the audio signal [70–72].

Preliminary studies have shown NitroMemantines to be highly neuroprotective in both *in vitro* and *in vivo* animal models. In fact, it appears to be substantially more effective than memantine. Moreover, because the memantine portion makes these drugs specific for the NMDA receptor and serves to target the NO group to the nitrosylation site on the receptor, NitroMemantines appear to lack the blood-pressure-lowering effect typical of nitroglycerin.

More research still needs to be performed on NitroMemantine drugs, but the fact that they chemically combine two clinically tolerated drugs (memantine and nitroglycerin) enhances their promise as second-generation memantine derivatives that are both clinically safe and neuroprotective.

## 22.6 SUMMARY

Necrosis- and apoptosis-mediated excitotoxic cell death is implicated in the pathophysiology of many neurologic diseases, including stroke, CNS trauma, dementia, glaucoma, polyglutamine (triple repeat) disorders, such as Huntington's disease and other neurodegenerative conditions. This type of excitotoxicity is caused, at least in part, by excessive activation of NMDA-type glutamate receptors. Intense insult, such as that occurring in the ischemic core after a stroke, leads to massive stimulation of NMDA receptors because of increased glutamate and energy failure leading to membrane depolarization, relief of $Mg^{2+}$ block of NMDA channels, and disruption of ionic homeostasis. The fulminant buildup of ions results in neuronal cell swelling and lysis (necrosis). In contrast, more moderate NMDA receptor hyperactivity, such as that occurring in the ischemic penumbra of a stroke and in many slow-onset neurodegenerative diseases, results in moderately excessive influx of calcium ions into nerve cells which, in turn, triggers free-radical formation and multiple pathways leading to the initiation of apoptotic-like damage [19,73]. Blockade of NMDA receptor activity prevents, to a large degree, both necrosis- and apoptosis-related excitotoxicity. NMDA receptor activity is also required for normal neural function, however, and so only those NMDA blockers that selectively reduce excessive receptor activation without affecting normal function will be clinically acceptable. We have shown that memantine is such a drug. Laboratory tests have demonstrated that memantine blocks excessive NMDA receptor activation but not normal, low-level activation. Importantly, the author, in collaboration with Vincent Chen (a graduate student at the time and currently a faculty member at the Burnham Institute and University of California at San Diego in La Jolla) reported that memantine had a relatively short dwell time in (and hence fast off-rate from) the NMDA-associated ion channel, in part explaining the drug's relatively low apparent affinity as an antagonist [9–11]. We realized that the relatively short dwell time/rapid off-rate from the channel was the predominant factor in determining the drug's clinical tolerability as well as its neuroprotective profile [9]. Most importantly, this mode of action meant that memantine blocked high (pathological) levels of glutamate at the NMDA receptor while relatively sparing the effects of low (physiological) levels of glutamate seen during normal neurotransmission because the drug does not accumulate in the channel during normal synaptic activity. The discovery that memantine, a low-affinity but still highly selective agent with a mechanism of uncompetitive antagonism, is neuroprotective yet clinically tolerated, triggered a paradigm shift in the history of drug development by the pharmaceutical industry [9–12]. Prior to that discovery, low-affinity drugs were thought to be inferior and not clinically useful. In particular, the relatively rapid off-rate from the NMDA receptor-associated ion channel of the memantine class of drugs largely accounts for its clinical tolerability as well as its low affinity. Clinical studies have borne out our hypothesis that low-affinity/fast off-rate memantine is a safe NMDA receptor antagonist in humans and beneficial in the treatment of neurological disorders mediated, at least in part, by excitotoxicity.

The NitroMemantines are second-generation NMDA receptor antagonists that may work even better than memantine by using the memantine binding site for the targeted delivery of NO to a second modulatory site on the NMDA receptor. Work is progressing rapidly in this area of investigation.

Clinical studies of the efficacy of memantine in the treatment of stroke, Alzheimer's disease, vascular dementia, HIV-associated dementia, glaucoma, and severe neuropathic pain, are currently under way; and there is every reason to expect the results to be positive, although this is, of course, not yet proven except in the case of moderate-to-severe Alzheimer's disease and possibly vascular dementia, in which phase III clinical trials have proven successful (Table 22.1) [13,14,74]. The efficacy of memantine in these neurodegenerative diseases and its ability to protect neurons in animal models of both acute and chronic neurologic disorders suggest that memantine and drugs acting in a similar manner could become very important new weapons in the fight against neuronal damage.

## ACKNOWLEDGMENTS

I thank my wonderful colleagues for helpful discussions and contributions to this work. I am especially grateful to H.-S. Vincent Chen, Yun-Beom Choi, and Jonathan S. Stamler for their collaborations. This work was supported in part by NIH grants P01 HD29587, R01 EY50477, and R01 EY09024.

## REFERENCES

1. Chen, G., Greenbard. P., and Yan, Z., Potentiation of NMDA receptor currents by dopamine D1 receptors in prefrontal cortex, *Proc. Natl. Acad. Sci. USA,* 101, 2596, 2004.
2. Lipton, S.A. and Rosenberg, P.A., Excitatory amino acids as a final common pathway for neurologic disorders, *N. Engl. J. Med.,* 330, 613, 1994.

3. Lipton, S.A. and Nicotera, P., Calcium, free radicals and excitotoxins in neuronal apoptosis, *Cell Calcium,* 23, 165, 1998.
4. Kemp, J.A. and McKernan, R.M., NMDA receptor pathways as drug targets, *Nat. Neurosci.,* 5 (Suppl.), 1039, 2002.
5. Lees, K.R. et al., Glycine antagonist (gavestinel) in neuroprotection (GAIN International) in patients with acute stroke: a randomised controlled trial. GAIN International Investigators, *Lancet,* 355, 1949, 2000.
6. Sacco, R.L. et al., Glycine antagonist in neuroprotection for patients with acute stroke: GAIN Americas: a randomized controlled trial, *JAMA,* 285, 1719, 2001.
7. Kemp, J.A., Kew, J.N., and Gill. R., in *Handbook of Experimental Pharmacology,* Jonas, P. and Monyer, H., Eds., Springer, Berlin, 1999, 101, 495.
8. Seif el Nasr, M. et al., Neuroprotective effect of memantine demonstrated in vivo and in vitro, *Eur. J. Pharmacol.,* 185, 19, 1990.
9. Lipton, S.A., Prospects for clinically tolerated NMDA antagonists: open-channel blockers and alternative redox states of nitric oxide, *Trends Neurosci.,* 16, 527, 1993.
10. Chen, H.S.-V. and Lipton, S.A., Mechanism of memantine block of NMDA-activated channels in rat retinal ganglion cells: uncompetitive antagonism, *J. Physiol. (Lond.),* 499, 27, 1997.
11. Chen, H.S.-V. et al., Open-channel block of N-methyl-D-aspartate (NMDA) responses by memantine: therapeutic advantage against NMDA receptor-mediated neurotoxicity, *J. Neurosci.,* 12, 4427, 1992.
12. Chen, H.S.-V. et al., Neuroprotective concentrations of the *N*-methyl-D-aspartate open-channel blocker memantine are effective without cytoplasmic vacuolation following post-ischemic administration and do not block maze learning or long-term potentiation, *Neuroscience,* 86, 1121, 1998.
13. Reisberg. B. et al., Memantine in moderate-to-severe Alzheimer's disease, *N. Engl. J. Med.,* 348, 1333, 2003.
14. Orgogozo, J.M. et al., Efficacy and safety of memantine in patients with mild to moderate vascular dementia: a randomized, placebo-controlled trial (MMM 300), *Stroke,* 33, 1834, 2002.
15. Lucas, D.R. and Newhouse, J.P., The toxic effect of sodium L-glutamate on the inner layers of the retina, *AMA Arch. Ophthalmol.,* 58, 193, 1957.
16. Olney, J.W., Glutamate-induced retinal degeneration in neonatal mice. Electron microscopy of the acutely evolving lesion, *J. Neuropathol. Exp. Neurol.,* 28, 455, 1969.
17. Olney, J.W. and Ho, O.L., Brain damage in infant mice following oral intake of glutamate, aspartate or cysteine, *Nature,* 227, 609, 1970.
18. Lipton, S.A., Molecular mechanisms of trauma-induced neuronal degeneration, *Curr. Opin. Neurol. Neurosurg.,* 6, 588, 1993.
19. Bonfoco, E. et al., Apoptosis and necrosis: two distinct events induced, respectively, by mild and intense insults with *N*-methyl-D-aspartate or nitric oxide/superoxide in cortical cell cultures, *Proc. Natl. Acad. Sci. USA,* 92, 7162, 1995.
20. Dreyer, E.B., Zhang, D., and Lipton, S.A., Transcriptional or translational inhibition blocks low dose NMDA-mediated cell death, *Neuroreport,* 6, 942, 1995.
21. Quigley, H.A. et al., Retinal ganglion cell death in experimental glaucoma and after axotomy occurs by apoptosis, *Invest. Ophthalmol. Vis. Sci.,* 36, 774, 1995.
22. Vorwerk, C.K. et al., Chronic low-dose glutamate is toxic to retinal ganglion cells. Toxicity blocked by memantine, *Invest. Ophthalmol. Vis. Sci.,* 37, 1618, 1996.
23. Dreyer, E.B. and Grosskreutz, C.L., Excitatory mechanisms in retinal ganglion cell death in primary open angle glaucoma (POAG), *Clin. Neurosci.,* 4, 270, 1997.

24. Dreyer, E.B. and Lipton, S.A., New perspectives on glaucoma, *JAMA*, 281, 306, 1999.
25. Naskar, R., Vorwerk. C.K., and Dreyer, E.B., Saving the nerve from glaucoma: memantine to caspaces, *Semin. Ophthalmol.,* 14, 152, 1999.
26. Zeevalk, G.D. and Nicklas, W.J., Evidence that the loss of the voltage-dependent $Mg^{2+}$ block at the *N*-methyl-D-aspartate receptor underlies receptor activation during inhibition of neuronal metabolism, *J. Neurochem.,* 59, 1211, 1992.
27. Rogawski, M.A. and Wenk, G.L., The neuropharmacological basis for the use of memantine in the treatment of Alzheimer's disease, *CNS Drug Rev.,* 9, 275, 2003.
28. Selkoe, D.J., Alzheimer's disease: genes, proteins, and therapy, *Physiol. Rev.,* 81, 741, 2001.
29. Wu, J., Anwyl, R., and Rowan, M.J., beta-Amyloid-(1-40) increases long-term potentiation in rat hippocampus in vitro, *Eur. J. Pharmacol.,* 284, R1, 1995.
30. Mattson, M.P. et al., beta-Amyloid peptides destabilize calcium homeostasis and render human cortical neurons vulnerable to excitotoxicity, *J. Neurosci.,* 12, 376, 1992.
31. Koh, J.Y., Yang, L.L., and Cotman, C.W., Beta-amyloid protein increases the vulnerability of cultured cortical neurons to excitotoxic damage, *Brain Res.,* 533, 315, 1990.
32. Topper, R. et al., Rapid appearance of beta-amyloid precursor protein immunoreactivity in glial cells following excitotoxic brain injury, *Acta Neuropathol. (Berl.),* 89, 23, 1995.
33. Harkany, T. et al., Beta-amyloid neurotoxicity is mediated by a glutamate-triggered excitotoxic cascade in rat nucleus basalis, *Eur. J. Neurosci.,* 12, 2735, 2000.
34. Couratier, P. et al., Modifications of neuronal phosphorylated tau immunoreactivity induced by NMDA toxicity, *Mol. Chem. Neuropathol.,* 27, 259, 1996.
35. Miguel-Hidalgo, J.J. et al., Neuroprotection by memantine against neurodegeneration induced by beta-amyloid(1-40), *Brain Res.,* 958, 210, 2002.
36. Tanila, H., Minkevicine, R., and Banjeree, P., Behavioral effects of subchronic memantine treatment in APP/PS1 double mutant mice modeling Alzheimer's disease, *J. Neurochem.*, 85 (Suppl. 1), 42, 2003.
37. Iqbal, K. et al., Memantine restores okadaic acid-induced changes in protein phosphatase-2A, CAMKII and tau hyperphosphorylation in rat, *J. Neurochem.,* 85 (Suppl. 1), 42, 2003.
38. Dawson, V.L. et al., Nitric oxide mediates glutamate neurotoxicity in primary cortical cultures, *Proc. Natl. Acad. Sci. USA,* 88, 6368, 1991.
39. Dawson, V.L. et al., Mechanisms of nitric oxide-mediated neurotoxicity in primary brain cultures, *J. Neurosci.,* 13, 2651, 1993.
40. Lipton, S.A. et al., A redox-based mechanism for the neuroprotective and neurodestructive effects of nitric oxide and related nitroso-compounds, *Nature,* 364, 626, 1993.
41. Tenneti, L. et al., Role of caspases in *N*-methyl-D-aspartate-induced apoptosis in cerebrocortical neurons, *J. Neurochem.,* 71, 946, 1998.
42. Yun, H.Y. et al., Nitric oxide mediates *N*-methyl-D-aspartate receptor-induced activation of p21ras, *Proc. Natl. Acad. Sci. USA,* 95, 5773, 1998.
43. Budd, S.L. et al., Mitochondrial and extramitochondrial apoptotic signaling pathways in cerebrocortical neurons, *Proc. Natl. Acad. Sci. USA,* 97, 6161, 2000.
44. Okamoto, S. et al., Dominant-interfering forms of MEF2 generated by caspase cleavage contribute to NMDA-induced neuronal apoptosis, *Proc. Natl. Acad. Sci. USA,* 99, 3974, 2002.
45. Johnson, J.W. and Ascher, P., Glycine potentiates the NMDA response in cultured mouse brain neurons, *Nature,* 325, 529, 1987.
46. McBain, C.J. and Mayer, M.L., *N*-methyl-D-aspartic acid receptor structure and function, *Physiol. Rev.,* 74, 723, 1994.

47. Lipton, S.A. et al., Cysteine regulation of protein function—as exemplified by NMDA-receptor modulation, *Trends Neurosci.,* 25, 474, 2002.
48. Koroshetz, W.J. and Moskowitz, M.A., Emerging treatments for stroke in humans, *Trends Pharmacol. Sci.,* 17, 227, 1996.
49. Hickenbottom, S.L. and Grotta, J., Neuroprotective therapy, *Semin. Neurol.,* 18, 485, 1998.
50. Lutsep, H.L. and Clark, W.M., Neuroprotection in acute ischaemic stroke. Current status and future potential, *Drugs R D,* 1, 3, 1999.
51. Rogawski, M.A., Low affinity channel blocking (uncompetitive) NMDA receptor antagonists as therapeutic agents—toward an understanding of their favorable tolerability, *Amino Acids,* 19, 133, 2000.
52. Palmer, G.C., Neuroprotection by NMDA receptor antagonists in a variety of neuropathologies, *Curr. Drug Targets,* 2, 241, 2001.
53. Chen, H.S., Rastogi, M., and Lipton, S.A., Q/R/N site mutations in the M2 region of NMDAR1/NMDAR2A receptors reveal a non-specific site for memantine action, *Soc. Neurosci. Abstr.,* 24, 342, 1998.
54. Bormann, J., Memantine is a potent blocker of *N*-methyl-D-aspartate (NMDA) receptor channels, *Eur. J. Pharmacol.,* 166, 591, 1989.
55. Blanpied, T.A. et al., Trapping channel block of NMDA-activated responses by amantadine and memantine, *J. Neurophysiol.,* 77, 309, 1997.
56. Sobolevsky, A.I.,, Koshelev S.G., and Khodorov, B.I., Interaction of memantine and amantadine with agonist-unbound NMDA-receptor channels in acutely isolated rat hippocampal neurons, *J. Physiol. (Lond.),* 512, 47, 1998.
57. Rammes, G. et al., The *N*-methyl-D-aspartate receptor channel blockers memantine, MRZ 2/579 and other amino-alkyl-cyclohexanes antagonise 5-HT(3) receptor currents in cultured HEK-293 and N1E-115 cell systems in a non-competitive manner, *Neurosci. Lett.,* 306, 81, 2001.
58. Reiser, G., Binmoller, F.J., and Koch, R., Memantine (1-amino-3,5-dimethyladamantane) blocks the serotonin-induced depolarization response in a neuronal cell line, *Brain Res.,* 443, 338, 1988.
59. Parsons, C.G., Danysz, W., and Quack, G., Memantine is a clinically well tolerated *N*-methyl-D-aspartate (NMDA) receptor antagonist — a review of preclinical data, *Neuropharmacology,* 38, 735, 1999.
60. Lipton, S.A., Memantine prevents HIV coat protein-induced neuronal injury in vitro, *Neurology,* 42, 1403, 1992.
61. Pellegrini, J.W. and Lipton, S.A., Delayed administration of memantine prevents *N*-methyl-D-aspartate receptor-mediated neurotoxicity, *Ann. Neurol.,* 33, 403, 1993.
62. Sucher, N.J., Lipton, S.A., and Dreyer, E.B., Molecular basis of glutamate toxicity in retinal ganglion cells, *Vision Res.,* 37, 3483, 1997.
63. Osborne, N.N., Memantine reduces alterations to the mammalian retina, in situ, induced by ischemia, *Visual Neurosci.,* 16, 45, 1999.
64. Tariot, P.N. et al., Memantine treatment in patients with moderate to severe Alzheimer's disease already receiving donepezil, *JAMA,* 291, 317, 2004.
65. Winblad, B. and Poritis, N., Memantine in severe dementia: results of the 9M-Best Study (Benefit and efficacy in severely demented patients during treatment with memantine), *Int. J. Geriatr. Psychiatry,* 14, 135, 1999.
66. Lipton, S.A. and Wang, Y.F., NO-related species can protect from focal cerebral ischemia/reperfusion, in *Pharmacology of Cerebral Ischemia,* Krieglstein, J. and Oberpichler-Schwenk, H., Eds., Wissenschaftliche Verlagsgesellschaft mbH, Stuttgart, 1996, 183.

67. Zurakowski, D. et al., Nitrate therapy may retard glaucomatous optic neuropathy, perhaps through modulation of glutamate receptors, *Vision Res.,* 38, 1489, 1998.
68. Lipton, S.A. et al., Redox modulation of the NMDA receptor by NO-related species, *Prog. Brain Res.,* 118, 73, 1998.
69. Choi, Y.B. et al., Molecular basis of NMDA receptor-coupled ion channel modulation by S-nitrosylation, *Nat. Neurosci.,* 3, 15, 2000.
70. Lipton, S.A. and Chen, H.S.-V., Paradigm shift in neuroprotective drug development: clinically tolerated NMDA receptor inhibition by memantine, *Cell Death Differ.,* 11, 18, 2004.
71. Lipton, S.A., Failures and successes of NMDA receptor antagonists: molecular basis for the use of open-channel blockers like memantine in the treatment of acute and chronic neurologic insults, *Neurorx.,* 1, 101, 2004.
72. Lipton S.A., Turning down, but not off, *Nature,* 428, 473, 2004.
73. Ankarcrona, M. et al., Glutamate-induced neuronal death: a succession of necrosis or apoptosis depending on mitochondrial function, *Neuron,* 15, 961, 1995.
74. Le, D.A. and Lipton, S.A., Potential and current use of *N*-methyl-D-aspartate (NMDA) receptor antagonists in diseases of aging, *Drugs Aging,* 18, 717, 2001.

## APPENDIX

### Scheme 1

$$\text{Channel} + \text{MEM} \leftrightarrow \text{Channel-MEM (blocked channel)} \qquad (1)$$

This simple bimolecular scheme predicts that the macroscopic blocking and unblocking actions of memantine (MEM) proceeds with exponential relaxation. The macroscopic pseudo-first-order rate constant of blocking ($k_{on}$) depends linearly on memantine concentration (as well as a constant A), and the macroscopic unblocking rate ($k_{off}$) is independent of memantine concentration ([MEM]).

$$k_{on} = A \bullet [\text{MEM}] \qquad (2)$$

$$k_{off}\text{: [MEM] independent} \qquad (3)$$

These predictions were borne out experimentally [10]. Both the macroscopic blocking and unblocking processes could be well fitted by a single exponential function. The macroscopic on-rate constant is the reciprocal of the measured time constant for onset ($\tau_{on}$) and is the sum of the pseudo-first-order blocking rate constant ($k_{on}$) and unblocking constant ($k_{off}$). The unblocking rate constant ($k_{off}$) is the reciprocal of the measured macroscopic unblocking time constant ($\tau_{off}$). These transformations lead to Equation 4 and Equation 5:

$$k_{on} = 1/\tau_{on} - 1/\tau_{off} \qquad (4)$$

$$k_{off} = 1/\tau_{off} \qquad (5)$$

The $k_{on}$ calculated from Equation 4 increased linearly with memantine concentration with a slope factor of $0.4 \pm 0.03 \times 10^6\ M^{-1}s^{-1}$ (mean ± SD), while the $k_{off}$

from Equation 5 remained relatively constant with a Y-axis intercept of $0.44 \pm 0.1\ s^{-1}$ [10]. A rapid method to validate this result was obtained by estimating the equilibrium apparent dissociation constant ($K_i$) for memantine action from the following equation:

$$K_i = k_{off}/(k_{on}/[MEM]) \qquad (6)$$

Here we found empirically that memantine was a relatively low-affinity (apparent affinity ~1 μ*M*) open-channel blocker of the NMDA receptor-coupled ion channel, and a major component of the affinity was determined by $k_{off}$ at clinically relevant concentrations in the low micromolar range.

# Index

## E



## F



## G

## K



## L

## O



## P